INTRODUCTION TO
AIRBORNE RADAR
SECOND EDITION

GEORGE W. STIMSON

SciTECH
PUBLISHING, INC.

SciTech Publishing Inc.
911 Paverstone Drive, Suite B
Raleigh, NC 27615
Phone: (919) 847-2434
Fax: (919) 847-2568
E-mail: info@scitechpub.com
Website: www.scitechpub.com

Acquisition and Product Development: Dudley R. Kay
Illustration and Page Layout: George Stimson and Shyam Reyes
Cover Design: Carolyn Allen—Intellisource Publishing and elaine kilcullen
Page Composition: Lehigh Press Colortronics
Printing: Overseas Printing, San Francisco, CA

Fifth printing
Printed in Hong Kong

Library of Congress Cataloging-in-Publication Data:

Stimson, George W.

      Introduction to Airborne Radar / George Stimson - 2nd ed
      xi, 576 p. : ill. (color) ; 29 cm
      includes index
      ISBN: 1-891121-01-4
      Subjects 1. Airplanes 2. Radar Equipment 3. Military Defense
      LC Classification: TL696.R2 S75 1998
      Dewey Class No.: 623/.7348 21

SciTech books may be purchased at quantity discounts for educational, business, or sales promotional use.

Members of the following professional associations may order directly from the associations.

The Institution of Electrical Engineers
Michael Faraday House
Six Hills Way, Stevenage, SG1 2AY, UK
Phone +44 (0) 1438 313311
Fax: +44 (0) 1438 313465
www.iee.org.uk
IEE Order No. RA 101

SPIE—The International Society for Optical Engineering
PO Box 10
Bellingham, WA 98227-0100
Phone: (360) 676-3290
Fax: (360) 647-1445
E-mail: bookorders@spie.org
SPIE Order No. PM56

This book is dedicated to **Meade A. Livesay** (left), veteran engineer, technical manager, and past President of the Hughes Radar Systems Group, who envisioned and commissioned the original writing of the book. He is seen here examining an advance copy of the first edition, with the author.

Preface

It is hoped that you will find this book as interesting and enjoyable to read as it was to write.

Key Features

As you will undoubtedly find, the book is unique in several respects.

First, beginning from scratch, it presents the wide range of airborne radar techniques in the form of an unfolding saga, not of individuals, but of radar concepts and principles. Each chapter tells a story, and the story flows naturally on from chapter to chapter.

Second, the book is designed to fulfill the needs of all who want to learn about radar, regardless of their technical backgrounds. It has sufficient technical depth and mathematical rigor to satisfy the instructor, the engineer, the professor. Yet, as long as a reader has a basic understanding of algebra and knows a little trigonometry and physics, the text painlessly takes the reader in bite-sized increments to the point of being able to talk on a sound footing with the radar experts.

Third, every technical concept is illustrated with a simple diagram immediately next to the text it relates to. Every illustration has a concise caption, which enables it to stand alone.

Fourth, to keep the text simple, where additional detail may be desired by some readers but not all, it is conveniently placed in a blue "panel" which one may skip, on a first reading, and come back to later on and examine at leisure. Exceptions, caveats, and reviewers comments are presented without detracting from the simplicity of the text in brief "side notes."

These features lead to the perhaps most unique aspect of the book. One can follow the development of each chapter by reading just the text, or just the illustrations and captions, or by seamlessly moving along between text and illustrations.

Yet another unique feature. Recognizing that people interested in airborne radar love airplanes, dispersed through the book are photos and renderings of radar-bearing aircraft, spanning the history of airborne radar from the Bristol Beaufighter of 1940 to the B-2 Bomber and F-22 fighter of today.

What's New

If you're familiar with the first edition, you may be wondering what's new in the second?

Prompted by the advent of "stealth," the daunting prospect of ever more sophisticated radar countermeasures, and the explosive growth of digital-processing throughput, which has made practical many radar techniques long considered "blue sky," 12 new chapters have been added. Briefly, they cover the following:

- Electronically steered array antennas (ESAs)— besides providing extreme beam agility, they're a "must" for stealth

- Antenna RCS reduction—also a crucial requirement of stealth

- Low-probability of intercept techniques (LPI) — besides greatly reducing vulnerability to countermeasures, they amazingly enable a radar to detect targets without its signals being usefully detected by an enemy

- Electronic countermeasures, counter countermeasures, and intelligence functions

- Multi-frequency operation and small-signal target detection—also essential in the era of stealth— plus space-time adaptive processing, true-time-delay beam steering, and 3-D SAR

- New modes and approaches to mode control that take advantage of the ESA's versatility

- Advanced airborne digital processing architectures—key to most of the above capabilities

- Detection and tracking of low-speed moving targets on the ground—an important topic missed in the first edition.

To illustrate the application of the basic radar principles, the book ends by briefly describing a dozen or so airborne radars currently in service in applications ranging from long-range surveillance to environmental monitoring.

Also warranting mention, the first three chapters have been extensively modified to provide a complete overview of virtually all of the basic principles and advanced features presented in the body of the book. These chapters may be useful in providing a "stand-alone" briefing on modern radar for students wanting a quick introduction to the subject.

Acknowledgements

Needless to say, I'm deeply grateful to the following engineers of the Hughes Aircraft Company (now a part of Raytheon) past and present, who have reviewed various sections of the book and contributed valuable suggestions, technical information, and insights.

For the first edition: *Eddie Phillips, Ben DeWaldt, Nate Greenblatt, Dave Goltzman, Kurt Harrison, Scott Fairchild, Verde Pieroni, Morris Swiger, Jeff Hoffner, John Wittmond, Fred Williams, Pete Demopolis, Denny Riggs,* and *Hugh Washburn.*

For the new chapters: *Doug Benedict, John Griffith, Don Parker, Steve Panaretos, Howard Nussbaum, Robert Rosen, Bill Posey, John Wittmond, Dave Sjolund, Lee Tower, Larry Petracelli, Robert Frankot,* and *Irwin Newberg.*

I am extremely grateful to *Merrill Skolnik* and *Russell Lefevre* (who reviewed an early draft of the second edition for the IEEE) for their encouragement and helpful suggestions.

Also, thanks are due to *Hugh Griffiths* of University College London and his colleagues, *Dr. David Belcher* and *Prof. Chris Oliver* of DERA Malvern, for the excellent SAR maps they provided; and to *Gerald Kaiser*, then professor at the University of Massachusetts-Lowell, who on his own initiative in anticipation of the second edition combed through the first from cover to cover to spot overlooked typos and other errors.

In addition, abundant thanks go to Hughes' ever helpful *Al Peña* for securing the negatives of the first edition for reuse in this edition.

Finally, special thanks to *Shyam Reyes*, for his invaluable aid with page composition and artwork, and to *Dudley Kay* and *Denise May* of SciTech, without whom the publication of this edition would not have been possible.

G.W.S., San Marino, California

B-2 STEALTH BOMBER

Having a range of 10,000 nmi with but one refueling, the B-2 can reach targets anywhere in the world quickly from bases in the United States.

Contents

F-22 STEALTH FIGHTER

With one pilot, two engines, and vector thrust, the low-observable F-22 air-dominance fighter is capable of supersonic flight without afterburner for long periods. Yet the pilot can precisely control its flight at all speeds from that of a piper cub to high supersonic levels.

PART I

Overview of Airborne Radar

BRISTOL BEAUFIGHTER (1940)

The first really successful radar-equipped fighter, flown by Flight Officer Ashfield, achieved its first kill on the night of November 7, 1940. The AI Mark IV radar detected airborne targets at a three to four mile range. The armament consisted of four 20 millimeter cannons and six 30 calibre machine guns.

Basic Concepts 1

Tapping the sidewalk repeatedly with his cane, a blind man makes his way along a busy street, keeping a fixed distance from the wall of a building on his right—hence also a safe distance from the curb and the traffic whizzing by on his left. Emitting a train of shrill beeps, a bat deftly avoids the obstacles in its path and unerringly homes in on a succession of tiny nocturnal insects that are its prey. Just as unerringly, the pilot of a supersonic fighter closes in on a possible enemy intruder, hidden behind a cloud bank a hundred and fifty miles away (Fig. 1). How do they do it?

Underlying each of these remarkable feats is a very simple and ancient principle: that of detecting objects and determining their distances (range) from the echoes they reflect. The chief difference is that, in the cases of the blind man and the bat, the echoes are those of sound waves, whereas in the case of the fighter, they are echoes of radio waves.

In this chapter, we will briefly review the fundamental radar[1] concept and see in a little more detail how it is applied to such practical uses as detecting targets and measuring their ranges and directions. We will then take up a second important concept: that of determining the relative speed or range rate of the reflecting object from the shift in the radio frequency of the reflected waves relative to that of the transmitted waves, the phenomenon known as the doppler effect. We will see how, by sensing doppler shifts, a radar can not only measure range rates but also differentiate between echoes from moving targets and the clutter of echoes from the ground and objects on it which are stationary. We will further learn how, rather than rejecting the echoes from the ground, the radar can use them to produce high resolution maps of the terrain (Fig. 2).

1. Looking out through a streamlined faring in the nose of a supersonic fighter, a small but powerful radar enables the pilot to home in on an intruder hidden behind or in a cloud bank a hundred and fifty miles away.

1. Radar = Radio Detection And Ranging.

2. Rather than rejecting echoes from the ground, as when searching for airborne targets, the radar may use them to produce real-time high-resolution maps of the terrain.

3

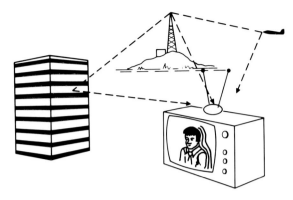

3. That radio waves are reflected by aircraft, buildings, and other objects is repeatedly demonstrated by the multiple images (ghosts) we sometimes see on TV screens.

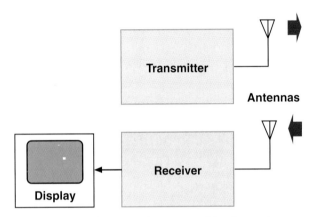

4. In rudimentary form, a radar consists of five basic elements.

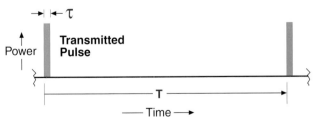

6. To keep transmission from interfering with reception, the radar usually transmits the radio waves in pulses and listens for the echoes in between.

Radio Detection

Most objects—aircraft, ships, vehicles, buildings, features of the terrain, etc.—reflect radio waves, much as they do light (Fig. 3). Radio waves and light are, in fact, the same thing—the flow of electromagnetic energy. The sole difference is that the frequencies of light are very much higher. The reflected energy is scattered in many directions, but a detectable portion of it is generally scattered back in the direction from which it originally emanated.

At the longer wavelengths (lower frequencies) used by many shipboard and ground based radars, the atmosphere is almost completely transparent. And it is nearly so even at the shorter wavelengths used by most airborne radars. By detecting the reflected radio waves, therefore, it is possible to "see" objects not only at night, as well as in the daytime, but through haze, fog, or clouds.

In its most rudimentary form, a radar consists of five elements: a radio transmitter, a radio receiver tuned to the transmitter's frequency, two antennas, and a display (Fig. 4). To detect the presence of an object (target), the transmitter generates radio waves, which are radiated by one of the antennas. The receiver, meanwhile, listens for the "echoes" of these waves, which are picked up by the other antenna. If a target is detected, a blip indicating its location appears on the display.

In practice, the transmitter and receiver generally share a common antenna (Fig. 5).

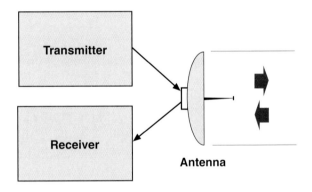

5. In practice, a single antenna is generally time-shared by the transmitter and the receiver.

To avoid problems of the transmitter interfering with reception, the radio waves are usually transmitted in pulses, and the receiver is turned off ("blanked") during transmission (Fig. 6). The rate at which the pulses are transmitted is called the *pulse repetition frequency* (PRF). So that the radar can differentiate between targets in different directions as well as detect targets at greater ranges, the antenna concentrates the radiated energy into a narrow beam.

To find a target, the beam is systematically swept through

the region in which targets are expected to appear. The path of the beam is called the *search scan pattern*. The region covered by the scan is called the *scan volume* or *frame*; the length of time the beam takes to scan the complete frame, the *frame time* (Fig. 7).

Incidentally, in the world of radar the term target is broadly used to refer to almost anything one wishes to detect: an aircraft, a ship, a vehicle, a man-made structure on the ground, a specific point in the terrain, rain (weather radars), aerosols, even free electrons .

Like light, radio waves of the frequencies used by most airborne radars travel essentially in straight lines. Consequently, for a radar to receive echoes from a target, the target must be within the line of sight (Fig. 8).

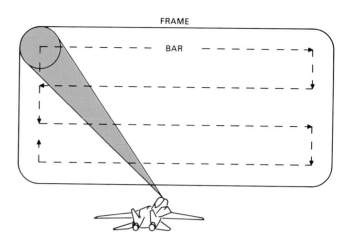

7. Typical search scan pattern for a fighter application. Number of bars and width and position of frame may be controlled by the operator.

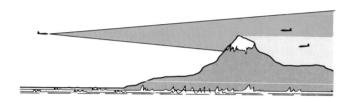

8. To be seen by most radars, a target must be within the line of sight.

Even then, the target will not be detected unless its echoes are strong enough to be discerned above the background of electrical noise that invariable exists in the output of a receiver, or, above the background of simultaneously received echoes from the ground (called *ground clutter*) which in some situations may be substantially stronger than the noise.

The strength of a target's echoes is inversely proportional to the target's range to the fourth power ($1/R^4$). Therefore, as a distant target approaches, its echoes rapidly grow stronger (Fig. 9).

The range at which they become strong enough to be detected depends upon a number of factors. Among the most important are these:

- Power of the transmitted waves
- Fraction of the time, τ/T, during which the power is transmitted
- Size of the antenna
- Reflecting characteristics of the target
- Length of time the target is in the antenna beam during each search scan
- Number of search scans in which the target appears
- Wavelength of the radio waves
- Strength of background noise or clutter

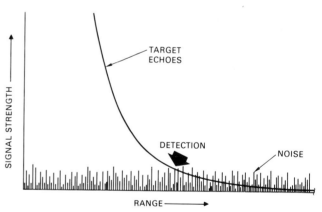

9. As a distant target approaches, its echoes rapidly grow stronger. But only when they emerge from the background of noise and/or ground clutter will they be detected.

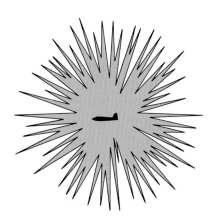

10. Since the target return scintillates and fades, and noise varies randomly, detection ranges must be expressed in terms of probabilities.

Much as the sunlight reflected from a car on a distant highway scintillates and fades, the strength of the echoes scattered in the radar's direction varies more or less at random (Fig. 10). Because of this and the randomness of the background noise, the range at which a given target is detected by the radar will not always be the same. Nevertheless, the probability of its being detected at any particular range (or by the time it reaches a given range) can be predicted with considerable certainty.

By optimizing those parameters over which one has control, a radar can be made small enough to fit in the nose of a fighter yet detect small targets at ranges on the order of a hundred miles. Radars of larger aircraft (Fig. 11) can detect targets at greater ranges.

11. Radars in larger aircraft (e.g. AWACS) can detect small aircraft at ranges out to 200 to 400 nmi.

Determining Target Position

In most applications, it is not enough merely to know that a target is present. It is also necessary to know the target's location—its distance (range) and direction (angle).

$$R = \frac{1}{2} \text{ (Round-Trip Time)} \times \text{(Speed of Light)}$$

$$= \frac{1}{2} \times \frac{10}{1,000,000} \text{ s} \times 300,000,000 \text{ m/s}$$

$$= 1.5 \text{ km}$$

12. Transit time is measured in millionths of a second (μs). A transit time of 10 μs corresponds to a range of 1.5 kilometers.

Measuring Range. Range may be determined by measuring the time the radio waves take to reach the target and return. Radio waves travel at essentially a constant speed—the speed of light. A target's range, therefore, is half the round-trip (two-way) transit time times the speed of light (Fig. 12). Since the speed of light is high—300 million meters per second—ranging times are generally expressed in millionths of a second (microseconds). A round-trip transit time of 10 microseconds, for example, corresponds to a range of 1.5 kilometers.

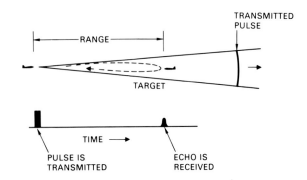

13. Usually, a target's range may be most easily determined by measuring the time between transmission of a pulse and reception of its echo.

The transit time is most simply measured by observing the time delay between transmission of a pulse and reception of the echo of that pulse (Fig. 13)—a technique called *pulse-delay ranging.* So that echoes of closely spaced targets won't overlap and appear to be the return from a single target, the width of the pulse, τ, is generally limited to a microsecond or less. To radiate enough energy to detect distant targets, however, pulses must often be made very much wider. This dilemma may be resolved by compressing the echoes after they are received.

One method of compression, called *chirp*, is to linearly increase the frequency of each transmitted pulse throughout its duration (Fig. 14). The received echoes are then passed through a filter which introduces a *delay* that decreases with increasing frequency, thereby compressing the received energy into a narrow pulse.

Another method of compression is to mark off each pulse into narrow segments and, as the pulse is transmitted, reverse the phase of certain segments according to a special code (Fig. 15). When each received echo is decoded, its energy is compressed into a pulse the width of a single segment.

With either technique, resolution of a foot or so may be obtained without limiting range. Resolutions of a few hundred feet, though, are more typical.

Radars which transmit a continuous wave (CW radars) or which transmit their pulses too close together for pulse-delay ranging, measure range with a technique called *frequency-modulation* (FM) ranging. In it, the frequency of the transmitted wave is varied and range is determined by observing the lag in time between this modulation and the corresponding modulation of the received echoes (Fig. 16).

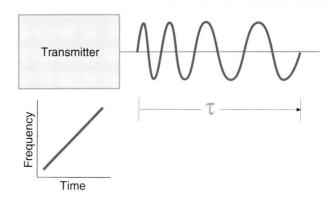

14. Chirp pulse compression modulation. The transmitter's frequency increases linearly throughout the duration, τ, of each pulse.

Transmitted Pulse

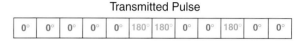

0°	0°	0°	0°	0°	180°	180°	0°	0°	180°	0°	0°

15. In binary phase-modulation pulse compression, the phases of certain segments of each transmitted pulse are reversed according to a special code. Decoding the received echoes compress them to the width of a single segment.

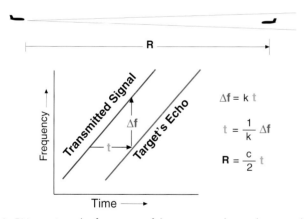

$$\Delta f = k\,t$$

$$t = \frac{1}{k}\,\Delta f$$

$$R = \frac{c}{2}\,t$$

16. In FM ranging, the frequency of the transmitted signal is varied linearly and the instantaneous difference, Δf, between the transmitter's frequency and the target echo's frequency is sensed. The round-trip transit time, t, to the target, hence the target's range, R, is proportional to this difference.

Measuring Direction. In most airborne radars, direction is measured in terms of the angle between the line of sight to the target and a horizontal reference direction such as north, or the longitudinal reference axis of the aircraft's fuselage. This angle is usually resolved into its horizontal and vertical components. The horizontal component is called *azimuth;* the vertical component, *elevation* (Fig. 17).

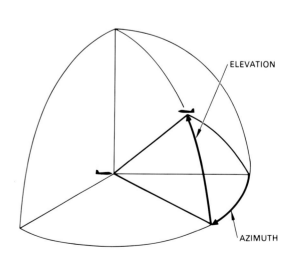

17. Angle between the fuselage reference axis and the line of sight to a target is usually resolved into azimuth and elevation components.

7

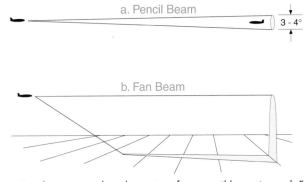

18. For detecting and tracking aircraft, a pencil beam is used. For long-range surveillance, mapping, or detecting targets on the ground, a fan beam may be used.

Where both azimuth and elevation are required, as for detecting and tracking an aircraft, the beam is given a more or less conical shape (Fig. 18a). This is called a *pencil beam*. Typically it is three or four degrees wide. Where only azimuth is required, as for long-range surveillance, mapping, or detecting targets on the ground, the beam may be given a fan shape (Fig. 18b).

Angular position may be measured with considerably greater precision than the width of the beam. For example, if echoes are received during a portion of the azimuth search scan extending from 30° to 34°, the target's azimuth may be concluded to be very nearly 32°. With more sophisticated processing of the echoes, such as used for automatic tracking, the angle can be determined more accurately.

Automatic Tracking. Frequently it is desired to follow the movements of one or more targets while continuing to search for more. This may be done in a mode of operation called *track-while-scan*. In it, the position of each target of interest is tracked on the basis of the periodic samples of its range, range rate, and direction obtained when the antenna beam sweeps across it (Fig. 19).

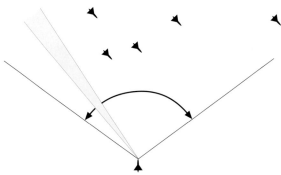

19. In track-while-scan, any number of targets may be tracked simultaneously on the basis of samples of each target's range, range rate, and direction obtained when the beam sweeps across it in the course of the search scan.

Track-while-scan is ideal for maintaining situation awareness. It provides sufficiently accurate target data for launching guided missiles, which can correct their trajectories after launch, and is particularly useful for launching missiles in rapid succession against several widely separated targets. But it does not provide accurate enough data for predicting the flight path of a target for a fighter's guns or of a tanker for refueling (Fig. 20). For such uses, the antenna is trained on the target continuously in a *single-target track* mode.

To keep the antenna trained on a target in this mode, the radar must be able to sense its pointing errors. This may be

20. For tasks requiring precision, such as predicting the flight path of a tanker in preparation for refueling, a single-target tracking mode is generally provided.

done in several ways. One is to rotate the beam so that its central axis sweeps out a small cone about the pointing axis (boresight line) of the antenna (Fig. 21). If the target is on the boresight line (i.e., no error exists), its distance from the center of the beam will be the same throughout the conical scan, and the amplitude of the received echoes will be unaffected by the scan. However, since the strength of the beam falls off toward its edges, if a tracking error exists, the echoes will be modulated by the scan. The amplitude of the modulation indicates the magnitude of the tracking error, and the point in the scan at which the amplitude reaches its minimum indicates the direction of the error.

In more advanced radars, the error is sensed by sequentially placing the center of the beam on one side and then the other of the boresight line during reception only, a technique called *lobing* (Fig. 22).

To avoid inaccuracies due to pulse-to-pulse fluctuations in the echoes' strength, more advanced radars form the lobes simultaneously, enabling the error to be sensed with a single pulse. In one such technique, called *amplitude-comparison monopulse*, the antenna is divided into halves which produce overlapping lobes. In another, called *phase-comparison monopulse*, both halves of the antenna produce beams pointing in the boresight direction. If a tracking error exists, the distance from the target to each half will differ slightly in proportion to the error θ_e. Consequently, the error can be determined by sensing the resulting difference in radio frequency phase of the signals received by the two halves (Fig. 23).

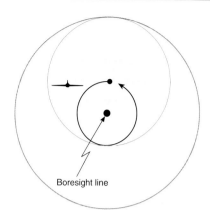

21. Conical scan. Angle tracking errors are sensed by rotating the antenna's beam about the boresight line and sensing the resulting modulation of the received echoes.

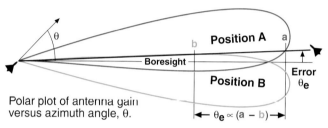

22. Lobing. For reception, antenna lobe is alternately deflected to the right and left of the boresight line to measure the angle-tracking error, θ_e.

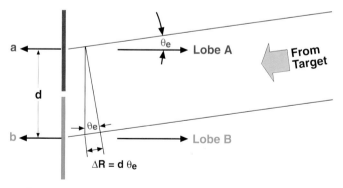

23. Phase comparison monopulse. Difference in distances from target to antenna's two halves, ΔR; hence (for small angles), the difference in phases of outputs a and b, is proportional to the tracking error, θ_e.

By continuously sensing the tracking error with either of these techniques and correcting the antenna's pointing direction to minimize the error, the antenna can be made to follow the target's movement precisely.

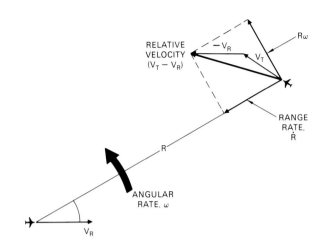

24. Target's relative velocity may be computed from measured values of range, range rate, and angular rate of line of sight.

25. A common example of the doppler shift. Motion of car crowds sound waves propagated ahead, spreads waves propagated behind.

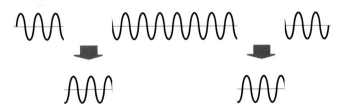

26. By cutting a radar's transmitted pulses from a continuous wave, the radio frequency phase of successive echoes from the same target will be coherent, enabling their doppler frequency to be readily measured.

While the target is being tracked in angle, its range and direction may be continuously measured. Its range rate may then be computed from the continuously measured range, and its angular rate (rate of rotation of the line of sight to the target) may be computed from the continuously measured direction. Knowing the target's range, range rate, direction, and angular rate, its velocity and acceleration may be computed as illustrated in Fig. 24.

For greater accuracy, both angular rate and range rate may be determined directly: Angular rate may be measured by mounting rate gyros sensitive to motion about the azimuth and elevation axes, on the antenna. Range rate may be measured by sensing the shift in the radio frequency of the target's echoes due to the doppler effect.

Exploiting the Doppler Effect

The classic example of the doppler effect is the change in pitch of a locomotive's whistle as it passes by. Today, a more common example is found in the roar of a racing car, which deepens as the car zooms by (Fig. 25).

Because of the doppler effect, the radio frequency of the echoes an airborne radar receives from an object is shifted relative to the frequency of the transmitter in proportion to the object's range rate. Since the range rates encountered by an airborne radar are a minuscule fraction of the speed of radio waves, the doppler shift—or *doppler frequency* as it is called—of even the most rapidly closing target is extremely slight. So slight that it shows up simply as a pulse-to-pulse shift in the radio frequency phase of the target's echoes. To measure the target's doppler frequency, therefore, the following two conditions must be met:

- At least several (and in some cases, a great many) successive echoes must be received from the target, and

- The first wavefront of each pulse must be separated from the last wavefront of the same polarity in the preceding pulse by a whole number of wavelengths— a quality called *coherence.*

Coherence may be achieved by, in effect, cutting the radar's transmitted pulses from a continuous wave (Fig. 26).

By sensing doppler frequencies, a radar can not only measure range rates directly, but also expand its capabilities in other respects. Chief among these is the substantial reduction, or in some cases complete elimination, of "clutter." The range rates of aircraft are generally quite different from the range rates of most points on the ground, as well as of rain and other stationary or slowly moving sources of unwanted return. By sensing doppler frequencies, therefore, a radar can differentiate echoes of aircraft from clutter and reject the clutter. This feature is called *moving target indication* (MTI). In some cases, it may also be called *airborne moving target indication* (AMTI) to differentiate it from the simpler MTI used in ground based radars.

MTI is of inestimable value in radars which must operate at low altitudes or look down in search of aircraft flying below them. The antenna beam then commonly intercepts the ground at the target's range. Without MTI, the target echoes would be lost in the ground return (Fig. 27). MTI can also be of great value when flying at higher altitudes and looking straight ahead. For even then, the lower edge of the beam may intercept the ground at long ranges.

A radar can similarly isolate the echoes of moving vehicles on the ground. In some situations where MTI is used, the abundance of moving vehicles on the ground can make aircraft difficult to spot. But echoes from aircraft and echoes from vehicles on the ground can usually be differentiated by virtue of differences in closing rates, due to the ground vehicles' lower speeds.

Where desired, by sensing the doppler shift, a radar can measure its own velocity. For this, the antenna beam is generally pointed ahead and down at a shallow angle. The echoes from the point at which the beam intercepts the ground are then isolated and their doppler shift is measured. By sequentially making several such measurements at different azimuth and elevation angles, the aircraft's horizontal ground speed can be accurately computed (Fig. 28).

Ground Mapping

The radio waves transmitted by a radar are scattered back in the direction of the radar in different amounts by different objects—little from smooth surfaces such as lakes[2] and roads, more from farm lands and brush, and heavily from most man-made structures. Thus, by displaying the differences in the intensities of the received echoes when the antenna beam is swept across the ground, it is possible to produce a pictorial map of the terrain, called a *ground map.*

Radar maps differ from aerial photographs and road maps in several fundamental respects: In the first place, because of the difference in wavelengths, the relative reflectivity of the various features of the terrain may be quite different for radio waves than for visible light. Consequently, what is bright in a photograph may not be bright in a radar map, and vice versa.

In addition, unlike road maps, radar maps contain shadows, may be distorted, and unless special measures are taken to improve azimuth resolution, may show very little detail.

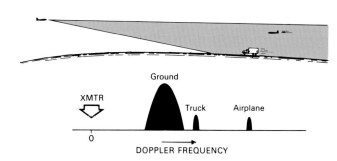

27. With MTI, echoes from aircraft and moving vehicles on the ground are separated from ground clutter on the basis of the differences in their doppler frequencies. Generally, echoes from aircraft and echoes from moving vehicles on the ground similarly may be differentiated as a result of the ground vehicles' lower speed.

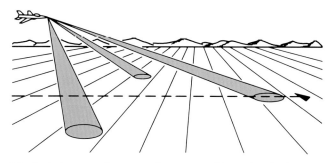

28. Radar's own velocity may be computed from doppler frequencies of three or more points on the ground at known angles.

2. This depends upon the lookdown angle. Water and flat ground directly below a radar produce very strong return.

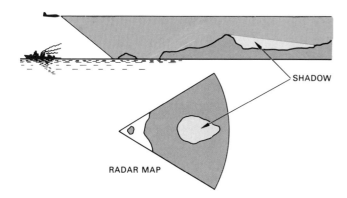

29. Shadows leave holes in radar maps. At steep lookdown angles, shadowing is minimized.

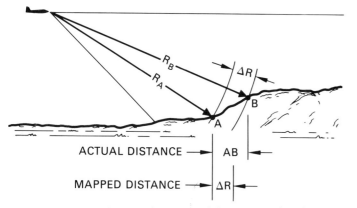

30. At steep lookdown angles, mapped distances are foreshortened. Except for distortion due to slope of the ground, foreshortening may be corrected before map is displayed.

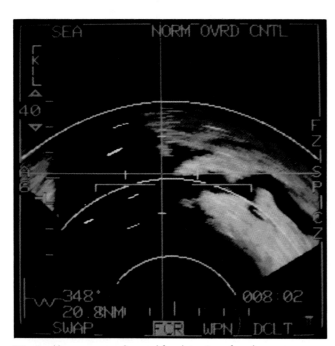

32. Real-beam map enhanced for detection of seaborne targets. Map was made by the radar of a fighter aircraft. Although azimuth resolution is limited, map can be highly useful. (Courtesy Northrop Grumman).

Shadows are produced whenever the transmitted waves are intercepted—in part or in whole—by hills, mountains, or other obstructions. The effect can be visualized by imagining that you are looking directly down on a relief map illuminated by a single light source at the radar's location (Fig. 29). Shadowing is minimal if the terrain is reasonably flat or if the radar is looking down at a fairly steep angle.

Distortion arises, however, if the lookdown angle is large. Since the radar measures distance in terms of slant range, the apparent horizontal distance between two points at the same azimuth is foreshortened (Fig. 30). If the terrain is sloping, two points separated by a small horizontal distance can, in the extreme, be mapped as a single point. Usually, the foreshortening can be corrected on the basis of the lookdown angle, before the map is displayed.

The degree of detail provided by a radar map depends upon the ability of the radar to separate (resolve) closely spaced objects in range and azimuth. Range resolution is limited primarily by the width of the radar's pulses.

By transmitting wide pulses and employing large amounts of pulse compression, the radar may obtain strong returns even from very long ranges and achieve range resolution as fine as a foot or so.

Fine azimuth resolution is not so easily obtained. In conventional (real-beam) ground mapping, azimuth resolution is determined by the width of the antenna beam (Fig. 31).

REAL-BEAM MAPPING

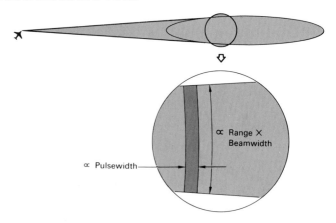

31. With conventional mapping, dimensions of resolution cell are determined by pulsewidth and width of the antenna beam.

With a beamwidth of 3°, for example, at a range of 10 miles azimuth resolution of a real-beam map may be no finer than half a mile (Fig. 32).

Azimuth resolution may be improved by operating at higher frequencies or by making the antenna larger. But if exceptionally high frequencies are used, detection ranges are reduced by atmospheric attenuation, and there are practical limitations on how large an antenna most aircraft can accommodate. However, an antenna of almost any length can by synthesized with a technique called *synthetic array radar* (or synthetic aperture radar), SAR.

SAR. Rather than scanning the terrain in the conventional way, with SAR the radar beam is pointed out to the side to illuminate the patch of ground of interest. Each time the radar radiates a pulse, it assumes the role of a single radiating element. Because of the aircraft's velocity, each such element is a little farther along on the flight path (Fig. 33). By storing the returns of a great many pulses and combining them—as a feed system combines the returns received by the radiating elements of a real antenna—the radar can synthesize the equivalent of a linear array long enough to provide azimuth resolution as fine as a foot or so (Fig. 34).

Moreover, by increasing the length of the synthesized array in proportion to the range of the area being mapped, the same fine resolution can be obtained at a range of 100 miles as at a range of only a few miles.

Moving targets tend to wash out in a SAR map because of their rotational motion. By taking advantage of it instead of the radar's forward motion, target images can be made, a technique called *inverse SAR* (ISAR).

Summary

By transmitting radio waves and listening for their echoes, a radar can detect objects day or night and in all kinds of weather. By concentrating the waves into a narrow beam, it can determine direction. And by measuring the transit time of the waves, it can measure range.

To find a target, the radar beam is repeatedly swept through a search scan. Once detected, the target may be automatically tracked and its relative velocity computed on the basis of either (a) periodic samples of its range and direction obtained during the scan or (b) continuous data obtained by training the antenna on the target. In the latter case, the target's echoes must be singled out in range and/or doppler frequency, and some means such as lobing must be provided to sense angular tracking errors.

Because of the doppler effect, the radio frequencies of the radar echoes are shifted in proportion to the reflecting object's range rates. By sensing these shifts, which is possible if the radar's pulses are coherent, the radar can measure target closing rates, reject clutter, and differentiate between ground return and moving vehicles on the ground. It can even measure its own velocity.

Since radio waves are scattered in different amounts by different features of the terrain, a radar can map the ground. With SAR, detailed maps can be made.

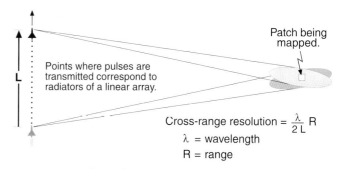

Cross-range resolution = $\frac{\lambda}{2L}$ R
λ = wavelength
R = range

33. SAR principle. With its antenna trained on a patch to be mapped, each time the radar transmits a pulse, it assumes the role of a single radiator. When the returns of a great many pulses are added up, the result is essentially the same as would have been obtained with a linear array antenna of length L. The mode illustrated here is called spotlight.

34. One-foot-resolution SAR map. Was made in real time in the spotlight mode from a long range, as indicated by radar shadows cast by trees. Regardless of the range, of course, radar maps always appear the same as if viewed from directly over head. (Crown copyright DERA Malvern)

13

MESSERSCHMITT Bf 110 G–4 (1941)

This was the first Luftwaffe radar-equipped fighter. It was equipped with a bulky antenna system, cutting the top speed of the Bf 110 by 25 mph. The Telefunken FuG 212 "Lichtenstein" radar presented targets from ranges of 600 feet out to 3–1/2 miles. The armament consisted of four 20 millimeter and four 7.9 millimeter guns firing forward and a 20 millimeter cannon fired vertically.

Approaches to Implementation

2

Having reviewed the basic radar concepts, we move on now to the practical consideration of their implementation While there is an endless variety of radar designs, we can get a rough idea of what is involved by considering three generic radars.

First is a radar of the sort used by the all-weather interceptors of the 1950s and 1960s, called simply a "pulsed" radar. In different configurations, it still is used today.

The second generic type is a far more capable one, called a "pulse-doppler" radar. It is the kind used in the current generation of conventional fighter and attack aircraft. In various forms, it too has a variety of applications.

The third generic type is a pulse-doppler radar tailored to meet the special requirements of stealth aircraft.

Generic "Pulsed" Radar

This radar (Fig. 1) is capable of automatic searching, single-target tracking, and real-beam ground mapping.

In the previous chapter, we learned that a pulsed radar consists of four basic functional elements: transmitter, receiver, time-shared antenna, and display. As you might expect, to implement even a simple practical radar, several other elements are also required. The more important of these are included in Fig. 2. The implementation of each of the elements shown in this figure is briefly outlined in the following paragraphs.

Synchronizer. This unit synchronizes the operation of the transmitter and the indicator by generating a continuous stream of very short, evenly spaced pulses. They designate the times at which successive radar pulses are to be transmitted and are supplied to the modulator and indicator.

1. Simple pulsed radar used in all-weather interceptors of 1950s and 1960s. In various forms, this generic type is in wide use even today.

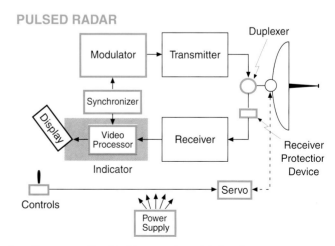

2. Elements outlined in blue must be added to the transmitter, receiver, antenna, and display of even a simple generic pulsed radar.

3. Magnetron transmitter tube.[1] Converts pulses of dc power to pulses of microwave energy. (Courtesy Litton Industries.)

1. Although you may not realize it, there is a good chance that you own a magnetron; their principal use today is in microwave ovens.

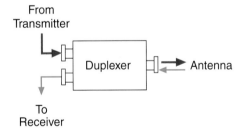

5. A duplexer is a device which passes the transmitter's high-power pulses to the antenna and the received echoes from the antenna to the receiver.

2. Active, gas-discharge switches, called TR (transmit-receive) and ATR (anti-transmit-receive) are also used.

Modulator. Upon receipt of each timing pulse, the modulator produces a high power pulse of direct current (dc) energy and supplies it to the transmitter.

Transmitter. This is a high-power oscillator, generally a magnetron (Fig. 3). For the duration of the input pulse from the modulator, the magnetron generates a high-power radio-frequency wave—in effect converting the dc pulse to a pulse of radio-frequency energy. (How it does this is illustrated in the panel on pages 18 and 19.) The wavelength of the energy is typically around 3 cm. The exact value may either be fixed by the design of the magnetron or tunable over a range of about 10% by the operator. The wave is radiated into a metal pipe (Fig. 4) called a waveguide, which conveys it the duplexer.

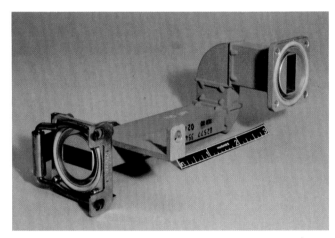

4. Representative waveguide: a metal pipe down which radio waves may be ducted. Width is usually about three quarters of the wavelength; height roughly half the wavelength.

Duplexer. This is essentially a waveguide switch (Fig. 5). Like a "Y" in a railroad track, it connects the transmitter and the receiver to the antenna. Unlike a railroad switch, however, the duplexer is usually a passive device which needn't be "thrown."[2] Sensitive to the direction of flow of the radio waves, it allows the waves coming from the transmitter to pass with negligible attenuation to the antenna, while blocking their flow to the receiver. Similarly, the duplexer allows the waves coming from the antenna to pass with negligible attenuation to the receiver, while blocking their way to the transmitter.

Antenna. In simple radars, the antenna generally consists of a radiator and a parabolically shaped reflector (dish), mounted on a common support. In the most rudimentary form, the radiator is little more than a horn-shaped nozzle on the end of the waveguide coming from the duplexer. The horn directs the radio wave arriving from the transmitter onto the dish, which reflects the wave in the form of a narrow beam (Fig. 6). Echoes intercepted by the dish are reflected into the horn and conveyed by the same waveguide back to the duplexer, thence to the receiver. (Instead of a dish antenna, some pulsed radars use a simple version of the planar array antenna described on page 28).

Generally, the antenna is mounted in gimbals, which allow it to be pivoted about both azimuth and elevation axes. In some cases, a third gimbal may be provided to isolate the antenna from the roll of the aircraft. Transducers on the gimbals provide the indicator with signals proportional to the displacement of the antenna about each axis.

Receiver Protection Device. Because of electrical discontinuities (mismatch of impedances) between the antenna and the waveguide conveying the radio waves to it, some of the energy of the radio waves is reflected from the antenna back to the duplexer. Since the duplexer performs its switching function purely on the basis of direction of flow, there is nothing to prevent this reflected energy from flowing on to the receiver, just as the radar echoes do. The reflected energy amounts to only a very small fraction of the transmitter's output. But because of the transmitter's high power, the reflections are strong enough to damage the receiver. To prevent the reflections from reaching the receiver, as well as to block any of the transmitter's energy that has leaked through the duplexer, a protection device is provided.

This device (Fig. 7) is essentially a high-speed microwave switch, which automatically blocks any radio waves strong enough to damage the receiver. Besides leakage and energy reflected by the antenna, the device also blocks any exceptionally strong signals which may be received from outside the radar—echoes received when the radar is inadvertently fired up in a hangar or is operated while facing a hangar wall at point blank range, or the direct transmission of another radar which happens to be looking directly into the radar antenna.

Receiver. Typically, the receiver is of a type called a superheterodyne (Fig. 8). It translates the received signals to a lower frequency at which they can be filtered and amplified more conveniently. Translation is accomplished by "beating" the received signals against the output of a low-power oscillator (called the *local oscillator* or LO) in a circuit called a *mixer.* The frequency of the resulting signal, called the *intermediate frequency* or IF, equals the difference between the signal's original frequency and the local oscillator frequency.

The output of the mixer is amplified by a tuned circuit (IF amplifier). It filters out any interfering signals, as well as

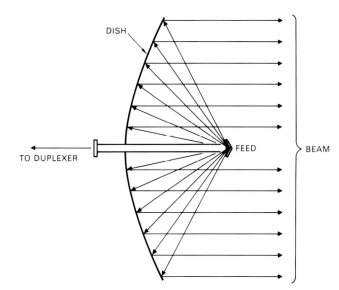

6.   Antenna for a simple pulsed radar consists of a single feed and a parabolic "dish" reflector, which forms the transmitted beam and reflects the returned echoes into the feed.

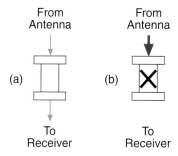

7.   Receiver protection device: (a) allows the weak echoes to pass from the duplexer to the receiver with negligible attenuation; but, (b) blocks any signals strong enough to damage the receiver.

RECEIVER

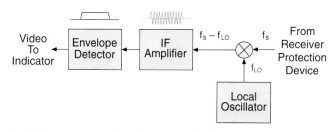

8.   The receiver translates the received radio waves (signal) to a lower frequency (IF), amplifies them, filters out signals of other frequencies, and produces a video output proportional to the received signal's amplitude.

THE VENERABLE MAGNETRON

Developed in the early years of World War II, the magnetron was the breakthrough that first made high-power microwave radars practical. Because of its comparatively low cost, small size, light weight, high efficiency, and rugged simplicity—plus its ability to produce high output powers with moderate input voltages—the magnetron has been widely used in radar transmitters ever since.

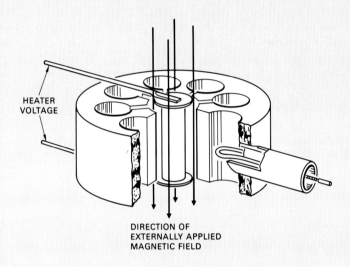

HEATER VOLTAGE

DIRECTION OF EXTERNALLY APPLIED MAGNETIC FIELD

The magnetron is one of a family of vacuum tube oscillators and amplifiers which take advantage of the fact that when an electron moves through a magnetic field whose direction is normal to the electron's velocity, the field exerts a force which causes the path of the electron to curve.

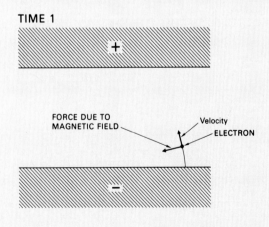

TIME 1

FORCE DUE TO MAGNETIC FIELD

Velocity
ELECTRON

+

−

The greater the electron's speed, the greater the curvature. (Because in these tubes the electric field that produces the electrons' motion is normal to the magnetic field, the tubes are called cross-field tubes.)

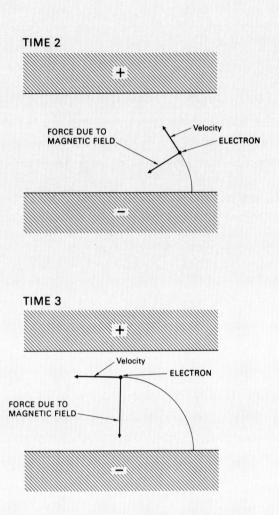

TIME 2

+

FORCE DUE TO MAGNETIC FIELD

Velocity
ELECTRON

−

TIME 3

+

Velocity
ELECTRON

FORCE DUE TO MAGNETIC FIELD

−

If we were to slice a magnetron in two, we would see that i consists of a cylindrical central electrode (cathode) ringed by a second cylindrical electrode (anode), with a gap (called the interaction space) in between.

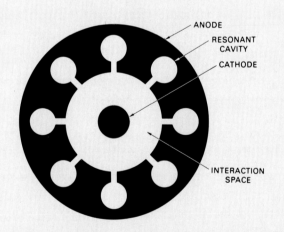

ANODE

RESONANT CAVITY

CATHODE

INTERACTION SPACE

Evenly distributed around the inner circumference of the anode are resonant cavities opening into the interaction space. The cathode is heated so that it emits electrons, which form a dense "cloud" around it. An externally mounted permanent magnet produces a strong magnetic field within the interaction space, normal to the axis of the electrodes.

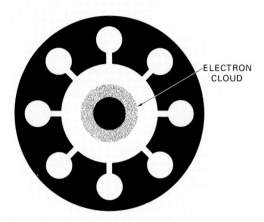

ELECTRON CLOUD

To cause the tube to generate radio waves, a strong dc voltage is applied between the electrodes—cathode negative, anode positive. Attracted by the positive voltage, the electrons accelerate toward the anode. But as the velocity of each electron increases, the magnetic field produces an increasingly strong force on the electrons, causing them to follow curved paths that carry them past the openings of the cavities.

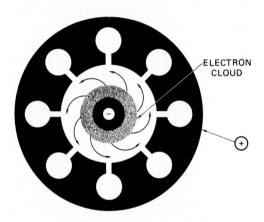

ELECTRON CLOUD

Much as a sound wave builds up in a bottle when you blow air across its mouth, an oscillating electromagnetic field (radio wave) builds up as a result of the electrons sweeping past the cavity openings. As with the sound wave, the frequency of the radio wave is the resonant frequency of the cavities.

It all starts with a minute, random disturbance which initiates an electromagnetic oscillation in one of the cavities. This oscillation propagates from cavity to cavity via the interaction space. The electric field of this incipient radio wave causes those electrons sweeping past the cavity openings during one peak of each cycle to slow down and move out toward the anode and those sweeping past during the other to speed up and move in toward the cathode. Consequently, the electrons quickly bunch up and form swirling spokes whose rotation is synchronized with the travel of the radio wave around the interaction space.

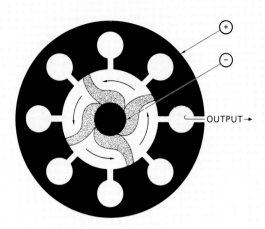

OUTPUT→

The electrons forming the spokes are gradually slowed down by their interaction with the traveling wave and in the process give up energy to the wave, thereby increasing its power. The slowing, of course, reduces the curvature of each electron's path, with the result that the electron soon reaches the anode. By the time it does, however, it has transferred to the radio wave up to 70 percent of the energy it acquired in being accelerated by the inter-electrode voltage. (What remains of the energy is absorbed as heat in the anode and must be carried away by the cooling system.) The spent electrons are returned to the cathode by the external power source. So the transfer of energy from the power source to the radio wave continues as long as the dc power is supplied.

Meanwhile, a tiny antenna inserted in one of the cavities bleeds the energy of the radio wave off into a waveguide which is the output "port" of the tube.

A magnetron's frequency may be varied over a limited range by changing the resonant frequency of the cavities through such techniques as lowering plungers into them.

Over the years a number of refinements have been made to the basic magnetron design. In one, a coaxial resonant output cavity is added.

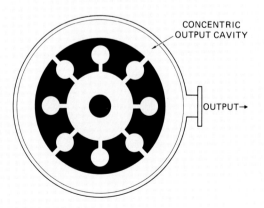

CONCENTRIC OUTPUT CAVITY

OUTPUT→

Energy is bled into it through slots in alternate cavities. The magnetron is tuned by changing the output cavity's resonant frequency.

the electrical background noise lying outside the band of frequencies occupied by the received signal.

Finally, the amplified signal is applied to a detector which produces an output voltage corresponding to the peak amplitude (or envelope) of the signal. It is similar to the signal that in a TV varies the intensity of the beam which paints the images on the picture tube. Consequently, the detector's output is called a video signal. This signal is supplied to the indicator.

Indicator. The indicator contains all of the circuitry needed to: (a) display the received echoes in a format that will satisfy the operator's requirements; (b) control the automatic searching and tracking functions; and (c) extract the desired target data when tracking a target.

Any of a variety of display formats may be used (see panel, on facing page). Only one of these, the B display will be described here.

For it, a video amplifier raises the receiver output to a level suitable for controlling the intensity of the display tube's cathode ray beam. The operator generally sets the gain of the amplifier so that noise spikes make the beam barely visible (Fig. 9). Target echoes strong enough to be detected above the noise will then produce a bright spot, or "blip." The vertical and horizontal positions of the beam are controlled as follows.

Each timing pulse from the synchronizer triggers the generation of a linearly increasing voltage that causes the beam to trace a vertical path from the bottom of the display to the top. Since the start of each trace is thus synchronized with the transmission of a radar pulse, if a target echo is received, the distance from the start of the trace to the point at which the target blip appears will correspond to the round-trip transit time for the echo, hence to the target's range. For this reason the trace is called the *range trace* and the vertical motion of the beam, the *range sweep.*

Meanwhile, the azimuth signal from the antenna is used to control the horizontal position of the range trace, and the elevation signal may be used to control the vertical position of a marker on the edge of the display, where an elevation scale is provided.

As the antenna executes its search scan, the range trace sweeps back and forth across the display in unison with the azimuth scan of the antenna. Each time the antenna beam sweeps across a target, a blip appears on the range trace, providing the operator with a plot of the range versus the azimuth of the target. (The typical location of the displays in a cockpit is shown in Fig. 10.)

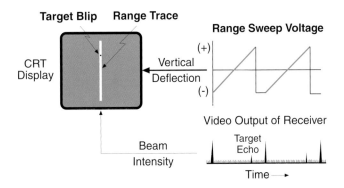

9. How range is displayed. Triggered by timing pulses from synchronizer, linear increase in vertical deflection voltage produces range sweep. Video output of receiver intensifies beam, producing target blip. (Strong video spikes are leakage of transmitted pulse through duplexer.)

10. Cockpit of a fighter/attack aircraft. Radar display is in upper right side of instrument panel. Combining glass for head-up display is in center of windscreen. Stored map for navigation is projected on display at lower center.

COMMON RADAR DISPLAYS

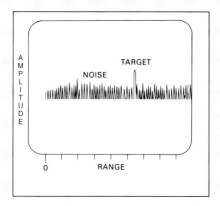

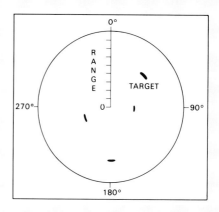

"A" Display. Plots amplitude of receiver output versus range on horizontal line, called a range trace. Simplest of all displays, but little used because it does not indicate azimuth.

PPI (Plan Position Indicator) Display. Targets displayed in polar plot centered on radar's position. Ideal for radars that provide 360 degree azimuth coverage.

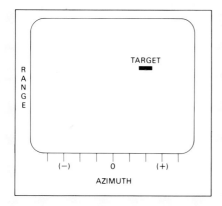

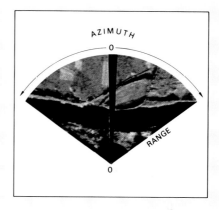

"B" Display. Targets displayed as blips on a rectangular plot of range versus azimuth. Widely used in fighter applications, where horizontal distortion near zero range is of little concern.

Sector PPI Display. Gives undistorted picture of region being scanned in azimuth. Commonly used for sector ground mapping.

PATCH MAP

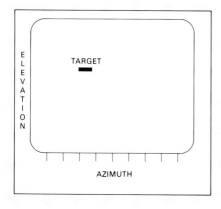

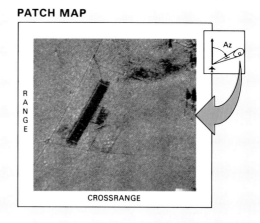

"C" Display. Shows target position on plot of elevation angle versus azimuth. Useful in pursuit attacks since display corresponds to pilot's view through windshield. Commonly projected on windshield as Head-Up Display.

Patch Map. In high resolution (SAR) ground mapping, a rectangular patch map is commonly displayed. This is a detailed map of a specific area of interest at a given range and azimuth angle. The range dimension of the patch is displayed vertically, the cross range dimension (i.e., dimension normal to the line of sight to the patch), horizontally.

ANTENNA SERVO

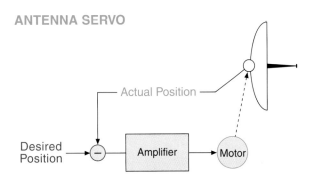

11. Antenna servo compares actual position of antenna with desired position, amplifies resulting error signal, and uses it to drive antenna in direction to reduce error to zero.

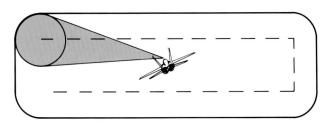

12. The antenna's search scan is stabilized in pitch and roll so that region searched will be unaffected by changes in aircraft attitude.

Antenna Servo. This unit positions the antenna in response to control signals which may be provided by any one of the following.

- The search scan circuitry in the indicator

- A hand control with which the operator can point the antenna manually

- The angle tracking system

A separate servo channel is provided for each gimbal. Their operation is illustrated in Fig. 11. The voltage obtained from the transducer on the gimbal is subtracted from the control signal, thereby producing an error signal proportional to the error in the antenna's position. This signal is then amplified and applied to a motor which rotates the antenna about the gimbal axis in such a way as to reduce the error to zero.

So that the search scan, which is usually much wider in azimuth than in elevation, will be unaffected by the attitude of the aircraft, stabilization may be provided (Fig. 12). If the antenna has a roll gimbal, the roll position of the antenna is compared with a reference provided by a vertical gyro and the resulting error signal is used to correct the roll position of the antenna.

Otherwise, the azimuth and elevation error signals are resolved into horizontal and vertical components on the basis of the reference provided by the gyro.

Power Supply. This element converts the power from the aircraft's typical 115 volt, 400 hertz primary power source to the various dc forms required by the radar. It first transforms the 400 hertz power to the standard voltages required; then converts them to dc, smooths them, and when necessary "regulates" them so they will remain constant in the face of changes in both the voltage of the primary power and the amounts of current drawn by the system. Though superficially mundane, elegant techniques have been devised to accomplish these tasks at a minimum cost in weight and dissipated power. (The antenna servo is generally operated directly off the 400 hertz supply and the relays off the aircraft's 28 volt dc supply.)

Automatic Tracking. Not all radars perform automatic tracking. Most of the simpler pulsed radars do not. Where automatic tracking is required, three additions must be made to the system just described. First, some means must be provided for isolating the target echoes in time (range). Second, a tracking scan such as the conical scanning or lobing described in the preceding chapter must be added to the antenna. Third, controls must be provided with which the operator can lock the radar onto the target's echoes.

For lock on, a pistol-grip hand control (Fig. 13) is generally designed so the operator can position a marker at any desired point on the range trace, and a button is provided with which he can tell the system that he has aligned the marker with the target he wishes to track. To lock onto a target, the operator takes control of the antenna with the hand control, aligns the antenna in azimuth so as to center the range trace on the target blip, adjusts the elevation of the antenna to maximize the brightness of the blip, runs the marker up the trace until it is just under the blip, and presses the lock-on button.

In the indicator, the circuit that controls the position of the marker on the display synchronizes the opening of an electronic switch, called a range gate, with the exact point in time after the start of the range sweep that an echo from the target will be received.

The gate stays open (switch closed) just long enough to allow the target echo to pass through and into the automatic tracking circuit. When the lock-on switch is depressed, control of the range gate is transferred to an automatic range tracking circuit (see panel below) which keeps the gate continuously centered on the target.

13. Hand control for a simple pulsed radar. Operator gains control of antenna by pressing trigger. For and aft motion controls position of range maker. Right and left motion controls antenna azimuth. Tilt switch on top controls elevation. Lock-on button is on side.

AUTOMATIC RANGE TRACKING

To control the timing of a range gate so it automatically follows (tracks) the changes in a target's range, a range tracking servo is provided.

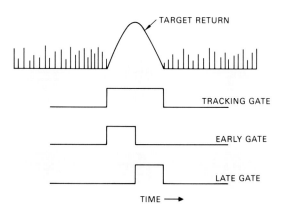

Typically, it samples the returns passed by the tracking gate with two secondary gates, each of which remains open only half as long as the tracking gate. One, called the early gate, opens

when the tracking gate opens, hence sampling the return passed by the first half of the tracking gate. The other, called the late gate, opens when the early gate closes and so samples the returns passed by the second half of the tracking gate.

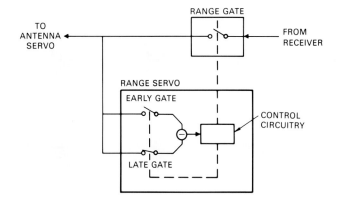

The range servo continuously adjusts the timing of the tracking gate so as to equalize the outputs of the early and late gates, thereby keeping the tracking gate centered on the target.

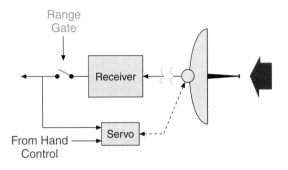

14. For automatically tracking a target, its echoes are isolated by closing an electronic switch (called the range gate) at the exact time each echo will be received.

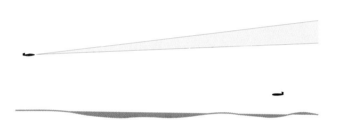

15. In early radars, ground clutter was avoided by keeping the radar beam from striking the ground, but this limited the radar's tactical capability.

Simultaneously, the tracking scan of the antenna is activated and control of the antenna servo (Fig. 14) is transferred to the automatic angle tracking system. It extracts signals proportional to the azimuth and elevation tracking errors from the output of the range tracker, and supplies these signals to the antenna servo.

Where extremely precise tracking is desired, rate integrating gyros (RIG) may be mounted on the antenna. They inertially establish azimuth and elevation axes to which the antenna servo is slaved, thereby holding the antenna solidly in the same position regardless of disturbances due to the aircraft's maneuvers. (This feature is called *space stabilizations.)*

The tracking error signals are smoothed and have corrections added to them to anticipate the effect of the aircraft's acceleration on the target's relative position. They are then applied to torque motors, which precess the gyros, thereby changing the directions of the reference axes they provide, so as to reduce the tracking errors to zero.

The principal shortcoming of the simple pulsed radar is that, since successive transmitted pulses are not coherent, it cannot easily differentiate between airborne targets and ground clutter. In early radars (Fig. 15), clutter was avoided simply by keeping the radar beam from striking the ground. But this seriously limited the radar's tactical ability.

In initial attempts to provide a lookdown capability, the radar detected the beat between the frequencies of the target echoes and the simultaneously received clutter (Fig. 16).

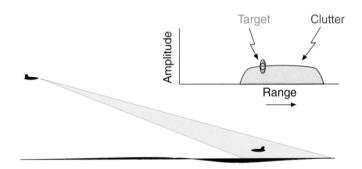

16. In initial attempts to provide a lookdown capability, the radar detected the "beat" between the frequency of the target echo and the simultaneously received clutter; performance was poor.

But since the clutter is generally spread over many frequencies, there were also beats between various clutter frequencies, as well as between these frequencies and the frequency of the target echoes. Hence, performance was marginal. These problems were completely circumvented with the advent of pulse-doppler operation.

Generic Pulse-Doppler Radar

Physically, this radar (Fig. 17) is no larger than many radars of the sort just described. Yet it provides a quantum improvement in performance. It can detect small aircraft at long ranges, even when their echoes are buried in strong ground clutter.

It can track them either singly or several at a time, while continuing to search for more. If desired, it can detect and track moving targets on the ground. And it can make real-time high-resolution SAR ground maps providing the same resolution at long ranges as at short. Moreover, besides these performance improvements the radar also achieves a quantum increase in reliability.

What makes the difference? The radar features three basic innovations:

- Coherence—enables detection of doppler frequencies

- Digital processing—ensures accuracy and repeatability

- Digital control—enables extreme flexibility

A simplified functional diagram of the radar is shown in Fig. 18. Comparing it with the corresponding diagram of the simple pulsed radar (Fig. 5), you will notice the following differences:

- Addition of a computer called the *radar data processor*

- Addition of a unit called the *exciter*

- Elimination of the synchronizer (its function is absorbed partly by the exciter but mostly by the data processor)

- Elimination of the modulator (its task is reduced to the point where it can be performed in the transmitter)

- Addition of a digital signal processor

- Elimination of the indicator (its functions are absorbed partly by the signal processor and partly by the data processor)

The added elements, as well as some important differences in the transmitter, antenna, and receiver, are briefly described in the following paragraphs.

Exciter. This element generates a continuous, highly stable, low-power signal of the desired frequency[3] and phase for the transmitter; and, precisely offset from it, local oscillator signals and a reference-frequency signal for the receiver.

17. No larger than many "pulsed" radars, the pulse-doppler radar has vastly greater capabilities.

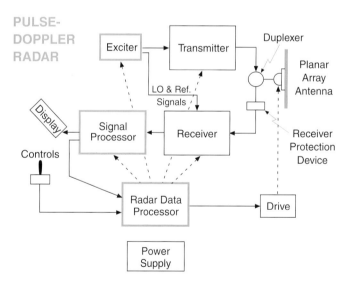

18. Principal elements of a pulse-doppler radar. Boxes with heavy borders were introduced with this generic system. Data processor controls all elements, verifies their operation, and isolates faults.

3. The frequency is selectable over a fairly wide range by the operator.

25

THE REMARKABLE GRIDDED TWT

The gridded traveling wave tube amplifier, or GTWT, is one of the key developments of the 1960's that made possible the truly versatile multimode airborne radar. With it, for the first time both the width and repetition frequency of a radar's high power transmitted pulses could not only be controlled precisely but be readily changed almost instantaneously to virtually any values within the power handling capacity of the tube. Added to these capabilities were those of the basic TWT: the high degree of coherence required for doppler operation; versatile, precise control of radio frequency; and the ability to conveniently code the pulse's radio frequency or phase for pulse compression.

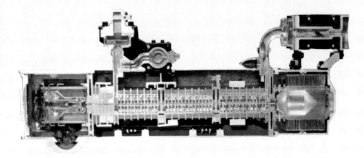

The Basic TWT. The TWT is one of a family of "linear beam" vacuum tube amplifiers (including the klystron), which convert the kinetic energy of an electron beam into microwave energy. In simplest form a TWT consists of four elements:

- Electron gun—produces the high-energy electron beam.

- Helix—guides the signal that is to be amplified.

- Collector—absorbs the unspent energy of the electrons, which are returned to the gun by a dc power supply.

- Electromagnet (solenoid)—keeps the beam from spreading as a result of the repulsive forces between electrons. (Often used instead is a chain of permanent magnets, called a periodic permanent magnet (PPM) since polarities of adjacent magnets are reversed.)

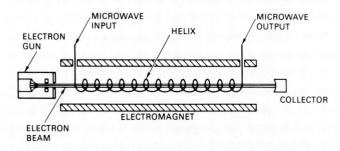

The microwave input signal is introduced at one end of the helix. Although the speed of the signal is essentially that of light, because of the greater distance the signal must cover in spiraling down the helix, its linear speed is slowed to the point where it travels slightly slower than the electrons in the beam. (For this reason, the helix is called a slow-wave structure.)

As the signal progresses, it forms a sinusoidal electric field that travels down the axis of the beam. Those electrons which

happen to be in positive nodes are speeded up by this field, and those in the negative nodes are slowed down. The electrons therefore tend to form bunches around the electrons at the nulls, whose speed is unchanged.

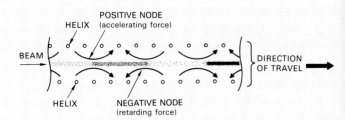

The traveling bunches in turn produce a strong electromagnetic field. Since it travels slightly faster than the signal, this field transfers energy from the electrons to the signal, thereby amplifying it and slowing the electrons. The longer the helix, the more the signal is amplified. In high gain tubes, attenuators called "severs" must be placed at intervals (of 20 to 35 dB gain) along the helix to absorb backward reflections which would cause self-oscillation. They reduce the gain somewhat (about 6 dB each) but have only a small effect on efficiency.

When the signal reaches the end of the helix, it is transferred to a waveguide which is the output port of the tube. The remaining kinetic energy of the electrons—which may amount to as much as 90 percent of the energy originally imparted by the gun—is absorbed as heat in the collector and must be carried away by cooling. Much of the unspent energy, though, can be recovered by making the collector negative enough (depressed collector) to decelerate the electrons before they strike it. (Kinetic energy is thus converted back to potential energy.)

High Power TWTs. Both the average and the peak power of helix TWTs are somewhat limited. As the *average* power is increased, an increasing number of electrons are intercepted by the helix, and it becomes difficult to remove enough of the resulting heat to avoid damage to the helix. As the required *peak* power is increased, the beam velocity must be increased, and a point is soon reached where the helix must be made too coarse to provide good interaction with the beam. In high power tubes, therefore, other slow wave structures are generally used. The most popular is a series of coupled cavities.

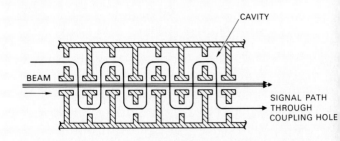

The Control Grid. While a pulsed output can be obtained by turning the tube "on" and "off", the pulses can be formed much more conveniently by interposing a grid between the cathode

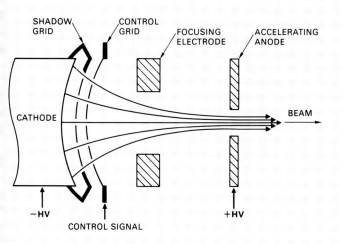

that emits the electrons and the anode whose positive voltage relative to the cathode accelerates them.

A low voltage control signal applied to this grid can turn the beam "on" and "off." To keep the grid from intercepting electrons and being damaged, it is placed in the shadow of a second grid which is electrically tied to the cathode. To eliminate all output between pulses, the low voltage microwave input signal may be pulsed.

Advantages. Besides the advantages listed earlier, the TWT can provide high-power outputs with gains of up to 10,000,000 or more and efficiencies of up to 50 percent. Low-power helix tubes have the added advantage of providing bandwidths of as much as two octaves (maximum frequency four times the minimum). In high-power tubes, though, where other slow-wave structures must be used, this is generally reduced to 5 to 20 percent, although some coupled cavity tubes having much greater bandwidths have been built.

Transmitter. The transmitter is a high-power amplifier of a type called a *gridded traveling-wave tube*, TWT (Fig. 19).

Keyed on and off to cut coherent pulses from the exciter's signal, it amplifies the pulses to the desired power level for transmission. As explained in the panel on the opposite page and above, the tube is turned on and off by a low-power signal applied to a control grid.

By appropriately modifying this signal, the width and repetition frequency of the high-power transmitted pulses can easily be changed to satisfy virtually any operating requirement.

Similarly, by modifying the exciter's low power signal, the frequency, phase, and power level of the high-power pulses can readily be changed, modulated, or coded for pulse compression (Fig 20).

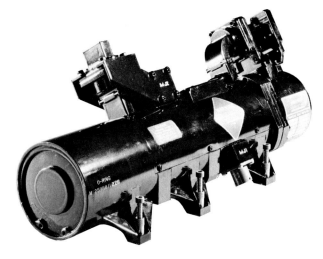

19. Gridded traveling-wave tube amplifies low-power wave from exciter to power required for transmission. Can readily be turned on and off with low-power control signal.

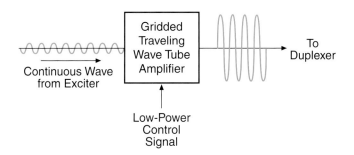

20. By keying the TWT with a low power control signal, the width and PRF of the high power pulses can readily be changed. And by modifying the low-power input provided by the exciter, the frequency, phase, and power of the pulses can readily be changed or modulated.

21. Planar array antenna. Radio waves are radiated though slots cut in a complex of waveguides behind the face of the antenna.

Antenna. The antenna is of a type called a planar array. Instead of employing a central feed that radiates the transmitted wave into a reflector, it consists of an array of many individual radiators distributed over a flat surface (Fig. 21). The radiators are slots cut in the walls of a complex of waveguides behind the antenna's face.

Though a planar array is more expensive than a dish antenna, its feed can be designed to distribute the radiated power across the array so as to minimize the radiated sidelobes, as is essential in some MTI modes. Also, the feed can readily be adapted to enable monopulse measurement of angle-tracking errors.

Receiver. This receiver (Fig. 22, bottom of page) differs in many respects from that described earlier. First, a low-noise preamplifier ahead of the mixer increases the power of the incoming echoes so that they can better compete with the electrical noise inherently generated in the mixer.

Second, more than one intermediate frequency translation is generally performed to avoid problems with image frequencies (see Chapter 5, page 64).

Third, the video detector is of a special type called a *synchronous detector* (Fig. 23). To detect doppler frequency shifts—which show up as pulse-to-pulse phase shifts—it beats the doppler-shifted received echoes against a reference signal from the exciter. Two bipolar video outputs are produced: the in-phase (I) and quadrature (Q) signals. Their amplitudes are sampled at intervals on the order of a pulse width.

The vector sum of the I and Q samples is proportional to the energy of the sampled signal: their ratio indicates the phase of the signal. The samples are converted into numbers by the analog-to-digital (A/D) converter and supplied to the signal processor.

Finally, to enable monopulse tracking, at least two parallel receiver channels must be provided.

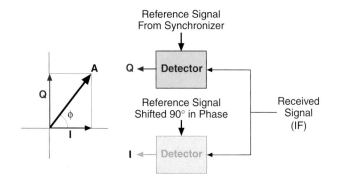

23. Synchronous detector. Vector sum of I and Q outputs is proportional to amplitude, A, of received signal. Ratio of outputs indicates the signal's phase, ϕ. Direction in which ϕ changes with time indicates whether the frequency of the signal is higher or lower than the reference frequency.

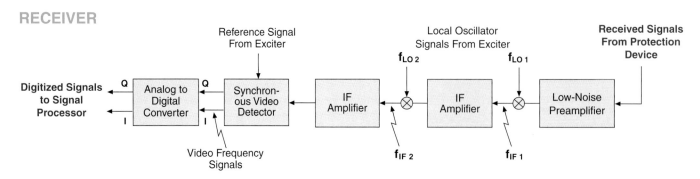

22. Receiver of generic pulse-doppler radar. To enable digital doppler filtering, synchronous video detector provides in-phase (I) and quadrature (Q) video outputs. To enable monopulse tracking, two receiver channels such as this must be provided.

Signal Processor. This processor (Fig. 24) is a digital computer specifically designed to efficiently perform the vast number of repetitive additions, subtractions, and multiplications required for real-time signal processing. Into it, the data processor loads the program for the currently selected mode of operation.

As required by this program, the signal processor (Fig. 25, bottom of page) sorts the incoming numbers from the A/D converter by time of arrival, hence range; stores the numbers for each range interval in memory locations called *range bins;* and filters out the bulk of the unwanted ground clutter on the basis of its doppler frequency. By forming a bank of narrowband filters for each range bin, the processor then integrates the energy of successive echoes from the same target (i.e., echoes having the same doppler frequency) and still further reduces the background of noise and clutter with which the target echoes must compete.

By examining the outputs of all the filters, the processor determines the level of the background noise and residual clutter, just as a human operator would by observing the range trace on an "A" display. On the basis of increases in amplitude above this level, it automatically detects the target echoes.

Rather than supplying the echoes directly to the display, the processor temporarily stores the targets' positions in its memory. Meanwhile, it continuously scans the memory at a rapid rate and provides the operator with a continuous bright TV-like display of the positions of all targets (Fig. 26). This feature, called *digital scan conversion,* gets around the problem of target blips fading from the display during the comparatively long azimuth scan time. The target positions are indicated by synthetic blips of uniform brightness on a clear background, making them extremely easy to see.

In the SAR ground mapping modes, the ground return is not clutter; rather it is signal, so it is not filtered out. To

24. Signal processor. Stored program for the selected mode of operation is automatically entered by the data processor.

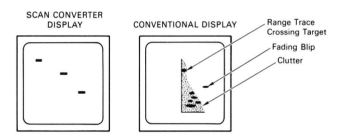

26. Scan converter provides continuous clean bright display of positions of all targets. In contrast, video signals drawn on conventional display-tube face by range trace, vary in brightness and rapidly fade away.

SIGNAL PROCESSOR

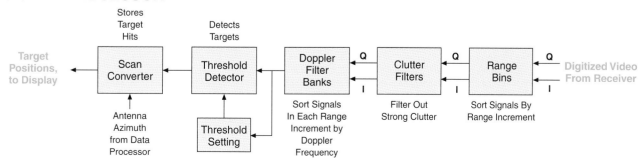

25. Signal processor sorts radar returns by range, storing them in range bins; filters out the strong clutter; then sorts the returns in each range bin by doppler frequency. Targets are detected automatically.

27. To provide a truly pictorial ground map, actual digital filter outputs are stored in the scan converter and continuously scanned for display.

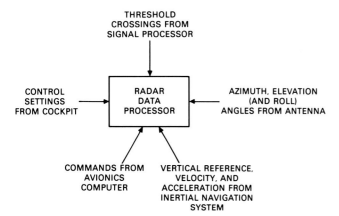

28. Principal inputs to the data processor.

29. B-2 bomber. Even a fairly large stealth aircraft has a radar cross section no larger than that of a bird.

provide fine range resolution without limiting detection range, the radar transmits wide pulses and employs large amounts of pulse compression. To provide fine azimuth resolution, the processor stores the returns of thousands of pulses from each range increment and integrates them to form very large banks of doppler filters having extremely narrow passbands. The filter outputs themselves are stored in the scan converter, which is scanned to produce a pictorial map on the radar display (Fig. 27).

Data Processor. A general-purpose digital computer, the data processor controls and performs routine computations for all units of the radar (Fig. 28). Monitoring the positions of selector switches on the control panel, it schedules and carries out the selection of operating modes, e.g., long-range search, track-while-scan, SAR mapping, close-in combat, etc. Receiving inputs from the aircraft's inertial navigation system, it stabilizes and controls the antenna during search and track. On the basis of inputs from the signal processor, it controls target acquisition, making it necessary for the operator only to bracket the target to be tracked, with a symbol on the display.

During automatic tracking, the data processor computes the tracking error signals in such a way as to anticipate the effects of all measurable and predictable variables—the velocity and acceleration of the radar bearing aircraft, the limits within which the target can reasonably be expected to change its velocity, the signal-to-noise ratio, and so on. This process yields extraordinarily smooth and accurate tracking.

Throughout, the data processor monitors all operations of the radar, including its own. In the event of a malfunction, it alerts the operator to the problem, and through built-in tests, isolates the failure to an assembly that can readily be replaced on the flight line.

Generic Radar for Stealth

In 1974 while reviewing the air battles of Vietnam and the Middle East, the U.S. Air Force concluded that in the future its aircraft would have great difficulty in getting through strong air defenses unless their detectability by radar could be reduced. Consequently, development was begun on what have come to be called low observable, or stealth, aircraft. Loosely speaking, a conventional fighter has a radar reflectivity—radar cross section (RCS)—comparable to that of a van. By contrast, even a fairly large stealth aircraft has an RCS no greater than that of a bird (Fig. 29).

What does that have to do with the design of radars for such aircraft?

Viewed broadside, the antenna of a conventional fighter's radar alone has an RCS many times that of the fighter. To put such an antenna in the nose of a stealth aircraft would be grossly counterproductive, to say the least. Furthermore, even if the aircraft managed to avoid being detected, the signals radiated by the radar would be intercepted by the enemy at long ranges, revealing both the aircraft's presence and its location. For these reasons, the first U.S. stealth fighter (Fig. 30) didn't even carry a radar.

Severe as these problems are, both can be acceptably resolved.

Reducing Antenna RCS. The first of several measures which must be taken to minimize the RCS of a radar's antenna is to mount it in a fixed position on the aircraft structure, tilted so that its face will not reflect radio waves back in the direction of an illuminating radar (Fig. 31).

The radar beam cannot then, of course, be steered mechanically. This requirement significantly influences the radar's front-end design. There are several possible approaches to nonmechanical beam steering.

The simplest and most widely used is the passive electronically steered array (ESA). It is a planar array antenna, in which a computer-controlled phase shifter is inserted in the feed system immediately behind each radiating element (Fig. 32). By individually controlling the phase shifters, the beam formed by the array can be steered anywhere within a fairly wide field or regard.[4]

A more versatile, but considerably more expensive, implementation is the active ESA. It differs from the passive ESA in having a tiny transmitter/receiver (T/R) module inserted behind each radiating element (Fig. 33). To steer the beam, provisions are included in each module for controlling both the phase and the amplitude of the signals the module transmits and receives.

30. Since no radar at the time had both a low-RCS antenna and a low probability of its signals being usefully intercepted by an enemy, the first U.S. stealth fighter, the F-117, was not equipped with a radar.

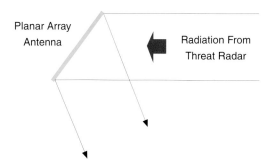

31. A first step in reducing the RCS of a radar antenna is to mount it in a fixed position, tilted so its face won't reflect radiation back to a radar.

4. Another name for this sort of antenna, commonly used in ground-based radars, is *phased array*.

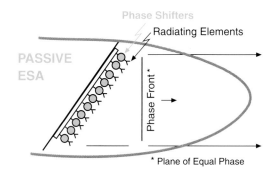

32. Passive electronically steered array antenna (ESA). By controlling the phase of the signals transmitted and received by each radiating element, the phase shifters can steer the radar beam anywhere within the field of regard.

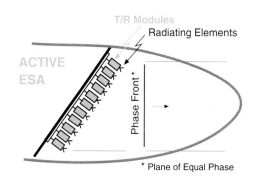

33. Active electronically steered array antenna (ESA). Transmit-receive (T/R) modules steer the radar beam by controlling the phase and amplitude of the signals radiated and received by each radiator.

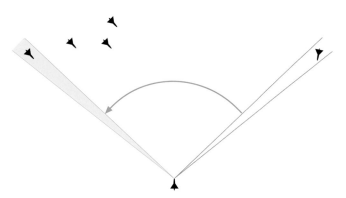

34. Since the radar beam has no inertia, with electronic steering it can be jumped anywhere within the field of regard in less than a millisecond.

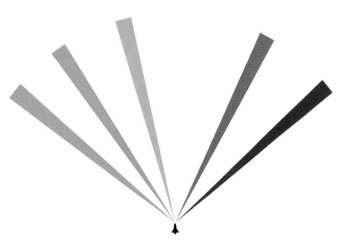

35. With an active ESA, the radar can even simultaneously radiate multiple, independently steerable beams on different frequencies.

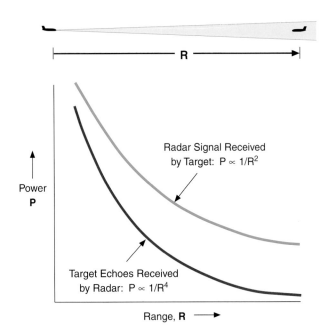

36. Handicap surmounted by a radar designed to have a low probability of its signals being usefully intercepted.

Another approach, still in its infancy, is photonic true-time-delay (TTD) beam steering. In it, the phase of the signals radiated and received by the individual T/R modules of an active ESA is controlled by introducing variable time delays in the elements' feeds, which are optical fibers. Their lengths, hence the time the signals take to pass through them, are varied by switching segments of fiber of selectable length into and out of each feed. This greatly broadens the span of frequencies over which the antenna can operate.

As will be explained in Chaps. 37 and 38, ESAs have many advantages. One of the more important is extreme beam agility (Fig. 34). Because the beam—as opposed to the conventional gimbaled antenna—has no inertia, it can, for example, interactively jump to one or another of several targets whenever and for whatever length of time is optimum for tracking it, without appreciably interrupting the beam's search scan. Among the special advantages of the active ESA is the ability to radiate multiple individually steerable beams on different frequencies (Fig. 35).

Avoiding Detection of the Radar's Signals. Keeping a radar's signals from being usefully intercepted by an enemy is especially challenging. As we saw in Chap. 1, because of the spreading of the radio waves occurring both on the way out to a target at a range, R, and on the way back to the radar, the strength of the target's echoes decreases as $1/R^4$. The strength of the radar's signals received by the target, however, decrease only as $1/R^2$ (Fig. 36).

To get around this huge handicap, an entire family of low-probability-of-intercept, LPI, features has been developed.

- Taking full advantage of the radar's ability to coherently integrated the target echoes

- Interactively reducing the peak transmitted power to the minimum needed at the time for target detection

- Spreading the radar's transmitted power over an immensely broad band of frequencies

- Supplementing the radar data with target data obtained from infrared and other passive sensors and offboard sources

- Turning the radar on only when absolutely necessary

Astonishing as it may seem, by combining these and other LPI techniques, the radar can detect and track targets without its signals being usefully intercepted by the enemy.

Meeting Stealth's Processing Requirements. Electronic beam steering and LPI, along with other advanced techniques, depend critically on immensely high digital processing throughputs. Despite the limited space available in high performance tactical aircraft, orders-of-magnitude increases in throughput have been realized through the use of very large scale integrated circuits, CMOS technology, and the distribution of processing tasks among a great many (up to a hundred or more) individual processing elements, operating in parallel and sharing bulk memories.

Further enhancing processing efficiency is *integrated* processing. Rather than providing separate processors for the aircraft's radar, electro-optical, and electronic warfare systems, a single integrated processor serves them all (Fig. 37). Size, weight, and cost are thereby reduced.

Further, with dramatically reduced memory costs, it has become possible to perform both signal and data processing in real time with commercial processing elements.

Summary

Illustrative of the various approaches to implementation are three generic designs: a simple "pulsed" radar, a "pulse-doppler" radar, and a radar for stealth.

The pulsed radar employs a magnetron transmitter, a parabolic-reflector antenna, and a superheterodyne receiver. Triggered by timing pulses from a synchronizer, a modulator provides the magnetron with pulses of dc power, which it converts into high-power microwave pulses. These are fed through a duplexer to the antenna. Echoes received by the antenna are fed by the duplexer through a protection device to the receiver, which amplifies and converts them into video signals for display on a range trace.

The pulse-doppler radar differs from the simple pulsed radar primarily in being coherent and largely digital. The transmitter, a gridded traveling-wave-tube amplifier, cuts pulses of selectable width and PRF from an exciter's low-power continuous wave—codable for pulse compression. The antenna is a planar array having a monopulse feed. The receiver features a low-noise preamplifier and a video detector, whose I and Q outputs are sampled at intervals on the order of a pulse width, digitized, and provided to a digital signal processor. It sorts them by range and doppler frequency, filters out the ground clutter, and automatically detects the target echoes, storing their locations in a memory continuously scanned to provide a TV-like display.

All operations of the radar are controlled by a digital computer (radar data processor), which loads the program for the selected mode of operation into the signal processor.

Implementation of a pulse-doppler radar for stealth differs from the foregoing principally in (a) having a fixed, low-RCS electronically steered antenna (ESA) and (b) incorporating features which minimize the possibility of its signals being usefully intercepted by an enemy.

37. A technician inserts a module into the integrated processor which jointly serves the radar, electro-optical, and electronic warfare systems of the F-22.

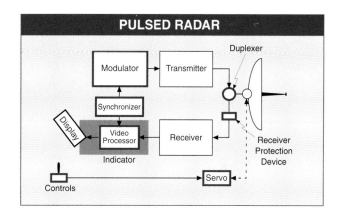

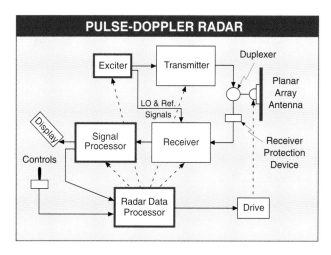

VOUGHT F4U–4N CORSAIR (1944)

This radar-equipped fighter was used very successfully by United States Navy and Marine pilots in World War II and the Korean conflict. Lt. Guy P. Bordelon became the only Navy Ace in Korea, achieving five kills with this aircraft at night.

Representative Applications 3

Having become acquainted with the basic radar principles and approaches to their implementation, in this chapter we'll briefly look at representative practical uses of airborne radar. Some of these—such as air-to-air collision avoidance, ice patrol, and search and rescue (Fig. 1)—are primarily civil applications. Others—such as early warning and missile guidance—are military. Still others—such as storm avoidance and windshear warning—are both.

1. Coast Guard helicopter, equipped with a multi-function radar having search, weather, and beacon modes.

Hazardous-Weather Detection

Three common threats to the safety of flight are turbulence, hail, and—particularly at low altitudes—windshears or microbursts, all of which are common products of thunder storms. One of the most common uses of airborne radar is alerting pilots to these hazards.

Storm Avoidance. If the radio frequency of a radar's transmitted pulses is appropriately chosen, the radar can see through clouds yet receive echoes from rain within and beyond them. The larger the rain drops, the stronger their echoes. So by sensing the rate of change of the strength of the echoes with range, the radar can detect thunder storms. And by scanning a wide sector ahead, the radar can display those regions in which hazardous weather and turbulence are apt to be encountered, hence should be avoided (Fig. 2).

Windshear Warning. Windshears are strong down drafts which can occur unexpectedly in thunder storms. At low altitudes the outflow of air from the core of the down draft can cause an aircraft to encounter an increasing headwind when flying into the down draft and a strong tail wind when emerging from it (Fig. 3). Without warning, this combination of conditions can cause an aircraft taking off or landing to crash.

Pulse doppler weather radars are sensitive not only to the intensity of the rainfall but also to its horizontal velocity, hence to the winds within a storm. By measuring the rate of change of the horizontal winds, these radars can detect a wind shear embedded in rain as much as 5 miles ahead, giving the pilot up to around 10 seconds of warning to avert it.

Navigational Aid

Among common navigational uses of airborne radar are marking the locations of remote facilities, assisting air traffic control, preventing air-to-air collisions, measuring absolute altitude, providing guidance for blind low altitude flight, and measuring the range and altitude of points on the ground ahead.

Marking Remote Facilities. For approaching helicopters and airplanes, the locations of off-shore drilling platforms, remote air fields, and the like may be marked with radar beacons. The simplest beacon—called a transponder—consists of a receiver, a low-power transmitter, and an omnidirectional antenna (Fig. 4). The transponder receives the pulses of any radar whose antenna beam sweeps over it and transmits "reply" pulses on a different frequency. Even

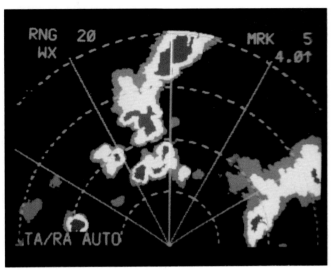

2.   Display of a weather radar employed on commercial airliners. Color coding indicates intensity of precipitation and turbulence.

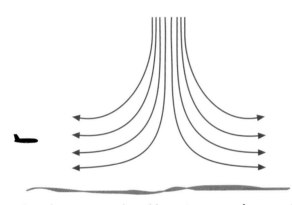

3.   Flow of air in a typical windshear. As an aircraft approaches the down draft, it encounters increasing head winds. As it emerges from the down-draft, it encounters strong tail winds.

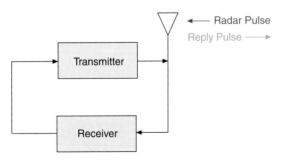

4.   Simple beacon transponder. Upon receiving a pulse from a radar, the transponder transmits a reply on another frequency.

though low powered, the replies are much stronger than the radar's echoes. And since their frequency is different from the radar's, they are not accompanied by clutter, but stand out clearly on the radar display.

A more capable beacon system (Fig. 5) includes an interrogator. It transmits coded interrogating pulses in response to which transponders return coded replies. The most common beacons of this sort are those of the air traffic control beacon system (ATRBS).

Assisting Air-Traffic Control. ATRBS transponders are carried on all but the smallest private aircraft. An ATRBS interrogator operates in conjunction with the air traffic control radar at every major airport. The interrogator's monopulse antenna is mounted atop the radar antenna, hence scans with it (Fig. 6), and the interrogator's pulses are synchronized with the radar's. Consequently, the operator can interrogate an incoming aircraft simply by touching its "blip" on the radar display with a light pen.

Ordinarily the interrogator uses only two of several possible codes. One requests the identification code of the aircraft carrying the transponder. The other requests the aircraft's altitude. Every beacon-equipped aircraft can thus be positively identified and its position accurately determined in three dimensions.[1]

Avoiding Air-to-Air Collisions. Another use of the ATRBS transponders is made by the traffic alert and collision avoidance system (TCAS II). Typically, integrated with an aircraft's weather radar, TCAS interrogates the air traffic control transponders in whatever aircraft happen to be within the search scan of the radar. From a transponder's replies, TCAS determines the aircraft's direction, range, altitude separation, and closing rate. Based on this information, TCAS prioritizes threats, interrogates high-priority threats at an increased rate, and if necessary give vertical and horizontal collision avoidance commands.

Measuring Absolute Altitude. In a great many situations, it is desirable to know an aircraft's absolute altitude.[2] Since beneath the aircraft there is usually a large area of ground at very nearly the same range (Fig. 7), a small low-power, broad-beam, downward-looking CW radar employing FM ranging can provide a continuous precise reading of absolute altitude. Interfaced with the aircraft's autopilot, the altimeter can ensure smooth tracking of the glide slope for instrument landings.

Altimeters may also be pulsed. For military uses, the probability of the altimeter's radiation being detected by an enemy is minimized by transmitting pulses at a very low

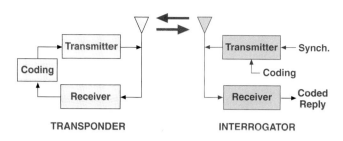

5. A complete radar beacon system. Interrogator is typically synchronized with a search radar, and the transponder's replies are shown on the radar's display.

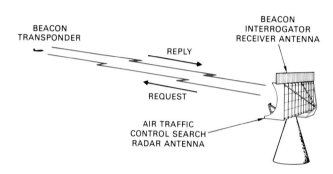

6. Antenna of ATRBS beacon interrogator is mounted atop antenna of air traffic control radar. Through coding of beacon pulses and replies, radar identifies approaching aircraft and obtains their altitudes and other flight data.

1. Sixteen million identification codes are available; so, every aircraft can be assigned a unique code.

2. Distance to the ground.

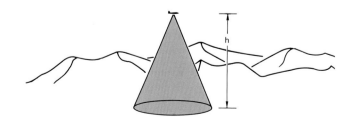

7. An aircraft's absolute altitude can be precisely determined by measuring the range to the ground beneath it with a small low-powered broad-beamed radar.

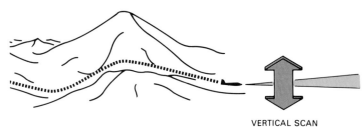

VERTICAL SCAN

8. For terrain following, a radar scans the terrain ahead vertically with a pencil beam.

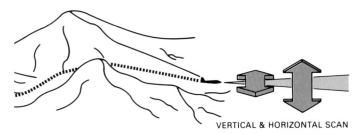

VERTICAL & HORIZONTAL SCAN

9. For terrain avoidance, the radar alternately scans terrain ahead vertically and horizontally.

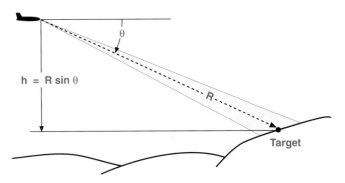

$h = R \sin \theta$

Target

10. Measurement of the range and the relative altitude of a point on the ground. The range the radar measures is that at which the elevation tracking-error signal is zero.

3. Damp out the earth's radius oscillation inherent in inertial systems.

PRF and employing large amounts of pulse compression to spread the pulses' power over a very wide band of frequencies.

Enabling Blind Low-Altitude Flight. To enable an attack fighter to avoid observation and enemy fire through "hedge-hopping" tactics, two basic radar modes have been developed: terrain following, and terrain avoidance.

In terrain following (Fig. 8), an aircraft's forward-looking radar scans the terrain ahead by sweeping a pencil beam vertically. From the elevation profile thus obtained vertical steering commands are computed. Supplied to the flight control system, they automatically fly the aircraft safely at terrain-skimming altitude.

Terrain avoidance (Fig. 9) is similar to terrain following, except that periodically the radar scans horizontally, enabling the aircraft not only to hug the ground but to fly around obstacles in its path. The aircraft is generally flown manually.

For pilotless aircraft a mode called TERCOM, for terrain contour mapping, is also available. It flies the aircraft on a precisely timed, preprogrammed ground-hugging trajectory along a known contour on a map. Ground clearance is measured with a very low power radar altimeter. Since it illuminates only the ground beneath the aircraft, the possibility of enemy detection is low. That may be further reduced by operating at frequencies for which atmospheric attenuation is high.

Forward Range and Altitude Measurement. On a bombing run over ground that is neither flat nor level, it is often necessary to precisely determine the range and altitude of the aircraft relative to the target. That can be done by training the radar beam on the target and measuring

(a) the antenna depression angle and

(b) the range to the ground at the center of the radar beam (Fig. 10).

This range may be identified by the return from it producing a nearly zero elevation tracking-error signal from a monopulse (or lobing) antenna.

Precision Velocity Update (PVU). As we saw in Chap. 1, by measuring the doppler frequency of the returns from three points on the ground ahead, a forward-looking radar can measure the radar's velocity. Such measurements can be used to 'update'[3] the aircraft's inertial navigation system. If the inertial system fails, the radar serves as a doppler navigator.

Ground Mapping

Radar ground-mapping applications are legion. They range from ice patrol and high resolution terrain mapping, to law enforcement and autonomous blind landing guidance, to name just a few.

Ice Patrol. One of the oldest civil radar mapping applications is charting passages through the ice in waters that freeze over during the winter. For this, patrol aircraft are equipped with real-beam mapping radars called SLARs, having long linear array antennas that look out on both sides of an aircraft. While the aircraft flies in a straight line, an optical scanner records the radar returns on film, thereby making a strip map of the passing scene.

Although SLAR resolution is limited, at short ranges it is quite adequate for mapping ice (Fig. 11). Moreover, because the radar is simple and its antennas are fixed, it is comparatively inexpensive.

High Resolution Terrain Mapping. For such applications as navigation, environmental monitoring, and geological exploration, SAR has the advantage of providing high resolution even at long ranges. Interferometric, 3-D SAR has proved especially useful for highly accurate, low cost terrain mapping (Fig 12). It also has military applications.

When mapping areas covered with dense tropical rain forests, a radar that transmits extremely short pulses may measure the distance to the ground under the canopy of trees—a technique called sounding.

Law Enforcement. Both SLAR and SAR have played important roles in oil-spill detection, fishery protection, and the interdiction of smugglers and drug traffickers. Since SAR can provide fine resolution at long ranges, it has the advantage of uncovering illicit activities without alerting the law breakers (Fig. 13).

11. Ice flow on Lake Erie, mapped by a real-beam side-looking array radar (SLAR) having long, fixed array antennas that look out on either side of the aircraft.

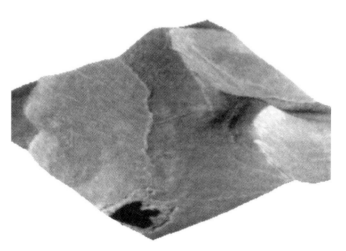

12. A representative interferometric, 3-D SAR map. (Crown copyright DERA Malvern)

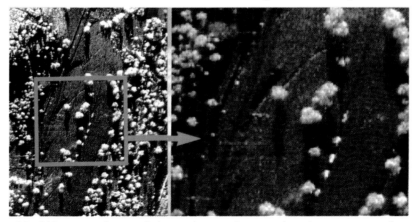

13. SAR map, such as might be used to interdict smugglers, shows a convoy of trucks on an off-road trail. As indicated by radar shadows of trees, map was made from a long stand-off range.

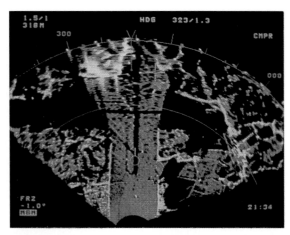

14. Forward-looking radar with monopulse antenna fills in gap in SAR map with real-beam mapping. Provides sufficient resolution to enable blind approaches at landing strips where navigation aids may not be available. (Courtesy Northrop Grumman)

15. Long-range, long-endurance unmanned reconnaissance aircraft, may relay 1-foot resolution SAR maps via satellite directly to users in the field.

16. The U. S. Navy carrier based E-2C Hawkeye early warning and sea surveillance aircraft has a large antenna housed in a circular radome (rotodome), which rotates continuously to provide 360° coverage.

Blind-Landing Guidance. The ground directly ahead of an aircraft cannot be mapped with SAR. So, for landing guidance, other techniques must be employed. One is to scan the narrow region ahead with a monopulse antenna. At the short ranges involved sufficiently fine azimuth resolution may be obtained to enable an aircrew to locate runways and markers (Fig. 14) and so make autonomous approaches to small or unimproved landing strips at night or in bad weather.

Reconnaissance and Surveillance

In military operations, airborne radar has proved invaluable for its ability to see through smoke, haze, clouds, and rain; to rapidly search vast regions; to detect targets at long ranges, and to simultaneously track a great many targets which may be widely dispersed.

We'll consider four representative applications here: long-range air-to-ground reconnaissance, early warning, air-to-ground battle surveillance, and balloon-borne low-altitude surveillance.

Long-Range Air-to-Ground Reconnaissance. Throughout the cold war, very high resolution SAR radars in the U-2 and later the higher flying TR-1 provided all-weather surveillance over the military buildup in the Soviet Union. During the war in the Persian Gulf, SAR also proved invaluable in pinpointing ground targets for fighters and bombers.

In the late 1990s, SAR radars were developed for both such missions in small pilotless reconnaissance aircraft capable of long-range endurance flight (Fig. 15). These radars may relay radar images of one-foot resolution via satellite directly to users in the field.

Early Warning and Sea Surveillance. An airborne radar can detect low-flying aircraft and surface vessels at far greater ranges than can a radar on the ground or the mast of a ship. Accordingly, to provide early warning of the approach of hostile aircraft and missiles and to maintain surveillance over the seas, radars are placed in high-flying loitering aircraft, such as the U.S. Navy Hawkeye and the U.S. Air Force AWACS (Figs. 16 and 17).

Because these aircraft are large and slow, the radars they carry can employ antennas large enough to provide high angular resolution while operating at frequencies low enough that atmospheric attenuation is negligible. And they can transmit very high powers.

Providing 360° coverage, they can detect low-flying aircraft out to the radar horizon—which at an altitude of

30,000 feet is more than 200 nautical miles—and detect higher altitude targets at substantially greater ranges. In addition, they can simultaneously track hundreds of targets.

Air-to-Ground Surveillance and Battle Management. Very much as AWACS provides surveillance over a vast air space, an airborne radar can also provide surveillance over a vast area on the ground.

Equipped with a long electronically steered side-looking antenna (Fig. 18), the U. S. Joint STARS radar detects and tracks moving targets on the ground with MTI and detects stationary targets with SAR.

17. The U.S. Air Force E-3 AWACS aircraft carries a high-power pulse doppler radar. Its 24-foot-long antenna is also housed in a rotodome.

18. Passive ESA of joint STARS radar is housed in a 24-foot-long radome. Radar performs SAR mapping and ground-moving target detection and tracking for surveillance and battle management.

Flying in a race-track pattern at an altitude of 35,000 feet a hundred miles behind a hostile border, the radar can maintain surveillance over a region extending a hundred or more miles into enemy territory. Through secure communication links, Joint STARS can provide fully processed radar data to an unlimited number of control stations on the ground.

Low Altitude Air and Sea Surveillance. A novel surveillance application of airborne radar arose in the U.S. war on drugs. The Customs Service undertook to implement a radar "fence" along the southern border of the U.S. by placing large-reflector, long-range surveillance radars in tethered balloons (Fig. 19).

Fighter/Interceptor Mission Support

The fighter/interceptor mission is twofold: to thwart attacks by aircraft and missiles, and to achieve control of the airspace over a given region. In both, the fighter's radar typically plays four vital roles: search, raid assessment, target identification, and fire control.

19. Aerostat carrying lightweight solid-state surveillance radar having a large parabolic reflector antenna. Tethered at 15,000 feet altitude, radar can detect small low altitude aircraft at ranges out to 200 miles. Aerostat can stay aloft for 30 days, remain operational in 70 mph winds, survive 90 mph winds.

20. Equipped with a high-power pulse-doppler radar, U.S. Navy F–14 air superiority fighter can provide surveillance over a huge volume of airspace. (Courtesy Grumman Aerospace Corp.)

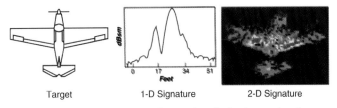

Target — 1-D Signature — 2-D Signature

21. 1–D and 2–D signatures of aircraft in flight obtained with a noncooperative target identification system.

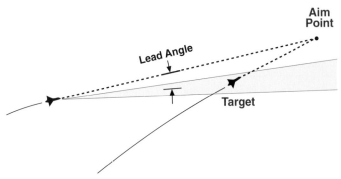

22. Lead-pursuit course for firing guns. Fighter's radar automatically locks onto target in an air-combat mode and tracks it in a single-target tracking mode.

Air-to-Air Search. The extent to which a fighter's radar must search for targets varies. At one extreme, the fighter may be "vectored" to intercept a target which has already been detected and is being precisely tracked. At the other extreme, the radar may be required to search a huge volume of air space for possible targets (Fig. 20).

Raid Assessment. Even if a radar has a narrow pencil beam, at long ranges it may not be able to resolve a close formation of approaching aircraft. Consequently, the fighter's radar is generally provided with a raid assessment mode.

In one version of this mode, the radar alternates between (a) track-while-scan to maintain situation awareness and (b) single target tracking of the suspect multiple target in a mode providing exceptionally fine range and doppler resolution.

Target Identification. To identify targets that are beyond visual range, some means of radar identification is generally desired.

The classical means is IFF, the World War II system upon which the civil-air-traffic-control beacon system was patterned. An IFF interrogator synchronized with the fighter's radar transmits interrogating pulses to which transponders carried in all friendly aircraft respond with coded replies. Despite use of sophisticated codes, the possibility of compromise is always present. So additional means of "noncooperative" target identification have been devised.

One of these is signature identification. It takes advantage of the unique characteristics of the echoes received from various aircraft to identify radar targets by type.

Another technique (Fig. 21) involves providing sufficiently fine range resolution that targets may be identified by their 1-D range profiles. Going a step further, by employing ISAR imaging, 2-D profiles may be provided.

Fire Control. Depending upon a target's range, the pilot may attack it with either the aircraft's guns or its guided missiles.

For firing guns, a selection of close-in combat modes may be provided in which the radar automatically locks onto the target in a single-target tracking mode and continuously supplies its range, range rate, angle, and angular rate to the aircraft's fire-control computer. The latter directs the pilot onto a lead-pursuit course against the target (Fig. 22) and at the appropriate range gives a firing command. Both steering instructions and firing command are presented on a head-up display; so the pilot need never take his eyes off the target.

Radar guided missiles, however, are often fired from beyond visual range. Representative examples are Phoenix and AMRAAM. Phoenix is a long-range missile used by the U.S. Navy F-14 air superiority fighter (Fig. 23). AMRAAM is a medium-range missile used by a wide variety of fighters. Both are generally launched while the fighter's radar is operating in a track-while-scan or search-while-track mode. Hence, several missiles may be launched in rapid succession and be in flight simultaneously against different targets.

Initially, Phoenix is guided inertially on a lofted trajectory. It then transitions to semi-active guidance, in which a radar seeker it carries homes on the periodic target illumination provided by the fighter's scanning radar. At close range, the seeker switches to active guidance, in which it provides its own target illumination.

AMRAAM (Fig. 24) is equipped with a command-inertial guidance system. It steers the missile on a preprogrammed intercept trajectory based on target data obtained by the fighter's radar prior to launch. If the target changes course after launch, target update messages are relayed to the missile by coding the radar's normal transmissions. Picked up by a receiver in the missile, the messages are decoded and used to correct the course set into the inertial guidance system.[4] For terminal guidance, the missile switches control to a short-range active radar seeker that it carries.

A third commonly used radar-guided missile is Sparrow. It is launched in a single-target-track mode and throughout its flight semiactively homes on the target illumination provided by the radar.

Air-to-Ground Weapon Delivery

Radar may play an important role in a wide variety of air-to-ground attacks. To illustrate, we'll look briefly at hypothetical missions of four different types: tactical-missile targeting, tactical bombing, strategic bombing, and ground-bases-defense suppression. In each, the basic strategy is to take advantage of radar's unique capabilities, while minimizing radiation from the radar.

Tactical-Missile Targeting. In this hypothetical mission, an attack helicopter lurks behind a hill overlooking a battlefield. With only the antenna pod of a short-range, ultra high-resolution (millimeter wave) radar atop the rotor mast showing (Fig. 25), the radar quickly scans the terrain for potential targets. Automatically prioritizing the targets it detects, the radar hands them off to a fire control system which fires small independently guided "launch and leave" missiles against them.

23. Long-range Phoenix missile is launched from F-14 air-superiority fighter. Missile homes semi-actively on periodic target illumination provided by fighter's scanning radar; converts to active guidance at close range.

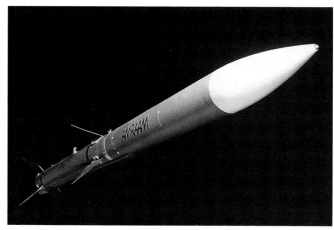

24. AMRAAM is inertially guided on preprogrammed intercept trajectory; receives update messages from radar if target maneuvers after launch. (Length. 12 ft.; range, 17+nmi)

4. If the missile is not in the radar beam at the time, the messages are received via the radar antenna's side lobes.

25. Small antenna of high-resolution millimeter-wave radar atop rotor mast enables attack helicopter to detect targets for its launch and leave missiles, while keeping out of sight from the battlefield.

Blind Tactical Bombing. In this hypothetical mission, a strike aircraft is guided by an inertial navigator on a terrain-skimming offset course to an area where a mobile missile launcher is believed to have been set up (Fig. 26, below). Upon reaching the area, the operator turns on the fighter's radar to update the navigator, then makes a single SAR map. With the map frozen on the radar display, the operator places a cursor over the target's approximate location. Turning the radar on again, he makes a detailed SAR map centered on the spot designated by the cursor.

Having identified the target, the operator places the cursor over it. Immediately, the pilot starts receiving steering instructions for the bombing run. At the optimum time, the bomb is automatically released.

By briefly breaking radio silence just three times, the radar has provided all the information needed to score a direct hit on the target, under conditions of zero visibility.

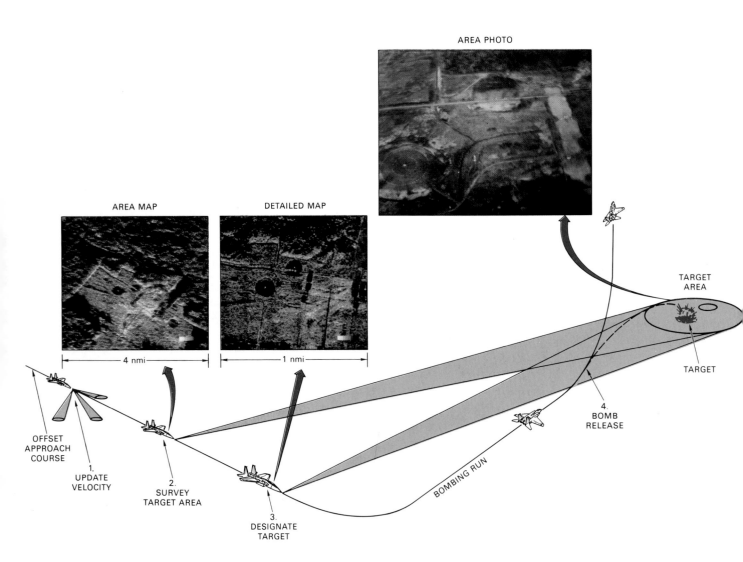

26. Representative blind bombing run. By turning radar on just three times, strike aircraft scores a direct hit from an offset approach course.

Precision Strategic Bombing. In this hypothetical mission, the flight crew of a stealth bomber, flying at some 20,000-feet altitude, turns the bomber's radar on just long enough to make a high-resolution map of an area where an enemy command center has been activated. This map, too, is frozen, but it is scaled to GPS coordinates. Upon identifying the target, the operator places a cursor over it, thereby entering the target's GPS coordinates into the GPS guidance system of a two thousand pound glide bomb.[6]

Automatically released at the optimum time, it glides out until it is almost directly over the target (Fig. 27), then dives vertically onto it with an accuracy of two or three feet.

Ground-Based-Defense Suppression. Ground-based enemy air-search radars and surface-to-air missile (SAM) sites, when radiating, may be put out of action with *high-speed anti-radiation missiles* (HARM).

In one hypothetical scenario, a specially equipped aircraft, lurking at low altitude outside the field of view of an enemy defense radar, determines its direction and range on the basis of data received via data link from other sources. The flight crew preprograms a HARM to search for the radar's signals. Launched in the direction of the radar, the missile soon acquires the radar's signals. Homing on them, it zooms in and destroys the radar before the enemy even realizes it is under attack.

Short-Range Air-to-Sea Search

Since its birth, airborne radar has played a key role in searching for both surface vessels and submarines.

Coast Guard aircraft typically are equipped with multifunction search and weather radars. Since virtually all vessels carry radar reflectors that return strong echoes, and since sea clutter is generally moderate compared to ground clutter, these radars, whether pulsed or pulse-doppler, can pick up small craft at long ranges.

While unable to see *beneath* the surface, radars providing fine resolution are widely used to detect periscopes and snorkels; ISAR is particularly useful for this.

Proximity Fuses

Another important application of airborne radar, which should not be overlooked, is proximity fuses (see panel alongside).

Conclusion

While we've looked only briefly at some of airborne radar's many applications, further information on radars for several of them is given in Part X.

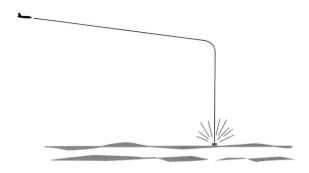

27. GPS guided bomb glides until it is almost directly over the target designated prior to launch on a SAR map made by the bomber's radar, then dives vertically onto it.

6. As a hedge against a GPS failure, alternate means of delivery are provided.

PROXIMITY FUSES
Then and Now

The earliest of these was the VT fuse of World War II. A tiny ultra-short-range CW radar, it detonated an artillery shell, when the return from the ground reached a predetermined amplitude, and an anti-aircraft shell, on the basis of the change in amplitude of the received signal as the shell approached an aircraft.

In guided missiles, much more sophisticated fuses are employed. They not only detect the presence of a target but time the detonation by such techniques as measuring the change in doppler frequency of the radar return as the missile approaches the target.

NORTHROP P–61A BLACK WIDOW (1944)

This was the first U.S. aircraft designed specifically as a night fighter. The Western Electric radar directed the firing of four 20 millimeter cannons and four 50 calibre machine guns. It entered squadron service in May 1944 and destroyed several enemy aircraft before the war ended in 1945.

PART II

Essential Groundwork

NORTH AMERICAN F–82F TWIN MUSTANG (1947)

At the start of the Korean war, the F–82F and G were the chief USAF night fighters. In 1950 an F–82G of the 68th Fighter Squadron achieved the first aerial victory of that war by downing a Russian built YAK–9 aircraft. The crew consisted of Lt. William Hudson, Pilot, and Lt. Carl Fraser, Radar Operator.

Radio Waves and Alternating Current Signals

Since radio waves and alternating current (ac) signals are vital to all radar functions, any introduction to radar logically begins with them. Indeed, many radar concepts which at first glance may appear quite difficult are simple when viewed in the light of a rudimentary knowledge of radio waves and ac signals.

In this chapter we will consider the nature of radio waves and their fundamental qualities.

Nature of Radio Waves

Radio waves are perhaps best conceived as energy that has been emitted into space. The energy exists partly in the form of an electric field and partly in the form of a magnetic field. For this reason, the waves are called electromagnetic.

Electric and Magnetic Fields. While neither field can be perceived directly, fields of both types are familiar to everyone. A common example of an electric field is that due to the charge which builds up between a cloud and the ground and produces lightning (Fig. 1). On a much smaller scale, another example of an electric field is that due to the charge which builds up on a comb on a particularly dry day, enabling the comb to attract a scrap of paper.

Examples of magnetic fields are equally common. At one extreme is the magnetic field that encircles the earth and to which compasses react. At the opposite extreme is the field surrounding a toy magnet, or the field produced by current flowing through the coil in a telephone ear piece, causing the diaphragm to vibrate and produce sound waves.

1. A common example of an electric field is that which builds up between a cloud and the ground.

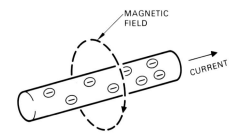

2.   Whenever an electric current flows,[1] a magnetic field is produced.

   1.   An electric current is a stream of charged particles, usually electrons.

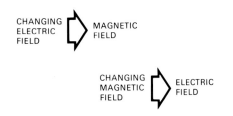

3.   Dynamic relationships giving rise to radio waves. If electric field varies sinusoidally, so will the magnetic field it produces. And if magnetic field varies sinusoidally, so will electric field it produces.

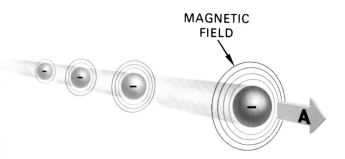

4.   Whenever an electric charge accelerates, a changing magnetic field is produced, and electromagnetic energy is radiated.

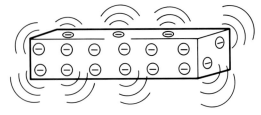

5.   Because of thermal agitation, everything around us radiates electromagnetic energy, a tiny portion of which is at radio frequencies.

In several important respects, the two types of fields are inextricably interrelated. For an electric current to flow—whether in a lightning bolt or in a telephone wire—an electric field must exist. And whenever an electric current flows (Fig. 2), a magnetic field is produced. The electromagnet is a common example.

If the fields vary with time, the interrelationship extends further. Any change in a magnetic field—increase or decrease in magnitude or movement relative to the observer—produces an electric field. We observe this relationship in the operation of electric generators and transformers. Similarly, although not so readily apparent, any change in an electric field produces a magnetic field. The effect is exactly the same as if an electric current actually flowed through the space in which the changing electric field exists.

Interestingly enough, the idea that a changing electric field might produce a magnetic field was conceived in the second half of the 19th century by James Clerk Maxwell. On the basis of this concept (Fig. 3) and the already demonstrated characteristics of electric and magnetic fields, he hypothesized the existence of electromagnetic waves and described their behavior mathematically (Maxwell's equations). Not until some 13 years later was their existence actually demonstrated (by Heinrich Hertz).

Electromagnetic Radiation. The dynamic relationship between the electric and magnetic fields—changing magnetic field produces an electric field and changing electric field produces a magnetic field—is what gives rise to electromagnetic waves. Because of this relationship, whenever a charge, such as that carried by an electron, accelerates—changes the direction or rate of its motion, hence changes the surrounding fields—electromagnetic energy is radiated (Fig. 4). The change in the motion of the charge causes a change in the surrounding magnetic field that is produced by the particle's motion. That change produces a changing electric field a bit further out, which in turn produces a changing magnetic field just beyond it, and on, and on, and on.

It follows that the sources of radiation are countless. As a result of thermal agitation, electrons in all matter are in continual random motion. Consequently, everything around us radiates electromagnetic energy (Fig. 5). Most of the energy is in the form of radiant heat (long wavelength infrared). But a tiny fraction invariably is in the form of radio waves. *Radiant heat, light, and radio waves are, in fact, all the same thing: electromagnetic radiation.* They differ only in wavelength.

In contrast to natural radiation, the waves radiated by a radar are produced by exciting a tuned circuit with a strong electric current. The waves, therefore, all have substantially the same wavelength and contain vastly more energy than that fraction of the natural radiation having the same wavelength.

How an Antenna Radiates Energy. The radiating element of most radar transmitters is generally buried at the origin of the system of waveguides that feed the radiation to the radar antenna. Consequently, we can get a clearer picture of how the radiation takes place by considering, instead, a simple elemental antenna in free space. For this purpose no better model can be found than the dipole used by Heinrich Hertz in his original demonstration of radio waves.

This antenna consists of a thin straight conductor, with flat plates like those of a capacitor at either end (Fig. 6). An alternating voltage applied at the center of the conductor causes a current to surge back and forth between the plates. The current produces a continuously changing magnetic field around the conductor. At the same time, the positive and negative charges that alternately build up on the plates as a result of the current flowing into and out of them produce a continuously changing electric field between the plates.

The fields are quite strong in the region immediately surrounding the antenna. And, as with the field of an electromagnet or the field between the plates of a capacitor, most of the energy each field contains returns to the antenna in the course of every oscillation.

But a portion does not. For the changing electric field between the plates produces a changing magnetic field just beyond it. That field in turn produces a changing electric field just beyond it; and so on.

Similarly, the changing magnetic field surrounding the conductor produces a changing electric field just beyond it; that field produces a changing magnetic field just beyond it, and so on.

By thus mutually interchanging energy, the electric and magnetic fields propagate outward from the antenna. Like ripples in a pond around a point where a stone has been thrown in (Fig. 7), the fields move on, long after the current that originally produced them has ceased. They and the energy they contain have escaped.

Visualizing a Wave's Field. Although the two fields can't be seen, both can be visualized quite easily.

The electric field may be visualized as the force it would exert on a tiny electrically charged particle suspended in

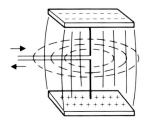

6.   Simple dipole antenna such as that used by Hertz to demonstrate radio waves.

7.   Like ripples on a pond, radio waves move on, long after the disturbance that produced them has ceased.

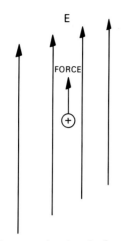

8.   Electric field is best visualized as the force it exerts on a
     charged particle.

the wave's path. The magnitude of the force corresponds to the field's strength (E); the direction of the force, to the field's direction. As in Fig. 8, the electric field is commonly portrayed as a series of solid lines whose directions indicate the field's direction and whose density (number of lines per unit of area in a plane normal to the direction) indicates the field strength.

The magnetic field may similarly be visualized as the force it would exert on a tiny magnet suspended in the wave's path. Again, the magnitude of the force corresponds to the field strength (H) and the direction, to the field's direction. This field is portrayed in the same way as the electric field, except that the lines are dashed (Fig. 9).

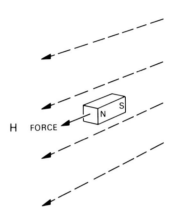

9.   Magnetic field is best visualized as the force it would exert on a
     tiny magnet.

Characteristics of Radio Waves

A radio wave has several fundamental qualities: speed, direction, polarization, intensity, wavelength, frequency, and phase.

Speed. In a vacuum, radio waves travel at constant speed—the speed of light, represented by the letter c. In the troposphere, they travel a tiny bit slower. Moreover, their speed varies slightly not only with the composition of the atmosphere, but with its temperature and pressure.

The variation, however, is extremely small—so small that for most practical purposes radio waves can be assumed to travel at a constant speed, the same as that in a vacuum. This speed is very nearly equal to 300×10^6 meters per second.

Direction. This is the direction in which a wave travels—the direction of propagation (Fig. 10). It is always perpendicular to the directions of both the electric and the magnetic fields. These directions, naturally, are always such that the direction of propagation is away from the radiator.

When a wave strikes a reflecting object, the direction of one or the other of the fields is reversed, thereby reversing

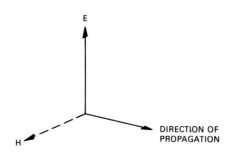

10. Direction of propagation is always perpendicular to directions
     of both electric and magnetic fields.

THE SPEED OF LIGHT AND RADIO WAVES

The speed of light in a nonmagnetic medium, such as the atmosphere, is

$$c = \frac{299.7925 \times 10^6}{(\kappa_e)^{1/2}} \text{ meters/s}^*$$

where κ_e is a characteristic, called the dielectric constant, of the medium through which the radiation is propagating. The dielectric constant for air is roughly 1.000536 at sea level.

Speed in the Atmosphere. The dielectric constant of the atmosphere varies slightly with the composition, temperature and pressure of the atmosphere. The variation is such that the speed of light is slightly higher at higher altitudes. The dielectric constant of the atmosphere also varies to some extent with wavelength. As a result, the speeds of light and radio waves are not quite the same, and the speed of radio waves is slightly different in different parts of the radio frequency spectrum.

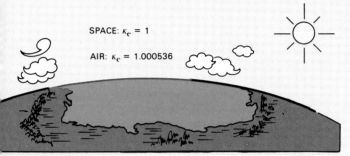

SPACE: $\kappa_e = 1$

AIR: $\kappa_e = 1.000536$

APPROXIMATE VALUES (c)	
300×10^6	Meter/s
984×10^6	Feet/s
162×10^3	Nautical miles/s
186×10^3	Statute miles/s

From Maxwell's equations, $c = (\mu\epsilon)^{-1/2}$, where $\mu = \mu_o \mu_m$, and $\epsilon = \epsilon_o \kappa_e$. But, $(\mu_o \epsilon_o)^{-1/2} = 299.7925 \times 10^6$ and, in a nonmagnetic medium, the permeability $\mu_m = 1$.

the direction of propagation. As will be made clear in the panel on the next page, which field reverses depends upon the electrical characteristics of the object.

Polarization. This is the term used to express the orientation of the wave's fields. By convention, it is taken as the direction of the electric field—the direction of the force exerted on an electrically charged particle. In free space, outside the immediate vicinity of the radiator, the magnetic field is perpendicular to the electrical field (Fig. 11), and, as just explained, the direction of propagation is perpendicular to both.

When the electric field is vertical, the wave is said to be vertically polarized. When the electric field is horizontal, the wave is said to be horizontally polarized.

If the radiating element emitting the wave is a length of thin conductor, the electric field in the direction of maximum radiation will be parallel to the conductor. If the conductor is vertical, therefore, the element is said to be vertically polarized (Fig. 12); if horizontal, the element is said to be horizontally polarized.

A receiving antenna placed in the path of a wave can extract the maximum amount of energy from it if the polarization (orientation) of the antenna and the polarization of the wave are the same. If the polarizations are not the same, the extracted energy is reduced in proportion to the cosine of the angle between them.

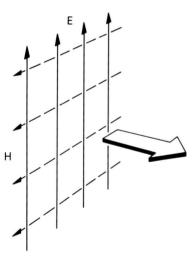

11. In free space, a wave's magnetic field is always perpendicular to its electric field. Direction of travel is perpendicular to both.

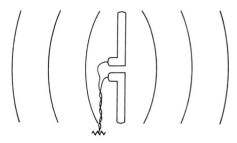

12. If the radiating element is vertical, the element is said to be vertically polarized.

53

REFLECTION, REFRACTION, AND DIFFRACTION

Any of three mechanisms may cause a radio wave to change directions: reflection (which makes radar possible), refraction, and diffraction—or a combination of the three.

Reflection from a Conductive Surface. When a wave strikes a conducting surface, its electric field is "short circuited." The resulting current causes the wave's energy to be reradiated, i.e., reflected.

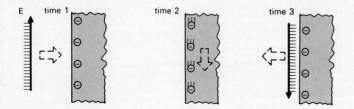

From a flat surface (irregularities small compared to a wavelength), reflection is mirror-like, hence is called specular. From an irregular or complex surface, such as that of an aircraft, reflection is diffuse, hence is called scattering.

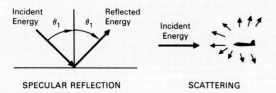

SPECULAR REFLECTION   SCATTERING

Reflection from a Nonconductive Surface. When a wave enters a nonconducting medium (such as Plexiglas) having a different dielectric constant from the medium through which the wave has been propagating, some of the wave's energy is reflected just as from a conducting surface. The reason is that the dielectric constant (κ_e) of the medium determines the division

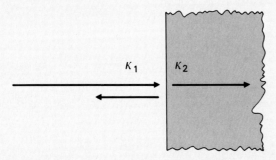

of energy between the wave's electric and magnetic fields. (In a vacuum, where $\kappa_e = 1$, the energy is divided equally between the two fields.) To adjust the balance to the new dielectric constant, some of the incident energy must be rejected. That energy is reflected.

Refraction. If the angle of incidence (θ_1) is greater than zero, when a wave enters a region of different dielectric constant the energy passing through is deflected, a phenomenon called refraction. The deflection increases with the angle of incidence and the ratio of the two dielectric constants, i.e., with the difference between the speeds in the two media.

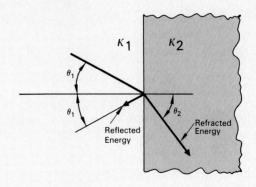

The reason is that the portion of the wave entering the new medium first travels briefly at a different speed than the portion entering next; that portion travels briefly at a different speed than the portion entering next; and so on. The ratio of the velocities in the two media is called the index of refraction.

Atmospheric Refraction. A form of refraction occurs in the atmosphere. Because of the increase in the speed of light (decrease in κ_e) with altitude, the path of a horizontally propagating wave gradually bends toward the earth. This phenomenon enables us to see the sun for a short time after it has set. It similarly enables a radar to see somewhat beyond the horizon.

Diffraction. A wave spreads around objects whose size is comparable to a wavelength and bends around the edges of larger obstructions. For a given size of obstruction, the longer the wavelength, the more significant the effect. That is why AM

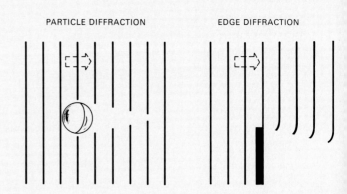

PARTICLE DIFFRACTION   EDGE DIFFRACTION

broadcast stations (operating at wavelengths of a few hundred meters) can be heard in the shadows of buildings and mountains whereas TV stations operating at wavelengths of only a few meters cannot. This phenomenon, called diffraction, stems from the fact that the energy at each point in a wave is passed on just as if a radiator actually existed at that point. The wave as a whole propagates in a given direction only because the radiation from all points in every wavefront reinforces in that direction and cancels in others. If the wavefronts are broken by an obstruction, cancellation at the edge of the wave is incomplete.

When a wave is reflected, the polarization of the reflected wave depends not only upon the polarization of the incident wave but upon the structure of the reflecting object. The polarization of radar echoes can, in fact, be used as an aid in discriminating classes of targets.

For the sake of simplicity, the discussion here has been limited to linearly polarized waves—waves whose polarization is the same throughout their length. In some applications, it is desirable to transmit waves whose polarization rotates through 360° in every wavelength (Fig. 13). This is called circular polarization. It may be achieved by simultaneously transmitting horizontally and vertically polarized waves which are 90° out of phase. In the most general case, polarization is elliptical—circular and linear polarization being extremes of elliptical polarization.

Intensity. This is the term for the rate at which a radio wave carries energy through space. It is defined as the amount of energy flowing per second through a unit of area in a plane normal to the direction of propagation (Fig. 14).[2]

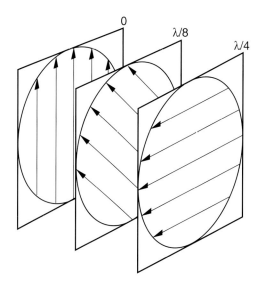

13. Polarization of a circularly polarized wave at points separated by 1/8 wavelength. Wave is produced by combining two equal-amplitude waves that are 90° out of phase.

2. Other terms for this rate are energy flux and power flow.

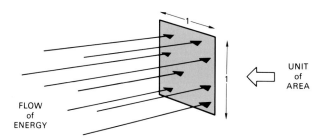

14. Intensity of a wave is the amount of energy flowing per second through a unit of area normal to the direction of propagation.

The intensity is directly related to the strengths of the electric and magnetic fields. Its instantaneous value equals the product of the strengths of the two fields times the sine of the angle between them. As previously noted, in free space outside the immediate vicinity of the antenna, that angle is 90°; so the intensity is simply the product of the two field strengths (EH).

Generally, what is of interest to us is not the instantaneous value of the intensity but the average value. If an antenna is interposed at some point in a wave's path, for example, multiplying the wave's average intensity at that point by the area of the antenna gives the amount of energy per second intercepted by the antenna (Fig. 15).

In an electrical circuit, the term used for the rate of flow of energy is power. Consequently, in considering the transmission and reception of radio waves, the term power density is often used for the wave's average intensity. (The two terms are equivalent.) The power of the received signal, then, is the power density of the intercepted wave times the area of the antenna.

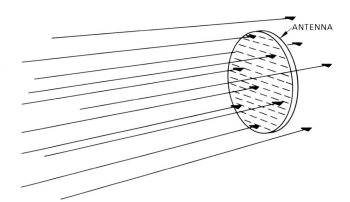

15. Power of received signal equals power density of intercepted wave times area of antenna. (Power density is another term for intensity.)

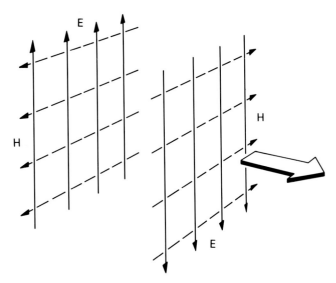

16. The fields of a radio wave, at points of maximum intensity, frozen in space. When intensities go through zero, directions of fields reverse.

3. A radio wave will have a pure sinusoidal shape, though, only if it is continuous and its peak amplitude, frequency, and phase are constant–i.e., the wave is unmodulated.

Wavelength. If we could freeze a linearly polarized radio wave and view its two fields from a distance, we would observe two things. First, the strength of the fields varies cyclically in the direction of the wave's travel. It builds up gradually from zero to its maximum value, returns gradually to zero, builds up to its maximum value again, and so on. (The fields in the planes of two successive maxima are shown in Fig. 16.) Second, we would see that each time the intensity goes through zero, the directions of both fields reverse.

The intensity of the fields is plotted versus distance along the direction of travel in Fig. 17. (It's plotted as negative when the directions of the forces exerted by the fields are reversed.) As you can see, the curve has an undulating shape very much like that of a shallow swell on the surface of the ocean. Assuming that the wave is continuous, the shape is the same as a plot of the sine of an angle versus the angle's size. Because of this, radio waves are referred to as sinusoidal, or sine waves.[3]

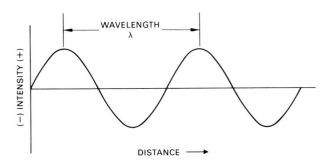

17. Variation in intensity of fields in direction of travel. Distance between crests is wavelength.

Referring again to Fig. 17, the distance between successive "crests" (or between points at which the intensity of the field goes through zero in the same direction) is the wavelength. The wavelength is usually represented by a lower case Greek letter lambda, λ, and expressed in meters, centimeters, or millimeters, depending upon its length.

Frequency. The frequency of a radio wave is directly related to the wavelength. To see the relationship, visualize if you will a radio wave traveling past a fixed point in space. The intensity of the electric and magnetic fields at this point increases and decreases cyclically as the wave goes by, just as the level of a buoy in the ocean rises and falls as a swell passes beneath it (Fig. 18).

If we place a receiving antenna in the wave's path and observe the voltage developed across the antenna terminals on an oscilloscope, we will see that it has the same shape (amplitude versus time) as our earlier plot of the intensity of the fields versus distance along the direction of travel.

18. Just as a buoy rises and falls when a swell passes under it, so the sstrengths of a radio wave's fields vary cyclically as the wave passes. Number of cycles per second is the frequency.

The number of cycles this signal completes per second is the wave's frequency.

Incidentally, the signal observed at the antenna terminals is similar to ordinary ac household power. The only difference is that it is generally far weaker and is usually of vastly higher frequency.

Frequency is usually represented by the lower case "f" and expressed in hertz, in honor of Heinrich Hertz. A hertz is one cycle per second. One thousand hertz is a kilohertz; one million hertz, a megahertz; one thousand megahertz, a gigahertz.

Since a radio wave travels at a constant speed, its frequency is inversely proportional to its wavelength. The shorter the wavelength—the more closely spaced the crests—the greater the number of them that will pass a given point in a given period of time; hence, the greater the frequency (Fig. 19).

The constant of proportionality between frequency and wavelength is, of course, the wave's speed. Expressed mathematically,

$$f = \frac{c}{\lambda}$$

where

f = frequency

c = speed of the wave (300×10^{6} meters/second)

λ = wavelength

With this formula, we can quickly find the frequency corresponding to any wavelength. A wave having a wavelength of 3 centimeters, for example, has a frequency of 10,000 megahertz.

Knowing the frequency we can find the wavelength simply by inverting the formula.

$$\lambda = \frac{c}{f}$$

Period. Another measure of frequency is period, T. It is the length of time a wave or signal takes to complete one cycle (Fig. 20).

If the frequency is known, the period can be obtained by dividing 1 second by the number of cycles per second.

$$\text{Period} = \frac{1}{f} \text{ second}$$

For example, if the frequency is 1 megahertz—i.e., the wave or signal completes one million cycles every second—

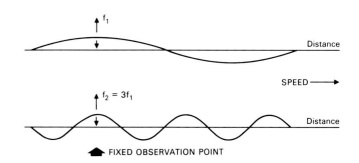

19. Since a radio wave travels at a constant speed, the shorter the wavelength, the higher the frequency.

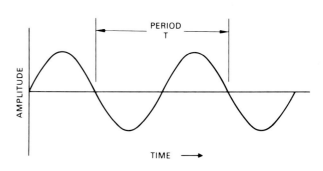

20. Period is length of time a signal takes to complete one cycle.

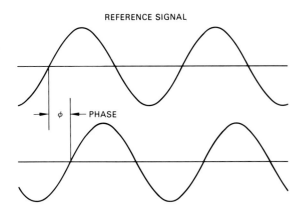

REFERENCE SIGNAL

φ ←— PHASE

21. Phase is the degree to which the cycles of a wave or signal
coincide with those of a reference signal of the same frequency.

it will complete one cycle in one-millionth of a second. Its period is one-millionth of a second: 1 microsecond.

Phase. A concept that is essential to understanding many aspects of radar operation is phase. Phase is the degree to which the individual cycles of a wave or signal coincide with those of a reference of the same frequency (Fig. 21).

Phase is commonly defined in terms of the points in time at which the amplitude of a signal goes through zero in a positive direction. The signal's phase, then, is the amount that these zero-crossings lead or lag the corresponding points in the reference signal.

This amount can be expressed in several ways. Perhaps the simplest is as a fraction of a wavelength or cycle. However, phase is generally expressed in degrees—360° corresponding to a complete cycle. If, for instance, a wave is lagging a quarter of a wavelength behind the reference, its phase is 360° x 1/4 = 90°.

Summary

Radio waves are radiated whenever an electric charge accelerates—whether due to thermal agitation in matter or a current surging back and forth through a conductor.

Their energy is contained partly in an electric field and partly in a magnetic field. The fields may be visualized in terms of the magnitude and direction of the forces they would exert on an electrically charged particle and a tiny magnet, suspended in the wave's path.

The polarization of the wave is the direction of the electric field. The direction of propagation is always perpendicular to the directions of both fields.

In free space at a distance of several wavelengths from the radiator, the magnetic field is perpendicular to the electric field, and the rate of flow of energy equals the product of the magnitudes of the two fields.

In an unmodulated wave, the intensity of the fields varies sinusoidally as the wave passes by. The distance between successive crests is the wavelength.

If a receiving antenna is placed in the path of a wave, an ac voltage proportional to the electric field will appear across its terminals. The number of cycles this signal completes per second is the wave's frequency. The length of time the signal takes to complete one cycle is its period.

Phase is the fraction of a cycle by which a signal leads or lags a reference signal of the same frequency. It is commonly expressed in degrees.

Some Relationships To Keep In Mind

- Speed of radio waves $= 300 \times 10^6$ m/s

  $= 300$ m/μs

- Wavelength $= \dfrac{300 \times 10^6}{\text{Frequency}}$

- Period $= \dfrac{1}{\text{Frequency}}$

Key to a Nonmathematical Understanding of Radar

5

One of the most powerful tools of the radar engineer—and certainly the simplest—is a graphic device called the *phasor*. Though no more than an arrow, the phasor is the key to a nonmathematical understanding of a great many seemingly esoteric concepts encountered in radar work: the formation of real and synthetic antenna beams, sidelobe reduction, the time-bandwidth product, the spectrum of a pulsed signal, and digital filtering, to name a few.

Unless you are already skilled in the use of phasors, *don't* yield to the temptation to skip ahead to chapters "about radar." Having mastered the phasor, you will be able to unlock the secrets of many intrinsically simple physical concepts which otherwise you may find yourself struggling to understand.

This chapter briefly describes the phasor. To demonstrate its application, the chapter goes on to use phasors to explain several basic concepts which are, themselves, essential to an understanding of material presented in later chapters.

How a Phasor Represents a Signal

A phasor is nothing more than a rotating arrow (vector); yet it can represent a sinusoidal signal completely (Fig. 1). The arrow is scaled in length to the signal's peak amplitude. It rotates like the hand of a clock and is positive in the counterclockwise direction, making one complete revolution for every cycle of the signal. The number of revolutions per second thus equals the signal's frequency.

The length of the projection of the arrow on a vertical line through the pivot point equals the peak amplitude

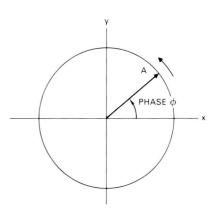

1. A phasor rotates counterclockwise, making one complete revolution for every cycle of the signal it represents.

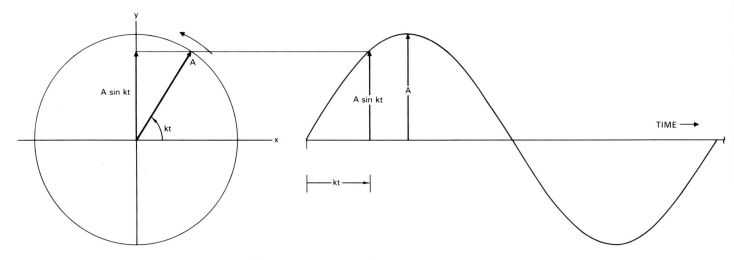

2.   For a sine wave, projection on y axis is signal's instantaneous amplitude.

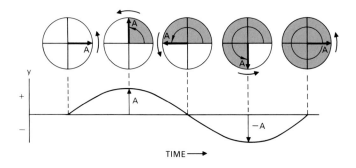

3.   As a phasor rotates, projection on y axis lengthens to maximum positive value, returns to zero, lengthens to maximum negative value, and returns to zero again.

times the sine of the angle between the arrow and the horizontal axis (Fig. 2, above). Consequently, if the signal is a sine wave, this projection corresponds to the signal's instantaneous amplitude.

As the arrow rotates (Fig. 3), the projection lengthens until it equals the arrow's full length, shrinks to zero, then lengthens in the opposite (negative) direction, and so on—exactly as the instantaneous amplitude of the signal varies with time.

If the signal is a cosine wave, the projection on the horizontal axis through the pivot corresponds to the instantaneous amplitude.

In the interest of simplicity, the arrow is drawn in a fixed position. It can be thought of as being illuminated by a strobe light that flashes "on" at exactly the same point in every cycle. That point is the instant the arrow would have crossed the horizontal axis had the signal the arrow represents been in phase with a reference signal of the same frequency (Fig. 4). In fact, the light of the strobe *is* the reference signal.

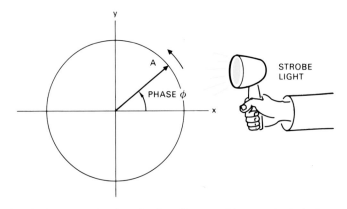

4.   A phasor can be thought of as illuminated by a strobe light that flashes "on" at the same time as a reference phasor would be crossing the x axis. Strobe provides the phase reference.

The angle the arrow makes with the horizontal axis, therefore, corresponds to the signal's phase—hence, the name phasor. If the signal is in phase with the reference, the phasor will line up with the horizontal axis (Fig. 5). If the signal is 90° out of phase with the reference—i.e., is in quadrature with it—the phasor will line up with the vertical axis. For a signal which leads the reference by 90°, the phasor will point up; for a signal that lags behind the reference by 90°, the phasor will point down.

Generally, the rate of rotation of a phasor is represented by the Greek letter omega, ω. While the value of ω can be expressed in many different units—e.g., in revolutions per second or degrees per second—it is most commonly expressed in radians per second. As you may recall, a radian is an angle which, if drawn from the center of a circle, is subtended by an arc the length of the radius. Since the circumference of a circle is 2π times the radius, the rate of rotation of a phasor in radians per second is 2π times the number of revolutions per second (Fig. 6). Thus,

$$\omega = 2\pi f$$

where f is the frequency of the signal, in hertz.

Representing individual signals graphically and concisely is not, of course, an end in itself. The real power of phasors lies in their ability to represent the relationships between two or more signals clearly and concisely. The following pages will briefly explain how phasors may be manipulated to portray (a) the addition of signals of the same frequency but different phases, (b) the addition of signals of different frequencies, and (c) the resolution of signals into in-phase and quadrature components. To illustrate the kind of insights which may be gained with phasors, several common but important aspects of radar operation will be used as examples: target scintillation, frequency translation, image frequencies, creation of sidebands, and the reason in-phase and quadrature channels are required for digital doppler filtering.

Combining Signals of Different Phase

To see how radio waves of the same frequency but different phases will combine, you draw two phasors from the same pivot point. Sliding one laterally, you add it to the tip of the other, then draw a third phasor from the pivot point to the tip of the second arrow. This phasor, which rotates counterclockwise in unison with the others, represents their sum (Fig. 7).

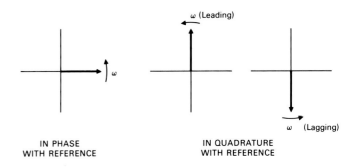

5. If the signal a phasor represents is in phase with the reference (strobe light), phasor will line up with x axis. If signal is in quadrature, phasor will line up with y axis.

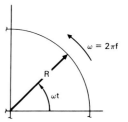

6. Rate of rotation, ω, is generally expressed in radians/second. Since there are 2π radians in a circle, $\omega = 2\pi f$.

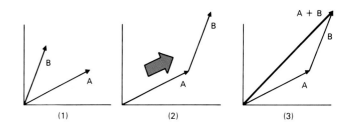

7. To add phasors A and B, you simply slide B to the tip of A. The sum is a phasor drawn from the origin to the tip of B.

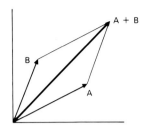

8. Phasors can also be added by constructing a parallelogram with them and drawing arrow from pivot to opposite corner.

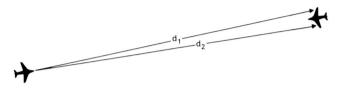

9. Situation in which a radar receives return primarily from two points on a target. Distances to the points are d_1 and d_2.

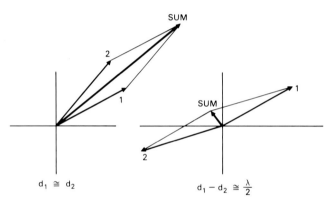

10. If distances d_1 and d_2 to the two points on the target are roughly equal, the combined return will be large; yet, if the distances differ by roughly half a wavelength, the combined return will be small.

You can also obtain the sum, without moving the second phasor, by constructing a parallelogram, two adjacent sides of which are the phasors you wish to add. The sum is a phasor drawn from the pivot point to the opposite corner of the parallelogram (Fig. 8).

To illustrate the value of such a seemingly superficial representation of the sum of two signals, we will use it to explain target scintillation.

Scintillation. Consider a situation where reflections of a radar's transmitted waves are received primarily from two parts of a target (Fig. 9). The fields of the reflected waves, of course, merge. To see what the resulting wave will be like under various conditions, we represent the waves with phasors.

To begin with, we assume that the target's orientation is such that the distances from the radar to the two parts of the target are almost the same (or differ by roughly a whole multiple of a wavelength). The two waves, therefore, are nearly in phase. As illustrated by the first diagram in Fig. 10, the amplitude of the resulting wave very nearly equals the sum of the amplitudes of the individual waves.

Next, we assume that the orientation of the target changes ever so slightly—as it might in normal flight—but enough so that the reflected waves are roughly 180° out of phase. The waves now (second diagram) largely cancel.

Clearly, if the phase difference is somewhere in between these extremes, the waves neither add nor cancel completely, and their sum has some intermediate value. Thus the sum may vary widely from one moment to the next. Recognizing, of course, that appreciable returns may be reflected from many different parts of a target, we can begin to see why a target's echoes scintillate and why the maximum detection range of a target can be predicted only in statistical terms.

What happens to the rest of the reflected energy when the waves don't add up completely? It doesn't disappear. The waves just add up more constructively in other directions for which the distances to the two parts of the target are such that the phases of the returns from them are more nearly the same.

Combining Signals of Different Frequency

The application of phasors is not limited to signals of the same frequency. Phasors can also be used to illustrate what happens when two or more signals of different frequency are added together or when the amplitude or phase of a signal of one frequency is varied—modulated—at a lower frequency.

To see how two signals of slightly different frequency combine, you draw a series of phasor diagrams, each showing the relationship between the signals at a progressively later instant in time. If you choose the instants so they are synchronized with the counterclockwise rotation of one of the phasors (i.e., if you adjust the frequency of the imaginary strobe light so it is the same as the frequency of one of the phasors), that phasor will occupy the same position in every diagram (Fig. 11).

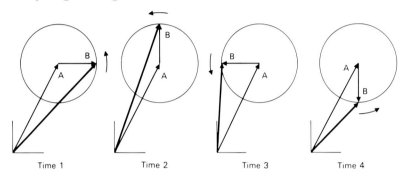

11. How signals of different frequencies combine. If strobe light is synchronized with rotation of phasor A, it will appear to remain stationary and phasor B will rotate relative to it.

The second phasor will then occupy progressively different positions. The difference from diagram to diagram corresponds to the difference between the two frequencies. (Usually, by indicating the relative rotation of the second phasor with a circle and/or a curved arrow in a single diagram, you can mentally visualize the effect of the difference in frequency.)

If the difference is positive—second frequency higher— the second phasor will rotate counterclockwise relative to the first (Fig. 12). If the difference is negative—second frequency lower—the second phasor will rotate clockwise relative to the first.

As the phasors slip into and out of phase, the amplitude of their sum fluctuates—is modulated—at a rate equal to the difference between the two frequencies. The phase of the sum also is modulated at this rate. It falls behind during one half of the difference-frequency cycle and slides ahead during the other half. As the phase changes, the rate of rotation of the sum phasor changes: the frequency of the signal is also modulated.

By representing signals of different frequencies in this way, many important aspects of a radar's operation can easily be illustrated graphically: image frequencies, creation of sidebands, and so forth.

Frequency Translation. As you may have surmised, since the amplitude of the sum of two phasors fluctuates at a rate

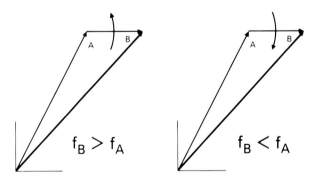

$f_B > f_A$ $f_B < f_A$

12. If the frequency of B is greater than that of A, phasor B will rotate counterclockwise relative to A. Otherwise it will appear to rotate clockwise.[1]

1. For larger frequency differences, these relationships do not necessarily hold. If a phasor's frequency is less than half the reference frequency or is between $1\frac{1}{2}$ and 2, $2\frac{1}{2}$ and 3, $3\frac{1}{2}$ and 4, etc. times the reference frequency, the phasor's apparent rotation will be reversed.

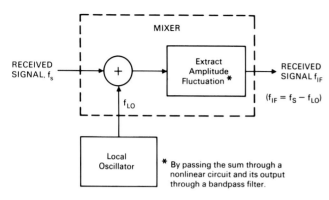

13. A received signal may be translated to a lower frequency f_{IF} by adding it to a local oscillator signal and extracting the amplitude modulation of the sum.

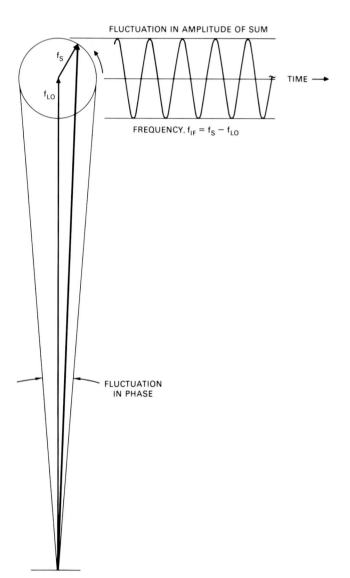

14. If local oscillator signal is stronger than received signal, fluctuation in amplitude of sum is virtually identical to received signal except for being shifted to f_{IF}.

equal to the difference between the rates of rotation of the phasors, you can readily shift a signal down in frequency by any desired amount. You simply add the signal to a signal of a suitably different frequency and extract the amplitude fluctuation.

We encounter this process all of the time. In the early stage of virtually every radio receiver, and a radar receiver is no exception, the received signal is translated to a lower "intermediate" frequency (Fig. 13). Translation is accomplished by "mixing" the signal with the output of a "local" oscillator, whose frequency is offset from the signal's frequency by the desired intermediate frequency (f_{IF}).

In one mixing technique, the signal is simply added to the local oscillator output, as in Fig. 14, and the fluctuation in the amplitude of the sum is extracted (detected).

In another mixing technique, the amplitude of the received signal itself is modulated by the local oscillator output. As will be explained shortly, amplitude modulation produces sidebands. In this case, the frequency of one of the sidebands is the difference between the frequencies of the received signal and local oscillator signal f_{IF}.

Image Frequencies. The phasor diagram of Fig. 15 (below) illustrates a subtler aspect of frequency translation. The same amplitude modulation will be produced by a signal whose frequency is above the local oscillator frequency as by one whose frequency is an equal amount below it. The phasors representing the two difference signals rotate in opposite directions, but the effect on the amplitude of the sum is essentially the same. It fluctuates at the difference frequency in either case.

Consequently, if a spurious signal exists whose frequency is the same amount below the local oscillator frequency as the desired signal is above it (or vice versa), both of the

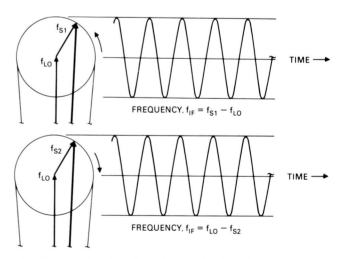

15. Amplitude modulation of sum by signals whose frequencies are above and below f_{LO} by the same amount.

signals will be translated to the same intermediate frequency. The spurious signal will thus interfere with the desired signal even though their original frequencies are separated by twice the intermediate frequency. The spurious signal is called an image and its frequency is called the image frequency (Fig. 16).

Another consequence of images is that noise occurring at the image frequency is added to the noise with which the desired signal must compete. As we shall see in a bit—also with the help of phasors—there are solutions to both of these image problems.

Creation of Sidebands. When phasors representing two signals of different frequency are added, the phase modulation of the sum can be eliminated completely by adding a third phasor, which is the same length as the second and rotates at the same rate relative to the first phasor but in the opposite direction (Fig. 17, below). If the counter-rotating phasors pass through the axis on the first phasor (vertical axis in Fig. 17) simultaneously, the phase modulation will cancel and only the amplitude of the sum will fluctuate. The sum will be a pure amplitude modulated, or AM signal—the same sort of signal one receives from an AM broadcast station when it is transmitting, say, a 400 hertz test signal.

As in the earlier examples of modulation, the frequency at which the amplitude of the sum is modulated is the difference between the frequency of either one of the counter-

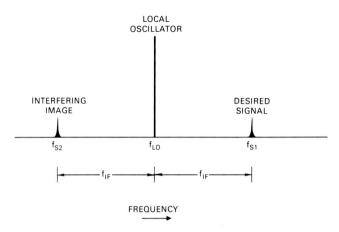

16. If operating frequency is higher than f_{LO}, the image frequency is $f_{LO} - f_{IF}$, and vice versa.

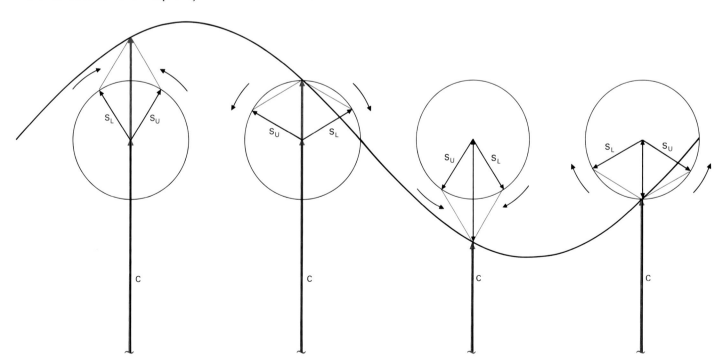

17. If two counter-rotating phasors, S_L and S_U, are added to a third phasor, C, and their phases and frequencies are such that all pass through the same axis together, their sum will be a pure amplitude modulated signal.

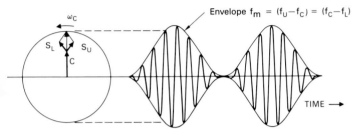

18. If amplitude of a carrier signal C is varied sinusoidally at rate, f_m, two new signals are produced, S_L and S_U.

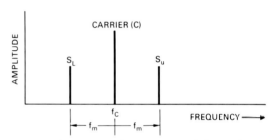

19. Since the frequencies of S_L and S_U are f_m hertz above and below f_c, they are called sidebands.

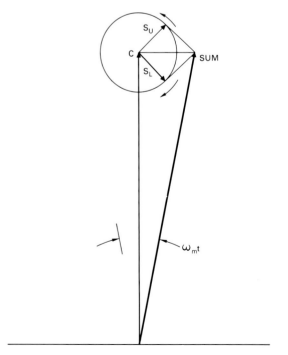

20. Frequency and phase modulation differ from amplitude modulation in that the phase of the sideband signals is shifted by 90°.

rotating phasors and the frequency of the "fixed" phasor. All three phasors, of course, rotate in unison with that phasor. But this rotation doesn't show up in the diagram because the imaginary strobe light which illuminates the phasors flashes "on" only once in every cycle of that phasor's rotation.

In some instances amplitude modulation is actually produced by generating the signals represented by the counter-rotating phasors separately and adding them to the signal that is to be modulated. Generally, though, it is the other way around. The signals represented by the counter-rotating phasors are the inevitable result of amplitude modulation.

As is illustrated by the phasor diagram of Fig. 18 and may be readily demonstrated with actual signals, whenever the amplitude of a signal of a given frequency (f_c) is modulated at a lower frequency (f_m), two new signals are invariably produced. One of these, represented by the phasor S_U in Fig. 18, has a frequency f_m hertz above f_c; the other, a frequency f_m hertz below it.

Since the frequencies of these signals lie on either side of f_c (Fig. 19), the signals are called sideband signals, or simply sidebands. Since the signal that is modulated carries the modulation—i.e., the modulation is added to and subtracted from the amplitude of this signal— it is called the carrier.

The light lines that join the crests of the modulated wave in Fig. 18 delineate what is called the *modulation envelope*. The frequency of the sidebands, you will notice, is the modulation frequency. The average separation of the sidebands from the baseline is the amplitude of the carrier.

Sidebands are similarly produced when the phase or frequency of a carrier signal is modulated. Only then, the phase relationship of the sidebands to the carrier is different (Fig. 20). If the percentage by which the phase or frequency is varied is large, many sideband pairs separated by multiples of the modulation frequency are created.

The generation of sidebands is an important consideration in the design of virtually every radar. As will be explained in detail in Chap. 9, for example, it is to avoid interference from sidebands due to the random fluctuation (noise modulation) of the output of the radar transmitter that one generally must employ pulsed transmission when the same antenna is used for both transmission and reception.

And, as will be explained in Chap. 23, it is the production of sidebands by the pulsed modulation of the transmitter that in some cases causes echoes from a target and a ground patch to be passed by the same doppler filter even though they have different doppler frequencies.

Resolving Signals into I and Q Components

Sometimes it is advantageous to resolve a signal into two components having the same frequency and peak amplitude but differing in phase by 90°. Since a cosine wave reaches its positive peak 90° before a sine wave does, the most convenient way of picturing the two components is as a sine wave ($A \sin \omega\tau$) and a cosine wave ($A \cos \omega\tau$). By convention, the cosine wave is called the in-phase or I component.[2] Since 90° is a quarter of a circle, the sine wave is called the quadrature or Q component.

If the signal is represented by a phasor, the instantaneous amplitude of the I component can be found simply by projecting the phasor onto the horizontal (x) axis. The instantaneous amplitude of the Q component can be found by projecting the phasor onto the vertical (y) axis (Fig. 21).

2. This convention was adopted because current passing through a resistance is in phase with the voltage across the resistance, whereas a current passing through a reactance either leads or lags behind the voltage by 90°.

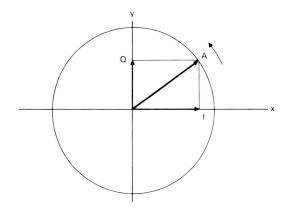

21. Instantaneous values of the I and Q components of a signal are obtained by projecting phasor representation of signal onto x as well as y axis.

For a phasor whose apparent rotation is counterclockwise—frequency of signal (represented by phasor) is higher than frequency of reference signal (strobe light)—the I component goes through its positive maximum 90° before the Q component. On the other hand, for a phasor whose apparent rotation is clockwise—frequency of signal represented by phasor is lower than that of reference—the Q component goes through its maximum in a positive direction 90° before the I component.

Distinguishing Direction of Doppler Shifts. One of the more striking examples of a requirement for resolving signals into I and Q components is found in radars that employ digital doppler filtering. For digital filtering, the IF output of the receiver must be converted to video frequencies. To preserve the sense (positive or negative) of a target's doppler shift once this conversion has been made, two video signals must be provided: one, corresponding to the cosine of the doppler frequency (I); the other, to the sine (Q). The reason is as follows.

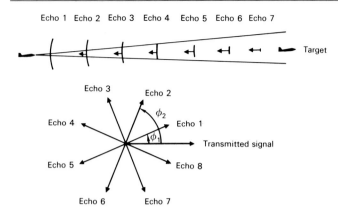

Echo 1   Echo 2   Echo 3   Echo 4   Echo 5   Echo 6   Echo 7

22. A target's doppler frequency shows up as a pulse-to-pulse shift in phase.

As will be explained in detail in Chap. 15, a target's doppler frequency shows up as a progressive shift in the radio frequency phase, ϕ, of successive echoes received from the target, relative to the phase of the pulses transmitted by the radar. This echo-to-echo phase shift is illustrated by the phasor diagram in Fig. 22.

By sensing the porgressive phase shift, the radar can produce a video signal whose amplitude fluctuates at the target's doppler frequency. The signal is illustrated for positive and negative doppler shifts in Fig. 23.

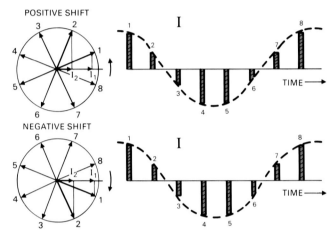

23. Video signal proportional to in-phase component of target echoes fluctuates at target's doppler frequency; but fluctuation is same for both positive and negative doppler shifts.

But as is clear from the figure, the fluctuations in the amplitude of this signal are the same for both positive and negative doppler shifts.

If both I and Q components of the phase shift are sensed, however, the difference between positive and negative doppler frequencies may be readily determined. For the fluctuation of the Q components will lag behind the fluctuation of the I component if the doppler shift is positive (Fig. 24). And it will lead the fluctuation of the I component if the doppler shift is negative.

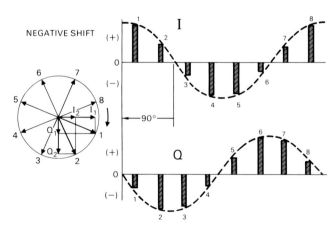

24. If the doppler frequency shift is positive and both I and Q video signals are provided, Q will lag I by 90°.

25. But if the doppler frequency shift is negative, Q will lead I by 90°.

KEY TO A MATHEMATICAL UNDERSTANDING OF RADAR

Phasors are also the key to a mathematical understanding of radar. For they enable one to visualize phase and frequency relationships in the domain of the complex variable.

Sine and cosine functions can be expressed in exponential as well as trignometric forms*

$$A \sin \omega t = \frac{A(e^{j\omega t} - e^{-j\omega t})}{2j}$$

$$A \cos \omega t = \frac{A(e^{j\omega t} + e^{-j\omega t})}{2}$$

The letter "j" in the exponential terms stands for $\sqrt{-1}$. Because $\sqrt{-1}$ alone cannot be evaluated, it is said to be an "imaginary" number. A variable having an imaginary part and a real part is called a complex variable.

Often, sinusoidal functions are more easily manipulated in the exponential form than in the trigonometric form. Yet, for many of us, the exponential terms, $e^{j\omega t}$ and $e^{-j\omega t}$, alone, have little physical meaning.

The functions they represent, however, can be visualized quite easily with phasors. For this, e^{j} is taken to mean rotation in a counterclockwise direction and e^{-j}, rotation in a clockwise direction.

The term $e^{j\omega t}$ then is represented by a phasor of unit length rotating counterclockwise at a rate of ω radians per second.

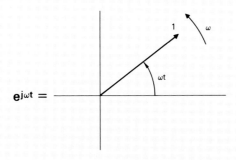

$$e^{j\omega t} =$$

The term $e^{-j\omega t}$ is similarly represented by a phasor of unit length rotating clockwise at a rate ω.

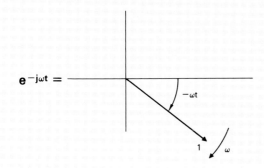

$$e^{-j\omega t} =$$

The sum $e^{j\omega t} + e^{-j\omega t}$ equals the sum of the projections of the two phasors onto the x axis. This sum, of course, is 2 cos ωt.

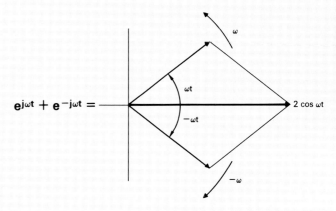

$$e^{j\omega t} + e^{-j\omega t} =$$

The difference $e^{j\omega t} - e^{-j\omega t}$ equals the projection of the first phasor on the y axis minus the projection of the second phasor on the y axis. This difference is 2 sin ωt.

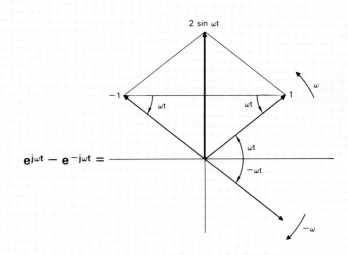

$$e^{j\omega t} - e^{-j\omega t} =$$

Using these basic relationships as building blocks and remembering the values of j raised to various powers,

$$j = \sqrt{-1}$$
$$j^2 = \sqrt{-1}\sqrt{-1} = -1$$
$$j^3 = -1\sqrt{-1} = -j$$
$$j^4 = (-1)(-1) = +1$$

one can easily visualize virtually any relationships involving the complex variable.

* The equivalence can be demonstrated by expanding the functions sin x, cos x, and e^{jx} into power series with Maclaurin's theorem.

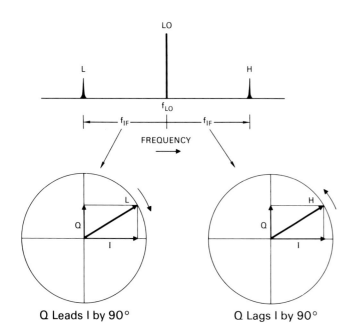

26. Quadrature component of mixer output will lead in-phase component if frequency of received signal is lower than f_{LO} and lag behind it if frequency of received signal is higher than f_{LO}.

Differentiating Between Signals and Images. Just as positive and negative doppler frequencies can be differentiated by resolving the received signals into I and Q components when they are converted from IF to video frequencies, so also, images can be differentiated from signals when the radar return is translated from the radar's operating frequency to IF. As can be seen from the phasor diagram of Fig. 26, if a signal's frequency is higher than the local oscillator frequency, the Q component of the mixer's output will lag 90° behind the I component. Yet, if the signal's frequency is lower than the local oscillator frequency, the Q component will lead the I component by 90°. By taking advantage of this difference, a receiver's mixer stage can be designed to reject images.

Summary

A powerful tool for visualizing phase and frequency relationships is the phasor. Its length corresponds to amplitude; its rate of rotation, to frequency; its angle, to phase. The phasor can be drawn in a fixed position by thinking of it as being illuminated by a strobe light which flashes on at the same point in every cycle. If the signal is in phase with the reference, it is drawn horizontally.

If signals of the same frequency are combined, the amplitude of the sum will depend on the relative phases of the signals. Because of this dependence, even a very slight change in target aspect can cause a target's echoes to scintillate.

If signals of different frequency are combined, their sum can be visualized by assuming the strobe is synchronized with the rotation of one of the phasors, causing it to appear fixed. The other then rotates at the difference frequency.

The amplitude and phase of the sum will be modulated at a rate equal to the difference between the frequencies. The phase modulation can be minimized by making the second signal much stronger than the first. By extracting the amplitude modulation, the first signal can be translated to the difference frequency. At the same time, however, a signal whose frequency is offset from that of the first signal by the same amount in the opposite direction (image) will also be translated to the difference frequency.

Whenever a carrier signal's amplitude is modulated, two sideband signals are produced. Their frequencies are separated from the carrier by the modulation frequency.

Resolution of a signal into in-phase (I) and quadrature (Q) components can be visualized by projecting the phasor representing the signal onto the x and y coordinates. Resolving the IF output of a receiver into I and Q components when it is converted to video enables a digital filter to differentiate between positive and negative doppler frequencies.

The Ubiquitous Decibel

The decibel—or dB, as it is called—is one of the most widely used tools of those who design and build radars. If you are already familiar with decibels, can readily translate to and from them, and feel at ease when the experts start throwing them about, then skip this short chapter. Otherwise, you will find the few minutes it takes you to read it well worthwhile.

What Decibels Are

The decibel is a logarithmic unit originally devised to express power ratios but used today to express a variety of other ratios, as well. Specifically,

$$\text{Power ratio in dB} = 10 \log_{10} \frac{P_2}{P_1}$$

where P_2 and P_1 are the two power levels being compared. For example, if P_2/P_1 is 1,000 then the power ratio in decibels is 30.

Origin. Named for Alexander Graham Bell, the unit originated as a measure of attenuation in telephone cable—the ratio of the power of the signal emerging from a cable to the power of the signal fed in at the other end. It so happened that one decibel almost exactly equaled the attenuation of one mile of standard telephone cable, the unit used until the decibel came along (Fig. 1). Also, one decibel relative to the threshold of hearing turned out to be very nearly the smallest ratio of audio-power levels that could be discerned by the human ear; so the dB was soon adopted in acoustics, too. From telephone communications, the dB was quite naturally passed on to radio communications; thence, to radar.

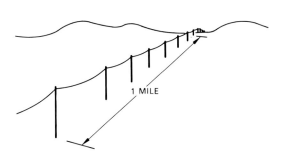

1. Conceived for use in communications, the decibel was the attenuation of one mile of standard telephone cable.

A SHORT COURSE IN LOGARITHMS

For years, logarithms to the base 10 were widely used to simplify the multiplication and division of large numbers. With the advent of the pocket calculator, however, logarithms to the base 10 became largely obsolete. Consequently, many people are today unfamiliar with them. For those who are (or have forgotten), this brief review is provided.

What a Logarithm Is. Suppose that a number has a value, N. Now suppose that n is the power to which 10 must be raised to equal N.

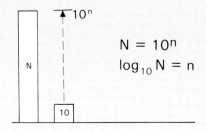

$$N = 10^n$$
$$\log_{10} N = n$$

The exponent n is the logarithm to the base 10 of N.

Logarithms of numbers that are whole multiples of 10 are whole numbers.

$$N = 10^3 \qquad \log_{10} N = 3$$
$$N = 10^2 \qquad \log_{10} N = 2$$

Since $10^1 = 10$ and $10^0 = 1$, logarithms of numbers between 10 and 1 are decimal fractions.

$$N = 10 \qquad \log_{10} N = 1$$
.   .
.   .
.   .
.   .
$$N = 1 \qquad \log_{10} N = 0$$

The logarithm of 2, for example, is 0.3.

Multiplying and Dividing with Logarithms. If two numbers are expressed as powers of the same number, say 10, they can be multiplied together by adding exponents.

$$10^3 \times 10^2 = 10^{(3 + 2)} = 10^5$$

And one number can be divided by the other by subtracting the exponent of the second from the exponent of the first.

$$10^3 \div 10^2 = 10^{(3 - 2)} = 10^1$$

Consequently, numbers expressed as logarithms can be multiplied or divided simply by adding or subtracting the logarithms.

$$\log_{10} 1000 \dots\dots\dots 3$$
$$+\log_{10} 100 \dots\dots\dots +2$$
$$\overline{\log_{10} (10^{(3 + 2)}) \dots\dots 5}$$

$$\log_{10} 1000 \dots\dots 3$$
$$-\log_{10} 100 \dots\dots -2$$
$$\overline{\log_{10} (10^{(3 - 2)}) \dots 1}$$

Similarly, a number can be raised to any power by multiplying logarithm by that power.

$$\log_{10} 1000 = 3$$
$$\log_{10} 1000^4 = 3 \times 4 = 12$$

And any root of a number can be taken by dividing its logarithm by that root.

$$\log_{10} 1000 = 3$$
$$\log_{10} 1000^{\frac{1}{4}} = 3 \div 4 = 0.75$$

These are the characteristics that made logarithms to the base 10 so useful, prior to the era of the pocket calculator.

Logarithm of a Number Expressed in Scientific Notation. Expressed in scientific notation, a number such as 200 is 2×10^2. The logarithm of the number, therefore, is the sum of the logarithms of the two parts.

$$200 = 2 \times 10^2$$
$$\log_{10} 200 = \log_{10} 2 + \log_{10} 10^2$$
$$0.3 + 2 = 2.3$$

To express a number as a logarithm, therefore, one needs to know only the logarithms of numbers between 1 and 10.

Converting from a Logarithm to dB. Going from the logarithm of a power ratio to the value of the ratio in dB is but a short step. You just multiply by 10. For example, if the number 200 was a power ratio ($P_2/P_1 = 200/1$), the ratio expressed in dB would be:

$$\log_{10} 200 = 2.3$$
$$10 \log_{10} 200 = 10 \times 2.3 = 23 \text{ dB}$$

Advantages. Several features of the decibel make it particularly useful to the radar engineer. First, since the decibel is logarithmic, it greatly reduces the size of the numbers required to express large ratios (Fig. 2).

A power ratio of 2 to 1 is 3 dB; yet a ratio of 10,000,000 to 1 is only 70 dB. Since the power levels encountered in a radar cover a tremendous range, the compression in the sheer size of numbers that decibels provide is extremely valuable.

Another advantage also stems from the decibel's logarithmic nature: two numbers expressed as logarithms can be multiplied simply by adding the logarithms. Expressing ratios in decibels, therefore, makes compound power ratios easier to work with. Multiplying 2500/1 by 63/1 in your head for example, isn't particularly easy. Yet when these same ratios are expressed in decibels, there is nothing to it: 34 + 18 = 52 dB (Fig. 3).

Similarly, with logarithms, the reciprocal of a number (one divided by the number) can be obtained simply by giving the logarithm a negative sign. By merely changing the sign of a ratio expressed in decibels, the ratio can instantly be turned upside down. If 157,500 is 52 dB, then $1/157,500$ is −52 dB (Fig. 4).

When it comes to raising ratios to higher powers or taking roots, these advantages are magnified. If a ratio such as 63 is expressed in decibels, you can square it by multiplying by two: 63^2 = 18 dB x 2 = 36 dB. You can take its fourth root by dividing by four: $\sqrt[4]{63}$ = 18 dB ÷ 4 = 4½ dB.

True, you can compress numbers by expressing them in scientific notation (e.g., 20,000,000 = 2 x 10^7). And you can quickly multiply, divide, and take roots of numbers of any size with a pocket calculator. But the decibel has the advantage of incorporating the power of 10 right in its value, thereby reducing the possibility of serious errors in keeping track of decimal places. And you can manipulate numbers that are expressed in decibels right in your head.

Furthermore, by tradition, many radar parameters are commonly expressed in decibels.

Perhaps the most compelling advantage is this. In the world of radar, where detection ranges vary as the one-fourth power of most parameters, target signal powers may vary by factors of trillions, and losses of 20 or 30 percent may be negligible, it is a lot easier to talk and think in terms of decibels than in terms of numbers expressed in scientific notation or ground out of a calculator.

To be able to throw decibels about as deftly as a seasoned radar engineer, you only need to know two things: (1) how to convert from power ratios to decibels and vice versa; (2) how to apply decibels to a few basic characteristics of a

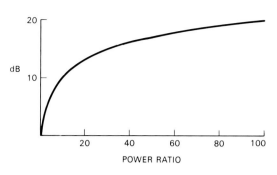

2. Being logarithmic, the decibel greatly reduces the size of the numbers required to express large power ratios.

$$\frac{2,500}{1} \times \frac{63}{1} = 157,500$$

$$34 \text{ dB} + 18 \text{ dB} = 52 \text{ dB}$$

3. Power ratios can be compounded simply by adding up their decibel equivalents.

$$52 \text{ dB} = \frac{157,500}{1} = 157,500$$

$$-52 \text{ dB} = \frac{1}{157,500} = 0.000006349$$

4. A power ratio can be inverted by changing the sign of its decibel equivalent.

radar. If you know the system, both things are surprisingly easy. And the system is really quite simple.

Converting from Power Ratios to dB

You can convert any power ratio (P_2 / P_1) to decibels, with any desired degree of accuracy, by dividing P_2 by P_1, finding the logarithm of the result, and multiplying by 10.

$$10 \log_{10} \frac{P_2}{P_1} = dB$$

Nevertheless, for the accuracy you will normally want, you don't need a calculator. With the method outlined below, you can do it all in your head—provided you have memorized a few simple numbers.

The first step is to express the ratio as a decimal number, in terms of a power of 10 (scientific notation). A ratio of 10,000/4, for example, is 2500. In scientific notation,

$$2500 = 2.5 \times 10^3$$

When converting to decibels, two portions of this expression are significant: the number 2.5, which we will call the basic power ratio; and the number 3, which is the power of 10.

Now, a ratio expressed in decibels similarly consists of two basic parts: (1) the digit in the "one's place" (plus any decimal fraction) and (2) the digit or digits to the left of the one's place. The digit in the one's place expresses the basic power ratio: 2.5, in the foregoing example. The digits, if any, to the left of the one's place express the power of 10: in this case, 3.

Incidentally, as you may already have observed, if the power ratio P_2/P_1 is rounded off to the nearest power of 10—e.g., $2.5 \times 10^3 \approx 10^3$—converting it to decibels is a trivial operation. The basic power ratio then is zero ($\log_{10}1 = 0$); so the decibel equivalent of P_2/P_1 is simply 10 times the power of 10, in this case, 30. Thus,

Power Ratio	Power of 10	dB
1	0	0
10	1	10
100	2	20
1000	3	30
10,000,000	7	70

The basic power ratio, of course, may have any value from 1 to (but not including) 10. So, the digit in the one's place can be any number from 0 through 9.999...

Table 1. Basic Power Ratios

Power Ratio	dB
1	0
1.26	1
1.6	2
2	3
2.5	4
3.2	5
4	6
5	7
6.3	8
8	9

Table 1 (opposite page) gives the basic power ratios for 0 to 9 dB. To simplify the table, all but the ratio for 1 dB have been rounded off to two digits. If you want to become adroit in the use of decibels, you should memorize these ratios.

Returning to our example, if we look up the decibel equivalent of the basic power ratio, 2.5, in Table 1 (or better yet, our memory) we find that it is 4 dB. So, expressed in decibels, the complete power ratio, 2.5 x 10^3, is 34 dB (Fig. 5).

Converting from dB to Power Ratios

To convert from decibels to a power ratio, you can also use a calculator. In this case, you divide the number of decibels by 10 to get the power of 10; then raise 10 to that power to get the power ratio.

$$\text{Power ratio} = 10^{dB/10}$$

But you can make the conversion just as easily in your head, using the procedure outlined in the preceding paragraphs in reverse.

Suppose, for example, you want to convert 36 dB to the corresponding power ratio. The digit in the one's place, 6, is the dB equivalent of a power ratio of 4. The digit to the left of the one's place, 3, is the power of 10. The power ratio, then, is 4 x 10^3 = 4,000 (Fig. 6).

As outlined here, the process may seem a bit laborious. But once you've tried it a few times, there is really nothing to it, if you remember the power ratios corresponding to decibels 1 through 9. An easy way to remember them is outlined in the panel on page 78.

Representing Power Ratios Less Than One

If 0 dB corresponds to a power ratio of one (1/1), how do you convert power ratios that are less than one to decibels? You use negative decibels, of course (Fig. 7). As previously noted, a ratio expressed in decibels can be inverted by putting a negative sign before it.

$$3 \text{ dB} = 2$$

$$-3 \text{ dB} = 1/2 = 0.5$$

What about a power ratio of zero? The smaller the power ratio is, the larger the number of negative decibels required to represent it. As a ratio approaches zero, the number of negative decibels increases without limit. For example, a power ratio of

$$0.000,000,000,000,000,001 = -180 \text{ dB}$$

There *is* no decibel equivalent of a power ratio of zero.

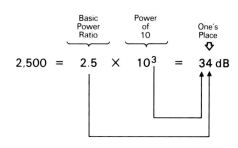

5. Conversion of a power ratio (2,500) to decibels.

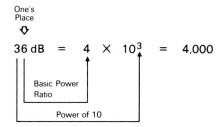

6. Conversion of decibels (36 dB) to a power ratio.

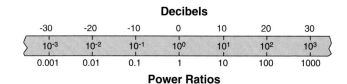

7. Negative decibels represent power ratios less than one; positive decibels, ratios greater than one; 0 dB, a ratio of 1.

Using Decibels

A common use of decibels in radar work is expressing power gains and power losses.

Gain is the term for an increase in power level. In the case of an amplifier—such as might raise a low power microwave signal to the desired level for radiation by an antenna—gain is the ratio of the power of the signal coming out of the amplifier to the power of the signal going into it.[1]

1. Assuming properly matched source and load impedances.

$$\text{Gain} = \frac{\text{Output power}}{\text{Input power}}$$

If the output power is 250 times the input power, the gain is 250. This ratio (250 to 1) is 24 dB (Fig. 8).

Loss is the term for a decrease in power. According to convention, it is the ratio of input power to output power—just the opposite of gain.

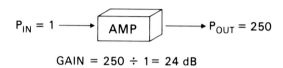

GAIN = 250 ÷ 1 = 24 dB

8. Gain is the ratio of output power to input power.

$$\text{Loss} = \frac{\text{Input power}}{\text{Output power}}$$

To illustrate, let us assume that the amplifier of the preceding example is connected to the antenna by a waveguide that absorbs some 20 percent of the power. The ratio of input to output power, therefore, is 10 to 8 (1.25), making the loss 1 dB (Fig. 9).

(Some people prefer to consider gain and loss as being synonymous and think only in terms of the ratio of output to input. Looked at this way, a 1 dB loss is a gain of −1 dB.)

Suppose, now, that we wish to find the total gain (G_T) between the input to the amplifier and the input to the antenna. To do this in terms of straight power ratios, we divide the gain of the amplifier (250) by the loss of the waveguides (1.25).

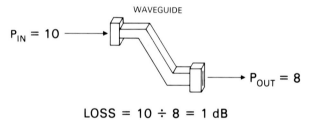

LOSS = 10 ÷ 8 = 1 dB

9. Loss is the ratio of input power to output power.

$$G_T = G_{AMP} \div L_{W.G.}$$
$$= 250 \div 1.25$$
$$= 200$$

On the other hand, to determine the gain and loss in dB (Fig. 10), we simply subtract the loss from the gain.

$$G_T = 24 \text{ dB} - 1 \text{ dB}$$
$$= 23 \text{ dB}$$

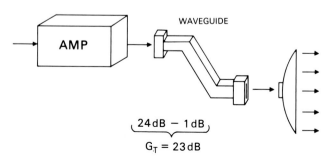

24 dB − 1 dB
G_T = 23 dB

10. In decibels, the overall gain is the gain minus the loss.

A decibel value of 23 dB is a power ratio of 200. So the answer is the same either way.

Following this same general procedure, we could take into account any number of gains and losses—additional stages of amplification inserted on either side of the original amplifier, losses in the antenna, losses in the radome, reductions in gain (losses) due to degradation in the field, etc. (Fig. 11). And by multiplying the total gain by the input power, we could quickly tell how much power would be supplied to the antenna.

$$G_T = (G_1 + G_2 + G_3 \dots) - (L_1 + L_2 + L_3 \dots)$$

11. Any number of gains and losses can readily be compounded.

Power Gain in Terms of Voltage

Sometimes it is convenient to express power in terms of voltages. The power dissipated in a resistance equals the voltage, V, applied across the resistance times the current, I, flowing through it: P = VI. But the current is equal to the voltage divided by the resistance: I = V/R. So the power is equal to (V²/R).

Accordingly, the power output of a circuit equals $(V_0)^2/R$, and the power input equals $(V_i)^2/R$. If the circuit's input and output impedances are the same, the gain is $(V_0)^2/(V_i)^2$ (Fig. 12). Expressed in decibels, then, the gain is

$$G = 10 \log_{10} \left(\frac{V_o}{V_i}\right)^2 = 20 \log_{10} \left(\frac{V_o}{V_i}\right)$$

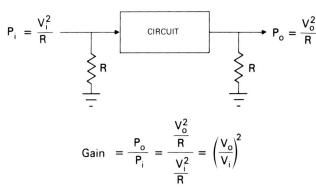

12. Expressed in terms of voltage, gain = $(V_o V_i)^2$, provided input and output resistances are the same.

Decibels as Absolute Units

While decibels were originally used only to express power ratios, they can also be used to express absolute values of power. All that is necessary is to establish some absolute unit of power as a reference. By relating a given value of power to this unit, that value can be expressed with decibels.

An often used unit is 1 watt. A decibel relative to 1 watt is called a dBW. A power of 1 watt is 0 dBW; a power of 2 watts is 3 dBW; a power of 1 kilowatt (10^3 watts) is 30 dBW (Fig. 13).

Another common reference unit is 1 milliwatt. A decibel relative to 1 milliwatt is called a dBm. The dBm is widely used for expressing small signal powers, such as the powers of radar echoes. They vary over a tremendous range. Echoes from a small, distant target may be as weak as –130 dBm, or less; while echoes from a short range target may be as strong as 0 dBm, or more. The dynamic range of echo powers is thus at least 130 dB. Considering that –130 dBm is 10^{-13}, or 0.000,000,000,000,1 millliwatt, the convenience of expressing absolute powers in dBm is striking (Fig. 14).

This advantage of decibels is so compelling that they have been applied to other variables than power. One example is radar cross section.

1 watt	= 0 dBW
2 watts	= 3 dBW
1 kilowatt	= 30 dBW

13. A decibel relative to 1 watt is called a dBW.

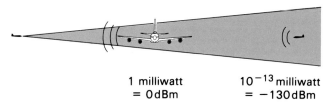

1 milliwatt = 0 dBm	10^{-13} milliwatt = –130 dBm

14. Power received from a large, short-range target can be 10,000,000,000,000 or more times that received from a small distant target. Advantage of expressing such powers in decibels relative to a milliwatt is obvious.

REMEMBERING THE BASIC POWER RATIOS

Whole Decibels. The table of equivalent power ratios has two characteristics that make it surprisingly easy to remember.

First, 3 dB corresponds almost exactly to a ratio of 2. Since adding decibels has the same effect as multiplying the ratios they represent, we can obtain the ratios for 6 dB and 9 dB directly from that for 3 dB.

$$3 \text{ dB} = 2$$

$$6 \text{ dB} = 3 \text{ dB} + 3 \text{ dB} = 2 \times 2 = 4$$

$$9 \text{ dB} = 6 \text{ dB} + 3 \text{ dB} = 4 \times 2 = 8$$

Second, 1 dB corresponds to about 1¼ (5/4). Since a negative sign inverts the ratio, −1 dB corresponds to 4/5 = 0.8. On the basis of these two ratios—1¼ and 0.8—we can determine all of the remaining ratios.

$$2 \text{ dB} = 3 \text{ dB} - 1 \text{ dB} = 2 \times 0.8 = 1.6$$

$$4 \text{ dB} = 3 \text{ dB} + 1 \text{ dB} = 2 \times 1\tfrac{1}{4} = 2.5$$

$$5 \text{ dB} = 6 \text{ dB} - 1 \text{ dB} = 4 \times 0.8 = 3.2$$

$$7 \text{ dB} = 6 \text{ dB} + 1 \text{ dB} = 4 \times 1\tfrac{1}{4} = 5$$

$$8 \text{ dB} = 9 \text{ dB} - 1 \text{ dB} = 8 \times 0.8 = 6.4$$

So, if you were stranded on a desert island and the batteries in your pocket calculator were dead, if you could remember just two ratios—those for 1 dB and 3 dB—you could reconstruct the entire table right in your head. You might starve, but you could talk in decibels until you did.

Oh yes . . . in case you forgot the ratio for 1 dB, you could find it by subtracting 9 dB from 10 dB.

$$1 \text{ dB} = 10 \text{ dB} - 9 \text{ dB} = 10 \div 8 = 1\tfrac{1}{4}$$

Fractions of a decibel. When you round off to the nearest whole decibel, the error in the power ratio is at most only 1 part in 7. While such accuracy is usually sufficient, greater precision is often required—as, for example, in compounding radar losses, which though small individually may be significant collectively. So, it is helpful to have a way of remembering the ratios for fractions of a decibel. A plot of decibels versus power ratio in the interval between 0 and 1 dB is practically a straight line. Since 0 dB corresponds to a ratio of 1 and 1 dB to a ratio of 1¼, the power ratio corresponding to any fraction of a decibel between 0 and 1 very nearly equals 1 plus ¼th of the fraction.

$$\text{Power ratio} = 1 + \frac{\text{Fraction of dB}}{4}$$

The ratio for ½ dB, for example, is 1 + ½/4 = 1⅛, or approximately 1.12.

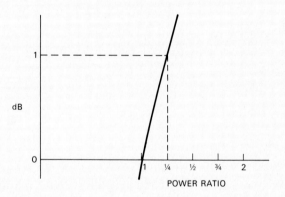

Remembering the "quarter-dB rule," you not only can round off to the nearest half dB, but could easily scratch out a table like this in the sand of a desert island.

$$0.8 \text{ dB} = 1.20$$
$$0.6 \text{ dB} = 1.15$$
$$0.5 \text{ dB} = 1.12$$
$$0.4 \text{ dB} = 1.10$$
$$0.2 \text{ dB} = 1.05$$

The radar cross section of a typical target can easily vary from 1 to 1000 square meters as the aspect of the target changes. A decibel relative to 1 square meter of radar cross section is called a dBsm (Fig. 15).

Another example is antenna gain. It is the ratio of the power per unit of solid angle radiated in a given direction to the power per unit of solid angle which would have been radiated had the same total power been distributed uniformly in all directions, i.e., isotropically. A decibel relative to isotropically radiated power is called a dBi.

Summary

The decibel was devised to express power ratios. Being logarithmic, it greatly compresses the numbers needed to express values having a wide dynamic range.

Decibels also make compounding ratios easy. Ratios can be multiplied by adding their decibel equivalents, divided (inverted) by giving them a negative sign, and raised to a power by multiplying them by that power.

A ratio expressed in dB can be thought of as consisting of two parts. The digit in the one's place expresses the basic ratio. The digit to the left of it is the power of 10. To translate from dB to a power ratio in your head, you convert the basic ratio; then, place a number of zeros to the right of it equal to the power of ten. To translate to decibels, you do the reverse.

Positive decibels correspond to ratios >1; zero decibels to a ratio of 1; negative decibels, to ratios <1. There is no decibel equivalent for a ratio of 0.

Decibels are commonly used to express gains and losses. Gain is output divided by input. Loss is input divided by output.

Referenced to absolute units, decibels are also used to express absolute values.

1 meter² = 0 dBsm   1,000 meters² = 30 dBsm

15. Because they vary widely in value, radar cross sections are conveniently expressed in decibels relative to 1 square meter.

Some Relationships To Keep In Mind

- Power ratio

$$dB = 10 \log_{10} \frac{P_2}{P_1}$$

- Power ratio in terms of voltages

$$dB = 20 \log_{10} \frac{V_2}{V_1}$$

- 1 dB = 1 ¼

- 3 dB = 2

- dBW = dB relative to 1 watt

- dBm = dB relative to 1 milliwatt

- dBsm = dB relative to 1 square meter of radar cross section

- dBi = dB relative to isotropic radiation

DOUGLAS F3D SKYNIGHT (1951)

This aircraft was the first jet and the first carrier-based fighter to be designed specifically as a night and all-weather fighter. Marine Major William T. Stratton Jr. and his Radar Operator, Sgt. Hans C. Hoglind, achieved the first jet-versus-jet night kill in history on November 3, 1952 against a YAK—15.

PART III

Radar Fundamentals

LOCKHEED F—94C STARFIRE (1953)

This aircraft was essentially an F—80, Shooting Star day fighter and was equipped with Hughes' first interceptor radar fire control system, the E1, guiding the firing of four 50 calibre machine guns. Later versions were equipped with a more powerful E—5 radar and fired 48 rockets.

Choice of Radio Frequency

7

A primary consideration in the design of virtually every radar is the frequency of the transmitted radio waves—the radar's *operating frequency*. How close a radar may come to satisfying many of the requirements imposed on it—detection range, angular resolution, doppler performance, size, weight, cost, etc.—often hinges on the choice of radio frequency. This choice, in turn, has a major impact on many important aspects of the design and implementation of the radar.

In this chapter, we will survey the broad span of radio frequencies used by radars and examine the factors which determine the optimum choice of frequency for particular applications.

Frequencies Used for Radar

Today, radars of various kinds operate at frequencies ranging from as low as a few megahertz to as high as 300,000,000 megahertz (Fig. 1).

At the low end are a few highly specialized radars: sounders that measure the height of the ionosphere, as well as radars that take advantage of ionospheric reflection to see over the horizon and detect targets thousands of miles away.

At the high end are laser radars, which operate in the visible region of the spectrum and are used to provide the angular resolution needed for such tasks as measuring the ranges of individual targets on the battlefield.

Most radars, however, employ frequencies lying somewhere between a few hundred megahertz and 100,000 megahertz.

To make such large values more manageable, it is customary to express them in gigahertz. One gigahertz, you

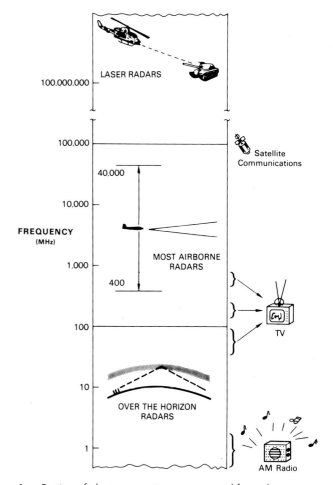

1. Portion of electromagnetic spectrum used for radar.

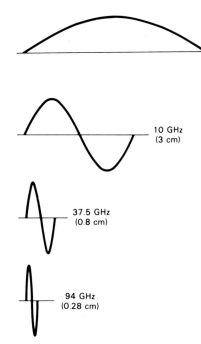

2.   Some wavelengths used by airborne radars, actual size.

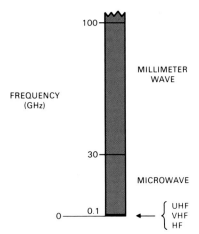

3.   Regions of the electromagnetic spectrum commonly used for radar, plotted on a linear scale.

UHF	300 – 900 MHz
VHF	30 – 300 MHz
HF	3 – 30 MHz

will recall, equals 1000 megahertz. A frequency of 100,000 megahertz, then, is 100 gigahertz.

As often as not, radar operating frequencies are expressed in terms of wavelength—the speed of light divided by the frequency (Fig. 2).

Incidentally, a convenient rule of thumb for converting from frequency to wavelength is *wavelength in centimeters = 30 divided by frequency in gigahertz.* The wavelength of a 10 gigahertz wave, for example, is $30 \div 10 = 3$ cm.

To convert from wavelength to frequency, you turn the rule around, interchanging wavelength and frequency: *frequency in gigahertz = 30 divided by wavelength in centimeters.* The frequency of a 3-cm wave is thus $30 \div 3 = 10$ GHz.

In English units, the rule is *wavelength in feet = 1 divided by frequency in gigahertz,* and vice versa. The wavelength of a 10 gigahertz wave is $1 \div 10 = 0.1$ foot.

Frequency Bands

Besides being identified by discrete values of frequency and wavelength, radio waves are also broadly classified as falling within one or another of several arbitrarily established regions of the radio frequency spectrum—high frequency (HF), very high frequency (VHF), ultra high frequency (UHF), and so on. The frequencies commonly used by radars fall in the VHF, UHF, microwave, and millimeter-wave regions (Fig. 3).

During World War II, the microwave region was broken into comparatively narrow bands and assigned letter designations for purposes of military security: L-band, S-band, C-band, X-band, and K-band. To enhance security, the designations were deliberately put out of alphabetical sequence. Though long since declassified, these designations have persisted to this day.

The K-band turned out to be very nearly centered on the resonant frequency of water vapor, where absorption of radio waves in the atmosphere is high. Consequently, the band was split up. The central portion retained the original designation. The lower portion was designated the Ku-band; the higher portion, the Ka-band. An easy way to keep these designations straight is to think of the "*u*" in Ku as standing for *under* and the *"a"* in Ka as standing for *above,* the central band.

In the 1970s, a complete new sequence of bands—neatly assigned consecutive letter designations from A to M—was devised for electronic countermeasures equipment (Fig. 4). Attempts were made to apply these designations to radars, as well. But largely because the junctions of the new bands occur at the centers of the traditional bands—about which many radars are clustered—these attempts proved abortive. In the U.S. the "new" band designations are generally used, as originally intended, only for counter measures.

If you haven't already done so, memorize the center frequencies and wavelengths of five of these radar bands:

Band	GHz	cm
Ka (above)	38	0.8
Ku (under)	15	2
X	10	3
C	6	5
S	3	10

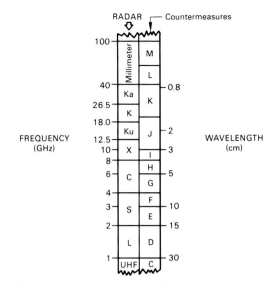

4. Radar and countermeasures band letter designations.

Influence of Frequency on Radar Performance

The best frequency to use depends upon the job the radar is intended to do. Like most other design decisions, the choice involves trade-offs among several factors—physical size, transmitted power, antenna beamwidth, atmospheric attenuation, and so on.

Physical Size. The dimensions of the hardware used to generate and transmit radio frequency power are in general proportional to wavelength. At the lower frequencies (longer wavelengths), the hardware is usually large and heavy. At the higher frequencies (shorter wavelengths), radars can be put in smaller packages and operate in more limited spaces, and they weigh correspondingly less (Fig. 5).

Transmitted Power. Because of its impact on hardware size, the choice of wavelength indirectly influences the ability of radar to transmit large amounts of power. The levels of power that can reasonably be handled by a radar transmitter are largely limited by voltage gradients (volts per unit of length) and heat dissipation requirements. It is not surprising, therefore, that the larger, heavier radars operating at wavelengths on the order of meters can transmit megawatts of average power, whereas millimeter-wave radars may be limited to only a few hundred watts of average power.

(Most often, though, within the range of available power

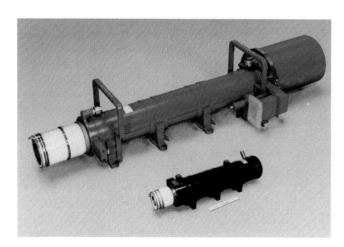

5. The physical size and power handling capacity of radio frequency components decreases with wavelength. Transmitter tube for 30 centimeter radar, top; transmitter tube for 0.8 centimeter radar, bottom.

85

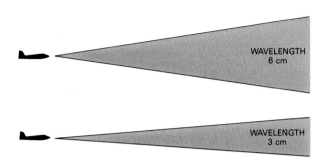

6. For same sized antenna, width of beam is proportional to wavelength.

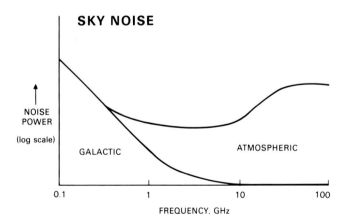

7. Ambient noise reaches a minimum somewhere between 0.3 gigahertz and 10 gigahertz, depending upon the level of the galactic noise, which varies with solar conditions.

the amount of power actually used is decided by size, weight, reliability, and cost considerations.)

Beamwidth. As will be explained in Chap. 8, the width of a radar's antenna beam is directly proportional to the ratio of the wavelength to the width of the antenna. To achieve a given beamwidth, the longer the wavelength, the wider the antenna must be. At low frequencies very large antennas must generally be used to achieve acceptably narrow beams. At high frequencies, small antennas will suffice (Fig. 6). The narrower the beam, of course, the greater the power that is concentrated in a particular direction at any one time, and the finer the angular resolution.

Atmospheric Attenuation. In passing through the atmosphere, radio waves may be attenuated by two basic mechanisms: absorption and scattering (see panel, right). The absorption is mainly due to oxygen and water vapor. The scattering is due almost entirely to condensed water vapor (e.g., raindrops). Both absorption and scattering increase with frequency. Below about 0.1 gigahertz, atmospheric attenuation is negligible. Above about 10 gigahertz, it becomes increasingly important.

Moreover, above that frequency, the radar's performance is increasingly degraded by weather clutter competing with desired targets. Even when the attenuation is reasonably low, if enough transmitted energy is scattered back in the direction of the radar, it will be detected. In simple radars which do not employ moving target indication (MTI), this return—called weather clutter—may obscure targets.

Ambient Noise. Electrical noise from sources outside the radar is high in the HF band. But it decreases with frequency (Fig. 7), reaching a minimum somewhere between about 0.3 and 10 gigahertz—depending upon the level of galactic noise, which varies with solar conditions. From there on, atmospheric noise predominates. It gradually becomes stronger and grows increasingly so at K-band and higher frequencies. In many radars internally generated noise predominates. But, when low-noise receivers are used to meet long range requirements, external noise can be an important consideration in the selection of frequency.

Doppler Considerations. Doppler shifts are proportional not only to closing rate but to radio frequency. The higher the frequency, the greater the doppler shift a given closing rate will produce. As will be made clear in later chapters, excessive doppler shifts can cause problems. In some cases, these tend to limit the frequencies that may be used. On the other hand, doppler sensitivity to small differences in clos-

86

ATMOSPHERIC ATTENUATION

Absorption. Energy is absorbed from radio waves passing through the atmosphere primarily by the gases comprising it. Absorption increases dramatically with the waves' frequency.

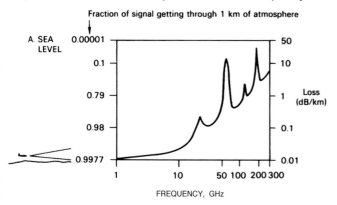

Fraction of signal getting through 1 km of atmosphere

Below about 0.1 gigahertz, absorption is negligible. Above 5 gigahertz, it becomes increasingly significant. Beyond about 20 gigahertz, it becomes severe.

Most of the absorption is due to oxygen and water vapor. Consequently, it not only decreases at the higher altitudes where the atmosphere is thinner, but decreases with decreasing humidity.

The molecules of oxygen and water vapor have resonant frequencies.

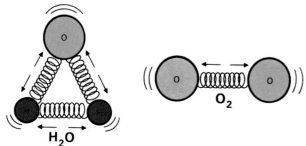

When excited at these frequencies, they absorb more energy. Hence the peaks in the absorption curve. The peaks are broadened by molecular collisions and so are sharper at high altitudes, where the atmosphere is less dense, but their frequencies are the same. (Plot B is drawn to the same scale as A; but is shifted down to encompass the lower curve.)

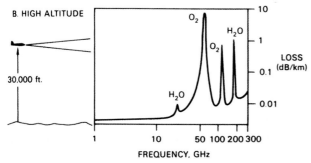

The peaks at 22 gigahertz and 185 gigahertz are due to water vapor. Those at 60 gigahertz and 120 gigahertz are due to oxygen. The regions between peaks are called windows.

Energy is also absorbed by particles suspended in the atmosphere, but their principal effect is scattering.

Scattering. Radio waves are scattered by particles suspended in the atmosphere. Scattering increases with the particles' dielectric constant and size relative to wavelength. Scattering becomes severe when the size is comparable to a wavelength.

The principal scatterers are raindrops and, to a lesser extent, hail (because of its much lower dielectric constant). Snowflakes, which contain less water and have lower fall rates, scatter less energy. Clouds, which consist of tiny droplets, scatter still less. Smoke and dust are negligible scatterers because of their small particle size and low dielectric constant.

Scattering becomes noticeable in the S-band (3 gigahertz). At those frequencies and higher, backscattering is sufficient to make rain visible.

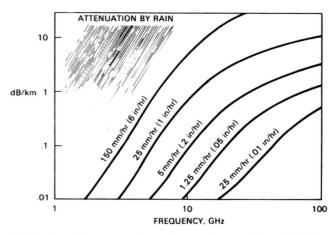

Both absorption and scattering by clouds are still negligible in the S-band. So meteorological radars operating there can measure rainfall rates without being hampered by attenuation due to clouds or by receiving enough backscatter from them to be confused with precipitation.

Above 10 gigahertz, scattering and absorption by clouds becomes appreciable. The attenuation is proportional to the amount of water in the clouds.

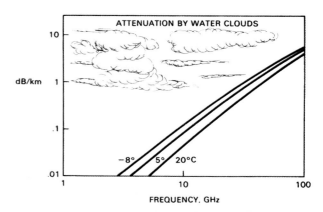

Attenuation increases with decreasing temperature since the dielectric constant of water is inversely proportional to temperature. Ice clouds, however, attenuate less because of the low dielectric constant of ice.

ing rate can be increased by selecting reasonably high frequencies.

Selecting the Optimum Frequency

From the preceding, it is evident that the selection of the radio frequency is influenced by several factors: the functions the radar is intended to perform, the environment in which the radar will be used, the physical constraints of the platform on which it will operate, and cost. To illustrate, let us consider some representative applications. To put the selection in context, we will consider not only airborne applications, but ground and shipboard applications, too.

Ground-Based Applications. These run the gamut of operating frequencies. At one extreme are the long-range multimegawatt surveillance radars. Unfettered by size limitations, they can be made large enough to provide acceptably high angular resolution while operating at relatively low frequencies. Over-the-horizon radars, as we've seen, operate in the HF-band where the ionosphere is suitably reflective. Space surveillance and early warning radars operate in the UHF and VHF bands, where ambient noise is minimal and atmospheric attenuation is negligible. These bands, however, are crowded with communication signals. So their use by radars (whose transmissions generally occupy a comparatively broad band of frequencies) is restricted to special applications and geographic areas.

Where such long ranges are not required and some atmospheric attenuation is therefore tolerable, ground radars may be reduced in size by moving up to L-, S-, and C-band frequencies or higher (Fig. 8).

Shipboard Applications. Aboard ships, physical size becomes a limiting factor in many applications. At the same time, the requirement that ships be able to operate in the most adverse weather puts an upper limit on the frequencies that may be used. This limit is relaxed however, where extremely long ranges are not required. Furthermore, higher frequencies must be used when operating against surface targets and targets at low elevation angles.

For, at grazing angles approaching zero, the return received directly from a target is very nearly cancelled by return from the same target, reflected off the water—a phenomenon called multipath propagation (Fig. 9). Cancellation is due to a 180° phase reversal occurring when the return is reflected. As the grazing angle increases, a difference develops between the lengths of the direct and indirect paths, and cancellation decreases. The shorter the wavelength, the more rapidly the cancellation disappears. For this reason, the shorter wavelength S- and X-band fre-

8. While ground-based radars commonly operate at lower frequencies, where long range is not important—as for this radar which traces the source of mortar fire—X-band may be used for small size.

9. At small grazing angles, return received directly from the target is very nearly cancelled by return reflected off the water.

quencies are widely used for surface search, detection of low flying targets, and piloting. (The same phenomenon is encountered on land when operating over a flat surface.)

Airborne Applications. In aircraft, the limitations on size are considerably more severe. The lowest frequencies generally used here are in the UHF and S-bands. They provide the long detection ranges needed for airborne early warning in the E2 and AWACS aircraft, respectively (Fig. 10). One look at the huge radomes of these planes, though, and it is clear why higher frequencies are commonly used when narrow antenna beams are required in smaller aircraft, such as fighters.

The next lowest-frequency applications are in the C-band. Radar altimeters operate here. Interestingly, the band was originally selected for this use because it made possible light, cheap equipment that could use a triode transmitter tube. These frequencies, of course, enable good cloud penetration. Because altimeters are simple, require only modest amounts of power, and do not need highly directive antennas, they can use these frequencies and still be made conveniently small.

Weather radars, which require greater directivity, operate in the C-band as well as in the X-band. The choice between the two bands reflects a dual trade-off. One is between storm penetration and scattering. If scattering is too severe, the radar will not penetrate deeply enough into a storm to see its full extent. Yet, if too little energy is scattered back to the radar, storms will not be visible at all. The other trade-off is between storm penetration and equipment size. C-band radars, providing better penetration, hence longer-range performance, are primarily used by commercial aircraft. X-band radars, providing adequate performance in smaller packages, are widely used by private aircraft.

Most fighter, attack, and reconnaissance radars operate in the X- and Ku-bands, with a great many operating in the 3-centimeter wavelength region of the X-band (Fig. 11).

The attractiveness of the 3-centimeter region is threefold. First, atmospheric attenuation, though appreciable, is still reasonably low—only 0.02 dB per kilometer for two-way transmission at sea level. Second, narrow beamwidths, providing high power densities and excellent angular resolution, can be achieved with antennas small enough to fit in the nose of a small aircraft. Third, because of their wide use, microwave components for 3-centimeter radars are readily available from a broad base of suppliers.

Where limited range is not a problem and both small size and high angular resolution are desired, higher frequencies may be used. Radars operating in the Ka-band, for

10. Operating in the S-band, AWACS radar provides early warning. But its antenna must be very large to provide desired angular resolution.

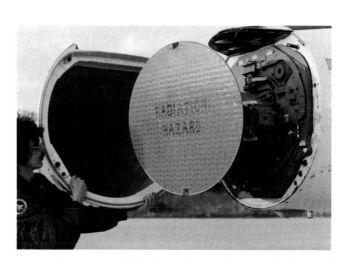

11. At X-band reasonably high angular resolution can be obtained with an antenna small enough to fit in the nose of a fighter.

12. Operating at 94 gigahertz, this tiny antenna of an air-to-air missile provides the same angular resolution as the much larger antenna pictured in Fig. 11.

Typical Frequency Selections

- Early warning radars.... UHF and S-band
- Radar altimeters .... C-band
- Weather radars............ C- and X-bands
- Fighter/Attack .... X- and Ku-bands

example, have been developed to perform ground search and terrain avoidance for some aircraft. But because of the high level of attenuation at these frequencies, to date there has been relatively little utilization of this band.

With the recent availability of suitable millimeter-wave power-generating components, radar designers are developing extremely small, albeit short-range, radars which take advantage of the atmospheric window at 94 gigahertz to give small air-to-air missiles high terminal accuracies (Fig. 12). At 94 gigahertz, a 3.8-inch antenna provides the same angular resolution as a 36-inch antenna would at 10 gigahertz (3 centimeters).

Summary

Radio frequencies employed by airborne radars range from a few hundred megahertz to 100 thousand megahertz, the optimum frequency for any one application being a trade-off among several factors.

In general, the lower the frequency, the greater the physical size and the higher the available maximum power. The higher the frequency, the narrower the beam that may be achieved with a given sized antenna.

At frequencies above about 0.1 gigahertz, attenuation due to atmospheric absorption—mainly by water vapor and oxygen—becomes significant. At frequencies of 3 gigahertz and higher, scattering by condensed water vapor—rain, hail, and, to less extent, snow—produces weather clutter. It not only increases attenuation, but in radars not equipped with MTI may obscure targets. Above about 10 gigahertz, absorption and scattering become increasingly severe, and attenuation due to clouds becomes important.

Noise is minimum between about 0.3 and 10 gigahertz, but becomes increasingly severe at 20 gigahertz and higher frequencies.

Doppler shifts increase with frequency, and this may also be a consideration.

Directivity and the Antenna Beam

8

The degree to which the antenna concentrates the radiated energy in a desired direction—*broadly referred to here as directivity*—is a key characteristic of virtually every airborne radar. Besides determining the radar's ability to locate targets in angle, directivity can vitally affect the ability to deal with ground clutter and is a major factor governing detection range.

In this chapter, we will learn how the energy radiated by an antenna is distributed in angle and examine the salient characteristics of the radiation pattern—beamwidth, gain, and sidelobes. We will then see how the sidelobes may be reduced; how fast, versatile beam positioning may be accomplished with electronic scanning; and how high angular resolution and angular measurement accuracy may be achieved. Finally, we will learn how the beam may be optimized for ground mapping.

Distribution of Radiated Energy in Angle

From common simplistic illustrations, it might be supposed that a radar antenna concentrates all of the transmitted energy into a narrow beam within which the power is uniformly distributed; that if a pencil beam were trained like a flashlight on an imaginary screen in the sky, it would illuminate a single round spot with uniform intensity. While this might be desirable, it is even less true of an antenna than of a flashlight.

Like all antennas, a pencil beam antenna radiates some energy in almost every direction. As illustrated in the three-dimensional plot of Fig. 1, most of the energy is concentrated in a more or less conical region surrounding the

1. Three-dimensional plot of the strength of the radiation from a pencil beam antenna.

2. A slice taken through 3-dimensional plot. Note the series of lesser lobes on either side of the mainlobe.

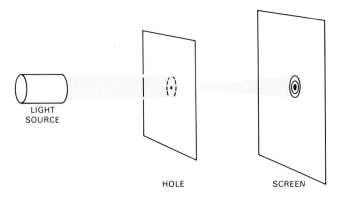

3. Lobular distribution of power is due to diffraction–the process that causes a beam of light projected through a tiny hole to spread and become fringed with concentric rings of light.

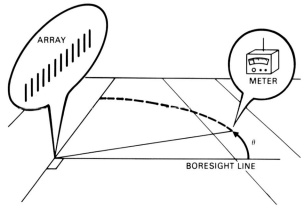

4. To determine the distribution of energy radiated by array, a field-strength meter is moved along arc of constant radius. The array consists of a row of closely spaced vertical radiators.

central axis, or boresight line, of the antenna. This region is called the mainlobe. If we slice the plot in two through the central axis of this lobe (the boresight line), we find that it is flanked on either side by a series of weaker lobes (Fig. 2). These are called sidelobes.

This lobular structure is due to diffraction—the phenomenon observed when a beam of light passes through a small circular hole (Fig. 3). The beam spreads and, if the light is all of one wavelength, becomes fringed with concentric rings of light of progressively decreasing intensity.

The phenomenon is most easily explained if we consider a type of horizontally oriented, one-dimensional antenna called a linear broadside array. It consists of a row of closely spaced radiators, each emitting in all azimuth directions a wave of the same amplitude, phase, and frequency. To measure the combined strength of these waves at various azimuth angles, we place a field strength meter far enough away that the lines of sight from the meter to all radiators are very nearly parallel (Fig. 4). Starting at a point on the perpendicular bisector of the array (boresight line), we move the meter along an arc of constant radius from the array center.

At any one point, the field strength depends upon the relative phases of the received waves. The relative phases, in turn, depend on the differences in distance to the individual radiators. These differences can best be visualized if we draw a line from one end of the array, perpendicular to the line of sight to the meter—the line AB in Fig. 5. The angle this line makes with the array equals the azimuth angle, θ, of the meter.

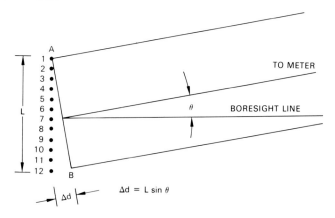

5. Line AB marks off differences in distance from individual array elements to meter. The angle that AB makes with array equals azimuth angle (θ) of meter.

Now, if θ is zero, the distance from the meter to all of the radiators is essentially the same. (The lines of sight to all radiators, remember, are essentially parallel.) The waves are in phase, and their fields add up to a large sum.

However, if θ is greater than zero, the distance to each successive radiator down the line is progressively greater. As a result, the phases of the received waves are all slightly different, and the sum is not as great.

As the azimuth angle increases, the differences in distance increase. A point is ultimately reached (Fig. 6) where the distance from the meter to the first radiator beyond the center of the array (No. 7) is half a wavelength greater than the distance to the radiator at the near end (No. 1). Consequently, the wave received from radiator No. 1 is cancelled by the wave received from radiator No. 7. The same is true of the waves received from radiators No. 2 and No. 8, and so on.

The sum of the waves received from all of the radiators, therefore, is zero. The meter has reached an azimuth angle where there is a null in the total radiation from the antenna.

If θ is increased further, the waves from the radiators at the ends of the array no longer cancel exactly, and the sum increases. As the difference in distance from the meter to the ends of the array approaches 1¹/₂ wavelengths, another peak is reached (Fig. 7). The waves from the radiators in the central portion of the array—Nos. 3 through 10—still cancel. But the waves from the radiators at either end—Nos. 1 and 2, and Nos. 11 and 12—add up to an appreciable sum. The meter is now in the center of the array's first sidelobe.

If θ is increased still further, the portion of the array for which cancellation occurs increases, and the same general process repeats. The meter thus moves through a succession of nulls and progressively weaker lobes.

The field strength measured in an excursion through several lobes on either side of the mainlobe is plotted versus azimuth angle in Fig. 8. The shape of this plot is approximated by the equation

$$E = K \frac{\sin x}{x}$$

where E is the field strength and x is proportional to θ. This is called a "sine-x-over-x" shape.

Actually, $x = \pi (L/\lambda) \sin \theta$. So x is directly proportional to θ only for small values of θ. As θ increases, sin θ becomes progressively less than θ, with the result that the higher-order sidelobes are spaced progressively farther apart.

The directivity of an array antenna has been explained here in terms of field strength, since that is both easily measured and easily visualized.

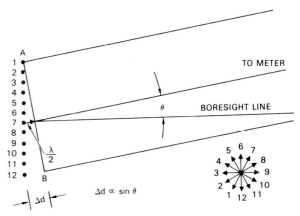

6. When distance from meter to radiator No. 7 becomes half a wavelength longer than distance to radiator No. 1, the signals received from these radiators cancel. So do all the others.

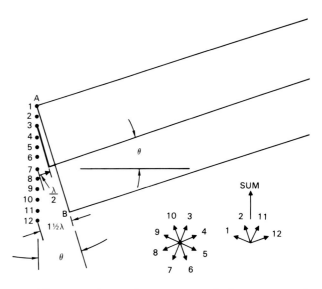

7. As difference in distance from meter to ends of array approaches 1¹/₂ wavelengths, only those signals from elements 3 through 10 cancel.

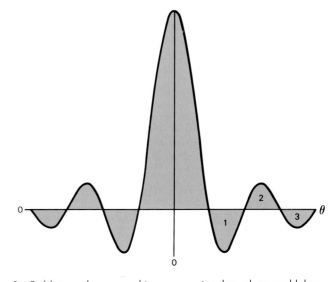

8. Field strength measured in an excursion through several lobes on either side of boresight line. (Radio frequency phase of odd numbered sidelobes—1, 3, etc.—is reversed; hence, these sidelobes are plotted as negatives.)

THE SIN X/X SHAPE

As the angle, θ, between the line of sight to a distant point and the boresight line of a linear array antenna increases, phasors representing the signals received from the individual radiators fan out and their sum decreases.

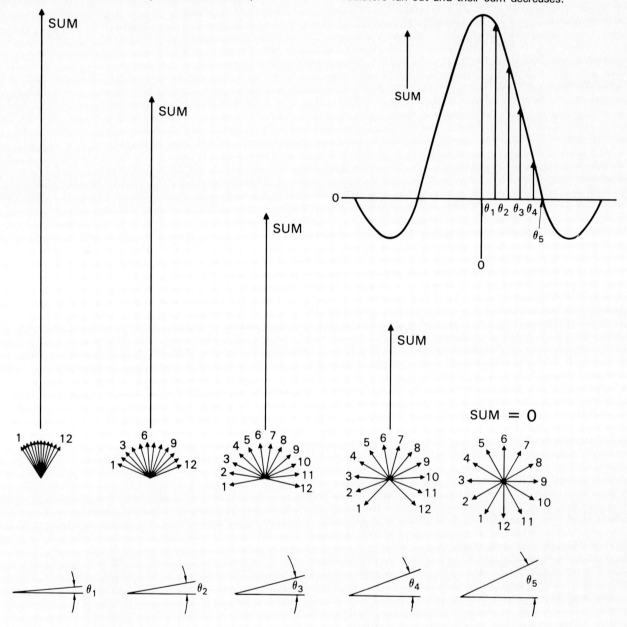

In a radar, however, what is important is the amount of energy radiated per unit of time: the *power* of the radiated waves (Fig. 9). Power is proportional to field strength squared. Expressed in terms of power, therefore, the equation for the distribution of the radiated energy in angle is

$$\text{Power} = \text{K'} \left(\frac{\sin x}{x}\right)^2$$

Two-dimensional planar arrays, such as are commonly used in airborne radars, consists essentially of a number of linear arrays stacked on top of one another.

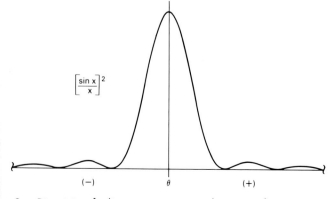

9.   Directivity of a linear array expressed in terms of power.

TWO COMMON TYPES OF AIRBORNE RADAR ANTENNAS

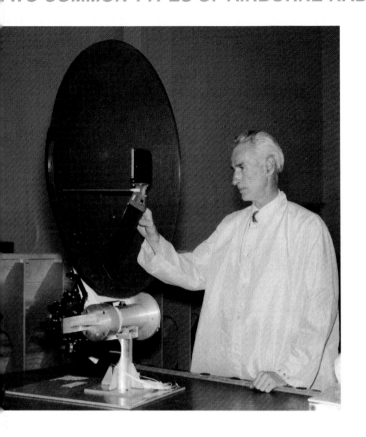

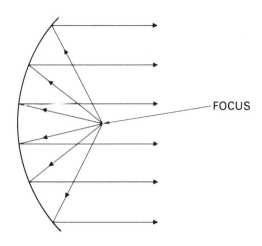

Parabolic reflector antenna—for years the most common type used in airborne radars. Feed located at focus of parabola directs radiation into dish, which reflects it. Curvature of parabola is such that distance from feed to dish to plane across mouth (aperture) of dish is the same for every path the radiation can take. Consequently, the phase of the radiation at every point in the plane of the aperture is the same, and a narrow pencil beam is formed. Antenna is simple and relatively inexpensive to fabricate.

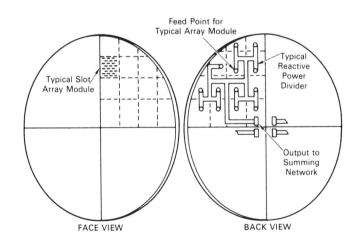

Planar array antenna—for an advanced fighter radar. Radiation of equal phase is emitted from 2-dimensional array of slots in face. Planar arrays provide relatively high aperture efficiency and low back radiation (spillover). By controlling excitation of slots with nondissipating attenuators on back of antenna, distribution of energy across aperture can be shaped to minimize sidelobes. Principal disadvantages are relatively narrow bandwidth ($\cong$ 10 percent) and higher cost. Also, circular polarization, if desired, is more difficult to obtain.

To give an antenna a circular or elliptical shape, the arrays above and below the central ones are progressively shortened. The total radiation from the antenna is the composite of the radiation from the individual arrays. Even if the radiation from every element were the same—which it never is—a plot of the total radiation would not have a simple sin x/x shape. Nevertheless, the general shape is much the same. (Incidentally, the shape for a uniformly illuminated circular array is exactly the same as the diffraction pattern mentioned earlier for light passing through a small round hole.[1])

Characteristics of the Radiation Pattern

A plot of the power (or field strength) of the radiation from an antenna in any one plane versus angle from the antenna's central axis is called a radiation pattern. In considering directivity, the power at the center of the mainlobe is taken as a reference and the power radiated in every other direction is taken in ratio to this value. The ratio is normally expressed in decibels and plotted in rectangular coordinates as in Fig. 10.

Since the pattern is usually not symmetrical about the center of the mainlobe, "cuts" must be taken through many different planes to describe an antenna's directivity fully. Also, patterns are generally measured in two polarizations: that for which the antenna was designed and the polarization at right angles to this—the cross polarization.

Generally, three characteristics of a radiation pattern are of interest: the width of the mainlobe, the gain of the mainlobe, and the relative strengths of the sidelobes.

Beamwidth. The width of the mainlobe is called the beamwidth. It is the angle between opposite edges of the beam. The beam is generally not symmetrical, so it is common to refer to azimuth beamwidth and elevation beamwidth.

Since the strength of the mainlobe falls off increasingly as the angle from the center of the beam increases, for any value of beamwidth to have meaning, one must specify what the edges of the beam are considered to be.

The edges are perhaps most easily defined as the nulls on either side of the mainlobe. However, from the standpoint of the operation of a radar (Fig. 11), it is generally more realistic to define them in terms of the points where the power has dropped to some arbitrarily selected fraction of that at the center of the beam. The fraction most commonly used is 1/2. Expressed in decibels, 1/2 is −3 dB. Beamwidth measured between these points, therefore, is called the 3-dB beamwidth.

1.   This pattern has a $J_1(x)$-over-x shape, where J_1 is the Bessel function of the first order.

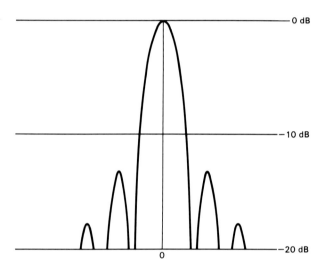

10.   Radiation pattern is normally plotted in rectangular coordinates in dB relative to gain at center of mainlobe.

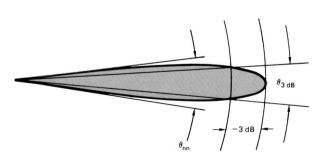

11.   Beamwidth is commonly measured between points where power has dropped to one half of maximum (−3 dB). Three dB beamwidth, θ_{3dB}, is roughly 1/2 null-to-null beamwidth, θ_{nn}.

Regardless of how it is defined, beamwidth is determined primarily by the size of the antenna's frontal area. This area is called the aperture. Its dimensions—width, height, or diameter—are gauged not in inches or centimeters, but in wavelengths of the radiated energy (Fig. 12).

The larger the appropriate dimension is in relation to the wavelength, the narrower the beam in a plane through that dimension will be. As we saw earlier, the nulls on either side of the mainlobe of a linear array occur at angles for which the distance from the observer to one end of the array is one wavelength longer than to the other end.

Therefore, for either a *linear array* or a *rectangular aperture* over which the illumination is uniformly distributed, the null-to-null beamwidth in radians is twice the ratio of the wavelength to the length of the array (Fig. 13).

$$\theta_{nn} = 2 \frac{\lambda}{L} \text{ radians}$$

where

λ = wavelength of radiated energy

L = length of aperture (same units as λ)

The 3-dB beamwidth is a little less than half the null-to-null width.

$$\theta_{3\,dB} = 0.88 \frac{\lambda}{L}$$

For a uniformly illuminated *circular aperture* of diameter d, the 3-dB beamwidth is a bit greater.

$$\theta_{3\,dB} = 1.02 \frac{\lambda}{d}$$

A circular antenna 60 centimeters in diameter, radiating energy of 3-centimeter wavelength, for example, has a beamwidth of 1.02 x 3/60 = 0.051 radian.

One radian equals 57.3° (Fig. 14). So the beamwidth in degrees is 0.051 x 57.3 = 2.9°.

If the antenna has tapered illumination, such as is typically used in radars for fighter aircraft, the beamwidth will be somewhat greater.

$$\theta_{3\,dB} = 1.25 \frac{\lambda}{d}$$

A 60-centimeter antenna with tapered illumination would thus have a beamwidth of about 3.6°.

A rule of thumb for estimating the beamwidths of tapered circular antennas is this: *at X-band, the 3-dB beamwidth is roughly 85° divided by the diameter in inches.*

The 3-dB beamwidth of a 20-inch diameter antenna is thus about 85° ÷ 20 = 4.25°. If the illumination is not

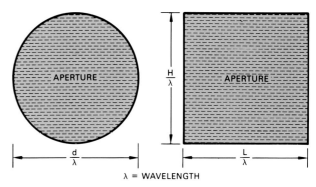

12. Beamwidth is determined primarily by dimensions of antenna aperture, in wavelengths.

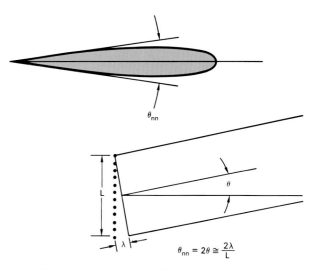

13. For a linear array, angle (in radians) from boresight line to first null equals ratio of wavelength to length of array. Null-to-null beamwidth is twice this angle.

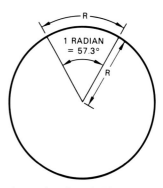

14. A radian is the angle subtended by an arc the length of the radius (R). The circumference of a circle equals $2\pi R$. Therefore, $2\pi = 360°$, and 1 radian = $360°/2\pi = 57.3°$.

X-Band Beamwidth Rule of Thumb	
For tapered illumination:	For untapered illumination:
$\theta_{3dB} \approx \dfrac{85°}{d}$	$\theta_{3dB} \approx \dfrac{70°}{d}$
d = diameter in inches	

tapered, 70° should be substituted for 85° in this rule. (See bottom of next page for an explanation of "tapering.")

Antenna Gain. The gain of an antenna is the ratio of the power per unit of solid angle radiated in a specific direction to the power per unit of solid angle that would have been radiated had the same total power been radiated uniformly in all directions—i.e., isotropically (Fig. 15).[2] An antenna thus has gain in almost every direction. In most directions, though, the gain is less than one, since the average gain in all directions is, by the law of conservation of energy, one.

The gain in the center of the mainlobe is thus a measure of the extent to which the radiated energy is concentrated in the direction the antenna is pointing. The narrower the mainlobe, the higher this gain will be.

The maximum gain that can be achieved with a given size antenna is proportional to the area of the antenna aperture in square wavelengths times an illumination efficiency factor. If the aperture were uniformly illuminated—a practically impossible condition, even if it were desired—the factor would equal one.

Actually, it ranges somewhere between 0.6 and 0.8 for planar arrays and may be as low as 0.45 for parabolic reflectors. In either case, for a given design, the factor tends to vary with the width of the band of frequencies the antenna is designed to pass. Typically, the greater the bandwidth, the lower the efficiency.

Because of the difficulty of determining the efficiency factor analytically, in practice the gain is determined experimentally and expressed in terms of an effective aperture area.

$$G = 4\pi \frac{A_e}{\lambda^2}$$

where

G = antenna gain at center of mainlobe

λ = wavelength of radiated energy

A_e = effective area of aperture (same units as λ^2)

Effective area is equal to physical area times aperture efficiency (which as noted above is virtually always less than 100 percent); so an alternate expression for antenna gain is

$$G = 4\pi \frac{A\eta}{\lambda^2}$$

where

A = physical area of aperture

η = aperture efficiency

2. Strictly speaking, the gain referred to here is directivity gain. More commonly, antenna gain connotes directivity gain less whatever power is lost in the antenna.

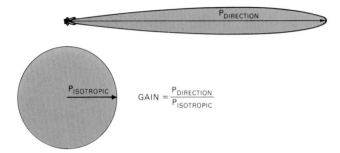

$$\text{GAIN} = \frac{P_{DIRECTION}}{P_{ISOTROPIC}}$$

15. Antenna directive gain is the ratio of power radiated in the direction of interest to power which would be radiated in that direction by an isotropic antenna, i.e., one that radiates waves of equal power in all directions.

Estimating Antenna Gain

X-Band Rule of Thumb:

$G \approx d^2 \eta$

d = diameter in cm
η = aperture efficiency

Example:

Diameter = 60 cm
Aperture efficiency = 0.7

$G \approx 60 \times 60 \times 0.7$
≈ 2520
$\approx 34 \text{ dB}$

General Rule of Thumb:

$G \approx 9 \, d^2 \eta$

d = diameter in wave-lengths

Example:

Wavelength = 3 cm
Diameter = 60 cm = 20 λ
Aperture efficiency = 0.7

$G \approx 9 \times (20)^2 \times 0.7$
≈ 2520
$\approx 34 \text{ dB}$

Sidelobes. An antenna's sidelobes are not limited to the forward hemisphere. They extend in all directions, even to the rear, for a certain amount of radiation invariably "spills over" around the edges of the antenna. Moreover, when the antenna is placed in a radome, the backward radiation is increased. For the radome scatters some energy from the mainlobe, much as the frosted glass of a light bulb diffuses light from the filament.

Nor are the sidelobes neatly defined, with sharp nulls in between. As can be seen from Fig. 16, the nulls tend to fill in.

For a uniformly illuminated circular aperture, the gain of the strongest (first) sidelobe is only about 1/64 that of the mainlobe. Stated in decibels, the first sidelobe has 18 dB less gain than the mainlobe—it is down 18 dB. The gain of the other sidelobes is substantially lower.

Nevertheless, in aggregate the sidelobes rob the mainlobe of a substantial amount of power. Because of the large solid angle they cover, roughly 25 percent of the total power radiated by a uniformly illuminated antenna is radiated outside the mainlobe.

Against most small targets even the strongest sidelobes are sufficiently weak that they can generally be ignored.[3] But against the ground, even the weakest sidelobes may produce considerable return. And, as will be explained in Chap. 22, buildings and other structures on the ground form corner reflectors which can return tremendously strong echoes, even when illuminated only by sidelobes.

In military applications, the sidelobes also increase both the radar's susceptibility to detection by an enemy and its vulnerability to jamming. Interference from a powerful noise jammer, for example, can be much stronger than the echoes of a small or distant target in the mainlobe. Consequently, it is generally desirable for the gain of the first sidelobes to be reduced to at least 80dB below that of the mainlobe.

Sidelobe Reduction. The degree to which the radiated power is concentrated into the mainlobe is called solid angle efficiency. To make it acceptably high, as well as to minimize problems of ground clutter and jamming, the gain of the sidelobes must generally be reduced. As explained earlier, the sidelobes are produced by radiation from the portion of the aperture near its edges. Consequently, they may be reduced by designing the antenna to radiate more power per unit area through the central portion of the aperture (Fig. 17). This technique is called illumination tapering. It increases the beamwidth somewhat, hence it reduces the peak gain of the mainlobe. But usually this is an acceptable price to pay for reduced sidelobes.

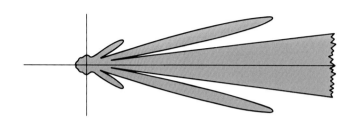

16. An antenna's sidelobes extend in all directions, even to the rear.

3. However, for some targets which in certain aspects reflect a large fraction of the incident energy back in the direction of the radar, sidelobe return can be substantial.

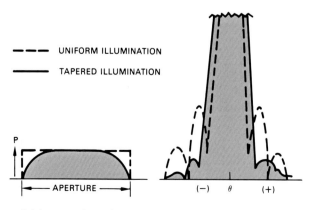

- - - - UNIFORM ILLUMINATION

──── TAPERED ILLUMINATION

17. Sidelobes may be reduced by tapering illumination at edges of aperture.

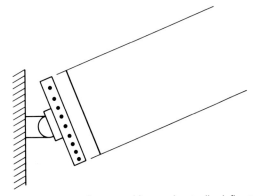

18. Beam is conventionally steered by mechanically deflecting the antenna.

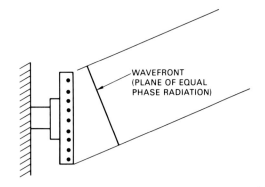

WAVEFRONT
(PLANE OF EQUAL
PHASE RADIATION)

19. With electronic steering, beam is steered by progressively shifting the phases of the signals radiated by the individual radiators.

Electronic Beam Steering

In most airborne radars, the antenna beam is positioned by physically moving the antenna through the desired azimuth and elevation angles (Fig. 18). An alternative method, possible with array antennas, is to differentially shift the phases of the radio waves emitted by the individual radiators. This technique is called electronic beam steering (or electronic scanning).

As with the simple linear array described earlier, the direction of maximum radiation from the array—i.e., direction of the mainlobe—is that for which the waves from all of the radiators are in phase. If the phases of the emitted waves are all the same, this direction is perpendicular to the plane of the array. However, if the phases are progressively shifted from one radiator to the next, the direction of maximum radiation will be correspondingly shifted (Fig. 19). By appropriately shifting the phases of the inputs to the individual radiators, therefore, the beam can be steered in any desired direction within a large solid angle.

Electronic steering has the advantage of being extremely flexible and remarkably fast. The beam can be given any shape, swept in any pattern at a very high rate, or jumped almost instantaneously to any position. It can even be split into two or more beams which radiate simultaneously on different frequencies and can be trained simultaneously on different targets (at the expense of a reduction in detection range).

Depending upon the application, electronic steering may be provided in one (Fig. 20) or two dimensions. Moreover, it may be combined with either mechanical beam steering or mechanical rotation of the antenna, as in the AWACS radar (see page 548).

20. Antenna for air-to-ground radar in which fan-shaped beam is electronically steered in azimuth. Antenna is carried in pod beneath aircraft; by rotating antenna about its longitudinal axis, it can be made to look out on either side of the aircraft.

Naturally, electronic steering also has disadvantages. Among these are increased complexity and degraded performance at large look angles.

The degradation in performance is due to foreshortening of the aperture when viewed from angles off dead center (Fig. 21). The length of the foreshortened dimension decreases as the cosine of the angle. The effect is negligible at small scan angles, but it becomes increasingly severe at large angles. The result of the foreshortening (effectively smaller aperture in the direction of illumination) is an increase in beamwidth and more importantly, a decrease in gain, limiting the maximum practical look angle to ±60°.

With mechanical steering, no such limitation occurs: the plane of the aperture is perpendicular to the direction of the mainlobe for all look angles.

Angular Resolution

The ability of a radar to resolve targets in azimuth and elevation is determined primarily by the azimuth and elevation beamwidths. This is illustrated simplistically by the two diagrams in Fig. 22.

In the first diagram, two identical targets, A and B, at nearly the same range are separated by slightly more than the width of the beam. As the beam sweeps across them, the radar receives echoes first from Target A, then from Target B. Consequently, the targets can easily be resolved.

In the second diagram, the same two targets are separated by less than the width of the beam. As the beam sweeps across them, the radar again receives echoes first from Target A. However, long before it stops receiving echoes from this target, it starts receiving echoes from Target B. The echoes from the two targets, therefore, meld together.

Superficially, angular resolution would appear to be limited to the null-to-null width of the mainlobe. But it is actually better than that, because the resolution depends not only upon the width of the lobe but on the distribution of power within it.

The graph in Fig. 23 is a plot of strength of the received signal as the mainlobe sweeps across an isolated target. When the leading edge of the lobe passes over the target, the echoes are so weak that they are undetectable. However, their strength increases rapidly. It reaches a maximum when the lobe is centered on the target, then drops to an undetectable value again as the trailing edge approaches the target.

This curve, you should note, is not the same shape as the radiation pattern plotted in similar coordinates, but is more sharply peaked. The reason is that the antenna's directivity applies equally to transmission and reception—a characteristic called reciprocity.

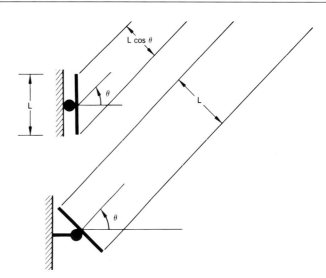

21. With electronic steering, length of aperture L decreases as cosine of look angle, θ. With mechanical steering, no such reduction occurs.

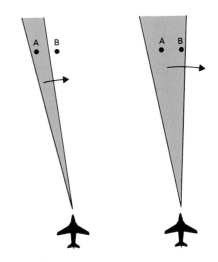

22. Ability to resolve targets in angle is determined primarily by antenna beamwidth. Targets can be resolved if beamwidth is less than their angular separation.

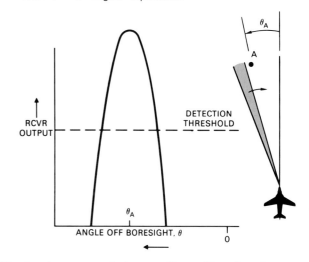

23. Angular accuracy is sharpened by peaking of receiver output as beam sweeps across target. Unless target echoes are very strong, azimuth angle over which return is detected is much less than null-to-null beamwidth, θ_{nn}.

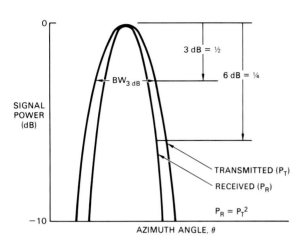

24. Since antenna's directivity is applied to both transmitted and received waves, the plot of received signal strength versus angle is more sharply peaked.

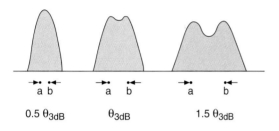

25. As separation between two closely spaced targets is increased, a notch develops in the plot of receiver output versus azimuth angle.

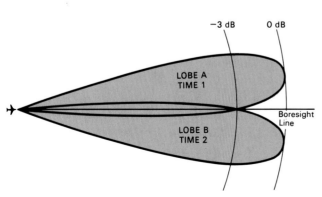

26. With sequential lobing, during reception the angular tracking error is determined by alternately placing mainlobe on one side and then the other of antenna boresight line.

To illustrate, suppose the position of a target is such that the power radiated in its direction is half that radiated in the center of the lobe (down 3 dB). When the target echoes are received, their power will again be cut in half. As a result, the received echoes will by only 1/4 as strong (down 6 dB) as when the target is in the center of the lobe (Fig. 24).

Because of this compounding, the plot or received signal power is narrower than the radiation pattern. And because the echoes received when the target is near the edges of the lobe are too weak to be detected (unless the target is at short range), the azimuth angle over which the target is detected is narrower than the null-to-null beamwidth.

The net effect of this narrowing on angular resolution is illustrated by the three plots of Fig. 25. They show a composite of the bell-shaped curves for two equally strong targets, A and B.

When the targets are closely spaced, the curves combine to produce a single broad hump. As the spacing increases, a notch develops in the top of this hump. The notch grows until the hump splits in two.

In practice, the notch becomes apparent at a target spacing of 1 to 1½ times the antenna's 3 dB beamwidth. The 3 dB beamwidth, therefore, has come to be used as the measure of the angular resolution of a radar.

Angle Measurement

The foregoing should not be taken to imply that the accuracy with which a radar can determine a target's direction is limited to the beamwidth. Since the amplitude of the received echoes varies symmetrically as the beam sweeps across a target, the direction of an isolated target can be determined to within a very small fraction of the beamwidth.

By stopping the antenna's search scan, target angle can be determined with still greater precision. One technique for accomplishing this is lobing.

Lobing. During reception, the center of the mainlobe is alternately placed on one side of the target and then the other (Fig 26). If the target is centered between lobes, the received echoes will be the same strength for both lobes. If it is not, the echoes will be stronger for one lobe than for the other.

Normally, the lobes are separated just enough to intersect at their half-power points. Since the slope of the radiation pattern in this region is relatively steep, a slight displacement of the target from a line through the crossover point results in a large difference in the strength of the echoes received through the two lobes (Fig. 27). By positioning the antenna to reduce this difference to zero (i.e., to eliminate the angular error), the antenna can be precisely lined up on the target.

Because the lobing is sequential, however, short-term changes in the strength of the target echoes—caused by scintillation or electronic countermeasures—can introduce large, spurious differences in the returns received through the two lobes and so degrade tracking accuracy. This problem may be avoided by designing the antenna to produce the lobes simultaneously, a technique called simultaneous lobing.

Since all the necessary angular tracking information is obtained from one reflected pulse, it is more commonly called monopulse operation.

Monopulse. Monopulse systems are of two general types. They differ both in regard to the direction of the lobes and in regard to the way the returns received through opposing lobes are compared.

The first type, called *amplitude comparison monopulse*, essentially duplicates sequential lobing with simultaneously formed lobes (Fig. 28).

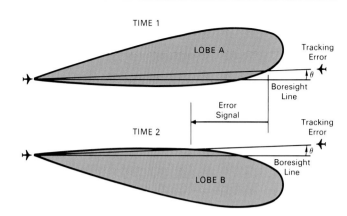

27. If the target is off the boresight line, return received through one lobe will be stronger than that received through the other. Magnitude of difference corresponds to magnitude of tracking error; sign of difference, to direction of error.

AMPLITUDE COMPARISON MONOPULSE

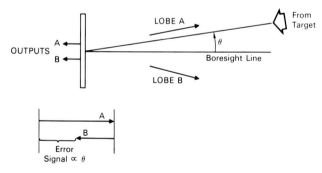

28. In essence, amplitude comparison monopulse duplicates sequential lobing in every respect except that return is received simultaneously through both lobes. Error signal is difference between outputs A and B.

Because the lobes point in slightly different directions, if a target is not on the boresight line of the antenna, the amplitude of the return received through one lobe differs from the amplitude of the return simultaneously received through the other lobe. The difference is proportional to the angular error.

By subtracting the output of one feed from the output of the other, an angular tracking error signal—often termed the *difference signal*—is produced. The sum of the two outputs—termed the *sum signal*—is used for range tracking.

**PHASE COMPARISON
MONOPULSE**

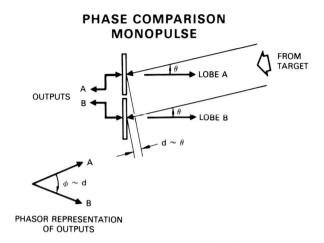

PHASOR REPRESENTATION
OF OUTPUTS

29. Phase-comparison monopulse. Since lobes of two antenna halves point in the same direction, amplitudes of outputs A and B are equal. But their phases differ by angle φ, which is proportional to angle error, θ.

The second type of monopulse is *phase-comparison.* In it, the array is divided into halves. The lobes produced by both halves point in the same direction. Consequently, the return received through one lobe has the same amplitude as that received through the other regardless of the angle of the target relative to the antenna boresight line. However, if an angular error exists, the *phases* of the returns will differ because of the difference in mean distance from the target to each half (Fig. 29).

An error signal proportional to the phase difference may be obtained by introducing a 180° phase shift in the output from one half and summing the two outputs. If no tracking error exists, the outputs cancel. If an angular error exists, the resulting phase difference partially offsets the external phase shift, and a difference output proportional to the tracking error is produced (Fig. 30).

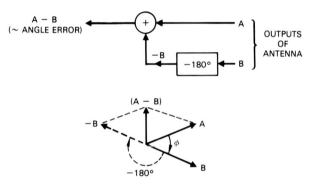

30. Phase difference between outputs of two antenna halves is converted to error signal by introducing 180° of phase shift in one output and adding the two together.

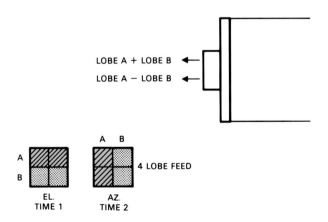

31. Monopulse antenna feed provides sum signal for range tracking; difference signals for angle tracking. Difference signals for azimuth and elevation tracking may be processed on a time-shared basis.

By combining the two outputs without the external phase shift, a sum signal is provided for range tracking.

For monopulse tracking in both azimuth and elevation, the antenna is typically divided into quadrants. The *azimuth difference signal* is obtained by separately summing the outputs of the two left quadrants and the two right quadrants and taking the difference between the two sums. The *elevation difference signal* is similarly produced by taking the difference between the sum of the outputs of the two upper quadrants and the sum of the outputs of the two lower quadrants.

Conventionally, three receiver channels would be provided: one, for the azimuth difference signal; a second, for the elevation difference signal; and a third, for the sum signal. The receiving system can, however, be simplified considerably, by alternately forming the azimuth and elevation difference signals (Fig. 31) and feeding them on a time-share basis through a single receiver channel.

HOW TO CALCULATE THE RADIATION PATTERN FOR A LINEAR ARRAY

If you wish, you can readily calculate the radiation pattern for a linear array consisting of any number of radiators having any spacing and any illumination taper.

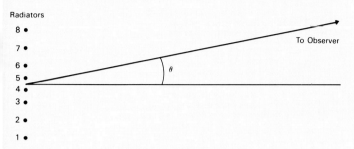

Radiators

For successive values of the angle, θ, you merely sum the contribution of the individual radiators to the total field strength in the direction, θ. If the array is symmetrical about its central axis, the summation only needs to be performed for half the array.

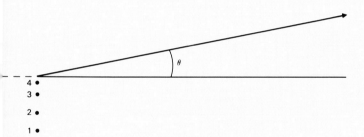

As was illustrated with phasors on page 93 of the text, the contribution of any given radiator, say No. 2, to the total field strength in a given direction is proportional to the amplitude of the signal (a_2) supplied to the radiator times the cosine of the phase of the radiation from this radiator relative to the radiation from the radiator at the center of the array (in this example a hypothetical central radiator).

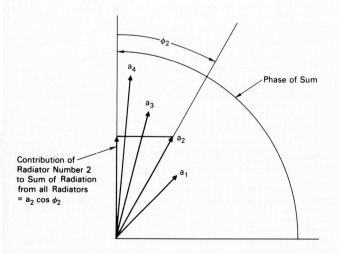

Contribution of Radiator Number 2 to Sum of Radiation from all Radiators = $a_2 \cos \phi_2$

The relative phase, of course, depends upon the difference Δd between the distance from radiator No. 2 to an observer (a long way off) in the direction θ and the distance from the *center* of the array to the same observer. That difference equals the distance of the radiator from the array center (d_2) times $\sin \theta$.

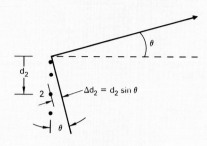

$\Delta d_2 = d_2 \sin \theta$

Dividing Δd_2 by the wavelength (λ) and multiplying by 2π yields the phase in radians.

$$\phi_2 = \frac{2\pi \, \Delta d_2}{\lambda} = \frac{2\pi \, d_2}{\lambda} \sin \theta$$

Thus, the contribution of radiator No. 2 to the total field strength in the direction θ is

$$E_2 \propto a_2 \cos \left(\frac{2\pi \, d_2}{\lambda} \sin \theta \right)$$

The total field strength, then, can be found by performing the following summation.

$$E_{Total} \propto \sum_{i=1}^{N/2} a_i \cos \left(\frac{2\pi \, d_i}{\lambda} \sin \theta \right)$$

By repeating the summation for values of θ from zero to 90°, you can obtain the radiation pattern for the array.

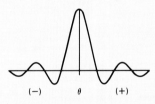

In case you're wondering how the above summation is related to the sin x/x equation given in the text, the relationship is direct. If we assume that the total excitation (A) is uniformly distributed over the length of the array (L), we can obtain the total field strength simply by integrating this same expression with respect to d over the length of the array.

$$E \propto \int_{-L/2}^{L/2} \frac{A}{L} \cos \left(\frac{2\pi d}{\lambda} \sin \theta \right) dd$$

$$\propto A \, \frac{\sin \left(\frac{\pi L}{\lambda} \sin \theta \right)}{\left(\frac{\pi L}{\lambda} \sin \theta \right)}$$

$$\propto A \, \frac{\sin x}{x}, \text{ where } x = \frac{\pi L}{\lambda} \sin \theta \simeq \frac{\pi L}{\lambda} \theta, \text{ for small } \theta.$$

PENCIL BEAM

32. For ground mapping, if the radar is at a low altitude, or the range interval being mapped is narrow, a pencil beam can be used. Otherwise, a fan beam is required.

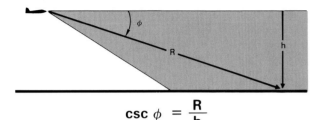

$$\csc \phi = \frac{R}{h}$$

$$\therefore R^2 \propto \csc^2 \phi$$

33. To illuminate ground at all ranges uniformly, power radiated at angle ϕ must be proportional to R^2; hence to the cosecant squared of the lookdown angle.

Useful Relationships To Remember

- For a circular uniformly illuminated X-band antenna of diameter d:

$$\theta_{3dB} = \frac{70°}{d} \quad \text{(d in inches)}$$

$$G = d^2 \eta \quad \text{(d in cm)}$$

(If illumination is tapered, substitute 85° for 70° in expression for beamwidth.)

- For circular, uniformly illuminated antenna and wavelength λ:

$$\theta_{3dB} = \frac{\lambda}{d} \text{ radians}$$

$$G = 9\left(\frac{d}{\lambda}\right)^2 \eta \qquad \text{(d and } \lambda \text{ in same units)}$$

- Angular resolution = θ_{3dB}

Antenna Beams for Ground Mapping

For ground mapping, the entire region being mapped must be illuminated by the antenna's mainlobe (Fig. 32). If the radar is operating at low altitudes, or if the range interval being mapped is relatively narrow, adequate illumination can be provided by a pencil beam. Otherwise, the antenna must radiate a fan-shaped beam.

Ideally, it is shaped so that the strength of the returns received from equivalent ground patches will be independent of their range. For that, the one-way gain of the antenna must be proportional to the square of the range, R, to the ground. This may be achieved by making the gain in the vertical plane proportional to the square of the cosecant of the look-down angle, ϕ (Fig. 33). Hence the beam is called a *cosecant-squared beam.*

It should be noted that multipurpose antennas, which are not exclusively designed for ground mapping, normally do not have a cosecant-squared beam but a pencil beam. In this case, reduction in strength of the return with range is compensated by increasing the receiver sensitivity with range, a process called sensitivity time control (STC) described in Chap. 25.

Summary

A directional antenna radiates a mainlobe surrounded by progressively weaker sidelobes. The width of the mainlobe (beamwidth) is inversely proportional to the width of the antenna aperture in wavelengths.

Antenna gain is the ratio of the power radiated in a specific direction to the power that would be radiated in that direction if the total power were radiated isotropically (uniformly in all directions). The gain on the axis of the mainlobe is proportional to the area of the aperture in square wavelengths.

Sidelobes rob the mainlobe of substantial power and are a source of undesirable ground clutter. Their gain can be reduced by radiating more power per unit area from the central portion of the aperture than from its edges.

Where extreme versatility and speed are required, the mainlobe of an array antenna may be steered by progressively shifting the phases of the waves radiated by successive radiating elements.

Angular resolution is determined by beamwidth. Angular measurement accuracy is much finer than the beamwidth and, in single-target tracking, can be made extremely fine through lobing. By designing the antenna to produce the lobes simultaneously (monopulse), angle tracking degradation due to short term variations in amplitude of the target return can be avoided.

Pulsed Operation

Radars are of two general types: continuous wave—called CW—and pulsed. A CW radar transmits continuously and simultaneously listens for the reflected echoes. A pulsed radar, on the other hand, transmits its radio waves intermittently in short pulses, and listens for the echoes in the periods between transmissions.

Pulsed radars fall into two categories: (1) those that sense doppler frequencies and (2) those that do not. The former have come to be called pulse-doppler radars; the latter, simply pulsed radars. Here, though, pulsed will be used in a general sense to refer to any radar that transmits pulses.

In this chapter, we'll consider the advantages of pulsed transmission, characteristics of the pulsed waveform, and effects of pulsed transmission on transmitted power and energy.

Advantages of Pulsed Transmission

With the exception of doppler navigators, altimeters, and VT proximity fuses, most airborne radars are pulsed. The chief reason is that with pulsed operation, one avoids the problem of the transmitter interfering with reception.

The transmitter's intended output—the signal—is not, of course, the problem. In doppler navigators, for example (Fig. 1) the doppler shift provides sufficient frequency separation to keep the transmitted signal from interfering with reception. And in altimeters (Fig. 2), where the doppler shift is usually near zero, interference from the transmitted signal is avoided by continuously shifting the transmitter's frequency. Because of the time the radio waves take to reach the ground and return, the frequency of the received signal lags behind the frequency of the transmitter; so, the signal is not interfered with.

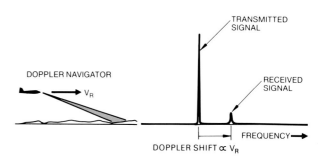

1. In applications where doppler frequency of return is large, doppler shift can keep transmitted signal from interfering with reception.

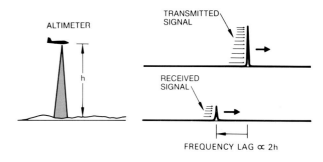

2. Where doppler shift may be negligible, transmitted signal can be kept from interfering with reception by continuously shifting transmitter frequency.

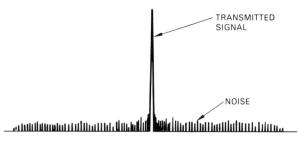

TRANSMITTED SIGNAL

NOISE

3. Noise sidebands blanket broad band of frequencies above and below transmitted signal and are vastly stronger than echoes of typical airborne targets. Interference from them can be avoided by transmitting pulses.

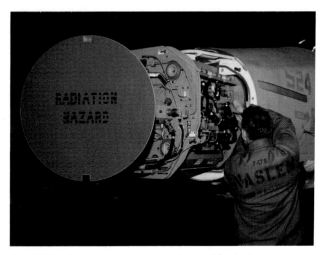

4. When the same antenna must be used for both transmission and reception, pulsed transmission avoids problem of transmitter noise leaking into receiver.

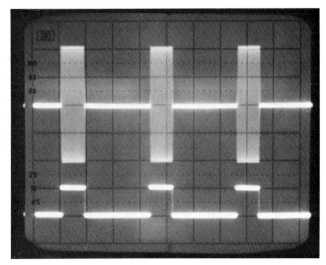

5. Basic characteristics of transmitter waveform.

The problem in most airborne applications is electrical *noise*. Unavoidably generated in every transmitter, it modulates the transmitter output. In so doing, it creates sidebands (see Chap. 5), which blanket a broad band of frequencies above and below the transmitter frequency (Fig. 3). Although the power of the noise sidebands may seem infinitesimal (and is negligible compared to that of echoes from the ground at short range), it can be many orders of magnitude stronger than the echoes from the average airborne target.

To keep the noise from interfering with reception, the receiver must be isolated from the transmitter. Adequate isolation can be obtained by physically separating the transmitter and receiver and providing separate antennas for each (as in ground and shipborne CW radars).

In airborne radars, however, because of space limitations it is usually necessary to use a single antenna for both transmission and reception (Fig. 4). When this is done, it is extremely difficult—hence costly—to prevent some of the noise in the transmitter output from leaking through the antenna into the receiver.

If the transmission is pulsed, neither the transmitted signal nor transmitter noise is a problem; the radar does not transmit and receive at the same time.

Pulsed operation has the further advantage of simplifying range measurement. If the pulses are adequately separated, a target's range can be precisely determined merely by measuring the elapsed time between the transmission of a pulse and the reception of the echo of that pulse.

Pulsed Waveform

Overall, the form of the radio waves radiated by a pulsed radar—the transmitted signal—is referred to as the transmitted *waveform* (Fig. 5). It has four basic characteristics:

- Carrier frequency

- Pulse width

- Modulation (if any) within or between the pulses

- Rate at which the pulses are transmitted (pulse repetition frequency)

Carrier Frequency. This is not always constant, but may be varied in different ways to satisfy specific system or operational requirements. It may be increased or decreased from one pulse to the next. It may be changed at random or in some specified pattern. It may even be increased or decreased in some prescribed pattern during each pulse—intrapulse modulation.

108

Pulse Width. This is the duration of the pulses (Fig. 6). It is commonly represented by the lower case Greek letter, τ. Pulse widths may range anywhere from a fraction of a microsecond to several milliseconds.

Pulse width may also be expressed in terms of physical length. That is, the distance, at any one instant, between the leading and trailing edges of a pulse as it travels through space. The length, L, of a pulse is equal to the pulse width, τ, times the speed of the waves. That speed is very nearly 1000×10^6 feet per second. Consequently, the physical length of a pulse (Fig. 7) is roughly 1000 feet per microsecond of pulse width.

$$\text{Pulse length} = 1000\,\tau \text{ feet}$$

where τ is the pulse width in microseconds.

Pulse length is of keen interest. For—without some sort of modulation within the pulse—it determines the ability of a radar to resolve (separate) closely spaced targets in range. The shorter the pulses (if not modulated for compression), the better the range resolution will be.

For a radar to resolve two targets in range with an unmodulated pulse, their range separation must be such that the trailing edge of the transmitted pulse will have passed the near target before the leading edge of the echo from the far target reaches the near target (Fig. 8). To satisfy this condition the range separation must be greater than half the pulse length.

As a measure of range resolution, therefore, radar designers have adopted a unit of pulse length, called the *radar foot*, which is twice the length of the conventional foot. The length of a 1-microsecond pulse is 500 radar feet, as opposed to 1000 conventional feet.

$$\text{Pulse length} = 500\,\tau \text{ radar feet}$$

As pulse length is decreased, the amount of energy contained in the individual pulses decreases. A point is ultimately reached where no further decrease in energy, hence in pulse width, is acceptable. Seemingly, this limitation puts a limit on the resolution a radar may achieve. That is not so.

Intrapulse Modulation. The limitation which minimum-pulse-length requirements impose on range resolution can be circumvented by coding successive increments of the transmitted pulse with phase or frequency modulation (Fig. 9). Each target echo will, of course, be similarly coded. By decoding the modulation when the echo is received and progressively delaying successive increments, the radar can, in effect, superimpose one increment on top

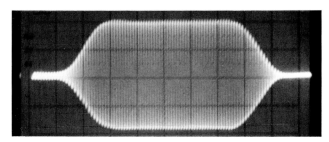

6. RF pulse as seen on oscilloscope. Pulse width is the duration of the pulse.

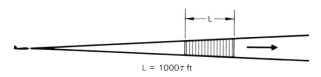

7. Pulse length is distance from leading to trailing edge of pulse as it travels through space.

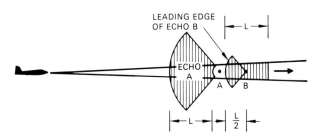

8. To resolve two targets A and B, with an unmodulated pulse of length L, their separation (AB) must be greater than L/2.

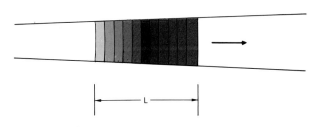

9. If successive segments of transmitted pulse are coded with intrapulse modulation, same resolution may be obtained as with a pulse the width of a single segment.

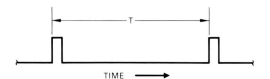

10. Number of pulses transmitted per second is pulse repetition frequency, PRF. Time between pulses is interpulse period, T.

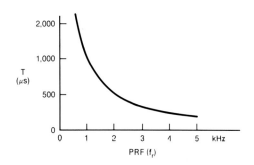

11. Interpulse period, T, decreases rapidly with increase in PRF.

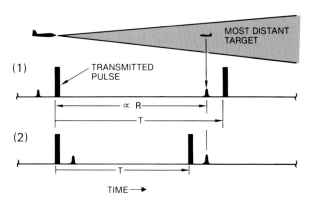

12. (1) If interpulse period, T, is long enough for all echoes from one pulse to be received before next pulse is transmitted, echoes may be presumed to belong to pulse that immediately precedes them. (2) But not if T is shorter than this.

1. Pulsed transmission is not, however, the only source of spectral lines.

of another. The resolution thus achieved is the same as if the radar had transmitted a pulse having nearly the same energy as the original pulse but the width of the individual increments. This technique is called *pulse compression*. We will take it up in detail in Chap. 13.

Pulse Repetition Frequency. This is the rate at which a radar's pulses are transmitted—the number of pulses per second (Fig. 10). It is referred to as the PRF and is commonly represented by f_r. The PRFs of airborne radars range anywhere from a few hundred hertz to several hundred kilohertz. For reasons to be discussed in subsequent chapters, during the course of a radar's operation, its PRF may be changed from time to time.

Another measure of pulse rate is the period between the start of one pulse and the start of the next pulse. This is called the *interpulse period* or the *pulse repetition interval*, PRI. It is generally represented by the upper case letter T.

The interpulse period (Fig. 11) is equal to one second divided by the number of pulses transmitted per second, f_r.

$$T = \frac{1}{f_r}$$

If the PRF is 100 hertz, for example, the interpulse period will be 1 / 100 = 0.01 second, or 10,000 microseconds.

The choice of PRF is crucial because it determines whether, and to what extent, the ranges and doppler frequencies observed by the radar will be ambiguous.

Range ambiguities arise as follows. A radar has no direct way of telling to which pulse a particular echo belongs. If the interpulse period is long enough for all of the echoes of one pulse to be received before the next pulse is transmitted, this doesn't matter: any echo can be assumed to belong to the immediately preceding pulse (Fig. 12). But, if the interpulse period is shorter than this, depending upon how much shorter it is, an echo may belong to any one of a number of preceding pulses. Thus, the ranges observed by the radar may be ambiguous. The higher the PRF, the shorter the interpulse period; hence, the more severe the ambiguities will be.

Doppler ambiguities arise because of the discontinuous nature of a pulsed signal. As will be explained in Chap. 17, a pulsed signal will pass through a filter (such as those which provide doppler frequency resolution in a radar signal processor) not only when the filter is tuned to the frequency corresponding to the signal's wavelength, but also when it is tuned above or below that frequency by multiples of the PRF. These frequencies are called *spectral lines* .[1]

Unfortunately, there is no direct way of telling which line a filter is passing. If the PRF is high enough, this doesn't matter. Any line of the received signal can be assumed to be the nearest spectral line of the transmitted signal, shifted by the target's doppler frequency. But, if the PRF is less than the doppler frequencies that may be encountered, depending on how much less it is, a line passed by a given filter may be any one of a number of lines of the transmitted signal, shifted by the target's doppler frequency (Fig. 13).

As with the observed ranges, therefore, depending on the PRF, the observed doppler frequencies may be ambiguous. In the case of doppler frequency, though, the relationship is reversed: the higher the PRF, the more widely spaced the spectral lines; hence, the less severe the ambiguities will be. (Range and doppler ambiguities and techniques for resolving them are discussed in Chaps. 12 and 21, respectively.)

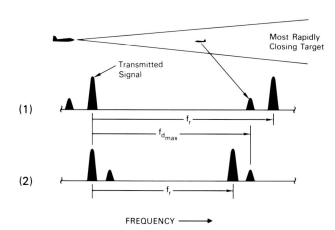

13. Spectral "lines" of a pulsed signal are spaced at intervals equal to PRF (f_r). (1) If f_r is greater than highest doppler frequency, $f_{d_{max}}$, any line to which a doppler filter is tuned may be presumed to be the next lower line of the transmitted signal shifted by the target's doppler frequency. (2) But not if f_r is less than $f_{d_{max}}$.

Output Power and Transmitted Energy

Before discussing the effect of pulsed transmission on output power and transmitted energy, it will be well to review the relationship between power and energy. As explained at some length in the panel on the next page, power is the rate of flow of energy (Fig. 14).

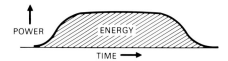

14. Power is rate of flow of energy. Backscattered energy is what a radar detects.

The amount of energy transmitted by a radar equals the output power times the length of time the radar is transmitting. Transmitted energy—(power) x (time)—does the work.

Two different measures are commonly used to describe the power of a pulsed radar's output: peak power and average power.

Peak Power. This is the power of the individual pulses. If the pulses are rectangular—that is, if the power level is constant from the beginning to the end of each pulse—peak power is simply the output power when the transmitter is on, or transmitting (Fig. 15). In this book, peak power is represented by the upper case P.

Peak power is important for several reasons. To begin with, it determines the voltages that must be applied to the transmitter.

Peak power also determines the intensities of the electromagnetic fields one must contend with: fields across insulators, fields in the waveguides that connect the transmitter to

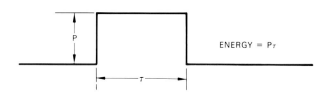

15. Peak power determines both voltage levels and energy per unit of pulse width.

THE VITAL DISTINCTION BETWEEN ENERGY AND POWER

Many of us use the terms power and energy loosely, often interchangeably. But if we are to understand the operation of a radar we must make a clear distinction between the two.

Energy is the capacity for doing work. It has many forms: mechanical, electrical, thermal, and so on. Work is accomplished by converting energy from one form to another.

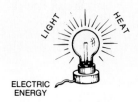

Take an electric lamp. It converts energy from electrical form to electromagnetic form. The result is light. Being inefficient, the lamp also converts a considerable amount of electrical energy to thermal energy. The result is heat, some of which is radiated in electromagnetic form.

Power is the rate at which work is done—the amount of energy converted from one form to another per second.

It is also the rate at which energy is transmitted—e.g., the amount of energy per second that a radar beams toward a target. The common units of power are the watt and the kilowatt (1000 watts).

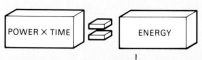

How much energy is converted or transmitted depends on how long the power is on. The common units of energy are the watt-second (joule) and the watt-hour (3600 watt-seconds).

A 25 watt lamp left on for 4 hours will convert 100 watt-hours of energy to light and heat—the same as a 100 watt lamp left on for only 1 hour.

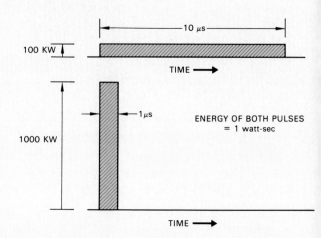

Similarly, a 100 kilowatt radar pulse having a duration of 10 will convey as much energy as a 1000 kilowatt pulse having duration of only 1 μs.

Equipment Rating. Although it is energy that is transmitted energy that does the work, most electrical apparatus is rated terms of power. Motors are rated in horsepower (746 watts = 1 hp). Radio transmitters are rated in watts or kilowatts.

The reason is that the power rating determines the energy handling capacity of the equipment and is a dominant factor its design.

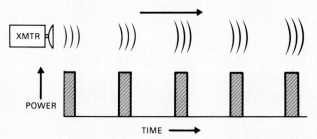

But we must not lose sight of the fact that the amount of ra frequency energy that a radar transmits toward a target equal the power of the transmitted waves times the duration of eac pulse, times the number of pulses.

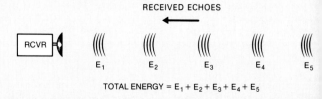

We must also remember that the extent to which the energy the received echoes can be used to detect the target depends the radar's ability to add up the energy contained in successiv echoes.

the antenna. If these fields are too intense, problems of corona and arcing will be encountered. Corona is a discharge which occurs when an electric field becomes strong enough to ionize the air; it is what makes high voltage power lines buzz (Fig. 16). Arcing occurs when ionization is sufficient for a conductive path to develop through the air. Both effects can result in a major loss of power, as well as in equipment damage. Consequently, there is an upper limit on the acceptable level of peak power.

Together, peak power and pulse width determine the amount of energy conveyed by the transmitted pulses. If the pulses are rectangular, the energy in each pulse equals the peak power times the pulse width.

$$\text{Energy per pulse} = P\tau$$

Usually, however, the energy in a train of pulses is what is important. This is related to average power.

Average Power. A radar's average transmitted power is the power of the transmitted pulses averaged over the interpulse period (Fig. 17). In this book, average power is represented by P_{avg}.

If a radar's pulses are rectangular, the average power equals the peak power times the ratio of the pulse width, τ, to the interpulse period, T.

$$P_{avg} = P \frac{\tau}{T}$$

For example, a radar having a peak power of 100 kilowatts, a pulse width of 1 microsecond, and an interpulse period of 2000 microseconds will have an average power of 100 x 1/2000 = 0.05 kilowatts, or 50 watts.

The ratio, t/T, is called the *duty factor* of the transmitter (Fig. 18). It represents the fraction of the time the radar is transmitting. If, for example, a radar's pulses are 0.5 microseconds wide and the interpulse period is 100 microseconds, the duty factor is 0.5 ÷ 100 = 0.005. The radar is transmitting five thousandths of the time it is in operation and is said to have a duty factor of 0.5 percent.

Average output power is important primarily because it is a key factor in determining the radar's potential detection range. The total amount of energy transmitted in a given period equals the average power times the length of the period, T.

$$\text{Transmitted energy} = P_{avg}T$$

16. Corona is what makes high voltage lines buzz. In a radar it can result in a major loss of power and equipment damage. (Courtesy Electro Power Research Inst.)

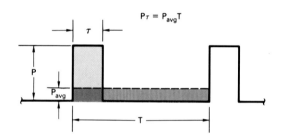

17. Average power is peak power times pulse width averaged over interpulse period.

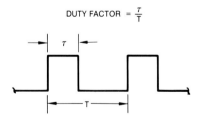

18. Duty factor is the fraction of time the radar is transmitting.

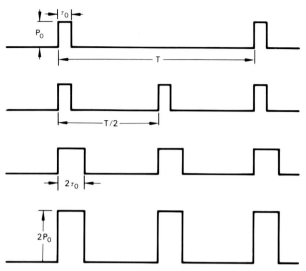

19. Three ways of increasing average power.

Some Relationships To Keep In Mind

- Pulse length $\approx 1000 \, \tau$ feet

- Range resolution $\approx 500 \, \tau$ feet

- Interpulse period, $T = \dfrac{1}{f_r}$

- Duty factor $= \dfrac{\tau}{T}$

- Average power, $P_{avg} = P \dfrac{\tau}{T}$

  (τ = pulse width in μs)
  (P = peak power)
  (T = interpulse period in μs)

In the interest of maximizing detection range, average power may be increased in any of three ways: by increasing the PRF, by increasing the pulse width, and by increasing the peak power (Fig. 19).

Average power is also of concern for other reasons. Together with the transmitter's efficiency, average power determines the amount of heat due to losses which the transmitter must dissipate. In turn, this determines the amount of cooling required. The average output power plus the losses determine the amount of input (prime) power that must be conditioned and supplied to the transmitter. Finally, the higher the average power, the larger and heavier the transmitter tends to be.

Summary

Because of the difficulty of preventing noise sidebands on the transmitted signal from leaking into the receiver, continuous wave (CW) transmission is generally practical against small targets only if separate antennas are used for both transmission and reception. Pulsed transmission avoids this problem—and provides a simple means of measuring range.

Basic characteristics of the transmitted waveform include radio frequency, pulse width, intrapulse or interpulse modulation, and pulse repetition frequency.

Radio frequency may be varied not only from pulse to pulse but within the pulses (intrapulse modulation).

Pulse width determines range resolution. By coding successive increments of each pulse with phase or frequency modulation and decoding the echoes, wide pulses can be transmitted to provide higher power output and the received pulses can be compressed (pulse compression) to provide fine resolution.

The PRF determines the extent of range and doppler ambiguities. The lower the PRF, the less severe the range ambiguities. The higher the PRF, the less severe the doppler ambiguities.

Peak power is the power of the individual pulses. The maximum usable peak power is generally limited by problems of arcing and corona.

Average power is peak power averaged over the interpulse period. The higher the peak power, pulse width, and PRF, the higher the average power will be.

Energy, not power, is what does the work. The energy in a pulse train equals the average power times the length of the train.

Detection Range 10

Generally, few things are of more fundamental concern to both designer and user alike than the maximum range at which a radar can detect targets. In this chapter, we will learn what determines that range.

We will begin by tracking down the sources of the electrical background noise against which a target's echoes must ultimately be discerned and finding what can be done to minimize the noise. We will then trace the factors upon which the strength of the echoes depends and examine the detection process. Finally, we'll see how, by integrating the return from a great many transmitted pulses, a radar can pull the weak echoes of distant targets out of the noise.

What Determines Detection Range

Airborne radars can be designed to detect targets at ranges of thousands of miles. As a rule though, they are designed to operate at much shorter ranges for at least one compelling reason: obstructions in the line of sight.

Radio waves of the frequencies used by airborne radars behave very much as visible light, except of course that they can penetrate clouds and are not scattered much by aerosols (tiny particles suspended in the atmosphere). They cannot penetrate liquids or solids very far. And although they bend slightly as a result of the increase in the speed of light with altitude and spread to some extent around obstructions, these effects are slight.

Consequently, no matter how powerful a radar is or how ingenious its design, its range is essentially limited to the maximum unobstructed line of sight. A radar cannot see through mountains, and it cannot see much at low altitudes or on the ground beyond the horizon.

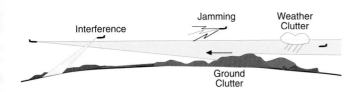

1. Just because a target is within the line of sight does not mean it will be detected. It may be obscured by competing clutter or man-made interference.

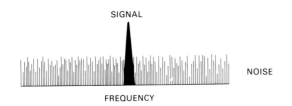

2. In the absence of clutter and interference, whether a target will be detected ultimately depends upon the strength of its echoes relative to the strength of the background noise.

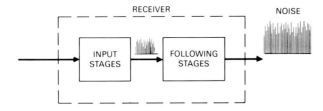

3. Amplified by full gain of receiver, noise generated in input stages swamps out that originating in following stages.

Just because a target is within the line of sight, however, does not mean that it will be detected (Fig. 1). Depending upon the operational situation, its echoes may be obscured by clutter returned from the ground, or (depending upon the wavelength and the weather) from rain, hail, or snow. At times, too, a target's echoes may be obscured by the transmissions of other radars, by jamming, or by other *electromagnetic interference* (EMI).

Clutter can largely be eliminated through doppler processing (MTI). And there are ways of dealing with most man-made interference, as well.

But, depending upon the strength of the transmitted waves, if the target is small or at long range, its echoes may still be obscured by the ever present background of electrical noise.

In a benign environment, then, whether a given target will be detected ultimately depends upon the strength of its echoes relative to the strength of the electrical background noise (Fig. 2)—the signal-to-noise ratio.

Electrical Background Noise

As the name implies, electrical noise is electrical energy of random amplitude and random frequency. It is present in the output of every radio receiver, and a radar receiver is no exception. At the frequencies used by most radars, the noise is generated primarily within the receiver itself.

Receiver Noise. Most of this noise originates in the input stages of the receiver. The reason is not that these stages are inherently more noisy than others but that, amplified by the receiver's full gain, noise generated there swamps out the noise generated farther along (Fig. 3).

Because the noise and the received signals are thus amplified equally (or nearly so), in computing signal-to-noise ratios, the factor of receiver gain can be eliminated by determining the signal strength at the input to the receiver and dividing the noise output of the receiver by the receiver's gain. Therefore, receiver noise is commonly defined as noise per unit of receiver gain.

$$\text{Receiver noise} = \frac{\text{Noise at output of receiver}}{\text{Receiver gain}}$$

This ratio can readily be measured in the laboratory by methods such as are outlined in the panel on the facing page.

Since the early days of radio it has been customary to describe the noise performance of a receiver in terms of a figure of merit called the *noise figure*, F_n. It is the ratio of the noise output of the actual receiver to the noise output of a

OW RECEIVER NOISE FIGURE IS MEASURED

Although you may never have to measure the noise figure of a
dar receiver, you may gain a better feel for its significance if
u have a general idea of how it is measured.

Measurement. Basically, this is a three-step process. First,
u connect a resistor across the input terminals of the receiver
d measure the output power, P_N.

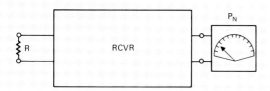

cond, you connect a noise generator to the input and increase
e noise power until the receiver output doubles.

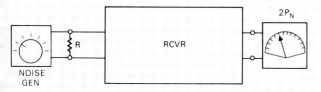

rd, you measure the power output of the noise generator, N_G.

What the Outputs Represent. We can see this best if we
present the receiver with an equivalent circuit consisting of an
der followed by an amplifier of gain, G. When only the resistor
connected to the input, the receiver output equals the gain (G)
nes the sum of the noise generated in the resistor (kT_0B) plus
e noise generated in the input stages of the receiver (N_R).

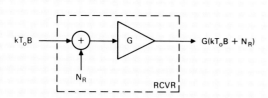

hen the noise generator is connected to the input, the noise
wer (N_G) is added to the sum.

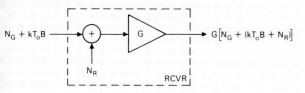

ce adding N_G to the input doubles the output, while
$_oB + N_R$) remains unchanged, it is clear that

$$N_G = kT_0B + N_R$$

This sum, divided by the noise generated in the resistor, is the
receiver noise figure, F_n. Therefore,

$$F_n = \frac{N_G}{kT_0B}$$

Value of the Resistor. As long as the resistor matches the
input resistance of the receiver, the value of R doesn't influence
the noise figure. We can see this by representing the resistor
with an equivalent circuit consisting of a voltage generator in
series with a resistance, R. The generator voltage (V) equals the
voltage of the noise thermally generated in the resistor.

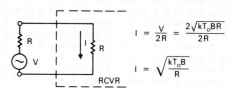

If the input resistance of the receiver equals R, then the current
(I) through the input resistance will be:

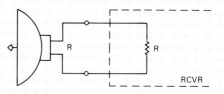

The power dissipated in the input resistance equals the square of
the current flowing through it times the resistance, so R cancels
out.

$$P = I^2R = \left(\sqrt{\frac{kT_0B}{R}} \right)^2 R = kT_0B$$

To extract the most power from the antenna the input resistance
of the receiver generally is made equal to the radiation resistance
of the antenna (R_r). (This resistance is the ratio of the voltage
applied across the antenna terminals when the antenna is
radiating to the component of the current flowing through the
terminals that is in phase with the voltage.)

Therefore, unless otherwise specified the value of the resistor
used in measuring the noise figure is assumed to equal the
radiation resistance of the antenna, hence can be ignored in
computing the noise figure.

hypothetical, "ideal" minimum-noise receiver providing equal gain.

$$F_n = \frac{\text{Noise output of actual receiver}}{\text{Noise output of ideal receiver}}$$

(Note that since the gains of both receivers are the same, F_n is independent of receiver gain.)

An ideal receiver, of course, would generate no noise whatsoever internally. The only noise in its output would be noise received from external sources. By and large, that noise has the same spectral characteristics as the noise resulting from thermal agitation in a conductor. Therefore, as a standard for determining F_n, the sources of external noise for both actual and ideal receivers can reasonably be represented by a resistor connected across the receiver's input terminals (Fig. 4). (A resistor is a conductor providing a specified resistance to the flow of current.)

Now, thermal agitation noise is produced by the continuous random motion of free electrons, which are present in every conductor. The amount of motion is proportional to the conductor's temperature above absolute zero. Quite by chance, at any one instant, more electrons will generally be moving in one direction than in another. This imbalance causes a random voltage proportional to the temperature to appear across the conductor (Fig. 5).

Thermal noise is spread more or less uniformly over the entire spectrum. So, the amount of noise appearing in the output of the ideal receiver is proportional to the absolute temperature of the resistor that is connected across its input terminals times the width of the band of frequencies passed by the receiver—the receiver bandwidth (Fig. 6).

The mean power—per unit of receiver gain—of the noise in the output of the hypothetical ideal receiver is thus,

$$\text{Mean noise power} = kT_0B \text{ watts}$$

(Ideal receiver)

where

k = Boltzmann's constant, 1.38×10^{-23} watt-second/°K

T_0 = absolute temperature of the resistor representing the external noise, °K

B = receiver bandwidth, hertz

Since the external noise is the same for both actual and ideal receivers, as long as everyone uses the same value for T_0 in determining the noise figure, the exact value is not critical. By convention, T_0 is taken to be 290°K, which is close to room temperature and conveniently makes kT_0 a round number (4×10^{-21} watt-second).

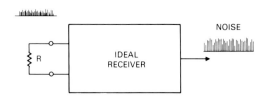

4. The only noise in the output of an ideal receiver would be that received from external sources. This is represented by thermal agitation in a resistor.

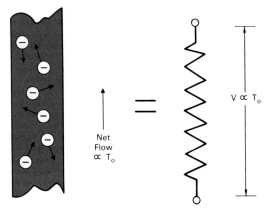

5. Because of thermal agitation, a random voltage proportional to absolute temperature appears across the electrical resistance of every conductor.

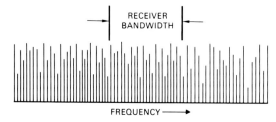

6. Noise in receiver output is proportional to bandwidth of receiver.

When the internally generated noise is considerably greater than the external noise (as it is in the vast majority of airborne radars in operation today), the noise figure, F_n, multiplied by the foregoing expression for mean noise power per unit of gain for an ideal receiver is commonly used to represent the level of background noise against which target echoes must be detected.

$$\text{Mean noise power} = F_n k T_0 B \text{ watts}$$

(Actual receiver)

This expression, it should be remembered, includes both a nominal estimate of the external noise (equivalent of the noise generated in a resistor at room temperature) and the accurately measured internally generated noise.

Although in many receivers, the internal noise predominates, it can be substantially reduced by adding a low-noise preamplifier ahead of the receiver's mixer stage and using a low-noise mixer (Fig. 7). The preamplifier increases the signal strength relative to the thermal noise originating in the subsequent stages, while contributing only a minimum amount of noise itself. When a low-noise front end is used, a more accurate estimate may have to be made of the noise received from sources ahead of the receiver.

Noise from Sources Ahead of the Receiver. As explained in Chap. 4, because of thermal agitation virtually everything around us radiates radio waves. The radiation is extremely weak. Nonetheless, it may be detected by a sensitive receiver and add to the noise in the receiver output. At the frequencies used by most airborne radars, the principal sources of this natural radiation are the ground, the atmosphere, and the sun (Fig. 8).

Radiation from the ground depends not only upon the temperature of the ground but on its "lossiness," or absorption. (The power of the radiated noise is proportional to the absolute temperature times the coefficient of absorption.) Thus, although a body of water may have the same temperature as a land mass, since water is a good conductor and the land usually is not, the water will radiate comparatively little noise. How much of the radiation that is received by the radar varies widely with the gain of the antenna and the direction in which it is looking. For example, far more noise is received when looking down at the warm earth than when looking off at a body of water which reflects the extreme cold of outer space.

The amount of noise received from the atmosphere depends not only upon the temperature and lossiness of the atmosphere but upon the amount of atmosphere the

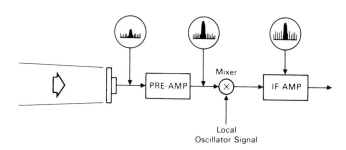

7. Receiver noise may be reduced substantially by providing a low-noise preamplifier ahead of the mixer and/or using a low-noise mixer.

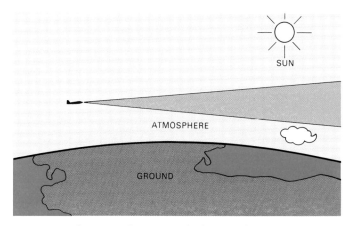

8. Principal sources of noise outside the aircraft. Amount received varies with antenna gain and direction.

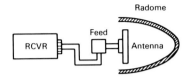

9.   Other sources of noise within the aircraft.

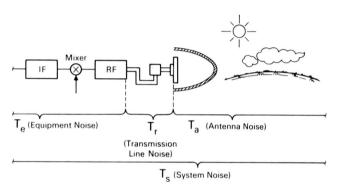

10.   When external noise is significant, noise from each source is assigned an equivalent noise temperature.[1]

1.   Since T_e does not include the noise of the input resistance, as $F_n T_o$ does, $T_e = T_o (F_n - 1)$.

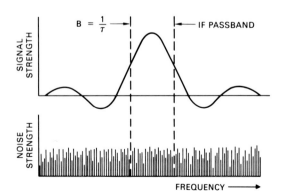

11.   Signal-to-noise ratio may be maximized by narrowing the passband of the IF amplifier to the point where only the bulk of the signal energy is passed.

antenna is looking through. Since the lossiness varies with frequency, the received noise also depends upon the radar's operating frequency.

Noise received from the sun varies widely with both solar conditions and the radar's operating frequency. Naturally, it is vastly greater if the sun happens to be in the antenna's mainlobe, as opposed to its sidelobes.

Within the aircraft carrying the radar, noise is radiated by the radome, the antenna, and the complex of waveguides connecting the antenna to the receiver (Fig. 9). Noise from these sources is likewise proportional to their absolute temperature times their loss coefficients.

As previously noted, noise from all of these external sources which falls within the receiver passband has essentially the same spectral characteristics as receiver noise. Consequently, when external noise is significant, the noise from each source, as well as the receiver noise, is usually assigned an equivalent noise temperature (Fig. 10). These temperatures are combined to produce an equivalent noise temperature for the entire system, T_s. The expression for noise power then becomes

$$\text{Mean noise power} = kT_sB$$

(All sources)

Competing Noise Energy. Whether noise is expressed in terms of receiver noise figure, F_nT_0, or equivalent noise temperature, T_s, it is noise *energy*, not power, with which a target's echoes must compete. As explained in the last chapter, power is but the rate of flow of energy. Noise energy is noise power times the length of time over which the noise energy flows—in this case, the duration of the period in which return may be received from any one resolvable increment of range. Therefore,

$$\text{Mean noise energy} = kT_sBt_n$$

where t_n is the duration of the noise.

For any given noise temperature and duration, the noise can be reduced by minimizing the receiver bandwidth, B. A common practice is to narrow the IF passband until it is just wide enough to pass most of the energy contained in the received echoes. This is called a *matched filter design* (Fig. 11).

Another way of looking at this is that the tuned circuits of the receiver IF amplifier integrate the received energy during the width, τ, of each received pulse. They thus accumulate the energy the pulse contains and reject the noise outside the pulse's bandwidth. The optimum bandwidth turns out to be very nearly equal to one divided by the pulse width, τ. When $1/\tau$ is substituted for B in the expression for mean noise, it becomes

$$\text{Mean noise energy} = \frac{kT_s t_n}{\tau}$$

(Matched filter design)

In doppler radars, bandwidth is further reduced by doppler filters, which follow the IF amplifier. (A separate filter is generally provided for every anticipated combination of resolvable range and doppler frequency.) As will be explained in Chap. 18, the passband of a doppler filter is approximately equal to $1/t_{int}$, where t_{int} is the time over which the filter adds up (integrates) the radar returns (Fig. 12).

Whereas τ is on the order of microseconds, t_{int} is on the order of milliseconds. Consequently, the passband of a doppler filter is on the order of 1/1000th of the width of the IF passband.

The integration time, t_{int}, is also the length of time over which the noise is received and integrated by the filter. When t_{int} is substituted for t_n and $1/t_{int}$ is substituted for $1/\tau$, the two terms cancel, leaving

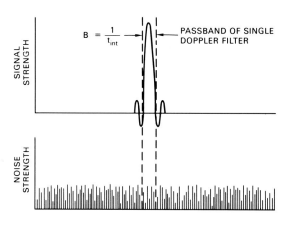

12. In a doppler radar the passband is further narrowed by doppler filtering.

$$\text{Mean noise energy} = kT_s$$

(Doppler radar)

One way of looking at the effect of a doppler filter on the noise is this. As the noise energy flows into the filter, the filter's passband (which is inversely proportional to integration time) simultaneously narrows. As a result, the level of the noise energy that accumulates in the filter is more or less independent of the length of the integration period.

Noise being random, of course, the level of the accumulated energy may vary widely from one integration period to another. But its mean value over a great many integration periods will be kT_s.

On average, therefore, for a target to be detected, enough energy must be received from it to noticeably raise the filter output above this mean level.

And that brings us to the question: what determines how much energy is received from a target; what is the energy of the signal?

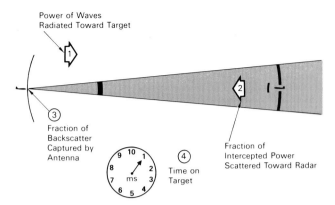

13. Factors which determine energy of target signal.

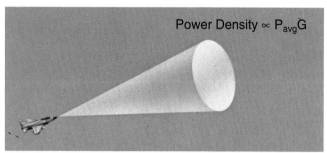

Power Density $\propto P_{avg}G$

14. Density of power radiated in target's direction is proportional to the average radiated power times the antenna gain in that direction.

2. Another term for power density is "power flux."

$A \propto R^2$

15. As waves travel out to a target, their power is spread over an increasingly large area.

Energy of the Target Signal

Four basic factors (Fig. 13) determine the amount of energy a radar will receive from a target during any one period of time that the antenna beam is trained on it:

- Average power—rate of flow of energy—of the radio waves radiated in the target's direction

- Fraction of the wave's power which is intercepted by the target and scattered back in the radar's direction

- Fraction of that power which is captured by the radar antenna

- Length of time the antenna beam is trained on the target

When the antenna is trained on a target, the power density of the radio waves radiated in the target's direction is proportional to the transmitter's average power output, P_{avg}, times the gain, G, of the antenna's mainlobe (Fig. 14). (Power density, you will recall, is the rate of flow of energy per unit of area normal to the wave's direction of propagation.)[2]

In transit to the target, the power density is diminished as a result of two things: absorption in the atmosphere, and spreading. Except at the shorter wavelength, attenuation due to absorption is comparatively small. For the moment it will be neglected, but not the reduction in power density due to spreading.

As the waves propagate toward the target, their energy spreads—like the substance of an expanding soap bubble—over an increasingly large area (Fig. 15). This area is proportional to the square of the distance from the radar. At the target's range, say R miles, the power density is only $1/R^2$ times what it was at a range of 1 mile.

The amount of power intercepted by the target equals the power density at the target's range times the geometric cross-sectional area of the target, as viewed from the radar (the projected area).

What fraction of the intercepted power is scattered back toward the radar depends upon the target's reflectivity and directivity. The reflectivity is simply the ratio of total scattered power to total intercepted power. The directivity—like the gain of an antenna—is the ratio of the power scattered in the direction of the radar to the power which would have been scattered in that direction had the scattering been uniform in all directions.

Customarily, a target's geometric cross-sectional area, reflectivity, and directivity are lumped into a single factor, called *radar cross section*. It is represented by the Greek letter sigma, σ, and is usually expressed in square meters. (See panel on right.)

RADAR CROSS SECTION

A target's radar cross section, σ, is most easily visualized as the product of three factors:

$$\sigma = \boxed{\text{Geometric Cross Section}} \times \boxed{\text{Reflectivity}} \times \boxed{\text{Directivity}}$$

Geometric Cross Section is the cross-sectional area of the target as viewed from the radar.

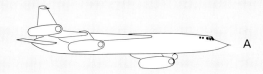

This area determines how much power the target will intercept.

$$P_{intercepted} = (A)(P_{incident})$$

where $P_{incident}$ is the power density of the incident waves.

Reflectivity is the term for the fraction of the intercepted power that is reradiated (scattered) by the target.

$$\text{Reflectivity} = \frac{P_{scatter}}{P_{intercepted}} = \frac{P_{scatter}}{(A)(P_{incident})}$$

(The scattered power equals the intercepted power less whatever power is absorbed by the target.)

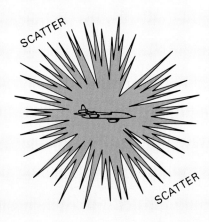

Directivity is the ratio of the power scattered back in the radar's direction to the power that would have been backscattered had the scattering been uniform in all directions, i.e., isotropically.

$$\text{Directivity} = \frac{P_{backscatter}}{P_{isotropic}}$$

Normally, $P_{backscatter}$ and $P_{isotropic}$ are expressed as power per unit of solid angle. $P_{isotropic}$ then equals $P_{scatter}$ divided by the number of units of solid angle in a sphere.

The unit of solid angle is the steradian. It is the angle subtended by an area on the surface of a sphere equal to the radius squared.

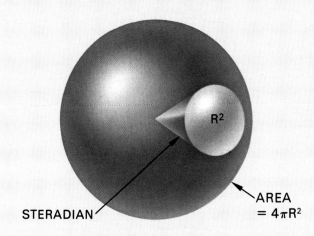

STERADIAN

AREA $= 4\pi R^2$

Since the area of a sphere is 4π times the radius squared, a sphere contains 4π steradians. Therefore:

$$\text{Directivity} = \frac{P_{backscatter}}{(1/4\pi)P_{scatter}}$$

A target may be thought of as consisting of a great many individual reflecting elements (scatterers).

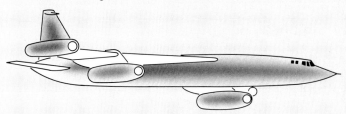

The extent to which the scatter from these combines constructively in the direction of the radar depends upon the relative phases of the backscatter from the individual elements. That in turn depends on the relative distances (in wavelengths) of the elements from the radar. Depending on the configuration and orientation of the target, the directivity may range anywhere from a small fraction to a large number.

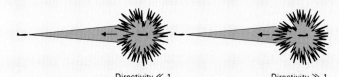

Directivity $\ll 1$ Directivity $\gg 1$

Complete Expression for σ. Expanded in terms of the factors outlined in the preceding paragraphs, the basic expression for radar cross section becomes:

$$\sigma = A \times \frac{P_{scatter}}{(A)(P_{incident})} \times \frac{P_{backscatter}}{(1/4\pi)(P_{scatter})}$$

Cancelling like terms and spelling out those that remain yields:

$$\sigma = 4\pi \frac{\text{Backscatter per steradian}}{\text{Power density of incident waves}}$$

This is the common form of the definition of radar cross section. It has the advantage of making radar equations easier to write. But expressing σ in terms of geometric cross section, reflectivity, and directivity is more illuminating since that shows the relationship between σ and the factors that determine its value.

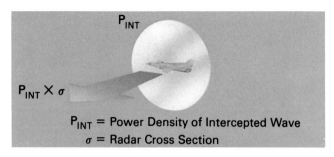

16. Density of power reflected in radar's direction equals density of intercepted wave times radar cross section.

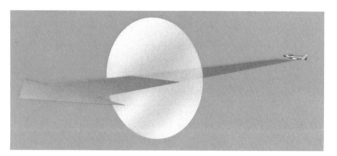

17. Reflected power undergoes equal amount of spreading in returning to radar.

The power density of the waves scattered back in the radar's direction, then, can be found by multiplying the power density of the transmitted waves when they reach the target by the target's radar cross section (Fig. 16). Since the directivity of a target can be quite high, for some target aspects the radar cross section may be many times the geometric cross-sectional area. For others, the reverse may be true.

As the waves propagate back from the target, they undergo the same geometrical spreading as on their way out. Their power density, which has already been reduced by a factor of $1/R^2$ is again reduced by $1/R^2$ (Fig. 17). The two factors are compounded, so the power density when the waves reach the radar is only $1/R^2 \times 1/R^2 = 1/R^4$ times what it would be if the target were at a range of only 1 mile (or whatever other unit of distance R is measure in).

To give you a feel for the magnitude of this difference, the relative strengths of the echoes from the same target in the same aspect at ranges of 1 to 50 miles are plotted below (Fig. 18). For a range of 1 mile, the strength is arbitrarily assumed to equal 1. At 50 miles, the relative strength is only 0.00000016—too small to be discernible in the figure.

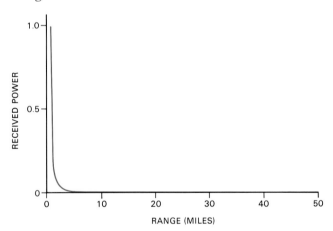

18. Reduction in strength of target echoes with range. Echoes from target at 50 miles are only 0.00000016 times as strong as echoes from same target at 1 mile range.

Incidentally, Fig. 18 dramatically illustrates why the receiver must be able to handle powers of vastly different magnitudes—i.e., have a wide dynamic range.

When the backscattered waves reach the antenna, it intercepts a fraction of their power. That fraction equals the power density of the waves times the effective area of the antenna, A_e (Fig. 19). The total amount of energy intercepted equals that product times the length of time the antenna is trained on the target, t_{ot}.

As we saw in Chap. 8, the area A_e takes into account the aperture efficiency of the antenna. For all of the intercepted

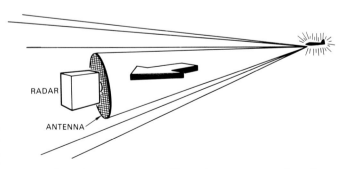

19. Reflected power intercepted by radar is proportional to effective area of antenna.

energy to be constructively added up by the antenna feed, of course, the target must be centered in the antenna's mainlobe.

If we multiply together the several factors which we have identified as governing the amount of energy received from a target, we get the following expression for received target signal energy.

$$\text{Signal energy} \cong K\,\frac{P_{avg}G\sigma A_e t_{ot}}{R^4}$$

where

K = factor of proportionality $(1/4\pi)^2$

P_{avg} = average transmitted power

G = antenna gain

σ = radar cross section of the target

A_e = effective area of the antenna

t_{ot} = time-on-target

R = range

This expression *roughly* indicates the total amount of energy that would be received by a radar during the antenna beam's time-on-target, t_{ot}. Whether all of the energy is actually utilized depends upon the radar's ability to integrate it.

In simple non-doppler radars, integration is performed by the display (e.g., by the phosphor that causes the image to persist on the face of the CRT) and by the eyes and the mind of the operator (Fig. 20). Because it takes place after detection, this integration is called *postdetection integration.*

In a doppler radar, integration is performed primarily by the signal processor's doppler filters before detection takes place. Provided the integration time, t_{int}, is made equal to t_{ot},[3] the above expression indicates the amplitude of the integrated target signal in the output of a filter at the end of each time-on-target. Whether the target will be detected, of course, depends upon the ratio of this amplitude to that of the integrated noise, discussed earlier.

To fully understand the relationship between signal-to-noise ratio and maximum detection range, though, we must know a little more about the actual detection process.

Detection Process

A small target, we will assume, is approaching a searching doppler radar from a very great distance. Initially, the target echoes are extremely weak—so weak they are completely lost in the background noise.

On first thought, one might suppose the echoes could be pulled out of the noise by increasing the gain of the receiver. But the receiver amplifies noise and echoes equally. Increasing its gain in no way alters the situation.

THE SCALE FACTOR K

If you're puzzled, the value of $4\pi^2$ for K in the equation for received signal energy, may be explained as follows:

$$\text{Density of power reaching target} = P_{avg}\,\frac{G}{\text{Area of a sphere of radius R}}$$

$$\text{Density of power returned to radar} = P_{reflected}\,\frac{\text{Directivity factor included in }\sigma}{\text{Area of a sphere of radius R}}$$

$$\text{Area of a sphere of radius R} = 4\pi R^2$$

Therefore: Both G and σ must be divided by $4\pi R^2$

$$\frac{G}{4\pi R^2} \times \frac{\sigma}{4\pi R^2} = \frac{G\sigma}{(4\pi)^2\,R^4}$$

Hence: $K = \dfrac{1}{(4\pi)^2}$

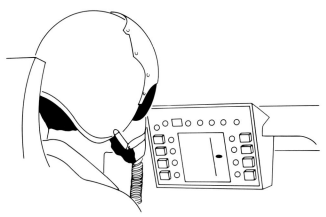

20. In non-doppler radars, integration takes place on the display and in the eyes and mind of the operator.

3. And provided the target is centered in the passband of one of the filters and the time on target coincides exactly with an integration period.

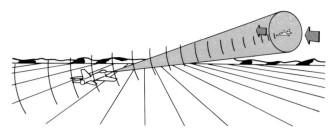

21. Each time the antenna beam sweeps across the target, a stream of echoes is received.

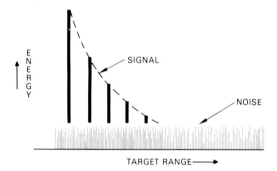

22. Received signal energy, for successive times-on-target. As range decreases, ratio of signal energy to noise energy increases.

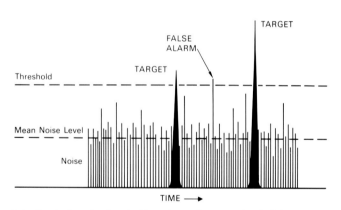

24. The higher the threshold is above the mean level of the noise, the lower the probability of a spike of noise crossing it and producing a false alarm.

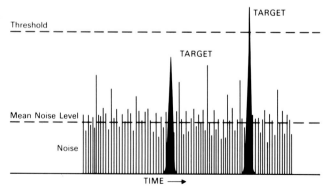

25. Yet, if the threshold is too high, some detectable targets may go undetected.

Each time the antenna beam sweeps over the target, a stream of pulses is received (Fig. 21). A doppler filter in the radar's signal processor adds up the energy contained in this stream. The target signal in the output of the filter thus corresponds fairly closely to the total amount of energy received during the antenna beam's time-on-target. This energy is indistinguishably combined with the noise energy that has accumulated in the filter during the same period.

As the target's range decreases, the strength of the integrated signal increases. The mean strength of the noise, on the other hand, remains about the same. Eventually, the signal becomes strong enough to be detected above the noise (Fig. 22).

In doppler radars, detection is performed automatically. At the end of every integration period, the output of each filter is applied to a separate detector. If the integrated signal plus the accompanying noise exceeds a certain threshold, the detector concludes that a target is present, and a bright, synthetic target blip is presented on the display. Otherwise, the display remains perfectly clear (Fig. 23).

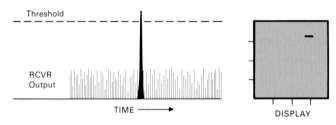

23. If the receiver output exceeds the detection threshold, a bright, synthetic blip appears on the display.

The completely random noise alone will occasionally exceed the threshold, and the detector will falsely indicate that a target has been detected (Fig. 24). This is called a *false alarm.* The chance of its occurring is called the *false-alarm probability.* The higher the detection threshold relative to the mean level of the noise energy, the lower the false-alarm probability will be, and vice versa.

Clearly the setting of the threshold is crucial. If it is too high (Fig. 25), detectable targets may go undetected. If it is too low, too many false alarms will occur. The optimum setting is just enough higher than the mean level of the noise to keep the false-alarm probability from exceeding an acceptable value. The mean level of the noise, as well as the system gain, may vary over a wide range. Consequently, the output of the radar's doppler filters must be continuously monitored to maintain the optimum threshold setting.

Generally, the threshold for each detector is individually set on the basis of both the probable noise level in the filter whose output is being detected (the "local" noise level) and

126

the average noise level in all of the filters (the "global" noise level). Typically, the local level is determined by averaging the outputs of a group (ensemble) of filters on either side of the one in question. Since most of these outputs will be due to noise, the average can be assumed to approximate the probable noise level in the bracketed filter.

The global noise level is determined by establishing a second, noise-detection threshold for every filter. This threshold is set far enough below the target-detection threshold so that in aggregate vastly more threshold crossings are made by noise spikes than by target echoes. By continually counting these crossings and statistically adjusting the count for the difference between the two thresholds, the false-alarm rate for the entire system can be determined.

Exactly how the local ensemble of filters is selected and how the average for the ensemble is weighted in comparison to the system false-alarm rate varies from system to system and mode to mode. As nearly as possible, however, the thresholds are set so as to maintain the false-alarm rate for each detector at the optimum value. If the rate is too high, the thresholds are raised; if it is too low, the thresholds are lowered. For this reason, the automatic detectors are called *constant false-alarm-rate* (CFAR) *detectors*.

Regardless of how close to optimum it is, the setting of the target detection threshold, relative to the mean level of the noise, establishes the minimum value of integrated signal energy, s_{det}, that, on average, is required for target detection (Fig. 26). Bear in mind, though, that because of the randomness of the noise energy about its mean value, the signal plus the accompanying noise will sometimes exceed the threshold even when the signal energy is less than s_{det}. Likewise, at other times it will fail to reach the threshold, even when the signal energy is greater than s_{det}.

Nevertheless, the range at which a given target's integrated signal becomes equal to s_{det} can be considered to be the maximum detection range (under the existing operating conditions) for that particular target.

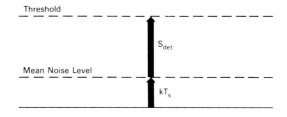

26. The setting of the threshold relative to the mean noise level establishes the minimum value of integrated signal, S_{det}, required for detection.

Integration and Its Leverage on Detection Range

Although implicit in the expression for signal energy (page 125), the immense importance of integration in pulling the weak echoes of distant targets out of noise is often overlooked. One can gain a valuable insight into this important process by performing a simple experiment.

Experimental Setup. To see how noise energy and signal energy integrate in a narrow bandpass (doppler) filter we set up a rudimentary radar to look for a test target at a given range and angle. Having trained the antenna in the

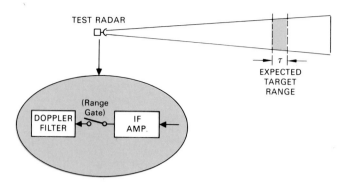

27. A rudimentary radar is set up to look for a target at a given range. Switch closes at the point in the interpulse period when return is received from the expected target's range.

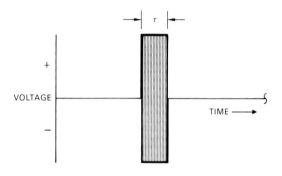

28. Having passed through the IF amplifier and been sliced very thin, each noise pulse looks much like target return.

expected target's direction, we turn on the receiver for a fixed period of time. Meanwhile, at that point in each interpulse period when return will be received from the expected target's range, we momentarily close a switch (range gate). It passes a slice of the receiver's IF output—one pulsewidth wide—to a narrowband filter (Fig. 27). The filter is tuned to the target's doppler frequency.

Noise Alone, Single Integration Period. Initially, we perform the experiment with no target present. When the range gate is closed, all the filter receives is a pulse of noise energy.

In this radar, as in most, the passband of the receiver's IF amplifier is just wide enough to pass the bulk of the energy in a target echo (matched filter design). Consequently, having passed through the IF amplifier and been sliced into narrow pulses, the noise looks much like target return (Fig. 28). The principal difference as seen by the doppler filter is this. Whereas the phase of the pulses received from a target is constant from pulse to pulse, the phase as well as the amplitude of the noise pulses varies randomly from pulse to pulse. We can see the variation in phase most clearly if we represent the pulses with phasors (Fig. 29).

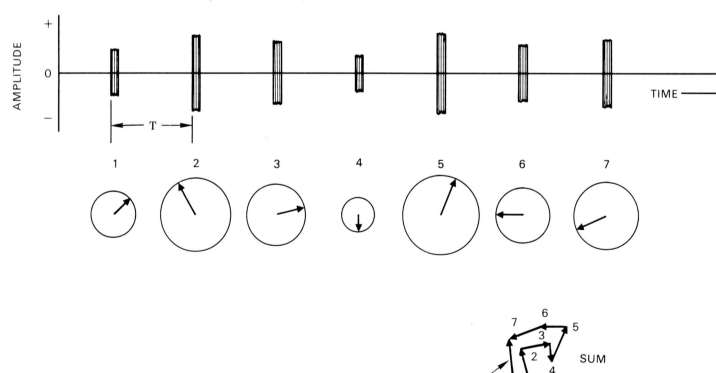

Noise Pulses (Output of IF Amplifier)

29. Phasor representation of noise pulses applied to doppler filter. Because of phase variation, amplitude of integrated noise is only a fraction of the sum of amplitudes of the individual pulses.

Now, the role of the filter is to further narrow the receiver passband by integrating the energy of successive pulses. What the filter does, in effect, is add up the phasors. In the case of noise, because of the randomness of phase, the pulses largely cancel.[4]

At the end of the integration period, the magnitude of the sum—the integrated noise, $\vec{N}$—is little different than the amplitude of a single noise pulse and only a fraction of the sum of the amplitudes of the individual pulses. The integration period, we will assume, corresponds to a single time-on-target.

Noise Alone, Successive Times-on-Target. We repeat the experiment a great many times—each repetition corresponding to a separate time-on-target. As expected, because of the randomness of the noise, the magnitude and phase of the energy that accumulates in the filter, $\vec{N}$, vary widely from one time-on-target to the next.

At the end of each time-on-target, the magnitude of the accumulated energy is "detected" (Fig. 30). That is, a voltage (video signal) proportional to the magnitude is produced. Incidentally, since the integration takes place before this detection, the integration is called *predetection integration.*

The video outputs for successive times-on-target are plotted in Fig. 31.

4. Actually, the phase of a target's returns varies from pulse to pulse in proportion to the target's doppler frequency. But as seen by a filter tuned to this frequency, the phase is constant.

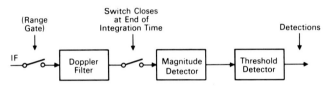

30. At the end of each integration period (time-on-target), the amplitude of the energy accumulated in the filter is detected and applied to a threshold detector.

INTEGRATED NOISE

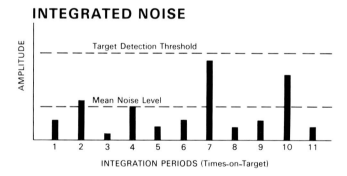

31. Outputs of the amplitude detector of Fig. 30 at end of successive times-on-target.

As you can see, over a number of integrated periods, the magnitude of the integrated noise varies randomly about a mean value. Though not illustrated here, the variation in phase is equally random.

Target Signal Only. We repeat the experiment; this time, with the target present but (through some magic) with the noise absent. Now each time the range gate is closed, the filter receives a pulse of energy from the target. Unlike the noise pulses, these all have the same phase.[5] When integrat-

5. During the very short time interval discussed here.

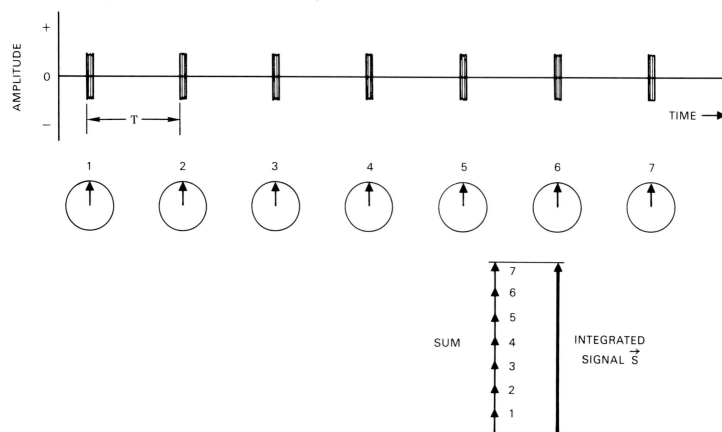

Signal Pulses (Output of IF Amplifier)

32. Phasor representation of signal pulses applied to doppler filter. Because phase is same from pulse to pulse, integrated signal is many times amplitude of individual pulses.

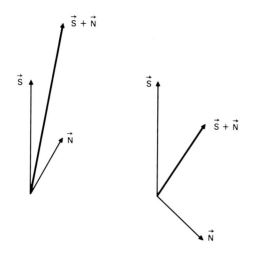

33. Amplitude of integrated signal plus noise varies widely, depending on amplitudes and phases of $\vec{S}$ and $\vec{N}$.

ed by the filter, they add constructively. At the end of each integration period, their sum (Fig. 32, above)—the magnitude of the integrated signal, $\vec{S}$—very nearly equals the sum of the amplitudes of the individual pulses.

How Signal and Noise Combine. Finally, we repeat the experiment several times with both target signal and noise present. Although they are indistinguishably mixed and so integrate simultaneously, we can visualize the result more clearly if we think of the signal and noise as being integrated separately and of their sums, $\vec{S}$ and $\vec{N}$, being vectorially added together at the end of the time-on-target. The magnitude of the vector sum, of course, depends not only upon the magnitudes of $\vec{S}$ and $\vec{N}$, but upon the phase angle between them (Fig. 33). If the noise is in phase with the signal, the two vectors will combine constructively; if the noise is 180° out of phase, they will combine destructively; and there are myriad possible combinations in between. For any one time-on-target, therefore, the magnitude of the energy that accumulates in the filter equals the magnitude of the integrated signal, $\vec{S}$, plus or minus some fraction of the magnitude of the integrated noise, $\vec{N}$.

Improvement in Signal-to-Noise Ratio. How predetection integration improves the signal-to-noise ratio should now be fairly clear. Whereas the noise energy that accumulates in the filter may vary widely from one integration period to another, the mean level of the noise energy is essentially independent of the integration time. The integrated signal energy (target return), on the other hand, increases in direct proportion to the integration time. By increasing the integration time, therefore, the signal-to-noise ratio can be increased significantly.

An individual target echo, for example, may contain only one thousandth as much energy as an individual noise pulse, yet after ten thousand pulses have been integrated, the signal may be considerably greater than the noise.

Indeed, the improvement in signal-to-noise ratio achievable through predetection integration is limited only by (1) length of the time-on-target, t_{ot}, or (2) the maximum practical length of the integration time, t_{int}, if that is less than t_{ot}, or (3) the length of time over which the target's doppler frequency remains close enough to the same value for the target echoes to be correlated by the filter (Fig. 34). The greater the improvement in signal-to-noise ratio, of course, the weaker the target echoes can be and still be detected; hence, the greater the detection range.

Postdetection Integration

Sometimes, the maximum practical integration time is a good deal less than the time-on-target. Take, for example, a situation where the doppler frequencies of expected targets may be subject to rapid change. Since the width of the filter passband is inversely proportional to the integration time (bandwidth $\cong 1/t_{int}$), making t_{int} as long as t_{ot} could narrow the passband to the point where the signal may very well move out of it long before the time-on-target ends (Fig. 35).

In such instances, rather than lose any of the signal, the integration time of the doppler filter is made short enough to provide the required bandwidth, and integration and video detection are repeated throughout the time-on-target (Fig. 36). The video outputs for successive integration periods are then added together (integrated) and their *sum* is applied to the threshold detector. This second integration process is fundamentally the same as that employed in non-doppler radars. Since it takes place after video detection, it is called *postdetection integration,* or PDI.

Once the output of a doppler filter (or the output of the IF amplifier in a non-doppler radar) has been converted to

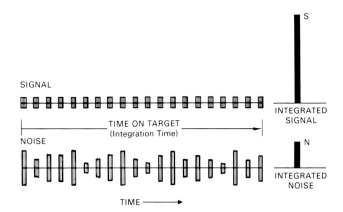

34. The improvement in signal-to-noise ratio is ultimately limited only by the time-on-target, provided target echoes remain correlated.

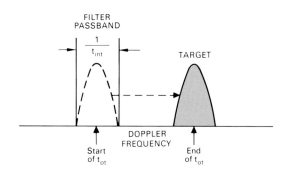

35. Situation in which target's doppler frequency changes radically during time-on-target, t_{ot}. If filter integration time, t_{int}, is made equal to t_{ot}, target will move out of passband before integration is finished.

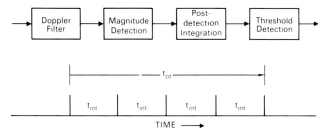

36. Problem is solved by dividing t_{ot} into a number of integration periods short enough to provide adequate doppler bandwidth and adding up filter outputs for entire time-on-target.

131

a video signal of single polarity, noise will no longer cancel when integrated. Rather, it will build up throughout the integration time in exactly the same way as the signal. Consequently, with PDI the mean signal-to-noise ratio cannot be increased. Nevertheless, an equivalent improvement in detection sensitivity may be achieved. To see why, we must look a little more closely at PDI.

Actually, PDI is nothing more than averaging. It has the same effect as passing the video signal through a low-pass (as opposed to bandpass) filter. You can visualize this most clearly by thinking of the video signal as consisting of a constant (dc) component, the amplitude of which corresponds to the mean level of the signal, plus a fluctuating (ac) component.

The amplitude of the dc component is unaltered by the averaging, but the amplitude of the ac component is reduced. The higher the frequency of the fluctuation and the greater the integration time (i.e., the larger the number of inputs averaged), the greater the reduction will be. Averaging improves detection sensitivity in two important ways.

First, it reduces the average deviation of the integrated noise energy. Consequently, without increasing the false-alarm probability, the target-detection threshold can be set closer to the mean noise level (Fig. 37). The integrated signal need not be as great to cross the threshold, and the weaker echoes of more distant targets can be detected.

The second improvement averaging makes is more subtle. As we just saw, when target return is received, the integrated signal is vectorially added to the integrated noise. Because of the randomness of the noise, as often as not, the noise will be out of phase with the signal and so will combine with the signal destructively.

However, when the integrated signal plus noise is averaged over many integration periods, the fluctuations due to the noise tend to cancel out, leaving only the signal. The possibility of missing an otherwise detectable target because of the destructive combination of it with noise is thus greatly reduced.

Together, these two effects of PDI can substantially reduce the signal-to-noise ratio required for detection. As illustrated in Fig. 38, the fluctuations in the noise and in the signal-plus-noise can, in the extreme, be reduced to the point where a signal can be detected when the mean signal-to-noise ratio is substantially less than one.

Sometimes, the equivalent of postdetection integration is approximated by using a so-called "m out of n" detection

NOISE ONLY

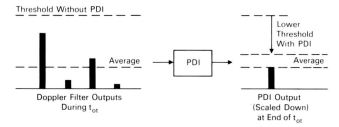

37. By averaging the noise outputs of a doppler filter during the time-on-target, PDI enables the target-detection threshold to be set much lower, without increasing the false-alarm probability.

PDI OUTPUTS FOR SIGNAL PLUS NOISE

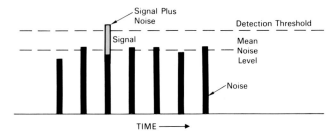

38. The two effects of PDI allow a signal to be detected even when the mean signal-to-noise ratio is less than one.

criteria. If the time-on-target spans n predetection integration periods, rather than requiring only one threshold crossing per time-on-target as a condition for detection, the signal processor requires m crossings (Fig. 39).

The chance of isolated noise spikes producing false alarms is thereby reduced. The detection threshold can be lowered without increasing the false-alarm probability, and more distant targets can be detected.

Summary

Since radio waves of the frequencies used by airborne radars travel essentially in straight lines, a target must be within the line of sight to be detected. Range may be further limited by clutter or man-made interference. Ultimately, it is determined by the signal-to-noise energy ratio.

The principal source of noise is thermal agitation in the input stages of the receiver. The noise energy is commonly expressed in terms of a figure of merit, F_n, relating it to an approximation of the external noise, provided by thermal agitation in a resistor connected across the receiver's input terminals. In the case of low-noise receivers, external noise sources become more significant, and noise is expressed in terms of an equivalent "system" noise temperature.

How much energy is received from a target depends primarily upon (1) the radar's average transmitted power, antenna gain, and effective antenna area; (2) the time-on-target; (3) the target's range, R, and radar cross section, σ—a factor which accounts for the size, reflectivity, and directivity of the target.

Most radars integrate the return received as the antenna scans across a target. If performed before video detection (predetection integration), the integration increases the signal-to-noise ratio in direct proportion to integration time. If performed after video detection (PDI), the integration accomplishes two things: (1) averages out the fluctuations in the noise, thereby reducing its peaks, and (2) averages out the destructive combination of the noise with the signal, thereby reducing the possibility of missing an otherwise detectable target.

For a target to be detected, the integrated signal must exceed a threshold set high enough to keep the probability of noise crossings acceptably low. In doppler radars, to maintain a constant, optimum false alarm rate (CFAR), the threshold setting of the magnitude detector for each doppler filter's output is based on the mean noise level in the outputs of an ensemble of adjacent filters, as well as on measurement of the mean noise level in the outputs of all the filters.

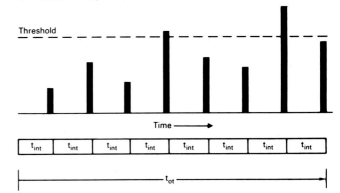

m = required number of threshold crossings
n = number of integration periods

39. Sometimes the equivalent of PDI is obtained by requiring m out of n threshold crossing in a time-on-target for a detection. Here m = 2 and n = 8.

Some Relationships To Keep In Mind

- Mean noise power = $F_n k T_0 B$ or $k T_s B$ watts

  F_n = Receiver noise figure

  T_0 = Noise temperature (nominally 290° K)

  k = Boltzmann's constant
  = 1.38×10^{-23} watt-second / °K

  B = Receiver bandwidth (hertz)

  T_s = System noise temperature (including internal + external noise)

- Mean noise energy = $k T_s B\, t_n$
  t_n = duration of the noise

- Mean noise energy = $\dfrac{k T_s t_n}{\tau}$
  (Matched filter)

- Mean noise energy = $k T_s$
  (Doppler radar)

- Signal energy = $K \dfrac{P_{avg}\, G\, \sigma\, A_e\, t_{ot}}{R^4}$

  $K = \dfrac{1}{(4\pi)^2}$

  P_{avg} = Average transmitted power, watts

  G = Antenna gain

  σ = Radar cross section of target

  A_e = Effective area of antenna

  t_{ot} = Time on target

  R = Range

NORTH AMERICAN F–86D SABRE (1952)

This was the first single-seat radar interceptor. It was an adaptation of the F–86E day fighter and featured the Hughes E–4 Radar Fire Control System firing 24 rockets from a retractable pod. Over 2500 of these aircraft were built with later versions bearing the designation F–86L and serving with Air National Guard units into the mid 1960's.

The Range Equation, What It Does and Doesn't Tell Us

11

I n the last chapter, we learned that within the line of sight, in the absence of interference and competing ground return, detection range is ultimately determined by the ratio of the energy received from a target—the signal—to the energy of the background noise. We identified the principal factors which determine the signal and noise energies and became acquainted with the detection process.

Building on that knowledge, in this chapter we will write a general equation for maximum detection range and analyze it to see how the individual factors we have identified influence the range. We will then narrow down to the special case of volume search. Finally, we will consider the statistical variation in detection range and see how it is accounted for.

General Range Equation

As we saw in the preceding chapter, when the radar antenna is trained on a target (Fig. 1), the energy received from the target during any one integration time is roughly

$$\text{Received signal energy} \cong \frac{P_{avg} G \sigma A_e t_{int}}{(4\pi)^2 R^4}$$

where

$$
\begin{aligned}
P_{avg} &= \text{average transmitted power} \\
G &= \text{antenna gain} \\
\sigma &= \text{radar cross section of target} \\
A_e &= \text{effective antenna area} \\
t_{int} &= \text{integration time} \\
R &= \text{range}
\end{aligned}
$$

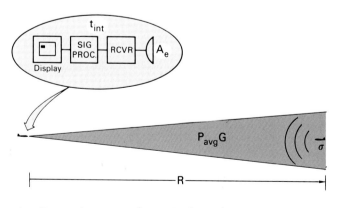

1.   Factors determining the received signal energy.

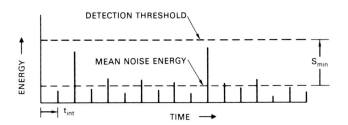

2. Integrated noise energy at end of successive integration times, t_{int}. On average, for a target to be detected integrated signal energy must equal S_{min}.

For the target to be detected, this energy plus the accompanying noise energy must exceed a certain threshold value. It is set just high enough above the mean noise level to reduce the probability of noise peaks crossing the threshold—false alarms—to an acceptably low value.

On average, the minimum energy that a signal must have to cross the detection threshold is the difference between the detection threshold and the mean level of the noise. This difference (Fig. 2) is commonly represented by the term S_{min}.

$$\text{Minimum detectable signal energy} = S_{min}$$

Assuming perfect integration, the maximum range at which a given target will be detected is the range at which the received signal energy becomes equal to S_{min}. Setting the expression for signal energy equal to S_{min} and solving for range, therefore, yields a simple equation for the maximum detection range.

$$R_{max} \cong \sqrt[4]{\frac{P_{avg}G\sigma A_e t_{int}}{(4\pi)^2 \, S_{min}}}$$

(Antenna trained on target)

As it stands, the equation applies only when the antenna is continuously trained on the target and the target is in the center of the mainlobe, i.e., when a target is being spotlighted. (Bear in mind that though the antenna may be continuously trained on the target, t_{int} is limited to the period of time that the phase of the target signal remains correlated.)

In search, the maximum integration time is limited to the time the antenna takes to sweep across the target—the time-on-target, t_{ot}. Moreover, the beam is actually centered on the target only for an instant, if centered at all. We can eliminate the first limitation simply by replacing t_{int} with t_{ot}. Temporarily, at least, we can get around the second limitation by pretending that the antenna gain is the same over the entire solid angle encompassed by the mainlobe and that, for the particular scan being used, the target is centered in the beam's path (Fig. 3).

Under these conditions, the equation gives the maximum detection range for a single search scan.

3. Simple range equation can be applied to search by pretending that transmitted energy is uniformly distributed over cross section of antenna beam and target is centered in beam's path.

$$R_{max} \cong \sqrt[4]{\frac{P_{avg}G\sigma A_e t_{ot}}{(4\pi)^2 \, S_{min}}}$$

(Single scan of antenna)

Incidentally, if we replace t_{ot} with the pulse width, τ, and P_{avg} with the peak power, P, the equation gives the range for single-pulse detection.

$$R_{max} \cong \sqrt[4]{\frac{PG\sigma A_e \tau}{(4\pi)^2 S_{min}}}$$

(Single pulse: non-doppler radar)

Assuming that postdetection integration is accounted for separately, this form of the equation applies to non-doppler radars (Fig. 4).

Omissions. Regardless of which of these forms we use, the equation is incomplete. Among the more obvious omissions are

- Absorption and scattering in the atmosphere (Fig. 5)

- Reduction in signal energy due to the target not necessarily being centered in the path of the scanning antenna beam (this is called *elevation beamshape loss*)

- The further reduction in signal energy as the beam sweeps across the target (Fig. 6) due to the fall-off in two-way antenna gain at angles off beam center (this is called *azimuth beamshape loss*)

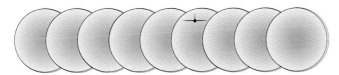

6.   Other factors not directly accounted for include possibility of target not being centered in beam's path and fall off in two-way gain of the antenna at angles off beam center.

- Losses due to imperfect IF-filter matching—some noise being unnecessarily passed and/or some signal energy being rejected (Fig. 7)

- Loss due to the target not necessarily being centered in a doppler filter

- Degradation of signal-to-noise ratio due to imperfect integration of the target return

- Effects of system degradation in the field

Nevertheless, the equation illustrates the relative contributions of what we have seen to be some of the more fundamental factors.[1]

A More Revealing Form of the Equation. The contribution of a couple of the factors represented by terms in the range equation can be seen more easily if we modify it

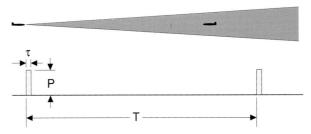

4.   Range equation can be applied to non-doppler radars by substituting pulse width, τ, for t_{int} and peak power $\bar{P}$, for P_{avg}. Postdetection integration must be accounted for separately.

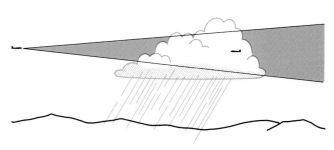

5.   One of the many important losses not accounted for by the simple range equation is atmospheric attenuation.

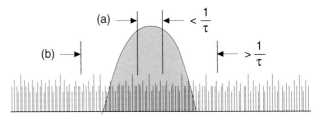

7.   IF–filter mismatch: (a) some signal being rejected that is stronger than accompanying noise; (b) some noise being passed that is stronger than accompanying signal.

1.   All omitted factors which reduce the signal-to-noise ratio are accounted for by including a loss factor, L, in the denominator.

137

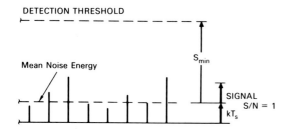

8. Since the detection threshold is related to mean noise level in a complex way, detection range can be expressed in terms of noise energy most simply by solving for the range at which the integrated signal-to-noise ratio is one.

slightly. First, since S_{min} is related in a fairly complex way to the mean noise energy, kT_s, if we back off from solving for the *maximum* detection range and solve merely for the range at which the integrated signal-to-noise ratio is one, we can substitute kT_s, directly for S_{min} (Fig. 8). Second, since antenna gain is proportional to effective antenna area divided by wavelength squared ($G \propto A_e/\lambda^2$), we can consolidate terms relating to the antenna by substituting A_e/λ^2 for G.

With these changes, the equation for a single search scan becomes

$$R_o \propto \sqrt[4]{\frac{P_{avg}A_e^2\sigma t_{ot}}{kT_s\lambda^2}}$$

(Single search scan: SNR = 1)

where R_o is the range at which the integrated signal-to-noise ratio (SNR) is one, λ is the wavelength, and the other terms are as previously defined.

What the Range Equation Tells Us

Incomplete as it is, the equation reveals a good deal not only about the effect of changing various parameters, but about some of the trade-offs which must be made in designing a radar.

Average Power. The equation tells us, for example, that increasing the power of the transmitter by a given factor increases the detection range by only about the fourth root of that factor. If we were to increase the power by, say, three times (Fig. 9), the detection range would increase by only about 30 percent ($R_2 = R_1\sqrt[4]{3} \cong 1.32$).

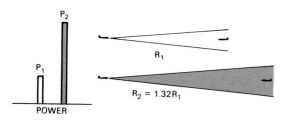

9. Tripling transmitter power would increase detection range by only 32 percent.

Noise. At the same time, the equation tells us that *de*creasing the mean level of the background noise (kT_s) by a given factor has the same effect as *in*creasing the average power by the same factor. If we could reduce the noise by 50 percent, for example, the detection range would increase by the same amount (Fig. 10) as if we had doubled the power, which is about 20 percent ($R_2 = R_1\sqrt[4]{2} \cong 1.19 R_1$).

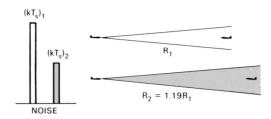

10. Decreasing system noise has the same effect on detection range as increasing power by the same factor.

Time-on-Target. The equation also enables us to predict the effect of changes in time-on-target, or integration time. Suppose that by slowing down the scan, we were to double the time-on-target. Provided the target return could still be integrated, this would have the same effect as doubling the power (Fig. 11).

Radar Cross Section. The equation further enables us to predict the differences in the ranges at which a given radar can detect targets of different sizes.

Suppose, for example, that the radar detects a target having a certain radar cross section at a range of 40 miles. Provided that the targets' aspect and directivity remaining the same, the radar should be able to detect a target having four times this radar cross section (Fig. 12) at a range of about 66 miles. ($R_2 = R_1 \sqrt[4]{4} \cong 40 \times 1.41 \cong 66$)

Antenna Size. Similarly, the equation enables us to predict the effects of changes in size of the antenna. Suppose the antenna is circular and we double its diameter. Assuming that the aperture efficiency remains the same, this increase would increase A_e by 2^2. ($A_e \propto d^2\eta$.) The range equation tells us that the increase in A_e (Fig. 13) would increase the range at which the radar might detect a given target by a factor of two, $R_2 = R_1 \sqrt[4]{(2^2)^2} = 2R_1$, provided we were spotlighting the target.

Doubling the antenna diameter, however, would cut the beamwidth in half. So, if the radar was searching for targets, we would have to slow down the antenna scan to maintain the same time-on-target. If we didn't, t_{ot} would be cut in half, and the range would be increased by a factor of only about 1.68. ($R_2 = R_1 \sqrt[4]{16 \times 0.5} \cong 1.68 R_1$)

Wavelength. Since wavelength squared is in the denominator of the equation, decreasing λ would *appear* to have the same effect on the radar's detection range as increasing the effective area of the antenna, A_e.

But here, an important limitation of our simple equation shows up. Depending upon what the original wavelength was and how much we decreased it, the first order effect of decreasing λ might be offset to a considerable extent by such factors as increased atmospheric absorption, one of the factors not accounted for in the equation.

Whereas the range equation indicates that decreasing the wavelength, λ, from 3 centimeters to 1 centimeter would increase the radar's detection range by about 70 percent (i.e., $R_2 = R_1 \sqrt[4]{1/(1/3)^2} \cong 1.73$), one look at a plot of atmospheric attenuation versus wavelength (Fig. 14) tells us that this is simply not so.

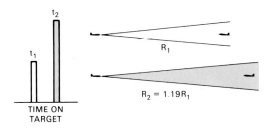

11. Doubling time-on-target would have the same effect as doubling transmitter power.

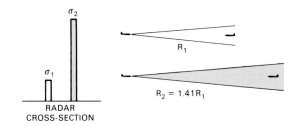

12. An increase in radar cross section has the same effect as a proportional increase in time-on-target.

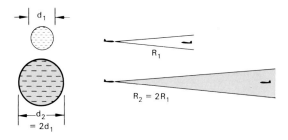

13. Doubling antenna diameter would double detection range, provided scan was slowed to provide same time-on-target.

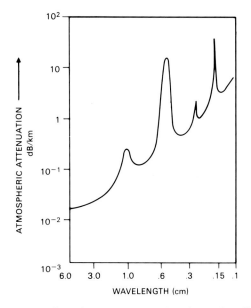

14. First order effect of decreasing wavelength may be offset by such factors as increased atmospheric attenuation.

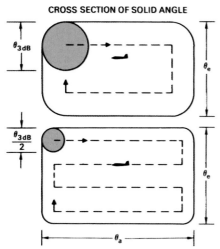

CROSS SECTION OF SOLID ANGLE

15. If beamwidth is reduced, scan must be slowed down to provide the same time-on-target, t_{ot}.

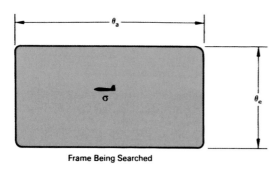

Frame Being Searched

16. During the scan frame time, the total backscattered energy is proportional to the ratio of the radar cross section (σ) of the target to the cross-sectional area $R^2(\theta_a\theta_e)$ of the solid angle scanned at the target's range, $\sigma/(R^2\theta_a\theta_e)$.

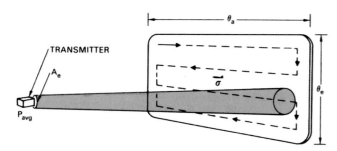

17. Fraction of backscatter intercepted by radar is proportional to effective antenna area divided by the range squared (A_e/R^2).

As with antenna size, decreasing λ would also decrease the beamwidth ($\theta_{3\,db} \propto \lambda/d$). In search, therefore, to keep the increase in range from being wiped out by a reduction in t_{ot}, the scan would have to be slowed down (Fig. 15).

Because the range equation we have been using doesn't account for the effect of changes in wavelength and antenna size on t_{ot}, it is not as illuminating as it might be for situations where a given volume of target space must be searched for a given period of time. For volume search, therefore, a slightly different form of the equation is commonly used.

Equation for Volume Search

To tailor the range equation to volume search, the time-on-target (t_{ot}) must be expressed in terms of (1) the length of time the antenna takes to complete one frame of the search scan, and (2) the size of the solid angle subtended by that frame. The scan frame time is represented by t_f; the solid angle, by the product of the azimuth and elevation angles spanning the frame, θ_a and θ_e, respectively. While this conversion is straightforward (see panel, bottom of next page), we can get a better physical feel for the fundamental relationships involved in volume search by starting from scratch with a simplified derivation.

Simplified Derivation. The total energy radiated during any one frame time, t_f, equals $P_{avg}t_f$. Assuming that the scan spreads the energy uniformly over the entire solid angle, the fraction of the energy that is intercepted by a target and scattered back toward the radar is proportional to the ratio of (a) the target's radar cross section to (b) the cross-sectional area of the solid angle of the search scan at the target's range (Fig. 16). The fraction of the backscattered energy captured by the radar antenna is proportional to A_e (Fig. 17).

For volume search, therefore, the simplified range equation can be rewritten as

$$R_o \propto \sqrt[4]{P_{avg}\, t_f \; \times \; \frac{\sigma}{\theta_a\, \theta_e} \; \times \; A_e}$$

where

t_f = frame time

θ_a = azimuth angle scanned

θ_e = elevation angle scanned

and the other terms are as previously defined.

Ignoring target radar cross section—over which we have no control—and rearranging, we get

$$R_o \propto \sqrt[4]{P_{avg}A_e \; \times \; \frac{t_f}{\theta_a\theta_e}}$$

What the Volume Search Equation Tells Us. From this simple equation, we can draw three important conclusions regarding detection range in volume search.

- Only through its secondary influences on atmospheric absorption, available average power, aperture efficiency, ambient noise, target directivity, and so on, does wavelength affect the range.

- For any combination of frame time and solid angle searched, range depends primarily upon the product, $P_{avg}A_e$ (Fig. 18).

- The greater the ratio of the frame time to the size of the volume searched, the greater the range will be.

Frame time, however, may be limited by required system reaction time—which is itself a function of detection range. And the size of the solid angle is dictated by the dispersion of anticipated targets. Therefore, the equation leads one to this general conclusion: To maximize detection range for volume search, use the highest possible average power and the largest possible antenna.

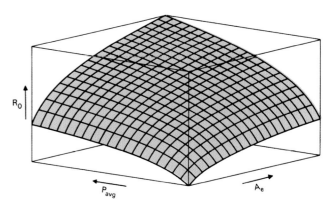

18. For any one combination of frame time and solid angle searched, detection range can be maximized by using the highest possible power (P_{avg}) and largest possible antenna (A_e).

TAILORING THE RANGE EQUATION TO VOLUME SEARCH

In adapting the radar equation to volume search, the antenna beam is conveniently thought of as having a uniform cross section of width $\theta_{3\,dB}$.

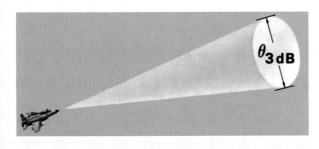

This beam is then thought of as jumping a beamwidth at a time through the solid angle that is to be searched.

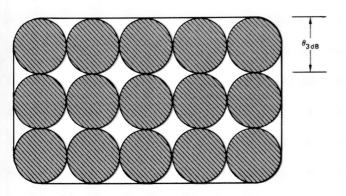

The number of such positions the beam occupies equals the cross-sectional area of the solid angle at unity range divided by the cross section of the beam at the same range.

$$\text{Number of beam positions} = \frac{\theta_a \theta_e}{(\theta_{3\,dB})^2}$$

Since the beam must complete its entire scan in one frame time (t_f), the length of time it dwells on any one target is

$$\text{Dwell time} = t_f \frac{(\theta_{3\,dB})^2}{\theta_a \theta_e}$$

Now, the beamwidth is proportional to the ratio of the wavelength, λ, to the diameter of the antenna, d. ($\theta_{3\,dB} \propto \lambda/d$.) And d, in turn, is proportional to the square root of the effective area of the antenna, $\sqrt{A_e}$. Thus, the dwell time, t_{ot} is

$$t_{ot} \propto t_f \frac{(\lambda/\sqrt{A_e})^2}{\theta_a \theta_e} = \frac{t_f \lambda^2}{\theta_a \theta_e A_e}$$

Substituting this expression for t_{ot} in the range equation given on page 138, we get:

$$R_o \propto \sqrt[4]{\frac{P_{avg} A_e^2 \sigma t_{ot}}{kT_s \lambda^2}} = \sqrt[4]{\frac{P_{avg} A_e^2 \sigma t_f \lambda^2}{kT_s \lambda^2 \theta_a \theta_e A_e}}$$

Cancelling like terms,

$$R_o \propto \sqrt[4]{\frac{P_{avg} A_e \sigma t_f}{kT_s \theta_a \theta_e}}$$

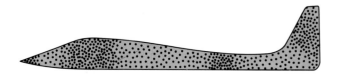

19. A target may be thought of as myriad, tiny reflectors. How their echoes add up depends upon their relative phases.

2. Generally, the sums tend to cluster around a median value.

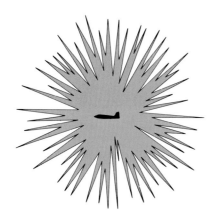

20. Polar plot of the radar cross section, σ, of a typical target. Note how widely σ varies with target aspect.

3. More precisely, like the light in a particular spectral line when a star is at a low elevation angle.

Even when all pertinent factors have been included in the radar equation, it cannot tell us with certainty at what range a given target will be detected. For not only background noise but radar cross section are continually fluctuating qualities.

Fluctuations in Radar Cross Section

The reason for these fluctuations can be seen if we think of the target as consisting of a large number of individual scatterers (Fig. 19). The extent to which the scatter from these adds up or cancels in the direction of the radar depends upon their relative phases. If the phases are more or less the same, the backscatter will add up to a large sum. If they are not, the sum may be comparatively small.[2]

The relative phases depend upon the instantaneous distances in wavelengths of the reflectors from the radar. Because of the round trip nature of the transmission, a difference in distance of 1/4 wavelength makes a difference in phase of 180°.

Since the wavelength may be very short, relatively small changes in target aspect, even vibration, can cause the target return to scintillate like the light from a star.[3] And since the configuration of many targets is radically different when viewed from different directions, larger changes in aspect may produce strong peaks or deep fades (Fig. 20). Over a period of time these variations will usually average out. But if the radar-bearing aircraft is approaching on a course that holds a target in the same relative aspect, a peak or fade may persist for some time.

During the early days of the all-weather interceptor, in fact, it was not uncommon to receive complaints from pilots who had picked up a target in a favorable aspect and locked onto it at long range, only to lose lock when the interceptor converted to a constant aspect attack course that happened to place the target in a deep fade. Being unfamiliar with the phenomenon, they thought the radar had malfunctioned.

Since the relative phases of the returns from the individual elements depend upon wavelength, the target aspects for which the fades occur will generally be slightly different for different wavelengths. One way of getting around the problem of target fading, therefore, is to switch periodically from one to another of several different radio frequencies, thereby providing what is called *frequency diversity*.

Detection Probability

Because of its randomness, detection performance against targets whose range is limited by thermal background noise is usually stated in terms of probabilities. For search, the most commonly used probability is *blip-scan ratio*, P_d. It is the probability of detecting a given target at a given range any time the antenna beam scans across the target (Fig. 21). It is also referred to as *single-scan* or *single-look probability*. The higher the probability specified, the shorter the range will be.

The notation used to represent the range is the letter "R" with a subscript indicating the probability. For instance, R_{50} represents the range for which the probability of detection is 50 percent; R_{90}, the range for which the probability is 90 percent.

How does one determine the range, say, for a probability of detection of 60 percent? There are five basic steps:

1. Decide on an acceptable system false-alarm rate.

2. Calculate the corresponding value of the false-alarm probability for the individual threshold detectors.

3. On the basis of the statistical characteristics of the noise, find the threshold setting that will limit the false-alarm probability to this value.

4. Determine the mean value of the integrated signal-to-noise ratio for which the signal plus the noise will have the specified probability of crossing the threshold (in this case 60 percent).

5. Compute the range at which this signal-to-noise ratio will be obtained.

That range is R_{60}.

What each step actually involves is outlined briefly in the following paragraphs.

Deciding on an Acceptable False-Alarm Rate. The average rate at which false alarms appear on the radar display—i.e., the number of false alarms per unit of time—is called the *false-alarm rate*, FAR. The mean time between false alarms is called the *false-alarm time*, t_{fa}. It, of course, is the reciprocal of the false-alarm rate.

$$t_{fa} = \frac{1}{FAR}$$

If false alarms occur only once every several hours, they will probably not even be noticed by the radar operator. Yet, if they occur at intervals on the order of a second, they may render the radar useless (Fig. 22). What is an acceptable false-alarm time depends upon the application. Since raising the detection threshold reduces the maximum detection range, where long range is desired the false-alarm time is usually made no longer than necessary to make the

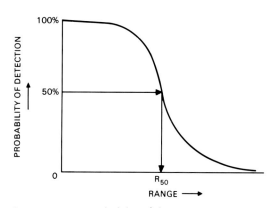

21. Blip-scan ratio is probability of detecting a given target at a given range on any one scan of the antenna.

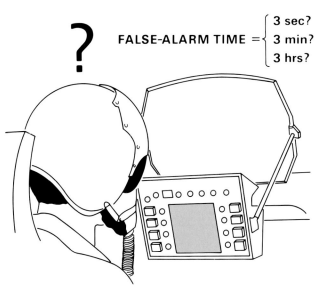

22. To the radar operator false-alarm probability has little direct meaning. Time between false alarms does.

radar easy to operate. In radars for fighter aircraft, for example, a false-alarm time of a minute or so is generally considered acceptable.

Calculating False-Alarm Probability. The mean time between false alarms is related to the false-alarm probability for the radar's threshold detectors by the following equation

$$t_{fa} = \frac{t_{int}}{P_{fa} N}$$

where

t_{fa} = average time between false alarms for the system

t_{int} = integration time of the radar's doppler filters (plus any PDI)

P_{fa} = false-alarm probability for a single threshold detector

N = number of threshold detectors

If you're puzzled, the analogy to the Lucky 8 Casino may help.

In this casino, whenever a roulette wheel spins an "8," an "eight-bell alarm" sounds, all bets at that wheel stay put, and everyone who has a bet down is served a free glass of champagne.

Before the casino opened, the question naturally arose: How often will the alarm go off?

Figuring that out was easy. There are 38 compartments in a wheel (Fig. 23); so, on average, the ball will light in Number 8 once every 38 spins. If three minutes elapsed between spins, the alarm would sound once every 38 x 3 = 114 minutes for each wheel. The casino would have 5 wheels; so the alarm would sound five times this often, or once every $114 \div 5 \cong 23$ minutes.

23. On average an 8 will be spun once every 38 spins. (Courtesy Las Vegas News Bureau)

$$1 \text{ wheel:} \quad 38 \frac{\text{spins}}{\text{alarm}} \quad x \quad 3 \frac{\text{min}}{\text{spin}} \quad = \quad 114 \frac{\text{min}}{\text{alarm}}$$

$$5 \text{ wheels:} \quad 114 \frac{\text{min}}{\text{alarm}} \quad \div \quad 5 \quad = \quad 23 \frac{\text{min}}{\text{alarm}}$$

Since the outcome of each spin is entirely random, the alarm would not, of course, sound at even intervals. There might be two or three alarms in a matter of minutes, or none for several hours. But, on average, the time between alarms would be 23 minutes.

With the exception of the champagne, the parallel between 8-bell alarms and false alarms on a radar display is direct. The probability of a wheel spinning an "8" corresponds to the false-alarm probability of one of the radar's threshold detectors; the time between spins, to the integration time of the radar's doppler filters; the number of roulette wheels, to the number of threshold detectors.

In general, a bank of doppler filters is provided for every resolvable increment of range (range gate), and a threshold detector is provided for every filter.

Substituting the product of the number of range gates (N_{RG}) times the number of doppler filters per filter bank (N_{DF}) for N in the equation at the top of the facing page and solving for P_{fa}, we get

$$P_{fa} = \frac{t_{int}}{t_{fa} \times N_{RG} \times N_{DF}}$$

Suppose, for example, that the filter integration time (t_{int}) is 0.01 second and the radar has 200 range gates with 512 doppler filters per bank (Fig. 24). To limit the false-alarm time (t_{fa}) to a minute and a half (90 seconds), we would have to set the threshold of each detector for a false-alarm probability of about 10^{-9}.

Setting the Detection Threshold. As explained in Chap. 10, the probability of noise crossing the target-detection threshold depends upon the setting of the threshold relative to the mean level of the noise. The higher the threshold is, the lower the probability of a crossing.

Just how high the threshold must be to keep P_{fa} from exceeding the specified value depends, of course, upon the statistical nature of the noise. Since the nature of thermal noise is well known and is essentially the same in all situations, determining the threshold setting that will yield a given false-alarm probability is comparatively simple. The statistical characteristics of the noise are usually represented by what is called a probability density curve (Fig. 25). This is a plot of the probability that the magnitude of the noise in the output of a narrowband filter will have a given amplitude at any one time.

The probability of the noise exceeding the detection threshold, V_T, equals the ratio of (1) the area under the curve to the right of V_T to (2) the total area under the curve, which is one since by definition the curve encompasses all possible magnitudes. This probability, of course, is the false-alarm probability, P_{fa}.

The area under the thermal-noise curve to the right of V_T in Fig. 25 is plotted versus the value of V_T in Fig. 26. With a curve like that, one can readily find the required threshold setting for any desired P_{fa}.

FALSE-ALARM CALCULATION

Problem
Determine the false-alarm probability of a radar's threshold detectors that will limit the system false-alarm time to no more than 90 seconds.

Conditions
Filter integration time . . . . . . t_{int} = 0.01 second
Number of range gates . . . N_{RG} = 200
Number of filters per bank . N_{DF} = 512

Calculation

$$P_{fa} = \frac{t_{int}}{t_{fa} \times N_{RG} \times N_{DF}}$$

$$P_{fa} = \frac{0.01}{90 \times 200 \times 512} \cong 1.09 \times 10^{-9}$$

24. Calculation of the detector false-alarm probability that will limit the time between system false alarms to 90 seconds.

THERMAL NOISE

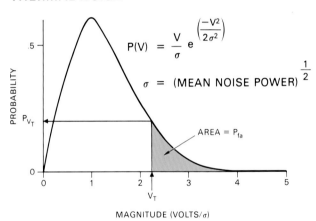

25. Probability density of thermal noise in the output of a narrowband filter. Mean noise power $\sigma^2 = kT_sB$, where kT_s is the integrated noise energy and B is the filter bandwidth, $1/t_{int}$.

FALSE – ALARM PROBABILITY

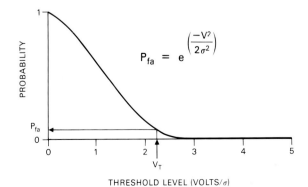

26. Area under the thermal-noise probability density curve of Fig. 25, to the right of the threshold voltage, V_T.

SIGNAL PLUS NOISE

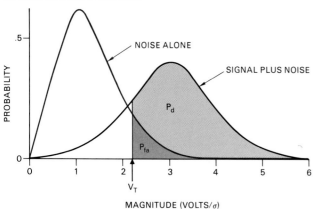

27. Probability density of filter output for a representative ratio of signal to noise. Area under curve to right of V_T is the probability of detection, P_d. Note that while increasing V_T decreases P_{fa}, it also decreases P_d.

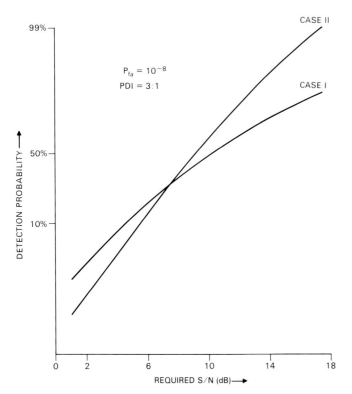

28. Standard curves base on simplified radar cross section models make coarse determination of required signal-to-noise ratio easy (the signal-to-noise ratio is evaluated at the output of the doppler filter).

Determining the Required Signal-to-Noise Ratio. The probability of a target signal plus the noise exceeding the threshold can similarly be determined. The probability density of the filter output for a representative signal-to-noise ratio is plotted in Fig. 27, along with a repeat of the probability density curve for the noise alone. As with P_{fa} in the case of noise alone, the area under the curve for signal plus noise to the right of V_T is the probability of detection, P_d. Unlike the fluctuation of the noise, the fluctuation of the signal—which we have seen is due to the variations in radar cross section—does not have a simple universal characteristic but varies from one target to the next and from one operational situation to another. Nevertheless, statistically it is possible to approximate radar cross sections having common characteristics quite accurately with standard mathematical models.

The required signal-to-noise ratio versus detection probability for a wide range of false-alarm probabilities has been calculated for a number of these models. Where specific radar cross section data is not available or a rigorous calculation is not required, curves based on these results make finding the required signal-to-noise ratio easy.

A commonly used set of curves are those based on the work of a man named Peter Swerling (Fig. 28). They apply to four different cases. Cases I and II assume a target made up of many independent scattering elements—as is a large (in comparison to the wavelength) complex target, such as an airplane. Cases III and IV assume a target made up of one large element plus many small independent elements—as is a small target of simple shape. Cases I and III assume that the radar cross section fluctuates only from scan to scan; Cases II and IV assume that it also fluctuates from pulse to pulse.

CASE	FLUCTUATIONS		SCATTERERS
	Scan–Scan	Pulse–Pulse	
I	X		Many Independent
II		X	
III	X		One Main
IV		X	

29. The four different cases to which the Swerling models apply.

With curves such as these, for almost any specified false-alarm probability, one can quickly find the integrated signal-to-noise ratio needed to provide any desired detection probability. Except for probabilities that are either very high or very low (and are poorly represented by simplified models) these curves are extremely useful.

Computing the Range. Having found the integrated signal-to-noise ratio needed to provide the desired detection probability, the range at which this ratio will be obtained can be computed with the equation derived earlier (top of page 138) for R_o. To adapt the equation for this use, the noise term (kT_s) is multiplied by the required signal-to-noise ratio.

$$R_{P_d} = \left[\frac{P_{avg} A_e^2 \sigma\, t_{ot}}{(4\pi)\, (S/N)_{req}\, kT_s\, \lambda^2 L} \right]^{1/4}$$

where R_{P_d} is the range for which the probability of detection is P_d, and $(S/N)_{req}$ is the required signal-to-noise ratio.

Cumulative Detection Probability

To account for the effects of closing rate, detection range is often expressed in terms of cumulative probability of detection: the probability that a given closing target will have been detected at least once by the time it reaches a certain range.

Cumulative probability of detection, P_c, is related to single-scan probability of detection, P_d, as follows.

$$P_c = 1 - (1 - P_d)^n$$

where n is the number of scans.

This equation may be readily understood. The term $(1-P_d)$ is the probability of the target *not* being detected in a given scan. This term to the nth power, $(1-P_d)^n$, is the probability of the target *not* being detected in n successive scans. One minus that probability is the probability of it *being* detected at least once in n scans.

If, for example, $P_d = 0.3$, the probability of the target not being detected in one scan would be $1 - 0.3 = 0.7$. The probability of it not being detected in ten scans would be $0.7^{10} = 0.03$. The probability of it being detected at least once in 10 scans, therefore, is $1 - 0.03 = 0.97$.

But determining the actual probability is not necessarily as straightforward as that. As the target closes, the value of P_d will increase. Also, a lot depends on how rapidly the target cross section (hence the signal) varies. If the rate is rapid enough for the variation to be essentially random from one scan to the next, over a period of several scans the variation will tend to cancel. If P_d for the range in question has a moderate value, as in the foregoing example, P_c will rapidly approach 100 percent.

SAMPLE RANGE COMPUTATION

Problem: Find the range at which the probability of a given pulse-doppler radar detecting a given target is 50%.

Characteristics Of The Radar:
Average Power (P_{avg})	**5** kW
Effective Area of Antenna (A_e). . . .	**4** ft²
Wavelength (λ)	**0.1** ft
Receiver Noise Figure (F_n)	**3** dB
Total losses (L)	**6** dB

Target: Fighter, viewed head-on, at constant look angle. Radar Cross Section, $\sigma = $ **10** ft².

Operating Conditions: Radar is searching a solid angle 100° wide in azimuth by 10° wide in elevation. The radar beam's time on target, $t_{ot} = $ **0.03** second.

Solution: Only two more values are needed to compute the range: the required signal-to-noise ratio, $(S/N)_{req}$, and the noise energy, KT_s.

Since the target is not very large and is viewed head-on from a constant angle, it's RCS will probably not fluctuate from pulse to pulse in the 0.03 second time on target. So, a rough estimate of $(S/N)_{req}$ can be obtained from Swerling's Case 1 curve in Figure 28. It indicates that for a probability of detection of **50%**, $(S/N)_{req}$ is about **10 dB**.

The value of kT_s, can be obtained by multiplying kT_0—which, as shown on page 118, equals 4×10^{-21}—by the receiver noise figure, 3dB (factor of 2). Thus, $kT_s = $ **8 x 10⁻²¹**.

Plugging the above values into the equation for R_{pd} (with the loss term, L, included in the denominator) yields:

$$R_{50} = \left[\frac{P_{avg} A_e^2 \sigma\, t_{ot}}{(4\pi)\, (S/N)_{req}\, kT_s\, \lambda^2\, L} \right]^{1/4}$$

$$R_{50} = \left[\frac{(5 \times 10^3) \times 4^2 \times 10 \times 0.03}{12.6 \times 10 \times (8 \times 10^{-21}) \times 0.1^2 \times 4} \right]^{1/4}$$

$$R_{P_d} = 8.78 \times 10^5 \text{ ft} = (8.78 \times 10^5)/(6076)$$
$$= 145 \text{ nmi}$$

THE MANY FORMS OF THE RADAR RANGE EQUATION

The many different forms of the equation for signal-to-noise ratio are easily confusing. To help you keep them straight in your mind, the constituent expressions for four of them are summarized here.

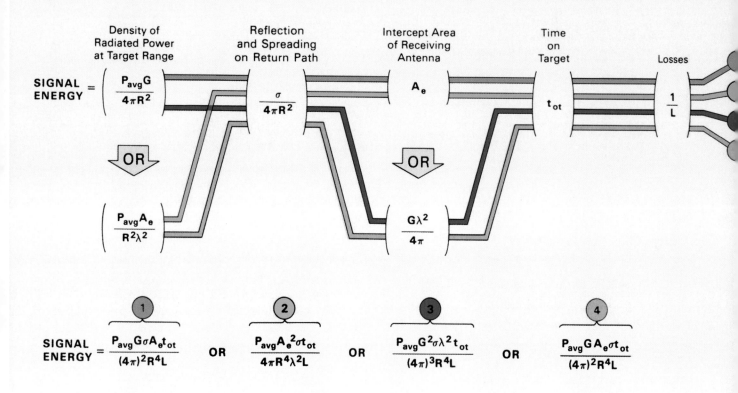

SIGNAL ENERGY $= \dfrac{P_{avg} G \sigma A_e t_{ot}}{(4\pi)^2 R^4 L}$ ① OR $\dfrac{P_{avg} A_e^2 \sigma t_{ot}}{4\pi R^4 \lambda^2 L}$ ② OR $\dfrac{P_{avg} G^2 \sigma \lambda^2 t_{ot}}{(4\pi)^3 R^4 L}$ ③ OR $\dfrac{P_{avg} G A_e \sigma t_{ot}}{(4\pi)^2 R^4 L}$ ④

NOISE ENERGY $=$

Matched Filter

(kT_s) OR $(kT_s B t_n)$ Ummatched Filter

OR

$(kT_o F_n)$ $(kT_o F_n B t_n)$

$A_c = \dfrac{G\lambda^2}{4\pi}$ = equivalent area of antenna

B = bandwidth

F_n = noise figure of receiver

G = antenna gain

k = Boltzmann's constant

L = losses

P_{avg} = average transmitter power

R = range

σ = radar cross section of target

t_n = duration of noise

T_o = ambient temperature (degrees Kelvin)

t_{ot} = time-on-target (dwell time)

T_s = system noise temperature (degrees Kelvin)

Area of sphere $= 4\pi R^2$

Antenna gain, $G = \dfrac{4\pi A_e}{\lambda^2}$

On the other hand, if there is little change in cross section from scan to scan and the target happens to be in a deep fade, the cumulative probability of detection for the same range may be quite low.

Summary

From the expressions for signal energy and noise energy a simple equation for detection range may be derived. It tells us that

- Range increases as the *¼th power* of average transmitted power, target radar cross section, and integration time

- Range increases as the *square root* of effective antenna area

- A reduction in noise is equivalent to a proportional increase in transmitted power

When adapted to the special case of volume search, the equation tells us that to the first order, range is independent of frequency and can be maximized by using the highest possible average power and the largest possible antenna.

Even when all secondary factors influencing signal-to-noise ratio have been accounted for, the range equation cannot tell us with certainty at what range a given target will be detected, since both noise and radar cross section fluctuate widely.

Consequently, detection range is usually specified in terms of probabilities. The most common probability for search is blip-scan ratio (also called single-scan or single-look probability). The range at which a given value of this probability may be achieved is determined by (1) establishing an acceptable false-alarm probability, (2) setting the target detection threshold just high enough to realize this probability, and (3) finding the signal-to-noise ratio for this setting that will provide the desired target detection probability, a process which may be simplified through the use of curves based on standard mathematical models of targets. The range at which that ratio will be achieved is then calculated with the range equation.

To account for the effect of closing rate in high-closing rate approaches, detection range may be expressed in terms of cumulative probability of detection—the probability that a given target will be detected at least once before it reaches a given range.

Some Relationships To Keep In Mind

- Range at which integrated signal-to-noise ratio is 1:

Spotlight
$$R_0 \propto \sqrt[4]{\frac{P_{avg} A_e^2 \sigma t_{ot}}{k T_s \lambda^2}}$$

Volume search
$$R_0 \propto \sqrt[4]{\frac{P_{avg} A_e \, \sigma \, t_f}{k T_s \theta_a \theta_e}}$$

- False-alarm time:

$$t_{fa} = \frac{1}{\text{Fale-alarm rate}} = \frac{t_{int}}{P_{fa} N}$$

N = number of threshold detectors

- False-alarm time for a single detector:

$$P_{fa} = \frac{t_{int}}{t_{fa} \times N_{RG} \times N_{df}}$$

N_{RG} = number of range gates

N_{df} = doppler filters per bank

- Range for which probability of detection is P_d :

$$R_{P_d} = \left[\frac{P_{avg} A_e^2 \, \sigma \, t_{ot}}{(4\pi) (S/N)_{req} k T_s \lambda^2 L} \right]^{1/4}$$

$(S/N)_{req}$ = required signal-to-noise ratio

- Cumulative probability of detection
$$P_c = 1 - (1 - P_d)^n$$

McDONNELL F2H–4 BANSHEE (1954)

This fighter equipped with a Hughes E–10 radar was the last of the Banshee
line. An earlier version, the F2H–3 with a Westinghouse APG–41 radar, was
used by the United States Navy during the Korean war.

Pulse Delay Ranging 12

By far the most widely used method of range measurement is pulse delay ranging. It is simple and can be extremely accurate. But since there is no direct way of telling for sure which transmitted pulse a received echo belongs to, the measurements are, to varying degrees, ambiguous.

In this chapter, we will look at pulse delay ranging more closely—learn how target ranges are actually measured and consider the nature of the ambiguities. We will see how ambiguities may be avoided at low PRFs, and resolved at higher PRFs. We will then consider ambiguities of a secondary type, called "ghosts," and see how these may be eliminated. Finally, we will look briefly at how range is measured during single-target tracking.

Basic Technique

When a radar's transmission is pulsed, the range of a target can be directly determined by measuring the time between the transmission of each pulse and reception of the echo from the target (Fig. 1). The round-trip time is divided by two to obtain the time the pulse took to reach the target. This time, multiplied by the speed of light, is the target's range. Expressed mathematically,

$$R = \frac{ct}{2}$$

where

R = range

c = speed of light

t = round-trip transit time

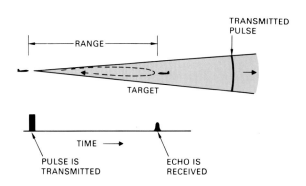

1. Range is determined by measuring the time between transmission of a pulse and reception of the target echo.

151

APPROXIMATE RANGING TIME

Unit of Distance	μs
1 nautical mile	12.4
1 statute mile	10.7
1 kilometer	6.67
1.5 kilometer	10.0

2. Rules of thumb for approximating round-trip ranging times.

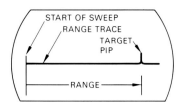

3. In simple analog radars, range is measured on the operator's display. Shown here is an A display of a World War II radar.

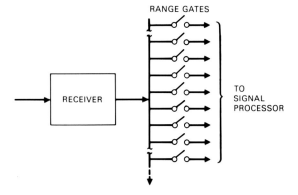

4. In sophisticated analog radars, range gates are sequentially opened (switch closed). Range is determined by noting which gate a target's echoes go through.

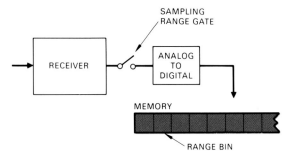

5. In digital radars, receiver output is periodically sampled by a range gate. Converted to a number, each sample is stored in a separate range bin.

A useful rule of thumb is *12.4 microseconds of round-trip transit time equals 1 nautical mile of range* (Fig. 2). If you wish to calculate ranges more accurately, the speed of light in various units of distance is given in Chap. 4.

Just how the range is actually measured varies with the type of radar.

Simple Analog Radars. In early radars, as well as many radars of today, range is measured right on the operator's display. This method is most graphically illustrated by the simple A display of World War II. For it, the electron beam of a cathode ray tube is repeatedly swept across the face of the tube (Fig. 3). It starts a new sweep each time the radar transmits a pulse, moves at a constant rate throughout the interpulse period, and "flies" back to the starting point again at the end of the period. Each sweep is called a range sweep; the line traced by the beam is called the range trace. When a target echo is received, it deflects the beam, causing a pip to appear on the range trace. The distance from the start of the trace to the pip corresponds to the time between transmission and reception, thus indicating the target's range.

Sophisticated Analog Radars. In these, range is measured in an analogous manner by applying the receiver output to a bank of switching circuits, called range gates (Fig. 4). The gates are opened sequentially at times corresponding to successive resolvable increments of range: first, Gate No. 1, then Gate No. 2, and so on. A target's range is determined by noting which gate, or adjacent pair of gates, its echoes pass through.

Enough range gates are provided to cover either the entire interpulse period or the portion of it corresponding to the range interval of interest.

Digital Radars. When digital signal processing is employed, range is essentially measured in the same way as in range-gated analog radars. The amplitude of the receiver's video output is periodically sampled by a range gate (Fig. 5). The samples are taken almost instantaneously. Each is held until the next sample is taken. During this interval, the amplitude of the sample is converted to a number. The numbers are temporarily stored in an electronic memory in positions called range bins. A separate bin is provided for each range increment within the interval of interest.

As noted in Chap. 2, to enable doppler filtering after the received signals have been converted to video, the receiver must provide both in-phase (I) and quadrature (Q) outputs (see page 28). Consequently, in digital doppler radars two numbers are stored for each range increment. Together, they correspond to the return passed by a single range gate in an analog system.

The choice of the sampling interval is generally a compromise. The larger the interval—i.e., the longer the time between samples—the less complex the system will be (Fig. 6). Yet, if the interval is greater than the duration (width) of the transmitted pulses, some of the signal will be lost when a target's echoes fall between sampling points. Moreover, the ability to resolve targets in range will be degraded.

To realize the full range-resolving potential of the pulses, as well as to enable more accurate range measurement, samples may be taken at considerably shorter intervals than the pulse width (Fig. 7). Range is then determined by interpolating between the numbers in adjacent range bins. If, for example, the numbers in two adjacent bins are equal, the target is assumed to be halfway between the ranges represented by the two bin positions. Depending on the sampling rate and the pulse width, the measurement can be quite precise.[1]

Using a comparatively high sampling rate also minimizes the loss in signal-to-noise ratio that occurs when a target's echoes fall partly in one sampling interval and partly in the next. This is called range-gate straddling loss.

Range Ambiguities

Pulse delay ranging works without a hitch as long as the round-trip transit time for the most distant target the radar may detect is shorter than the interpulse period. But if the radar detects a target whose transit time exceeds the interpulse period, the echo of one pulse will be received after the next pulse has been transmitted, and the target will appear, falsely, to be at a much shorter range than it actually is.

Nature of the Ambiguities. To get a more precise feel for the nature of the ambiguities, let us consider a specific example. Suppose the length of the interpulse period, T, corresponds to a range of 50 nautical miles, and echoes are received from a target at 60 miles (Fig. 8). The transit time for this target will be 20 percent greater than the interpulse period (60/50 = 1.2). Consequently, the echo of Pulse No. 1 will not be received until 0.2T microsecond after Pulse No. 2 is transmitted. The echo of Pulse No. 2 will not be received until 0.2T microsecond after Pulse No. 3 is transmitted, and so on.

If the difference between the time an echo is received and the time the immediately preceding pulse was transmitted is used as the measure of range, the target will appear to

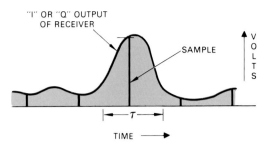

6. Video signal is generally sampled at intervals on the order of a pulse width, τ.

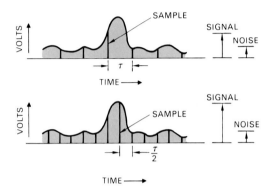

7. To enable more accurate measurement and minimize loss of signal-to-noise ratio, samples may be taken at intervals shorter than a pulse width; range is then computed by interpolating between samples.

1. If pulse compression is used, the intervals must be shorter than the compressed pulse width.

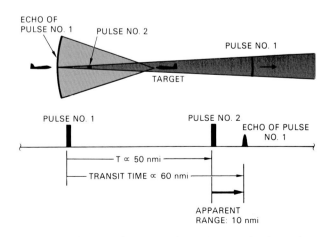

8. If interpulse period corresponds to 50 nautical miles and transit time to 60 nautical miles, range will appear to be only 10 nautical miles.

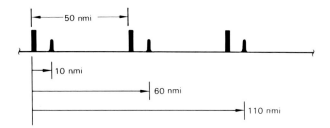

9. There is no direct way of telling whether true range is really 10 nautical miles, or 60 nautical miles, or 110 nautical miles, or . . . .

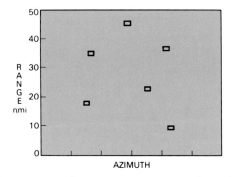

10. The true range of any target appearing on this radar display *may* be greater than 50 nautical miles. Ergo, all ranges are ambiguous.

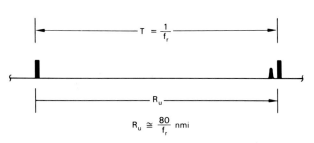

11. Longest range from which unambiguous return may be received, R_u, corresponds to interpulse period, T.

be at a range of only 10 miles (0.2 x 50). In fact, there will be no direct way of telling whether the target's true range is 10 miles, or 60 miles, or for that matter, 110 or 160 miles (Fig. 9). In short, the observed range will be ambiguous.

Not only that, but as long as there is a possibility of detecting targets at ranges greater than 50 nautical miles, the observed ranges of all targets detected by the radar will be ambiguous—even though their true ranges may be less than 50 miles. Put another way, if the range indicated by *any* target blip on the radar display can be greater than 50 miles, the range indicated by *every* target blip is ambiguous. There is no telling which of the blips represents a target at the greater range (Fig. 10). Therefore, range is almost always ambiguous. This point is often overlooked.

The extent of the range ambiguities in the return from a single target are commonly gauged by the number of interpulse periods spanned by the transit time. That is, by whether the target's echoes are received during the first, second, third, fourth, etc., interpulse period following transmission of the pulses that produced them. An echo received during the first interpulse period is called a single-time-around echo. Echoes received during subsequent periods are called multiple-time-around echoes, or MTAEs.

Maximum Unambiguous Range. For a given PRF, the longest range from which single-time-around echoes can be received—hence the longest range from which *any* return may be received without the observed ranges being ambiguous—is called the maximum unambiguous range (or simply unambiguous range). It is commonly represented by R_u. Since the round-trip transit time for this range equals the interpulse period,

$$R_u = \frac{cT}{2}$$

where

R_u = maximum unambiguous range

c = speed of light

T = interpulse period

Since the interpulse period is equal to one divided by the PRF (f_r), an alternative expression is

$$R_u = \frac{c}{2f_r}$$

A useful rule of thumb is R_u *in nautical miles equals 80 divided by the PRF in kilohertz* (Fig. 11). For a PRF of 10 kilohertz, for example, R_u would be 80 ÷ 10 = 8 nautical miles.

In metric units, R_u *equals 150 kilometers divided by the PRF in kilohertz.*

Strategy to Follow. What one does about range ambiguities depends both upon their severity and on the penalty that must be paid for mistaking a distant target for a target at closer range. The severity, in turn, depends upon the maximum range at which targets are apt to be detected and on the PRF. Often, the PRF is determined by considerations other than range measurement, such as providing adequate doppler resolution for clutter rejection. The penalty for not resolving an ambiguity, of course, depends upon the operational situation.

Obviously, the possibility of ambiguities could be eliminated altogether by making the PRF low enough to place R_u beyond the maximum range at which *any* target is apt to be detected (Fig. 12). However, since targets of large radar cross section may be detected at very great ranges, it may well be impractical to set the PRF this low, even when a comparatively low PRF is acceptable.

On the other hand, for the expected conditions of use, the probability of detecting such large targets may be slight, and the consequences of sometimes mistaking them for targets at closer range may be of no great importance.

Eliminating Ambiguous Return

If targets at greater ranges than R_u are of no concern to us, we can solve the problem of ambiguities simply by rejecting all return from beyond R_u (Fig. 13). This may sound like a neat trick, but it can be accomplished quite easily.

One technique is PRF jittering. It takes advantage of the dependence of the apparent ranges of targets beyond R_u on the PRF.

Since the echoes received from these targets are not due to the transmitted pulses that immediately precede them, any change in PRF—hence in R_u—will change the targets' apparent ranges (Fig. 14). On the other hand, since the echoes received from targets within R_u *are* due to the pulses that immediately precede them, changes in PRF will not affect these targets' apparent ranges.

Therefore, by transmitting at one and then the other of two different PRFs on alternate integration periods, any targets at ranges greater than R_u can be identified and rejected. The ranges of all targets appearing on the display, then, will be unambiguous.

Naturally, one pays a price for this improvement. As explained in Chap. 10, the time-on-target is generally limited. Since it must be divided between the two PRFs, the total potential integration time is cut in half (Fig. 15). This reduces the maximum detection range.

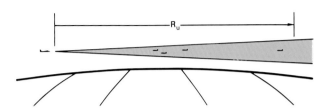

12. Ambiguities can be avoided completely only by making R_u, greater than the range at which *any* target may be detected.

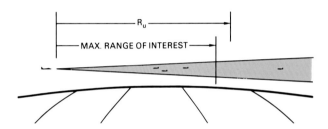

13. If R_u is greater than maximum range of interest, problem of ambiguities can be solved by eliminating all return from ranges greater than R_u.

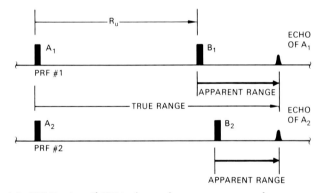

14. PRF jittering. If PRF is changed, apparent range of a target beyond R_u will change—identifying range as ambiguous.

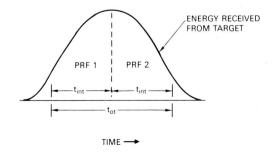

15. The penalty for PRF jittering: potential integration time is cut in half, reducing detection sensitivity.

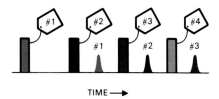

TIME ⟶

16. By tagging transmitted pulses, we can tell which pulse each echo belongs to. But except for frequency modulation, tagging has proved impractical.

2. Echoes being received in part or in whole when the radar is transmitting and the receiver is blanked.

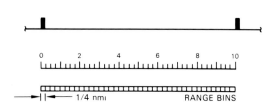

17. To span 10 nautical miles ranging interval, a bank of 40 range bins is provided. Each represents a range increment of 1/4 nmi.

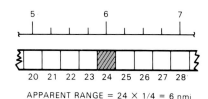

APPARENT RANGE = 24 × 1/4 = 6 nmi

18. A target appears in bin No. 24—apparent range, 6 nmi.

Resolving Ambiguities

For reasons having nothing to do with ranging, the PRF may have to be made so high that the maximum range of interest is longer than R_u—often many times so. The radar must then be able to resolve range ambiguities.

Tagging Pulses. Superficially, it might seem that the easiest way to resolve the ambiguities would be to "tag" successive transmitted pulses (Fig. 16). That is, change (modulate) their amplitude, width, or frequency in some cyclical pattern. By looking for corresponding changes in the target echoes, one could then tell which transmitted pulse each echo belongs to and thereby resolve the ambiguities.

But for one reason or another—problems of mechanization, in the case of amplitude modulation; eclipsing[2] and range gate straddling, in the case of pulse width modulation—only one of these approaches has as yet proved practical: frequency modulation (see Chap. 8). For air-to-air applications, even this approach has serious limitations.

PRF Switching. The resolution technique commonly used is a simple extension of PRF jittering, called PRF switching. It goes a step beyond jittering by taking account of how *much* a target's apparent range changes when the PRF is changed. Knowing this and the amount the PRF has changed, it is possible to determine the number of whole times, n, that R_u is contained in the target's true range.

Determining n. How this is done is best illustrated by a hypothetical example. We will assume that for other reasons than ranging, a PRF of 8 kilohertz has been selected. Consequently, the maximum unambiguous range, R_u, is 80 ÷ 8 = 10 nmi. However, the radar must detect targets out to ranges of at least 48 miles—nearly 5 x R_u—and undoubtedly it will detect some targets at ranges beyond that, as well.

The apparent ranges of all targets will, of course, lie between 0 and 10 nautical miles (Fig. 17). To span this 10-mile interval, a bank of 40 range bins has been provided. Each bin position represents a range interval of 1/4 mile (10 miles ÷ 40 bins = 1/4 mile per bin).

A target is detected in bin No. 24. The target's apparent range is 24 x 1/4 = 6 miles (Fig. 18). On the basis of this information alone, we know only that the target could be at any one of the following ranges:

6 nmi
10 + 6 = 16 nmi
10 + 10 + 6 = 26 nmi
10 + 10 + 10 + 6 = 36 nmi
10 + 10 + 10 +10 + 6 = 46 nmi
10 + 10 + 10 + 10 + 10 + 6 = 56 nmi

To determine which of these is the true range, we switch to a second PRF. To keep the explanation simple, we will assume that this PRF is just enough lower than the first to make R_u ¼ mile longer than it was before (Fig. 19).

What happens to the target's apparent range when the PRF is switched will depend upon what the target's true range is. If the true range is 6 miles, the switch will not affect the apparent range. The target will remain in bin No. 24.

But if the true range is greater than R_u, for every whole time R_u is contained in the target's true range, the apparent range will decrease by ¼ mile: the target will move one bin position to the left in Fig. 20. For the PRFs used here, n equals the number of bins the target shifts.

Computing the Range. We can find the true range, therefore, by (1) counting the number of bin positions the target moves, (2) multiplying this number by R_u, and (3) adding the result to the apparent range.

Suppose the target moves from bin No. 24 (apparent range, 6 miles) to bin No. 21, a jump of three bins (Fig. 21). The target's true range, then, is $(3 \times 10) + 6 = 36$ miles.

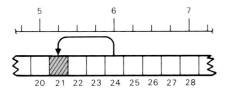

21. If target jumps 3 bins, true range is $(3 \times 10) + 6 = 36$ nmi.

General Relationships. From the foregoing, we can draw the following conclusions. The number of whole times, n, that R_u is contained in a target's true range equals the change in apparent range when the PRF is switched, divided by the change in R_u for the two PRFs.

$$n = \frac{\Delta R_{apparent}}{\Delta R_u}$$

The true range is n times R_u plus the apparent range.

$$R_{true} = nR_u + R_{apparent}$$

Eliminating Ghosts

When PRF switching is used, a secondary sort of ambiguity, called ghosting, is sometimes encountered. It may occur when two targets are detected simultaneously—i.e., at the same azimuth and elevation angles—and their range rates are so nearly equal that their echoes cannot be separated on the basis of doppler frequency (Fig. 22). Under this condition, when the PRF is switched and one or both targets move to different range bins, we may not be able to tell

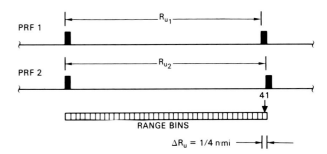

19. PRF is changed to increase R_u by ¼ nmi.[3]

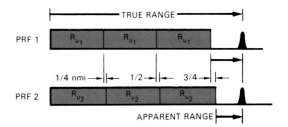

20. For every whole time R_u is contained in true range, apparent range will decrease ¼ nmi when the PRF is switched.

3. A practical system would not be mechanized with PRFs so closely spaced. The principle, though, is the same.

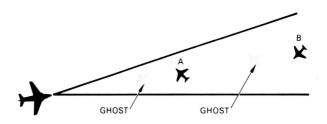

22. If more than one target is detected at the same angle and the targets are not resolvable in doppler frequency, a problem of ghosts will occur.

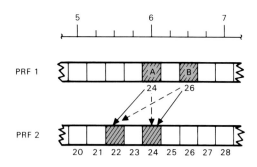

23. When PRF was switched, did A move to bin No. 22 and B to bin No. 24, or, did A stay put?

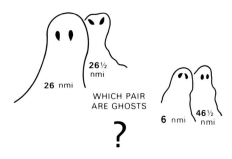

24. Each target shown in Fig. 23 has two possible true ranges.

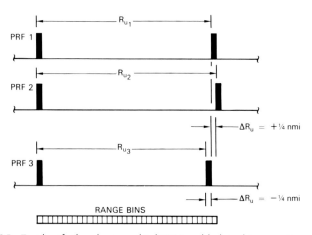

25. To identify the ghosts, a third PRF is added. In this case, it decreases R_u by 1/4 nautical mile.

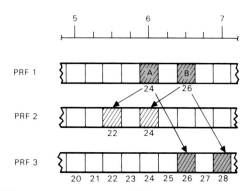

26. When radar is switched to PRF 3, targets jump to bins 26 and 28. The value of n for both targets must be 2.

which target has moved to which bin. Each target will appear to have two possible ranges. One is the true range; the other, in radar jargon, is a ghost.

Example of Ghosts. Figure 23 shows two targets, A and B, in the same bank of range bins as used in the preceding example. When the radar is transmitting at the first PRF, the targets are two bins apart: A is in bin No. 24 (apparent range, 6 miles); B is in bin No. 26 (apparent range, 6 1/2 miles). When we switch to the second PRF, the targets appear in bins No. 22 and No. 24. But we have no direct way of telling whether A and B have both moved to the left two bins, or, whether A has merely stayed put and B has moved four bins to the left and is in bin No. 22.

Each target thus has two possible true ranges (Fig. 24). If both A and B have moved two bin positions, the true ranges are

$$\text{Target A: } (2 \times 10) + 6 = 26 \text{ nmi}$$
$$\text{Target B: } (2 \times 10) + 6\tfrac{1}{2} = 26\tfrac{1}{2} \text{ nmi}$$

On the other hand, if A stayed put and B moved four bin positions, the true ranges are

$$\text{Target A: } (0 \times 10) + 6 = 6 \text{ nmi}$$
$$\text{Target B: } (4 \times 10) + 6\tfrac{1}{2} = 46\tfrac{1}{2} \text{ nmi}$$

One of the two pairs of ranges are ghosts.

Identifying Ghosts. The ghosts may be identified by switching to a third PRF (Fig. 25). To simplify the explanation, we'll assume that PRF No. 3 is just enough higher than PRF No. 1 to decrease R_u by 1/4 mile—i.e., shorten it by one range bin (from 40 to 39 bins). Accordingly, when PRF No. 3 is used, for every whole time R_u is contained in either target's true range, the target will appear one position to the *right* of the bin it occupied when PRF No. 1 was used. This is the same number of positions it appeared to the *left* of that bin when PRF No. 2 was used.

Let's say for example, that we switch to PRF No. 3 and the targets appear in bins 26 and 28. Which of the two pairs of ranges are ghosts?

As you can see from the figure (Fig. 26), bin 26 is two positions to the right of the bin A originally occupied. Likewise, bin 28 is two positions to the right of the bin B originally occupied. Since, when we switched earlier to PRF No. 2, one target appeared two positions to the left of the bin A originally occupied and the other target appeared two positions to the left of the bin B originally occupied, we conclude that n = 2 for both targets. Their true ranges are 26 miles and 26 1/2 miles. The other pair of ranges are ghosts.

It may be instructive to consider where the targets would have appeared when we switched to PRF No. 3 had the first pair of ranges been ghosts and the second pair—6 miles and 46½ miles—been the true ranges. In that case (Fig. 27), since n = 0 for 6 miles, target A would have stayed put. Since n = 4 for 40 miles, target B would have moved 4 positions to the *right*—the same distance (for these particular PRFs) that it must have moved to the *left* when earlier we switched to PRF No. 2.

How Many PRFs?

From what has been said so far, it might appear that no more than three PRFs would ever be required: one for measuring ranges, another for resolving range ambiguities, a third for deghosting simultaneously detected targets. This is not so, however.

Number of PRFs for Resolving Ambiguities. Depending on how great the detection ranges are and how high and widely spaced the PRFs are, more than one PRF (besides the first) may be required to resolve ambiguities. Figure 28 illustrates why.

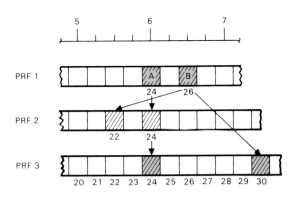

27. If A's true range had been 6 miles, it would have stayed put when the radar was switched to PRF 3, and B would have jumped four positions to the right.

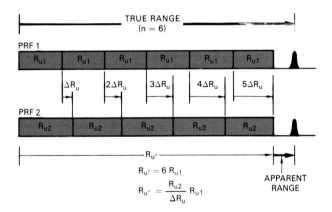

28. Range for which ambiguities can no longer be resolved by switching between two PRFs. Since $5R_{u2} = 6R_{u1}$, apparent range does not change when PRF is switched. R_u', is maximum unambiguous range for this combination of PRFs.

The true range in that example includes six whole multiples of the unambiguous range for PRF No. 1 (n = 6). This is clear. But the difference in the unambiguous ranges for the two PRFs (ΔR_u) is such that five times the unambiguous range for PRF No. 2 exactly equals six times the unambiguous range for PRF No. 1. Consequently, for the target range assumed here (Fig. 29), when the PRF is switched the apparent range remains the same, just as though n = 0.

If the true range were long enough to make n = 7 or more, the apparent range would again change when the PRF was switched, but the change then would only indicate

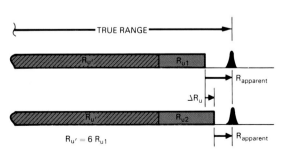

29. If true range is increased beyond R_u', apparent range will change when PRF is switched, but (in this case) only by amount corresponding to (n − 6).

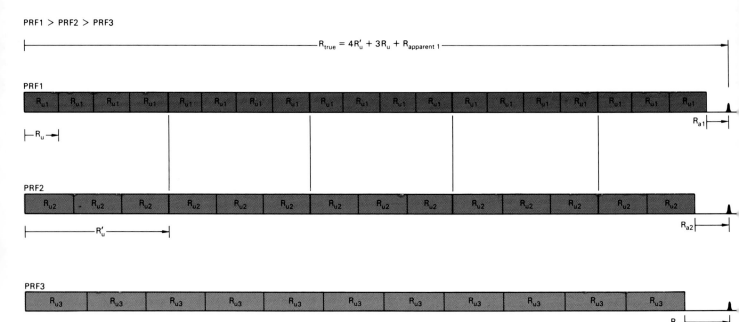

PRF1 > PRF2 > PRF3

30. Just as adding a second PRF increases the unambiguous range from R_u to R_u', adding a third PRF increases it to R_u''. For any one combination of R_{u1}, R_{u2}, R_{u3}, and $R_{apparent}$, there is only one possible value of the true range. It is uniquely indicated by the values of the three apparent ranges, R_{a1}, R_{a2}, and R_{a3}.

how much n exceeds 6. This particular combination of PRFs extends the maximum unambiguous range to six times the unambiguous range for PRF No. 1, but no farther (Fig. 30).

In fact, a more general expression for the true range than that given earlier might be

$$\text{True range} = n'R_u' + nR_u + R_{apparent}$$

where R_u' is the unambiguous range for the combination of the two PRFs and n' is the number of whole times R_u' is contained in the true range. To find the value of n' we must switch to a third PRF.

With the aid of a diagram like Fig. 30, it can be shown that for every additional PRF the unambiguous range for the combination increases by the ratio of (a) R_u for the added PRF to (b) the difference between that value of R_u and the value for the preceding PRF (Fig. 31). Thus, if the unambiguous ranges for three PRFs taken individually are 3, 4, and 5 miles, the unambiguous range for the combination is $3/1 \times 4/1 \times 5/1 = 60$ miles. How many PRFs are required for resolving range ambiguities, then, depends upon the desired maximum unambiguous range and the values of R_u for the individual PRFs.

$$\text{Unambiguous Range (Multiple PRFs)} = R_{u_1} \underbrace{\left(\frac{R_{u_2}}{\Delta R_{u_1}}\right)}_{R_{u_2} - R_{u_1}} \underbrace{\left(\frac{R_{u_3}}{\Delta R_{u_2}}\right)}_{R_{u_3} - R_{u_2}} \underbrace{\left(\frac{R_{u_4}}{\Delta R_{u_3}}\right)}_{R_{u_4} - R_{u_3}} \cdots$$

1st PRF 2nd PRF 3rd PRF 4th PRF

31. For each additional PRF, the unambiguous range for the combination is increased by the ratio of the unambiguous range, R_u, for the added PRF to the difference between R_u for that PRF and R_u for the preceding PRF.

Number of PRFs for Deghosting. More PRFs may also be required for deghosting. To deghost *all* possible combinations of the observed ranges of more than two simultaneously detected targets, an additional PRF must be provided for each additional target. Thus, if a single PRF suffices to resolve range ambiguities, a radar employing N PRFs can uniquely measure the ranges of (N − 1) simultaneously detected targets.

The Trade-off. As with PRF jittering, one pays a price for PRF switching. Each additional PRF not only reduces the integration time—hence reduces detection range—but increases the complexity of mechanization. The number of PRFs actually used, therefore, is a compromise between these costs and the cost of occasionally having to contend with ambiguous ranges and unresolved ghosts (Fig. 32).

The optimum number of PRFs naturally varies with the application. For most of the fighter applications in which the required PRFs are low enough to make PRF switching practical, one additional PRF generally suffices for resolving ambiguities and another for deghosting—making a total of three.

32. The number of PRFs actually used is always a compromise.

Single-Target Tracking

During single-target tracking, range measurement is simplified in two respects.

First, only two adjacent range gates must be provided (Fig. 33). The time delay between the transmission of a pulse and the opening of these gates is automatically adjusted to equalize the output of the two gates, thereby centering them on the target. By measuring this delay, the target's apparent range may be precisely determined.

Second, once the ambiguities in the target's range have been resolved, no further resolution is necessary. Accurate track can be kept of the true range simply by keeping continuous track of the changes in apparent range.

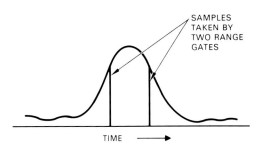

33. For single-target tracking, only two range gates are needed. By positioning them to equalize their outputs, they are centered on a target.

Summary

With pulse delay ranging, range is determined by measuring the time between transmission of a pulse and reception of an echo. In rudimentary radars, the measurement is made on the range trace of the display. In sophisticated analog radars it is made by opening a succession of range gates. Digital radars accomplish the equivalent by periodically sampling the receiver output, converting the samples to numbers, and storing them in a bank of range bins.

The range for which the round-trip transit time equals the interpulse period is called the maximum unambiguous range, R_u. A target at greater range will appear to have a

Some Relationships To Keep In Mind

- Ranging time:

  $$12.4 \ \mu s \ = \ 1 \ \text{nmi of range}$$

- Maximum unambiguous range:

  $$R_u \ (\text{nmi}) \ = \ \frac{80}{\text{PRF (kHz)}}$$

- When PRF switching is used to resolve range ambiguities:

  $$R_{true} \ = \ nR_u \ + \ R_{apparent}$$

  $$n \ = \ \frac{\Delta R_{apparent}}{\Delta R_u}$$

range equal to its true range minus some multiple of R_u. As long as there is a possibility of detecting any targets at ranges greater than R_u, all observed ranges are ambiguous.

What one does about range ambiguities depends upon their severity and the penalty for ambiguous measurements.

If the PRF can be set low enough to make R_u greater than the maximum range of interest, ambiguities can be avoided by discarding the return from those targets beyond R_u. These can be identified by jittering the PRF and looking for a corresponding jitter in the apparent target ranges.

If higher PRFs are required, ambiguities must be resolved. This can be done by switching between two or more PRFs and measuring the changes, if any, in the apparent ranges.

If two or more targets are detected simultaneously, each target may appear to have two possible ranges, one of which is a ghost. Ghosts can be eliminated by switching to additional PRFs.

Besides increasing complexity, using more than one PRF decreases detection range. The optimum number of PRFs is a compromise between these costs and the cost of occasionally having to contend with unresolved ambiguities and ghosts.

Pulse Compression 13

Ideally, if we wanted both long detection range and fine range resolution, we would transmit extremely narrow pulses of exceptionally high peak power. But there are practical limits on the level of peak power one can use. To obtain long detection ranges at PRFs low enough for pulse delay ranging, fairly wide pulses must be transmitted.

One solution to this dilemma is pulse compression. That is, transmit internally modulated pulses of sufficient width to provide the necessary average power at a reasonable level of peak power; then, "compress" the received echoes by decoding their modulation.

This chapter explains the two most common methods of coding—linear frequency modulation and binary phase modulation. It also briefly describes a third method, polyphase modulation.[1]

Linear Frequency Modulation (Chirp)

Because of its parallel to the chirping of a bird, this method of coding was called "chirp" by its inventors. Since it was the first pulse compression technique, some people still use the terms chirp and pulse compression synonymously.

Basic Concept. With chirp, the radio frequency of each transmitted pulse is increased at a constant rate throughout its length (Fig. 1). Every echo, naturally, has the same linear increase in frequency.

The received echoes are passed through a filter. It introduces a time lag that *de*creases linearly with frequency at exactly the same rate as the frequency of the echoes *in*creases. Being of progressively higher frequency, the trailing portions of an echo take less time to pass through than

1. All methods of pulse compression are essentially matched filtering schemes in which the transmitted pulses are coded and the received pulses are passed through a filter whose time-frequency characteristic is the conjugate (opposite) of the coding.

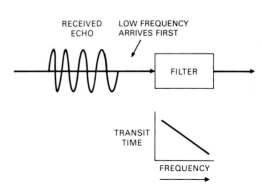

1. With chirp, transmitter frequency is increased linearly throughout pulse. Echo is passed through filter that introduces time lag inversely proportional to frequency.

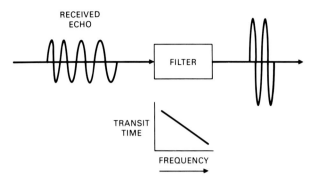

2. Since trailing portions of echo take less time to pass through filter, successive portions tend to bunch up: Amplitude of pulse is increased and width is decreased.

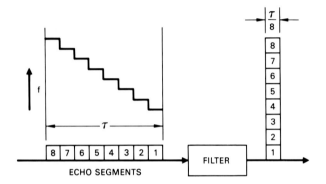

3. Linear frequency modulated pulse can be thought of as being made up of segments of progressively higher frequency. In going through filter, second segment catches up with first; third, with second; etc.

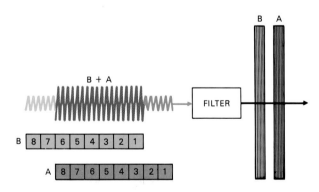

4. Echoes from closely spaced targets, A and B, are merged but, because of coding, separate in output of filter.

the leading portion. Successive portions thus tend to bunch up. Consequently, when the pulse emerges from the filter its amplitude is much greater and its width much less than when it entered (Fig. 2). The pulse has been compressed.

Filtering may be done with an analog device—such as an acoustical delay line—or digitally. Depending on the mechanization, the frequency can either be increasing, as described here, or decreasing, in which case the delay increases with frequency.

Incremental-Frequency Explanation. What actually happens when an echo passes through the filter can be visualized most easily if we think of the echo as consisting of a number of segments of equal length and progressively higher frequency (Fig. 3). In fact, in one form of pulse coding—incremental frequency modulation—the transmitted wave is modulated in exactly this way. The first segment, having the lowest frequency, takes longest to get through the filter. The second segment takes less time than the first; the third less time than the second, etc.

The increments of frequency are such that the difference in transit time for successive segments just equals their width. If the segments are 0.1 microsecond wide, the first segment takes 0.1 microsecond longer to go through than the second; it, in turn, takes 0.1 microsecond longer to go through than the third, etc. As a result, in passing through the filter, the second segment catches up with the first; the third segment catches up with the second, the fourth catches up with the third, and so on. All segments thus combine and emerge from the filter at one time. The output pulse is only a fraction of the width of the received echo; yet, has many times its peak power.

How Range Resolution Is Improved. Figure 4 shows what happens when the echoes from two closely spaced targets pass through the filter. Since the range separation is small compared to the pulse length, the incoming echoes are merged indistinguishably. In the filter output, however, they appear separately—staggered by the target's range separation.

It seems like magic, until you consider the coding. Because of it, each segment of the echo from the near target emerges from the filter at the same time as the first segment of this echo. And each segment of the echo from the far target emerges at the same time as the first segment of that echo. The difference between these times, of course, is the length of time the leading edge of the transmitted pulse took to travel from the first target to the second and back.

The range resolution is thus improved by the ratio of the width of the individual segments to the total width of the pulse.

Range Sidelobes. If we look at an output pulse closely (Fig. 5), we will see that it is preceded and followed by a series of lesser pulses. These are called *range sidelobes.* They are half the width of the compressed pulse, have the same shape as antenna sidelobes, and are equally undesirable.

As with the compression process, the source of the range sidelobes can be visualized most easily in terms of incremental frequency modulation.

In Chap. 16, it will be demonstrated that the energy of a single short pulse, such as an individual segment of an incremental-frequency-modulated pulse, is not concentrated at a single frequency—as might be expected from the uniform spacing of the wavefronts. Rather, it is spread over a broad band, extending equally above and below that frequency. The envelope of this spectrum (plot of amplitude versus frequency) has a sin x/x shape.

As each segment of the uncompressed pulse passes through the compression filter, the energy of the higher frequency spectral sidelobes travels faster than the energy of the main spectral lobe, and the energy of the lower frequency spectral sidelobes travels slower. Consequently, when the segments emerge in unison from the filter, their combined amplitude—plotted against time—has the same sin x/x shape as the spectra of the individual segments (Fig. 6).

Like antenna sidelobes, range sidelobes can be reduced to acceptable amplitudes. The reduction is accomplished by designing the filter to do the equivalent of illumination tapering in an antenna—i.e., taper the amplitude of the uncompressed pulse at its leading and trailing ends. The compressed pulse widens a bit, just as the antenna beam does; but, again, this is a small price to pay for the reduction achieved in the sidelobes.

Stretch-Radar Decoding. For a narrow range swath, linear frequency modulation may be conveniently decoded by a technique called stretch radar.

With this technique (described in detail in the panel on the next page), pulse delay time (range) is converted to frequency. As a result, the return from any one range has a constant frequency, and the returns from different ranges may be separated with a bank of narrowband filters, implemented with the highly efficient fast Fourier transform (see Chap. 20).

Incidentally, stretch radar is similar to the FM ranging technique used by CW radars. The principal differences are that instead of transmitting pulses the CW radar transmits continuously, and the period over which the transmitter's frequency changes in any one direction is many times the

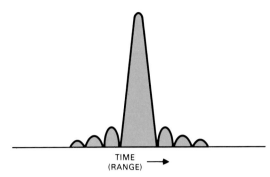

5. Plot of filter output versus time has same shape as antenna radiation pattern. Sidelobes can be reduced through weighting.

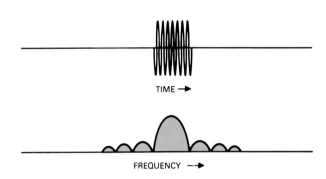

6. Energy of a single short pulse, such as a segment of an incremental-frequency-modulated pulse, is spread over a band of frequencies. Envelope has sin x/x shape.

STRETCH RADAR DECODING OF CHIRP

For a narrow range swath, such as is mapped by a synthetic array radar, linear frequency modulation is commonly decoded by a technique called stretch radar.

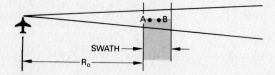

As the return from the swath is received, its frequency is subtracted from a reference frequency that increases at the same rate as the transmitter frequency. The reference frequency, however, increases continuously throughout the entire period in which the return is received.

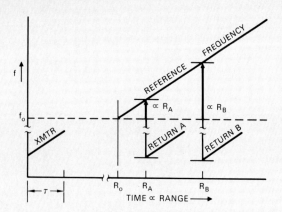

Consequently, the difference between the reference frequency and the frequency of the return from any one point on the ground is constant. Moreover, as can be seen from the above figure, if we subtract the reference frequency's initial offset, f_o, from the difference thus obtained, the result is proportional to the range of the point from the near edge of the swath. Range is thus converted to frequency.

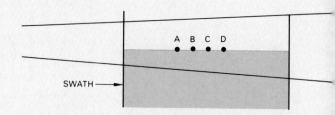

To see how fine resolution is thereby achieved, consider the returns from four closely spaced points after the subtraction has been performed.

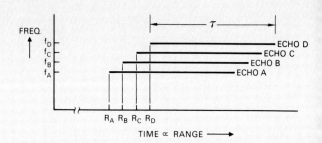

Although the returns were received almost simultaneously, the slight stagger in their arrival times has resulted in clearly discernible differences in frequency.

As indicated in the figure below, the continuously changing reference frequency may be subtracted at one of three points in the receiving system. One is the mixer, which converts the radar returns to the receiver's intermediate frequency (IF). A second point is the synchronous detector, which converts the output of the IF amplifier to video frequencies. A third point is in the signal processor, after the video has been digitized.

To sort the difference frequencies, the video output of the synchronous detector is applied to a bank of narrowband filters implemented with the highly efficient fast Fourier transform.

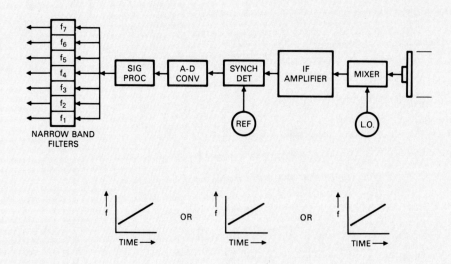

round-trip ranging time. Range is determined by measuring the instantaneous difference between the frequencies of the transmitted and received signals.

Pulse Compression Ratio. The extent to which the received pulses are compressed—i.e., the ratio of the uncompressed width, τ, to the compressed width, τ_{comp}—is called the pulse compression ratio. With incremental frequency modulation this is the ratio of τ to the width of the modulation increments (Fig. 7). But what is it when the modulation is strictly linear? And what determines how much the transmitter frequency must be increased over the length of the transmitted pulse?

To answer these questions, it is necessary to consider an important characteristic of the pulse compression filter which until now we have ignored: frequency sensitivity. If returns received simultaneously from two slightly different ranges are to be separated on the basis of the difference in their frequencies, besides providing a delay proportional to frequency, a second requirement must also be satisfied. The frequency difference must be large enough for the signals to be resolved by the filter.

As will be made clear in Chap. 18, the frequency resolution of a filter increases with the duration of the signals passing through it—in this case, the width of the uncompressed pulses (Fig. 8).

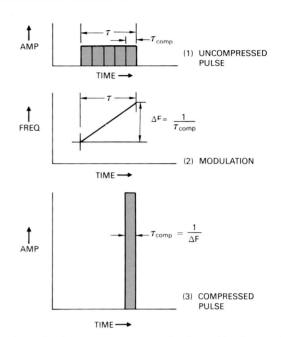

7. Relationship between uncompressed pulse width, chirp modulation, and compressed pulse width.

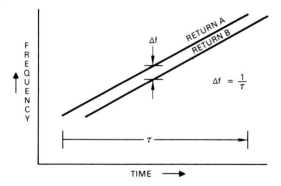

8. For a filter to resolve two simultaneously received returns, the instantaneous difference in their frequencies (Δf) must at least equal 1 divided by their duration (τ).

For the pulses to be resolved, the frequency difference, Δf, must at least equal one divided by the uncompressed pulsewidth, τ.

$$\Delta f = \frac{1}{\tau}$$

If Δf is the minimum resolvable frequency difference, the compressed pulsewidth, τ_{comp}, is the period of time in which the frequency of the uncompressed pulse changes by Δf (Fig. 9). If the frequency of the uncompressed pulse changes at a rate of $\Delta f / \tau_{comp}$ hertz per second, then the

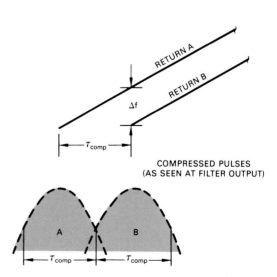

9. If the minimum resolvable frequency difference is Δf, the time in which the frequency of the uncompressed pulse changes by Δf will be the width of the compressed pulse τ_{comp}.

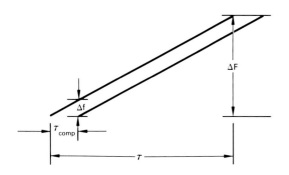

10. Ratio of uncompressed to compressed pulse widths equals ratio of total change in frequency of compressed pulse (ΔF) to minimum resolvable frequency difference (Δf).

total change in frequency, ΔF, over the duration of the uncompressed pulse will be this rate times the uncompressed pulsewidth, τ.

As is apparent from the geometry of Fig. 10, the pulse compression ratio, τ / τ_{comp}, equals the ratio of ΔF to Δf.

$$\text{Pulse compression ratio} = \frac{\tau}{\tau_{comp}} = \frac{\Delta F}{\Delta f}$$

Substituting $1/\tau$ for Δf, we find that the pulse compression ratio equals the uncompressed pulsewidth times ΔF.

$$\text{Pulse compression ratio} = \tau \Delta F$$

The quantity $\tau \Delta F$ is called the *time-bandwidth product*.

This simple relationship—pulse compression ratio equals time-bandwidth product—tells us a lot. To begin with, for a given uncompressed pulse width, τ, the compression ratio increases directly as ΔF. Conversely, for a given value of ΔF, the ratio increases directly as the uncompressed width, τ.

If we set the time-bandwidth product equal to τ / τ_{comp}, τ cancels out,

$$\tau \Delta F = \frac{\tau}{\tau_{comp}}$$

and we find that

$$\tau_{comp} = \frac{1}{\Delta F}$$

The width of the compressed pulse is determined entirely by the change in transmitter frequency over the duration of the transmitted pulse—the greater the frequency change, the narrower the compressed pulse.

Finally, when solved for ΔF, this last equation tells us that the total change in transmitter frequency must be

$$\Delta F = \frac{1}{\tau_{comp}}$$

To get a feel for the relative values involved, let us consider a representative example of chirp. Assume that to provide adequate average power, the width of a radar's transmitted pulses must be 10 microseconds. To provide the desired range resolution (5 feet) a compressed pulsewidth of 0.01 microsecond is required. The pulse compression ratio, therefore, must be

$$\frac{\tau}{\tau_{comp}} = \frac{10}{0.01} = 1000$$

To achieve a compressed pulsewidth of 0.01 microsecond (10^{-8} s), the change in transmitter frequency (ΔF) over the duration of each transmitted pulse must be $1/10^{-8} = 10^8$ hertz.

Since the duration of the uncompressed pulse is 10 microseconds (10^{-5}s), the rate of change of the transmitter frequency will be $10^8/10^{-5} = 10^{13}$, or 10,000 gigahertz (GHz)!

This, incidentally, explains why stretch decoding is practical only for relatively narrow range intervals. The ranging time for an interval of 50 nautical miles, for instance, is 12.4 x 50 = 620 μs. If the receiver local-oscillator frequency were shifted at a rate of 10,000 gigahertz per second throughout that time (Fig. 11), the total shift would be 10,000 x 620 x 10^{-6} = 6.2 GHz! Even at Ku-band frequencies (15 gigahertz), such a large shift is far from practical.

Relative Merits of Chirp. Linear frequency modulation has the advantage of enabling very large compression ratios to be achieved. In addition, it is comparatively simple. No matter when a pulse is received or what its exact frequency is, it will pass through the filter equally well and with the same amount of compression.

The principal disadvantage is a slight ambiguity between range and doppler frequency. If the frequency of a pulse has been, say, increased by a positive doppler shift, the pulse will emerge from the chirp filter a little sooner than if there were no such shift. The radar will have no way of telling whether this difference is due to a doppler shift or to the echo being reflected from a slightly greater range. However, since the doppler shifts typically encountered are very much less than the increment, ΔF, over which the frequency of the individual transmitted pulses is swept, ambiguity is generally not a problem.

Binary Phase Modulation

As the name implies, in this type of coding, the radio frequency phase of the transmitted pulses is modulated, and the modulation is done—as in incremental frequency modulation—in finite increments. Here, though, only two increments are used: 0° and 180°.

Basic Concept. Each transmitted pulse is, in effect, marked off into narrow segments of equal length. The radio frequency phase of certain segments is shifted by 180°, according to a predetermined binary code. This is illustrated for a 3-segment code in Fig. 12. (So you can readily discern the phases, the wavelength has been arbitrarily

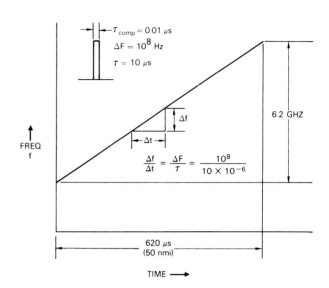

11. If stretch processing were used over a 50 mile range interval to decode a 10 microsecond pulse modulated for 1000:1 compression ratio, receiver local oscillator would have to be swept over an impractical 6.2 gigahertz.

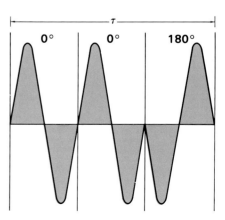

12. Binary phase coding of a transmitted pulse. Pulse is marked off into segments; phases of certain segments (here, No. 3) are reversed.

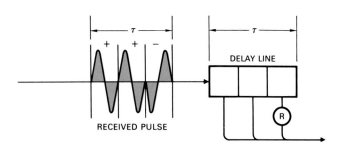

13. Received pulses are passed through tapped delay line. Separate tap is provided for each segment of pulse.

increased to the point where each segment contains only one cycle.)

A common shorthand method of indicating the coding on paper is to represent the segments with + and − signs. An unshifted segment is represented by a +; a shifted segment, by a − sign. The signs making up the code are referred to as digits.

The received echoes are passed through a delay line (Fig. 13) which provides a time delay exactly equal to the duration of the uncompressed pulses, τ. Thus, as the trailing edge of an echo enters the line, the leading edge emerges from the other end. The delay line may be implemented either with an analog device or digitally.

Like the transmitted pulses, the delay line is divided into segments. An output tap is provided for each segment. The taps are all tied to a single output terminal. At any one instant, the signal at this terminal corresponds to the sum of whatever segments of a received pulse currently occupy the individual segments of the line.

Now, in certain of the taps, 180° phase reversals are inserted. Their positions correspond to the positions of the phase-shifted segments in the transmitted pulse. Thus, when a received echo has progressed to the point where it completely fills the line, the outputs from all of the taps will be in phase (Fig. 14).

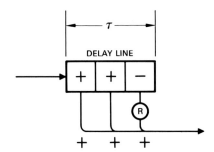

14. Phase reversal, R, is so placed that when a pulse completely fills the delay line, outputs from all taps will be in phase.

Their sum will then equal the amplitude of the pulse times the number of segments it contains.

To see step by step how the pulse is compressed, consider a simple three-segment delay line and the three-digit code, illustrated earlier.

Suppose an echo from a single point target is received. Initially, the output from the delay line is zero. When Segment No. 1 of the echo has entered the line, the signal at the output terminal corresponds to the amplitude of this

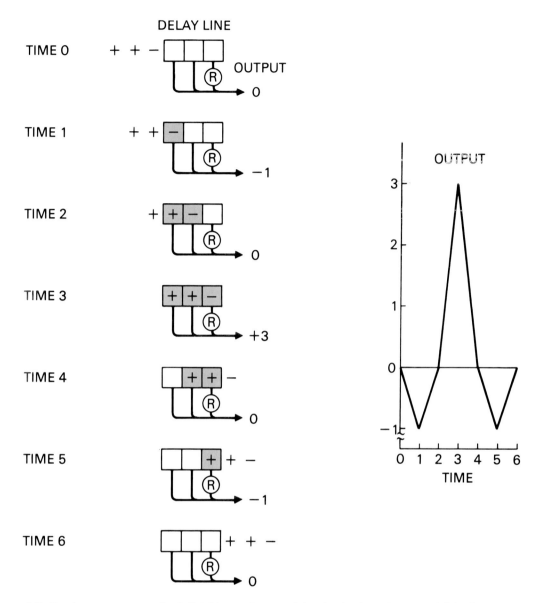

15. Step-by-step progress of a 3-digit binary phase modulated pulse through a tapped delay line.

segment (Fig. 15). Since its phase is 180°, the output is negative: −1.

An instant later, Segment No. 2 has entered the line. Now the output signal equals the sum of Segments No. 1 and No. 2. Since the segments are 180° out of phase, however, they cancel: The output is 0.

When Segment No. 3 has entered the line, the output signal is the sum of all three segments. Segment No. 1, you will notice, has reached a point in the line where the tap contains a phase reversal. The output from this tap, therefore, is in phase with unshifted Segment No. 2. The phase of Segment No. 3 also being unshifted, the combined output of the three taps is three times the amplitude of the individual segments: +3.

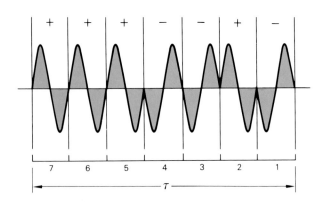

16. A seven-digit binary phase code.

As Segments No. 2 and No. 3 pass through the line, this same process continues. The output drops to zero, then increases to minus one, and finally returns to zero again.

A somewhat more practical example is shown in Fig. 16. This code has seven digits. Assuming no losses, the peak amplitude of the compressed pulse is seven times that of the uncompressed pulse, and the compressed pulse is only one-seventh as wide.

To see why the code produces the output it does, transfer the code to a sheet of paper and slide it across the delay line plotted in Fig. 17 (below), digit by digit, noting the sum of the outputs for each position. (A minus sign, −, over a tap with a reversal ® in it, becomes a +; and a + becomes a −.) You should get the output shown in the figure.

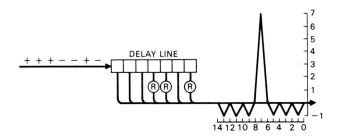

17. Output produced when seven-digit phase code is passed through tapped delay line with phase reversals in appropriate taps.

N	BARKER CODES
2	+ − OR (+ +)
3	+ + −
4	+ − + + OR (+ − − −)
5	+ + + − +
7	+ + + − − + −
11	+ + + − − − + − − + −
13	+ + + + + − − + + − + − +

Note. Plus and minus signs may be interchanged (+ + − changed to − − +); order of digits may be reversed (+ + − changed to − + +). Codes in parentheses are complementary codes.

18. Barker codes come very close to the goal of producing no sidelobes. But the largest code contains only 13 digits.

Sidelobes. Ideally, for all positions of the echo in the line—except the central one—the outputs from the same number of taps would have phases of 0° and 180°. The outputs would then cancel, and there would be no range sidelobes.

One set of codes, called the Barker codes, comes very close to meeting this goal (Fig. 18). Two of these have been used in the examples. As you have seen, they produce sidelobes whose amplitudes are no greater than the amplitude of the individual segments. Consequently, the ratio of mainlobe amplitude to sidelobe amplitude, as well as the pulse compression ratio, increases with the number of segments into which the pulses are divided—i.e., the number of digits in the binary code.

Unfortunately, the longest Barker code contains only 13 digits. Other binary codes can be made practically any length, but their sidelobe characteristics, though reasonably good, are not quite so desirable.

Complementary Barker Codes. It turns out that the four-digit Barker code has a special feature which enables us not only to eliminate the sidelobes altogether but to build codes of great length.

172

This code, and also the two-digit code, have complementary forms. Corresponding sidelobes produced by the two forms have opposite phases. Therefore, it we alternately modulate successive transmitted pulses with the two forms of the code—and appropriately switch the locations of the phase reversals in the outputs of the delay line, for alternate interpulse periods—when the returns from successive pulses are integrated the sidelobes cancel (Fig. 19).

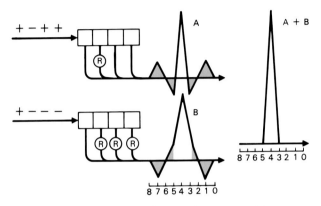

19. Echoes with complementary phase coding, received from same target during alternate interpulse periods. When echoes are integrated, time sidelobes cancel.

More importantly, by chaining the complementary forms together according to a certain pattern, we can build codes of almost any length. As illustrated in Fig. 20, the two forms of the four-digit code are just such combinations of the two forms of the two-digit code; and these are just such combinations of the two fundamental binary digits, + and −.

Unlike the unchained Barker codes, the chained codes produce sidelobes having amplitudes greater than one. But since the chains are complementary, these larger sidelobes—like the others—cancel when successive pulses are integrated.

Limitations of Phase Coding. The principal limitation of phase coding is its sensitivity to doppler frequencies. If the energy contained in all segments of a phase-coded pulse is to add up completely when the pulse is centered in the delay line, while cancelling when it is not, very little shift in phase over the length of the pulse can be tolerated, other than the 180° phase reversal due to the coding.

As will be explained in Chap. 15, a doppler shift is actually a continuous phase shift. A doppler shift, of, say, 10 kilohertz amounts to a phase shift of 10,000 x 360° per second, or 3.6° per microsecond. If the radar's pulses are as much as 50 microseconds long (Fig. 21), this shift will itself equal 180° over the length of the pulse, and performance will deteriorate. For the scheme to be effective, either the

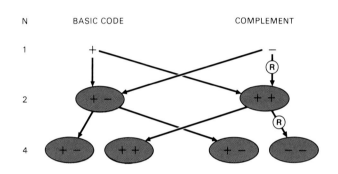

20. How complementary codes are formed. Basic two-digit code is formed by chaining basic binary digit (+) to its complement (−), Complementary two-digit code is formed by chaining basic binary digit (+) to its complement with sign reversed (+). Basic four-digit code is formed by chaining basic two-digit code to complementary two-digit code. Complementary four-digit code is formed by chaining basic two-digit code to complementary two-digit code with sign reversed, and so on.

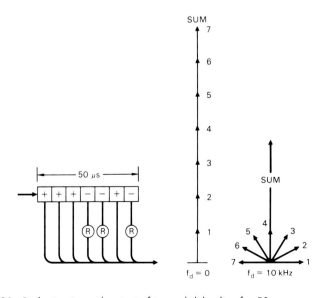

21. Reduction in peak output of tapped delay line for 50 microsecond, phase-coded pulse, resulting from doppler shift of 10 kilohertz.

173

2. This constraint may in some cases be circumvented through "doppler tuning," a technique whereby the doppler shift is largely removed before the pulses are passed through the compression filter.

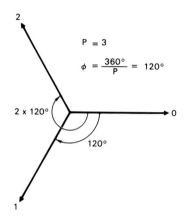

22. Phase increments for a Frank code in which number of phases, P, is three.

$\Delta\Phi \;=\; (G - 1) \times (P - 1) \times \phi°$

$G \;=\;$ group

$P \;=\;$ number of phases

$\phi \;=\;$ basic phase increment

doppler shifts must be comparatively small or the uncompressed pulses reasonably short.[2]

Polyphase Codes

Phase coding is not, of course, limited to just two increments (0° and 180°). Codes employing any number of different, harmonically related phases may be used—e.g., 0°, 90°, 180°, 270°. One example is a family called Frank codes.

The fundamental phase increment, ϕ, for a Frank code is established by dividing 360° by the number of different phases to be used in the code, P. The coded pulse is then built by chaining together P groups of P segments each. The total number of segments in a pulse, therefore, equals P^2

In a three-phase code (Fig. 22), for example, the fundamental phase increment is 360° ÷ 3 = 120°, making the phases 0°, 120°, and 240°. The coded pulse consists of three groups of three segments—a total of 9 segments.

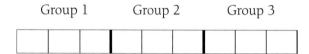

Phases are assigned to the individual segments according to two simple rules. (1) The phase of the first segment of every group is 0°. That is, 0° __ __, 0° __, __, 0° __, __. (2) The phases of the remaining segments in each group increase in increments of

$$\Delta\Phi = (G - 1) \times (P - 1) \times \phi°$$

where G is the group number and ϕ is the basic increment.

For a three-phase code (P = 3, ϕ = 120°, P − 1 = 2), then $\Delta\Phi = (G − 1) \times 2\phi$. So the phase increment in Group 1 is 0°, the phase increment for Group 2 is 2ϕ, and the phase increment for Group 3 is 4ϕ.

Written in terms of ϕ, the nine digits of the code for P = 3 thus are

Group 1	Group 2	Group 3
0, 0, 0,	0, 2ϕ, 4ϕ,	0, 4ϕ, 8ϕ

Substituting 120° for ϕ and dropping multiples of 360°, the code becomes

Group 1	Group 2	Group 3
0°, 0°, 0°,	0°, 240°, 120°,	0°, 120°, 240°

174

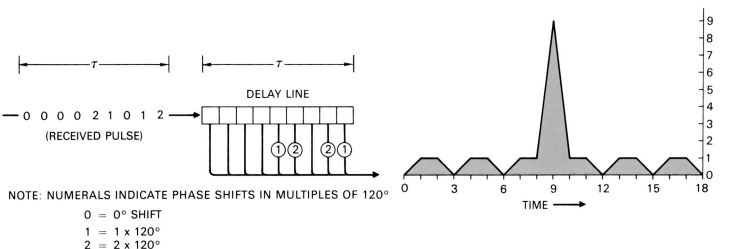

NOTE: NUMERALS INDICATE PHASE SHIFTS IN MULTIPLES OF 120°

    0 = 0° SHIFT
    1 = 1 x 120°
    2 = 2 x 120°

23. Processing of Frank codes is similar to that of binary codes. Phase shifts introduced in taps complement shifts in corresponding segments of coded pulse. If phase of a segment is shifted by 1 x 120°, corresponding tap adds a shift of 2 x 120°, making total shift when pulse fills line equal 3 x 120° = 360°.

Echoes are decoded by passing them through a tapped delay line (or the digital equivalent) in the same way as binary-phase-coded echoes (Fig. 23). The only difference: the phase shifts in the taps have more than one value.

For a given number of segments, a Frank code provides the same pulse compression ratio as a binary phase code and the same ratio of peak amplitude to sidelobe amplitude as a Barker code. Yet, by using more phases (increasing P), the codes can be made any length. As P is increased, however, the size of the fundamental phase increment decreases, making performance more sensitive to externally introduced phase shifts and imposing more severe restrictions on uncompressed pulse width and maximum doppler shift.

Summary

The commonly used pulse compression techniques are linear frequency modulation (chirp) and binary phase coding.

In chirp, the frequency of each transmitted pulse is continuously *increased* or *decreased*. Received pulses are passed through a filter, which introduces a delay that *decreases* or *increases* with frequency. Successive increments of a pulse, therefore, bunch up. Width of the compressed pulse is 1/ΔF, where ΔF is the total change in frequency. The pulse has sidelobes, but they can be acceptably reduced by tapering the amplitude of the uncompressed pulse.

When only a narrow range swath is of interest, chirp can be decoded with stretch radar techniques, whereby range is converted to frequency. Differences in frequency are resolved by a bank of fixed-tuned filters, implemented with the efficient fast Fourier transform.

Chirp has the advantage of providing large compression ratios and being simple.

In binary phase modulation, each pulse is marked off into segments, and the phase of certain segments is reversed. Received pulses are passed through a tapped delay line having phase reversals in corresponding taps. The output pulse is the width of the segments. It, too, has sidelobes.

With Barker codes the mainlobe-to-sidelobe ratio equals the pulse-compression ratio, but the longest code is only 13 digits. Sidelobes can be eliminated by alternately transmitting complementary forms of the four-digit code. These can be chained to any length. But, if doppler shifts are large, performance deteriorates unless pulses are reasonably short.

Polyphase—e.g., Frank—codes can also be used, but they are even more sensitive to doppler frequency.

FM Ranging

14

If enough PRFs can be provided to resolve the growing number of range ambiguities that arise as the PRF is increased, pulse delay ranging can be employed successfully even at fairly high PRFs. However, a point is ultimately reached where the echoes return so "thick and fast" it is virtually impossible to resolve the ambiguities (Fig. 1). Range, if required, must then be measured indirectly, as in CW radars. The most common indirect method is linear frequency modulation, or FM, ranging.[1]

This chapter briefly describes the principle of FM ranging. It explains how doppler frequency shifts, which would otherwise introduce gross measurement errors, are taken into account and how a problem of ghosting similar to that encountered in PRF switching is handled. Finally, it briefly considers the accuracy which may be obtained with FM ranging.

Basic Principle

With FM ranging, the time lag between transmission and reception is converted to a frequency difference. By measuring it, the time lag—hence the range—is determined.

In simplest form, the process is as follows. The radio frequency of the transmitter is increased at a constant rate. Each successive transmitted pulse thus has a slightly higher radio frequency. The linear modulation is continued for a period at least several times as long as the round-trip transit time for the most distant target of significance (Fig. 2). Over the course of this period, the instantaneous difference between the frequency of the received echoes

1. If the PRF is increased beyond a certain point, it becomes impractical, if not impossible, to resolve range ambiguities. Ranging time must then be measured indirectly.

1. Another form of FM ranging which has advantages in some applications employs sinusoidal modulation.

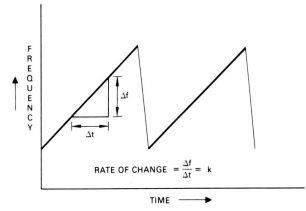

$$\text{RATE OF CHANGE} = \frac{\Delta f}{\Delta t} = k$$

2. In simplest form, FM ranging involves changing the transmitter frequency at a constant rate. Length of slope is generally many times maximum round-trip transit time.

and the frequency of the transmitter is measured. The transmitter is then returned to the starting frequency, and the cycle is repeated.

Just how the measured frequency difference is related to a target's range is illustrated in Fig. 3 for a static situation. That is a situation, such as a tail chase, where the range rate is zero.

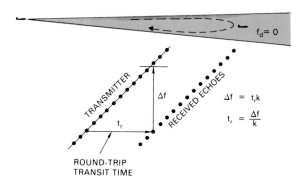

3. Difference between the frequency of an echo and the frequency of the transmitter at the time an echo is received (Δf) is proportional to transit time (t_r).

In this figure, the radio frequency of both the transmitter and the echoes received from a target are plotted versus time. The dots on the plot of transmitter frequency represent individual transmitted pulses. The horizontal distance between each of these dots and the dot representing the received target echo is the round-trip transit time. The vertical distance between the echo dot and the line representing the transmitter frequency is the difference, Δf, between the frequency of the echo and the frequency of the transmitter when the echo is received.

As you can see, this difference equals the rate of change of the transmitter frequency—hertz per microsecond—times the round-trip transit time. By measuring the frequency difference and dividing it by the rate (which we already know), we can find the transit time.

Suppose, for example, that the measured frequency difference is 10,000 hertz and the transmitter frequency has been increasing at a rate of 10 hertz per microsecond. The transit time is

$$t_r = \frac{10,000 \text{ Hz}}{10 \text{ Hz/}\mu s} = 1,000 \ \mu s$$

Since 12.4 microseconds of round-trip transit time correspond to one nautical mile of range, the target's range is equal to $1000 \div 12.4 = 81$ nautical miles.

Accounting for the Doppler Shift

Actually, the process is more complicated than just outlined, for the range rate rarely is zero. The frequency of a target echo is not equal solely to the frequency of the transmitted pulse that produced it, but to that frequency plus the target's doppler frequency. To find the transit time, we must add the doppler frequency, f_d, to the measured frequency difference (Fig. 4).

Including a Constant-Frequency Segment. As you may have surmised, the doppler frequency can be found by interrupting the frequency modulation at the end of each cycle and transmitting at a constant frequency for a brief period. During this period, the difference between the echo frequency and the transmitter frequency will be due solely to the target's doppler frequency. By measuring that difference (Fig. 5) and adding it to the difference measured during the sloping segment, we can find the transit time.

Alternate, Two-Slope Cycle. It turns out that the doppler frequency can be added just as easily by employing a two-slope modulation cycle. The first slope is the same as the rising-frequency slope just described. Once it has been traversed, the frequency is decreased at the same rate until the starting frequency is again reached (Fig. 6 below). The cycle is then repeated.

If the target is closing—i.e., has a positive doppler frequency, f_d—the difference between the frequency of the transmitter and the frequency of the received echoes will be decreased by f_d during the rising-frequency segment and increased by f_d during the falling-frequency segment. (The

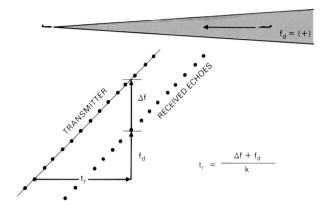

4. Frequency difference, Δf, between transmitter and received echoes is reduced by target's doppler frequency, f_d. To find transit time, f_d must be added to Δf.

$$t_r = \frac{\Delta f + f_d}{k}$$

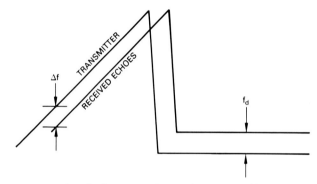

5. Target's doppler frequency (f_d) may be measured by adding a constant frequency segment to the modulation cycle.

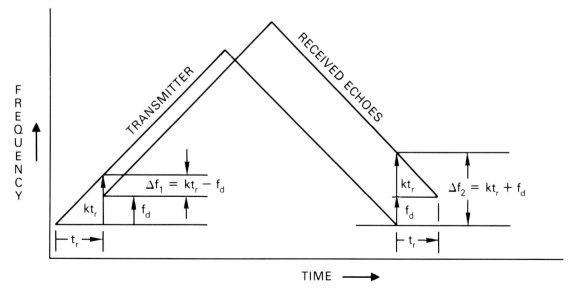

6. With two-slope modulation, frequency difference is decreased by f_d during rising slope and increased by f_d during rising slope and increased by f_d during falling slope.

reverse will be true if the target is opening.) Consequently, if the frequency differences for the two segments are added, the doppler frequency will cancel out.

$$\Delta f_1 = kt_r - f_d$$
$$\Delta f_2 = kt_r + f_d$$
$$\overline{\Delta f_1 + \Delta f_2 = 2kt_r + 0}$$

The sum, then, will be twice the frequency difference, kt_r, due to the round-trip transit time. The latter can be found by dividing the sum by twice the rate of change of the transmitter frequency.

$$t_r = \frac{\Delta f_1 + \Delta f_2}{2k}$$

where

t_r = round-trip transit time

Δ_{f_1} = difference between transmitter and echo frequencies during rising-frequency segment

Δf_2 = difference between transmitter and echo frequencies during falling-frequency segment

k = rate of change of transmitter frequency

Again, knowing the transit time, we can readily calculate the target range.

Suppose the target used in the previous example (k = 10 Hz/µs, kt_r = 10 kHz) had a doppler frequency of 3 kilohertz. During the rising-frequency segment, the measured frequency difference would have been 10 − 3 = 7 kHz. During the falling-frequency segment, it would have been 10 + 3 = 13 kHz. Adding the two differences and dividing by 2k (20 Hz / µs) gives the same transit time, 1000 microseconds, as when the doppler frequency was zero.

Although in both this example and the illustrations the doppler frequency is positive, the equation works just as well for negative doppler frequencies.

Eliminating Ghosts

If the antenna beam encompasses two targets at the same time, a problem of ghosting may be encountered, as with PRF switching. There will be two frequency differences during the first segment of the modulation cycle and two during the second (Fig. 7). This is true, of course, regardless of whether both segments are sloped or one is sloped and the other is not.

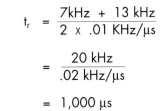

$$t_r = \frac{7kHz + 13\ kHz}{2 \times .01\ KHz/\mu s}$$

$$= \frac{20\ kHz}{.02\ kHz/\mu s}$$

$$= 1,000\ \mu s$$

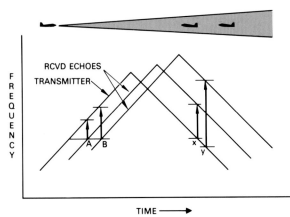

7. If two targets are detected simultaneously, two frequency differences will be measured during each segment of the cycle.

Why Ghosts? Although we can tell from the continuity in the plots of frequency versus time which frequency differences belong to the same target, the continuity is not visible to the radar. As will be explained in detail in Chap. 18, for a radar to discern small frequency differences such as are normally encountered in FM ranging, it must receive echoes from a target for an appreciable period of time. In essence, all the radar observes are two frequency differences at the end of the first segment and two (probably different) frequency differences at the end of the second segment.

This is illustrated in Fig. 8. There, the first two differences are referred to as A and B; the second two, as x and y. Without some further information it is impossible to tell for sure how these differences should be paired—whether A and x pertain to the same target or A and y.

Identifying Ghosts. In applications where ghosts are apt to be encountered, they may be eliminated by adding another segment to the modulation cycle—much as they are by adding another PRF when PRF switching is used.

A representative three-slope cycle consists of equal increasing and decreasing frequency segments—such as we just considered—plus a constant-frequency segment (Fig. 9). The latter, of course, provides a direct measure of the targets' doppler frequencies.

There is, however, no direct way of pairing the measured doppler frequencies with A and B or x and y, either. But knowing the doppler frequencies, the correct pairing of A and B with x and y can quickly be found. Just as doppler frequency cancels out when we *add* the frequency differences for positively and negatively sloping segments, so transit time cancels out when we *subtract* the differences. The result then is twice the doppler frequency.

$$\Delta f_2 = k t_r + f_d$$
$$\underline{- (\Delta f_1 = k t_r - f_d)}$$
$$\Delta f_2 - \Delta f_1 = 0 + 2 f_d$$

Therefore, by subtracting A (or B) from x (or y) and comparing the results with the measured doppler frequencies, we can tell which of the two possible pairings is correct (Fig. 10). If, we say, $(x - A)$ is twice one of the measured doppler frequencies, then the pairing should be as follows:

x with A

y with B

Otherwise, y should be paired with A and x with B.

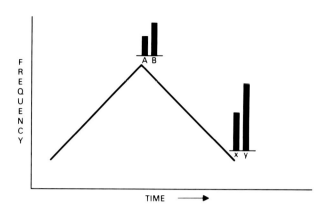

8.   All the radar sees are two frequency differences at the end of each segment. Radar has no way of telling whether A should be paired with x or y.

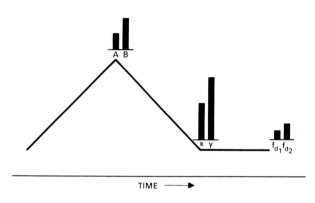

9.   The problem is solved by adding a third segment in which doppler frequencies are separately measured.

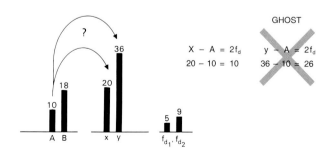

10.   Knowing the two doppler frequencies, the radar can readily tell whether x and A or y and A should be paired.

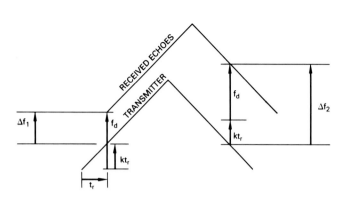

11. When doppler frequency (f_d) is greater than frequency difference due to ranging time (kt_r), echo frequencies are higher than transmitter frequency during the rising frequency segment.

Doppler Frequency Greater Than kt_r. In the illustrations shown so far, the doppler frequency has been less than the frequency difference due to the ranging time, kt_r. While this is true of applications such as altimetry, it is not true of air-to-air applications. For these, the rate of change of the transmitter frequency is generally made low enough so that the maximum value of kt_r will be only a small fraction of the highest doppler frequency normally encountered. In that case, a plot of the frequency of the echoes from a closing target during the rising-frequency portion of the modulation cycle appears as in Fig. 11.

The relationships between the measured frequency differences for two or more simultaneously detected targets can then be seen more clearly if the differences for each segment of the cycle are plotted on separate horizontal scales—one above the other—as in Fig. 12 (below). The differences for the rising-frequency segment (A and B in the figure) appear on the negative half of the frequency scale; and the differences for the falling-frequency segment (x and y), on the positive half. The differences can be paired by drawing horizontal arrows, between them, of lengths corresponding to the doppler frequencies measured in the third segment of the cycle.

3 SLOPES; 2 TARGETS

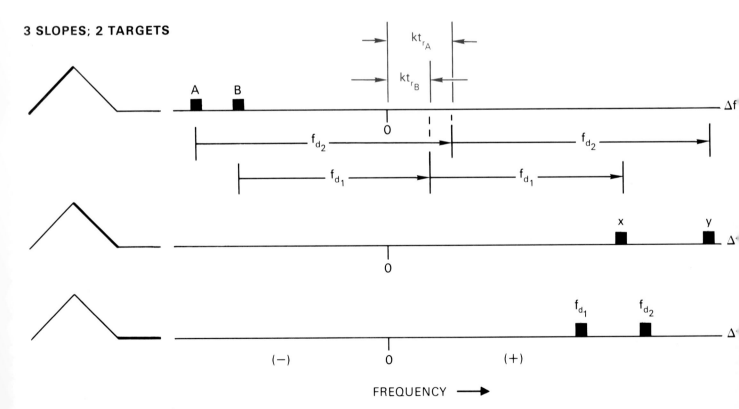

12. Relationships between frequency differences measured during rising and falling slopes can be seen more clearly if plotted on separate line for each slope.

In this case, we find that y is separated from A by two lengths of the arrow, f_{d_2}. These frequency differences, therefore, belong to the same target. The point where they abut corresponds to the frequency difference due to the transit time for the target, kt_{r_A}:

$$A + f_{d_2} = kt_{r_A}$$

Similarly, x is separated from B by two lengths of the arrow, f_{d_1}, and the point where they abut corresponds to the frequency difference due to the transit time for the target, kt_{r_B}.

Three Targets Detected Simultaneously. Figure 13 (below) plots the measured frequency differences for three targets. They, too, can be paired easily. Comparing C with x, y, and z, we find that it is separated from z, by $2f_{d_3}$. There are still two possible combinations of A and B with x and y: A with x and B with y, or A with y and B with x. But, with C out of the way, we can readily tell which of these are ghosts—just as we did when only two targets were detected to begin with.

Certain combinations of ranges and doppler frequencies may occur, however, for which more than one pairing of A, B, and C with x, y, and z are possible. One of these is

SLOPES; 3 TARGETS

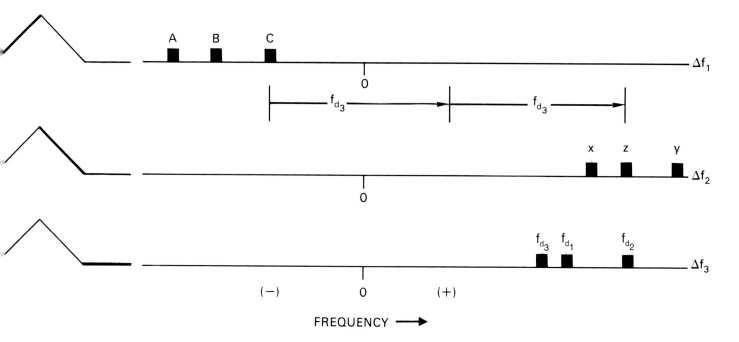

13. When three targets are detected simultaneously, once one combination of frequency differences has been paired, the others may be paired in the same way as when only two targets are detected.

3 SLOPES; 3 TARGETS—3 GHOSTS

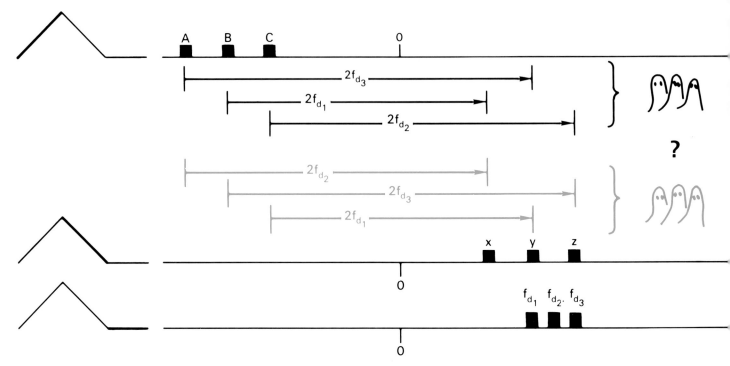

14. With only three slopes, certain combinations of three targets will leave unresolved ghosts, but these combinations are rare.

illustrated in Fig. 14 (above). The frequency differences shown there can readily be paired as follows.

$$A + 2f_{d_3} = y$$
$$B + 2f_{d_1} = x$$
$$C + 2f_{d_2} = z$$

But a second pairing is also possible.

$$A + 2f_{d_2} = x$$
$$B + 2f_{d_3} = z$$
$$C + 2f_{d_1} = y$$

The ranges indicated by one or the other of these pairings are ghosts, and with only three PRFs, we cannot tell which. As the number of simultaneously detected targets increases, the number of these potential ghost-producing combinations, though small, goes up.

They can be eliminated by adding more slopes to the cycle. As with PRFs in pulse delay ranging, if N is the number of slopes, all possible combinations of N − 1 simultaneously detected targets can be deghosted. But the problem of ghosts is generally much less severe with FM ranging. For, in situations where it is normally used, neither range nor doppler frequency is ambiguous.

Recognizing that there will always be some possibility of encountering unresolved ghosts, a three-slope modulation cycle usually suffices.

Performance

The accuracy of FM ranging depends upon two basic factors: the rate, k, at which the transmitter frequency is changed, and the accuracy with which the frequency differences are measured.

The greater k is, the greater the frequency difference that a given transit time will produce. The greater this difference and the greater the accuracy with which frequency can be measured, the more accurately the range will be determined.

Frequency Measurement Accuracy. This increases with the length of time t_{int} over which the measurement is made—the length of the segments of the modulation cycle (Fig. 15).

In search operation, the length of the segments is limited by the length of time the antenna beam takes to scan across a target: time on target, t_{ot}.

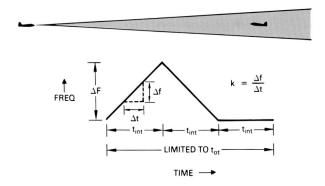

15. Frequency measurement accuracy is limited by time-on-target, t_{ot}.

Since the time-on-target is generally fixed by other considerations, the steepness of the sloping segments of the cycle—the rate, k—becomes the controlling factor for range measurement accuracy.

Steepness of Slope, k. In applications such as low-altitude altimeters, k can be made sufficiently high to provide extremely precise range measurements (Fig. 16).

However, as will be explained in detail in Chap. 27, in air-to-air applications, the value of k is severely limited. As k is increased, ground return—which may be received from ranges out to hundreds of miles—is smeared over an increasingly broad band of frequencies. A point is quickly reached, where the clutter blankets the targets, even though

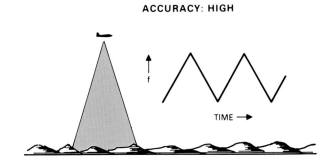

16. In altimeters, slope of modulation curve can be made steep enough to provide highly accurate range measurement.

185

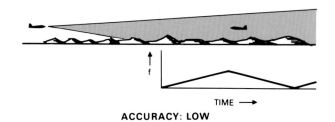

ACCURACY: LOW

17. For air-to-air applications, slopes must be made shallow to avoid smearing the spectrum of the ground return. The result is low accuracy.

the doppler frequencies of targets and clutter may be quite different (Fig. 17).

Because of the limitations on k, in these applications FM ranging is fairly imprecise. Whereas pulse-delay ranging yields accuracies on the order of feet, FM ranging yields accuracies on the order of miles.

Reduction in Detection Sensitivity. As with PRF switching, FM ranging reduces the detection sensitivity from what it would be with no ranging. The reduction is primarily due to the fact that if a target is to be detected at all, it must be detected independently on each segment of the modulation cycle. The time-on-target for each segment is the total time-on-target divided by the number of segments. Also, each time the slope of the transmitter frequency curve is changed, the return from a considerable number of transmitted pulses must be discarded.

In many situations, though, the reduction in detection range is an acceptable price to pay for being able to range.

Summary

With FM ranging, the time lag between transmission and reception is converted to a frequency shift. By measuring this shift, the range is determined. Typically, the transmitter frequency is changed at a constant rate. The change is continued over a considerable period of time so the frequency difference can be accurately measured.

To cancel the contribution of the target's doppler frequency to the measured frequency difference, a second measurement is made. This is done either while transmitting at a constant frequency or while changing the transmitter frequency in the opposite direction. The second measurement is then subtracted from the first.

To resolve ambiguities occurring when two targets are detected simultaneously, a third measurement may be made.

For long range applications, FM ranging is more complicated and generally less accurate than pulse-delay ranging and reduces the radar's detection range. But it enables the high PRF waveform to be mechanized while still ranging.

PART IV

Pulse Doppler Radar

DOUGLAS F4D SKYRAY (1956)

This bat-winged Navy interceptor set world records for speed and climb.
The Westinghouse APG—50 (AERO—13) radar directed the firing of 20
millimeter cannons and 2.75 inch rockets.

Doppler Effect 15

By sensing doppler frequencies, a radar not only can measure range rates, but also can separate target echoes from clutter, or produce high resolution ground maps. Since these are important functions of many of today's airborne radars, one of the keys to understanding their operation is a good understanding of the doppler effect.

Accordingly, in this chapter, we will look at the doppler shift more closely—first, in terms of the compression or expansion of wavelength and, second, in terms of the continuous shift of phase. We will then pinpoint the factors which determine the doppler frequencies of the return from both aircraft and the ground. Finally, we will consider the special case of the doppler shift of a target's echoes as observed by a semiactive missile.

Doppler Effect and Its Causes

The doppler effect is a shift in the frequency of a wave radiated, reflected, or received by an object in motion. As illustrated in Fig. 1, a wave radiated from a point source is compressed in the direction of motion and is spread out in the opposite direction. In both cases, the greater the object's speed, the greater the effect will be. Only at right angles to the motion is the wave unaffected. Since frequency is inversely proportional to wavelength, the more compressed the wave is, the higher its frequency is, and vice versa. Therefore, the frequency of the wave is shifted in direct proportion to the object's velocity.

In the case of a radar, doppler shifts are produced by the relative motion of the radar and the objects from which the radar's radio waves are reflected. If the distance between

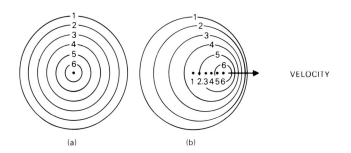

1. A wave radiated from a point source when stationary (a) and when moving (b). Wave is compressed in direction of motion, spread out in opposite direction, and unaffected in direction normal to motion.

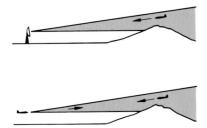

2. With a ground-based radar, relative motion is due entirely to the target's motion. With airborne radar, it is due to motion of both radar and target.

the radar and a reflecting object is decreasing, the waves are compressed. Their wavelength is shortened and their frequency is increased. If the distance is increasing, the effect is just the opposite.

With ground-based radars, any relative motion is due entirely to movement of the radar's targets. Return from the ground has no doppler shift (Fig. 2). Differentiating between ground clutter and the echoes of moving targets, therefore, is comparatively easy.

With airborne radars, on the other hand, the relative motion may be due to the motion of either the radar or the targets, or both. Except in such aircraft as hovering helicopters, the radar is always in motion. Consequently, both target echoes and ground return have doppler shifts. This greatly complicates the task of separating target echoes from ground clutter. A radar can differentiate between the two only on the basis of differences in the magnitudes of their doppler shifts.

Before discussing that, however, let's take a close look at how the shift actually occurs.

Where and How the Doppler Shift Takes Place

If both radar and target are moving, the radio waves may be compressed (or stretched) at three points in their travel: transmission, reflection, and reception.

The compression in wavelength occurring in the simple case of a radar closing on a target, head-on, is illustrated in Fig. 3 at the top of the facing page.

In these simplified diagrams, the slightly curved vertical lines represent planes (viewed edge-on) at every point on which the phase of the wave's fields is the same. These planes are called *wavefronts*. Those shown here, we'll say, are planes on which the fields have their maximum intensity in a positive direction—they represent wave "crests." Two successive wavefronts are shown at each of the points in question. So that you can keep track of the wavefronts easily, they are color coded—wavefront No. 1, red; wavefront No. 2, blue.

For the sake of readability, the diagrams have not been drawn to scale. In reading them, you must keep two things in mind. First, the wavelength—spacing between successive wavefronts of the same phase—is generally only a small fraction of the length of the aircraft. Second, since the speed of light is 162,000 nautical miles per second, in a given period of time the aircraft would travel only a minuscule fraction of the distance traveled by the waves.

The diagram at top left in Fig. 3 illustrates the compression in wavelength occurring when a wave is transmitted.

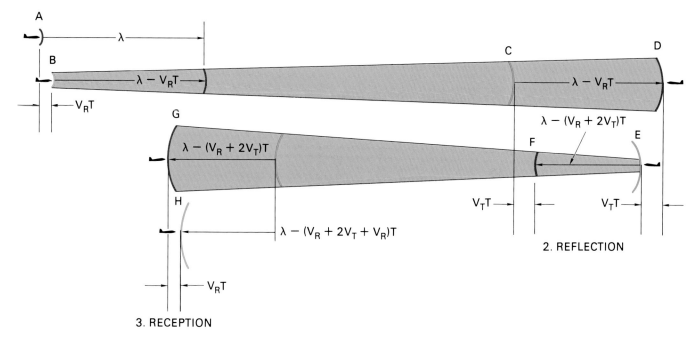

1. TRANSMISSION

2. REFLECTION

3. RECEPTION

TOTAL COMPRESSION $\propto 2(V_R + V_T)T$

3.   Compression in wavelength occurs during transmission, reflection, and reception.

The radar is a point A when it transmits wavefront No. 1. By the time it transmits wavefront No. 2, it has advanced to point B, decreasing the wavelength by a distance equal to the velocity of the radar (V_R) times the time between transmission of the two wavefronts. That time, of course, is the period of the wave (T). The space between wavefronts as the wave travels out to the target, therefore, is ($\lambda - V_R T$).

The diagram on the right in Fig. 3 illustrates the compression occurring when the wave is reflected by the target. When wavefront No. 1 is reflected, the target is at point D and wavefront No. 2 is at point C. By the time wavefront No. 2 is reflected, the target has advanced to point E, shortening the distance the wavefront has had to travel from point C to reach the target by an amount equal to the velocity of the target, V_T, times the period, T. Meanwhile, the reflection of wavefront No. 1 has traveled an equal distance (D to F). But the target's advance has reduced the separation between this reflected wavefront and the reflection of wavefront No. 2, which is just now leaving the target, by $V_T T$.

The space between wavefronts of the reflected wave as it travels back to the radar, therefore is $\lambda - (V_R + 2V_T)T$.

The third diagram in Fig. 3 illustrates the reception of the two wavefronts by the radar. The radar is at point G when it receives wavefront No. 1. Wavefront No. 2 is one compressed wavelength away. But by the time this wavefront is received, the radar has advanced to point H. Thus,

191

during reception, the wavelength is still further compressed by a distance $V_R T$—the same as during transmission.

In all, the wavelength is compressed by twice the sum of the two velocities times the period of the transmitted wave, T.

$$\text{Total compression} = 2(V_R + V_T)T$$

Since T is very short, the compression is extremely slight. For an X-band radio wave and values of V_R and V_T of 600 knots, the compression is only about 5 millionths of an inch. Nevertheless, since the radio frequency of an X-band wave is very high (10 gigahertz), the resulting frequency shift is more than 40 kilohertz.

Magnitude of the Doppler Frequency

Although we can get a physical feel for the doppler effect by observing the compression in wavelength due to the relative motion of a radar and a target, we can calculate the doppler frequency much more simply on the basis of the shift in phase of the received wave.

Frequency, a Continuous Phase Shift. While not generally though of in this way, a change in the frequency of a wave is tantamount to a continuous shift in phase. This is illustrated in Fig. 4. It shows a one-second sample of two waves, A and B. Their frequencies are 10 hertz and 11 hertz, respectively. At 11 hertz, B completes one more cycle per second than A. In other words, every second, the phase of B relative to A advances 360°. Since A completes 10 cycles every second, the gain in phase per cycle of A is 360° ÷ 10 = 36°.

If we wish to shift the frequency of B down to 10 hertz, therefore, we can do so simply by inserting a time delay equivalent to 36° of phase between successive wavefronts (Fig. 5).

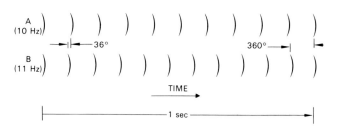

4. Wavefronts of two waves of slightly different frequency. Frequency difference is tantamount to a continuous shift in phase—here, 36° per cycle.

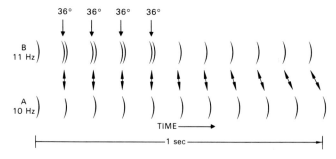

5. By inserting 36° phase shift between wavefronts, frequency is decreased by 1 hertz from 11 hertz to 10 hertz. When insertion is discontinued, wave reverts to its original frequency.

As long as we continue shifting the wave's phase, the frequency shift will persist. But if we stop, B will revert to its original frequency. By shifting phase in the opposite direction—i.e., decreasing the time between wavefronts, we can similarly *increase* a wave's frequency.

So it is with the doppler shift in the signal a radar receives from a target. Only in this case, phase is not shifted through the arbitrary insertion or removal of increments of time between wavefronts.

Rather, it is shifted as the result of the continuous change in the time the radio waves take to travel from the radar to the target and back—the change in the round-trip transit time.

Phasor Representation of Doppler Frequency. As with so many radar concepts, the shift in phase may be most easily visualized with phasors. The phase of the received wave relative to the transmitted wave is portrayed by a simple phasor diagram in Fig. 6. Phasor T represents the transmitted wave; phasor R, the received wave. (To make the relationship between the two phasors easier to visualize, we'll assume that the radar transmits continuously, though that is not necessary). At any one instant, the phase of the received wave, R, lags that of the transmitted wave, T, by the round-trip transit time, t_{rt}. If t_{rt} were zero, there would be no lag, and the two phasors would coincide. If t_{rt} were half the period of the transmitted wave, R would lag half a revolution behind T.

Let us suppose, more realistically, that t_{rt} is 100,000 times the period of the transmitted wave plus some fraction, ϕ (Fig. 7). The rotation of R, though 100,000 complete revolutions behind the rotation of T, will be out of phase with it by only the fraction of a complete revolution (cycle), ϕ.

Now, if the transit time is constant (range rate = 0), the phase lag, too, will be constant and the angle ϕ will remain the same. The two phasors, therefore, will rotate at the same rate. The frequencies of the transmitted and received signals will be the same.

However, if the transit time decreases slightly, the total phase lag will decrease, reducing the angle ϕ. If the decrease continues (decreasing range), R will rotate counterclockwise relative to T (Fig. 8). The frequency of the received wave will be greater than that of the transmitted wave.

Essentially, the same thing happens if the transit time increases (positive range rate). The only difference is that then the phase lag increases, and R (though still rotating counterclockwise in absolute terms) rotates clockwise relative to T. The frequency of the received wave is less than that of the transmitted wave (Fig. 8).

In either event, the difference in frequency between the transmitted and received waves—the target's doppler frequency, f_d—is proportional to the rate of change of ϕ.

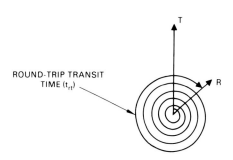

6. Phase of received wave lags that of transmitted wave by round-trip transit time.

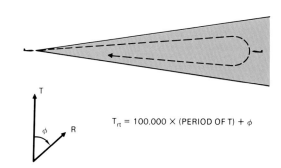

$$T_{rt} = 100,000 \times (\text{PERIOD OF T}) + \phi$$

7. If round-trip transit time is 100,000 times the period of the transmitted wave plus a fraction ϕ, R will be out of phase with T by only the fraction ϕ.

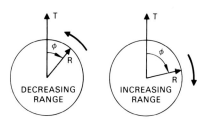

8. If range decreases, ϕ will decrease, causing R to rotate counterclockwise relative to T, and so have a higher frequency.

Equation for f_d Derived. If the rate of change of the phase angle, $\dot{\phi}$, is measured in whole revolutions per second (1 revolution = 2π radians = 360° = 1 whole cycle per second), the doppler frequency in hertz equals $\dot{\phi}$. Since phasor R makes one revolution relative to phasor T every time the round-trip distance (d) to the target changes by one wavelength (λ), the doppler frequency equals the rate of change of d in wavelengths.

$$f_d = -\frac{\dot{d}}{\lambda}$$

The minus sign accounts for the fact that, if $\dot{d}$ is negative (closing target), the doppler frequency is positive.

Since d is twice the target's range (d = 2R), the rate of change of d (Fig. 9) is twice the range rate ($\dot{d} = 2\dot{R}$). The target's doppler frequency, therefore, is twice the range rate divided by the wavelength.

$$f_d = -2\frac{\dot{R}}{\lambda}$$

where

f_d = doppler frequency, hertz

$\dot{R}$ = range rate, feet (or meters) per second

λ = transmitted wavelength, same units as R

Since wavelength equals the speed of light divided by the frequency of the wave (Fig. 10), an alternative expression for doppler frequency is

$$f_d = -2\frac{\dot{R}f}{c}$$

where f is the frequency of the transmitted wave and c is the speed of light.

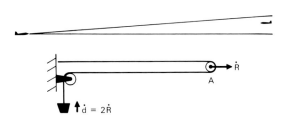

9. As illustrated by this simple mechanical analogy, the round-trip distance from radar to target changes at twice the range rate. If pulley A moves to right at rate $\dot{R}$, weight moves up at rate $\dot{d}$, which is twice $\dot{R}$.

$$\lambda = cT$$

$$\text{But }\; T = \frac{1}{f}$$

$$\therefore\;\; \lambda = \frac{c}{f} \;\text{ and }\; \frac{1}{\lambda} = \frac{f}{c}$$

10. Reciprocal of wavelength ($1/\lambda$) is equal to f/c.

DOPPLER SHIFT IN A NUTSHELL

For every half wavelength per second that a target's range decreases, the radio frequency phase of the received echo advances by the equivalent of one whole cycle per second.

$$\therefore f_d = \frac{-\dot{R}}{\lambda/2} = \frac{-2\dot{R}}{\lambda}$$

where

f_d = doppler shift (positive for decreasing R)

$\dot{R}$ = radial component of relative velocity

λ = wavelength

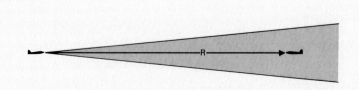

Doppler Frequency of an Aircraft

With either of these expressions, you can quickly and accurately calculate the doppler frequency of any target for any radar. Take an X-band radar (Fig. 11). Its wavelength is 0.1 foot. Suppose the radar is closing on a target at 1000 feet per second ($\dot{R}$ = –1000 fps). The target's doppler frequency is (–2 x –1000) / 0.1 = 20,000 Hz, or 20 kHz.

If the wavelength were only half as long—0.05 foot, instead of 0.1 foot—the same closing rate would produce twice the doppler shift—40 kilohertz, instead of 20 kilohertz.

The equations apply equally to targets whose range is increasing. In this case, f_d has a negative sign, signifying that the radio frequency of the echoes is f_d hertz less than the transmitter frequency.

A simple rule of thumb for estimating doppler frequencies for X-band radars is *1 knot of range rate produces 35 hertz of doppler shift* (Fig. 12). By this rule, a target whose closing rate is 600 knots would have a doppler frequency of 600 x 35 = 21 kHz. Turning the rule around, a target whose doppler frequency is 7 kilohertz would have a range rate of 7000 ÷ 35 = 200 knots.

For other wavelengths, you simply scale the constants to the wavelength: 10.5 hertz per knot for S-band (λ = 10 cm); 21 hertz, for C-band (λ = 5 cm); etc.

Since for X-band (λ = 0.1 foot, another useful rule of thumb is *1000 feet per second of range rate produces 20 kilohertz of doppler shift*.

A target's range rate, of course, depends upon the velocities of both the radar and the target. For a radar approaching a target head-on (Fig. 13a) the range rate is simply the numerical sum of the magnitudes of the two velocities.

$$\dot{R} = - (V_R + V_T)$$

Consequently,

$$f_d = -2 \; \frac{\dot{R}}{\lambda} = 2 \; \frac{V_R + V_T}{\lambda}$$

For a target tail-on (Fig. 13b), the rate is the difference between them. If the radar's velocity is greater than the target's, the range rate will be negative (decreasing range). If the radar's velocity is less than the target's, the range rate will be positive (increasing range). If the two velocities are equal, the rate will be zero.

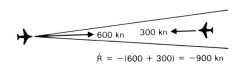

$$f_d = -2 \left(\frac{-1000}{0.1} \right) = 20 \text{ kHz}$$

11. With this expression you can easily calculate the doppler frequency of any target.

DOPPLER FREQUENCIES FOR X-BAND

Closing Rate	f_d (Hz)
1 knot	35
1 mile/hour	30
1 kilometer/hour	19
1000 fps	20 X 10³

12. Rules of thumb for estimating doppler frequency.

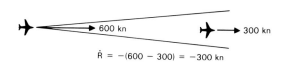

13a. For a target approaching nose-on, range rate is sum of the magnitudes of aircraft velocities.

13b. For tail-on approach, range rate is the difference between them.

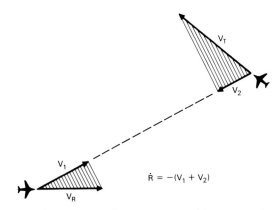

14. In general, range rate of a target is sum of the magnitudes of the projections of radar velocity and target velocity on line of sight to target.

For the more general case where the velocities are not colinear, the range rate is the sum of the projections of the radar velocity and the target velocity on the line of sight to the target. As illustrated in Fig. 14, if the projection of the target velocity is toward the radar, the range will be decreasing. But if it is not, whether the range is decreasing or increasing depends upon the relative magnitudes of the two projections—as in the colinear tail-on case.

A target's doppler frequency, therefore, can vary widely depending on the operational situation. In nose-on approaches, it is always high. In tail-on approaches, it is generally low. In between, its value depends upon the look angle and the direction the target is flying.

Doppler Frequency of Ground Return

The doppler frequency of the return from a patch of ground is also proportional to the range rate divided by the wavelength. The only difference: the range rate of a patch of ground is due entirely to the radar's own velocity (Fig. 15).

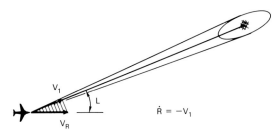

15. Range rate of a ground patch is the magnitude of the projection of radar velocity on line of sight to patch.

Therefore, the projection of the radar's velocity on the line of sight to the patch can be substituted for $-\dot{R}$. For a ground patch dead ahead, this projection equals the radar's full velocity, V_R. For a ground patch directly to the side or directly below, the projection is zero. In between, it equals V_R times the cosine of the angle, L, between V_R and the line of sight to the patch.

The doppler frequency of the return from a patch of ground, therefore, is

$$f_d = 2\ \frac{V_R \cos L}{\lambda}$$

where

$$
\begin{aligned}
f_d &= \text{doppler frequency of ground patch, hertz}\\
V_R &= \text{velocity of radar, feet (meters) per second}\\
L &= \text{angle between } V_R \text{ and line of sight to patch}\\
\lambda &= \text{transmitted wavelength, same units as in } V_R
\end{aligned}
$$

Suppose, for example, the velocity and wavelength are such that $2V_R/\lambda = 10{,}000$ and return is received from a patch at an angle of 60°. The cosine of 60° being 0.5, the doppler frequency is 10,000 x 0.5 = 5 kHz.

If the angle, L, is resolved into its azimuth and elevation components, the term cos L in the above equation must be replaced by the product of the cosines of the azimuth and elevation angles of the patch (Fig. 16)

$$f_d = 2 \frac{V_R \cos \eta \cos \varepsilon}{\lambda}$$

where

η = azimuth angle of patch

ε = lookdown angle of patch

As a rule, ground return is received, not from a single small patch, but from a great many patches at a great many different angles. The return therefore covers a broad spectrum of frequencies. This spectrum is discussed in Chap. 22.

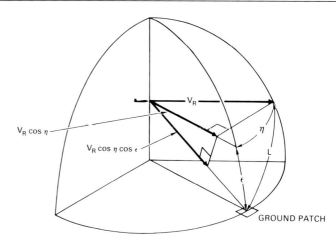

16. Radar velocity V_R, is projected onto line of sight to ground patch in terms of azimuth angle, η, and depression angle, ε. Projection of V_R onto line of sight then equals $V_r \cos \eta \cos \varepsilon$.

Doppler Frequency Seen by a Semiactive Missile

A semiactive missile, you may recall, homes on the scatter from a target which is illuminated by a radar carried in the launch aircraft. Therefore, the doppler frequency of the target as seen by the missile may be quite different from that seen by the illuminating radar.

This is illustrated for a simple colinear case in Fig. 17. The distance, d, from radar to target to missile changes at a rate equal to the radar velocity plus two times the target velocity plus the missile velocity (V_M).

$$\dot{d} = -(V_R + 2 V_T + V_M)$$

The missile velocity, V_M, equals the radar velocity plus the incremental velocity of the missile relative to the radar ($V_M = V_R + \Delta V_M$). With this substitution,

$$\dot{d} = -(2V_R + 2V_T + \Delta V_M)$$

The range rate of the target relative to the radar is $\dot{R} = -(V_R + V_T)$, and the doppler frequency is $-\dot{d}/\lambda$. Therefore, expressed in terms of relative velocities, the target's doppler frequency as seen by the missile is

$$f_{d_M} = \frac{-2\dot{R} + \Delta V_M}{\lambda}$$

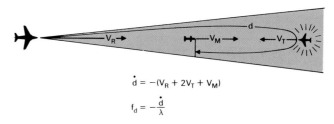

$$\dot{d} = -(V_R + 2V_T + V_M)$$

$$f_d = -\frac{\dot{d}}{\lambda}$$

17. Doppler frequency of target as seen by semiactive missile is proportional to rate of change of distance, d, from radar to target to missile.

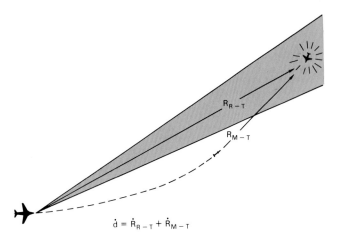

$$\dot{d} = \dot{R}_{R-T} + \dot{R}_{M-T}$$

18. Rate of change of distance to missile is sum of range rate of target relative to radar plus range rate of target relative to missile.

The foregoing equation, of course, applies only if the velocities are all colinear and the missile is on the line of sight from the radar to the target.

In the more general case, the rate at which the distance from radar to target to missile changes equals the range rate of the target relative to the radar plus the range rate of the target relative to the missile (Fig. 18). The latter is the sum of the projections of V_M and $-V_T$ on the line of sight from missile to target.

Initially, f_{d_M} may be comparatively high. However, as the attack progresses, f_{d_M} may fall off considerably—particularly if the missile is drawn into a tail chase, as shown in Fig. 19.

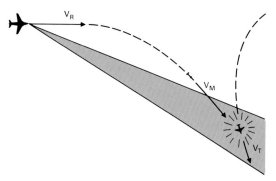

19. Target's doppler frequency as seen by missile may decrease as attack progresses, particularly if missile is drawn into a tail chase.

Some Relationships To Keep In Mind

• Doppler frequency of a target:

$$f_d = -2\,\frac{\dot{R}}{\lambda}$$

Where $\dot{R}$ = range rate
$\quad\quad\ \ \lambda$ = wavelength

• Doppler Frequency of a ground patch:

$$f_d = -2\,\frac{V_R \cos L}{\lambda}$$

Where V_R = radar's velocity
$\quad\quad\quad$ L = look angle to the patch

• Doppler shifts at X band for common velocities:
$\quad$ 1 knot = 35 Hz of doppler shift
$\quad$ 1000 fps = 20 Hz

Summary

In the case of radar echoes, the doppler effect can be visualized as the crowding (or spreading) of wavefronts due to motion of the reflecting object relative to the radar. Since frequency is tantamount to a continuous phase shift, the resulting shift in frequency is equal to the rate (wavelengths per second) at which the round-trip distance traveled by the radio waves is changing—i.e., twice the range rate divided by the wavelength.

Range rate of a moving target is determined by the velocities of the radar and target and by the angle of the line of sight to the target relative to the direction of the radar's velocity. In nose-on approaches the range rate is usually greater than the radar's velocity; in tail-on approaches, less.

Range rate of a patch of ground is determined solely by the radar's velocity and the angle to the patch. Since return may be received from ground patches in many directions, the ground return generally covers a broad band of frequencies.

Spectrum of Pulsed Signal

16

In Chap. 9, we considered the effect of pulsed transmission on a transmitter's power output. But we did not consider its crucial effect on the spectra of the transmitted and received signals—i.e., on the distribution of their energy over the range of possible radio frequencies.

It so happens that, if a radio wave of constant wavelength is transmitted in short pulses, it can be detected by a receiver at more than one radio frequency. As seen by the tuned circuit of a receiver that can be tuned continuously over the complete radio frequency spectrum, the wave's power is spread over a broad band of frequencies. This is true regardless of how "sharply" the circuit is tuned.

On the surface, this behavior is perplexing. For if we plot one or more complete cycles of the wave and measure the spacing between zero crossings or even the rate of change of the wave's amplitude with time, we observe no change whatsoever in frequency as a result of having chopped the wave into pulses. No matter how we define frequency—whether in terms of wavelength, or period, or rate of change of phase with time—as seen by us, the wave has only one frequency.

The reason for the difference between our perception and the receiver's will be explained in the next chapter. In this chapter, we will accept that difference as a fact and concern ourselves only with the nature of the spectrum observed by the receiver. By performing a few simple experiments, we will determine how the spectrum of a pulsed signal is influenced by the pulse width, PRF, and duration of the signal. Along the way, we will learn what coherence is and why it is so vital in a doppler radar.

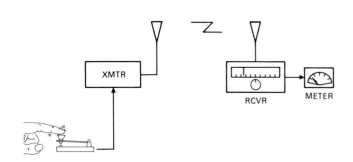

1. To determine the effect of pulse modulation on radio frequency, we perform a series of simple experiments with a microwave transmitter and receiver.

1. The receiver's passband is only one hertz wide; outside this band, it's sensitivity is negligible.

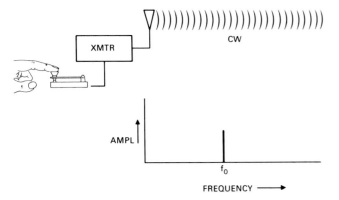

2. A continuous-wave signal produces an output from the receiver only when it is tuned to a single frequency.

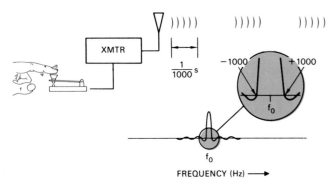

3. A train of independent pulses having a pulse width of 0.01 second and a constant PRF produces a receiver output that is continuous over a band of frequencies 2000 hertz wide.

Illustrative Experiments

To get a feel for the relationships in question, we will perform a series of simple experiments—in our mind's eye, of course. All require just two pieces of equipment (Fig. 1).

One is a microwave transmitter, which for the initial experiments consists simply of an oscillator. Its output signal, we'll assume, has a constant amplitude and a constant, highly stable wavelength. A key is provided with which we can turn the transmitter on or off at any desired instant.

The other equipment is a microwave receiver with which we can detect the transmitted signal, in much the same way as one "tunes-in" a radio station on a broadcast receiver. This receiver, however, is far more selective[1] and can be tuned over a much broader band of frequencies. A meter indicates the amplitude of the receiver's output.

Bandwidth

To find what determines the bandwidth of a pulsed signal, we perform two experiments.

Experiment No. 1: CW Signal. In this, the control experiment, we transmit a continuous wave at a frequency, f_o, and slowly tune—a hertz at a time—through the receiver's frequency range in search of the transmitted signal (Fig. 2).

As anyone might have predicted, the signal produces a strong output from the receiver at a single point on our hypothetical radio dial: the frequency f_o. Though we search the entire tuning range, we find no trace of the signal at any other frequency. If we plot the amplitude of the receiver output versus frequency, it appears as a vertical line.

Experiment No. 2: Stream of Independent Pulses. In our second experiment, we periodically key the transmitter "on" and "off" so that it transmits a continuous stream of pulses having a constant PRF (Fig. 3). It should be noted, however, that although the keying is as precise as we can make it, the radio frequency phases of successive pulses are not the same, but vary randomly from pulse to pulse.

Each pulse is exactly 1/1000th of a second long. While 1/1000th of a second (1000 microseconds) is a very short time, bear in mind that it is on the order of a thousand times longer than the pulses of a great many radars.

Because of the signal's lower average power (the transmitter is "on" only a fraction of the time), the receiver output is not as strong as before. But it still occurs at the same point, f_o, on the dial. The plot of receiver output versus frequency, however, is not quite as sharp as before. In fact, if we expand it, we see that it is continuous over a band of frequencies extending from 1000 hertz below f_o to 1000 hertz above it. The null-to-null bandwidth of 2 kilohertz.

The signal also produces an output in a succession of contiguous bands above and below this one. Within these bands, which are half as wide as the central one, the output is very much weaker, becoming more so, the farther the bands are removed from f_o. The plot of receiver output versus frequency (Fig. 4) has, in fact, the same sin x / x shape as the radiation pattern of a uniformly illuminated linear-array antenna. Although these spectral sidelobes are important, for the time being will ignore them and concern ourselves only with the central band.

Now, the width of this central band might be determined by either the PRF or the pulse width, or by both.

To see if it is the PRF, we repeat the experiment at several progressively lower PRFs. But, except for a reduction in receiver output due to the lower duty factor, the receiver output is unchanged. For a signal of the sort our simple transmitter puts out, the PRF does not affect the spectrum.

Carrying this finding to a logical extreme in which the interpulse period is stretched to days, we further conclude that the spectrum of a single pulse is exactly the same as that of a stream of independent pulses. Bandwidth, we conclude, is not determined by the PRF. What about pulse width?

To find the relationship between bandwidth and pulse width, we repeat the experiment several times using progressively narrower pulses. The final pulse width is 1/1,000,000th of a second (1 microsecond).

The result of narrowing the pulses is striking. As the pulse width decreases, the bandwidth increases tremendously (Fig. 5). For the final pulse width—1 microsecond—the band extends from 1,000,000 hertz below f_o to 1,000,000 hertz above it. The total bandwidth, from null to null, is 2 megahertz!

A frequency of 2 megahertz is 2 divided by 1 millionth of a second. Similarly, a frequency of 2 kilohertz is 2 divided by 1/1000th of a second. Consequently, we conclude that the null-to-null width of the spectral lobe of a stream of independent pulses is

$$BW_{nn} = \frac{2}{\tau}$$

where

$\quad$ BW_{nn} = null-to-null bandwidth

$\quad$ τ = pulse width

The null-to-null bandwidth of a half-microsecond pulse, for example, is $2 \div 0.5\ \mu s = 4$ MHz (Fig. 6).

But this raises a serious question. If at X-band the doppler shift is only 35 kilohertz per thousand knots of closing rate, the doppler frequencies encountered by most airborne radars will be no more than a few hundred kilohertz.

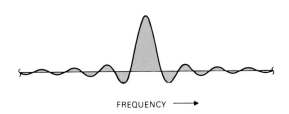

4. Plot of receiver output versus frequency has sin x / x shape. Sidelobes half the width of the central lobe and continuously diminishing in amplitude extend above and below mainlobe.

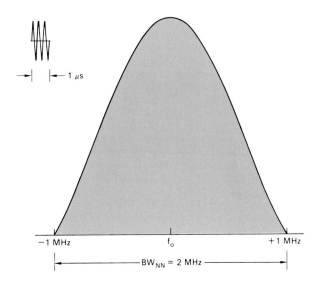

5. For a 1 microsecond pulse width, the null-to-null bandwidth of the central lobe is 2 megahertz.

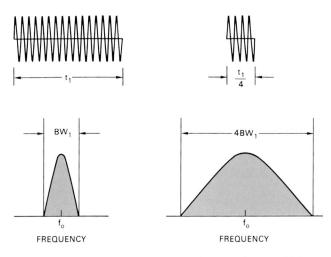

6. The narrower the pulses, the wider the central spectral lobe.

If a pulsed signal has a null-to-null bandwidth on the order of a few megahertz—roughly 10 times the highest doppler shift—then how can a pulsed radar ever detect doppler frequencies? The answer is that it can't, unless the received pulses are, in some way, coherent.

Coherence

By coherence is meant a consistency, or continuity, in the phase of a signal from one pulse to the next. There are many forms of coherence. That used almost universally is illustrated in Fig. 7. In it, the first wavefront in each pulse is separated from the last wavefront of the same polarity in the preceding pulse by some integral number of wavelengths. For example, if the wavelength is exactly 3 centimeters: the separation may be 3,000,000 or 3,000,003 or 3,000,006 centimeters etc.; but not, say, 3,000,001 or 3,000,0033.15.

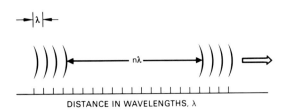

7. Common form of coherence. First wavefront in second pulse is separated from last wavefront of same phase in first pulse by a whole number of wavelengths.

In the preceding experiment, you will recall, the pulses were formed by keying the transmitter (oscillator) "on" for each pulse. Although the keying was fairly precise, the radio frequency phases of the individual pulses—their "starting" phases—varied at random from pulse to pulse. The transmitted signal was *not* coherent.

This is not surprising, when you consider that the period of, say, an X-band signal is only 1/10,000th of a microsecond and a degree of phase is only 1/3,600,000th of a microsecond.

Achieving Coherence. With a somewhat more elaborate transmitter, coherence can readily be achieved. The type of transmitter most commonly used in doppler radars is called a *master oscillator–power amplifier* (Fig. 8). It consists of an oscillator, which produces a low-power signal of highly stable wavelength, and an amplifier, which amplifies the signal to the power level needed for transmission.

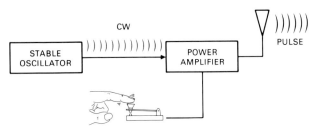

8. Coherent pulse train may be produced with master oscillator-power amplifier. Oscillator runs continuously; amplifier is keyed "on" to produce pulses.

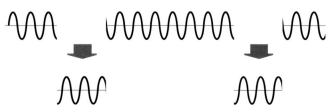

9. Pulses of master oscillator-power amplifier are in effect cut from a continuous wave; hence are coherent.

The oscillator runs continuously; the power amplifier is keyed on and off to produce the pulses. Although the keying is no more precise than in the simple, noncoherent (random starting phase) transmitter, the radio frequency phases of successive pulses are exactly the same as if the pulses had been cut from a continuous wave (Fig. 9). The separation between the last wavefront in one pulse and the first wavefront of the same polarity in the next pulse is thus always exactly equal to a whole number of wavelengths. The pulses are coherent.

Experiment No. 3: Effect of Coherence. To see what effect coherence has on the bandwidth of a pulsed signal, we perform Experiment No. 2 again. But this time, we use a master oscillator–power amplifier transmitter.

The effect of changing to coherent transmission is remarkable, to say the least (Fig. 10). Whereas, with noncoherent transmission, the signal's central spectral lobe is spread over a broad band of frequencies, with coherent transmission, it peaks up almost as sharply as the continuous wave did. There is, however, one important difference. Instead of appearing at only one point on the radio dial, the coherent pulsed signal appears at many different points. Its spectrum, in fact, consists of a series of evenly spaced lines.

Comparing this spectrum with the corresponding spectrum for the noncoherent signal—same PRF and same pulse width—we observe two things. First, at those frequencies where the coherent signal produces an output, it is a great deal stronger than the output produced by the noncoherent signal, evidently because the energy has been concentrated into a few narrow lines. Second, the "envelope" within which these lines fit (Fig. 11) has the same shape (sin x/x) and the same null-to-null width ($2/\tau$) as the spectrum of the noncoherent signal.

Suspecting that the spacing of the lines is related to the PRF, we repeat the experiment several times, at progressively higher PRFs. As the PRF is increased, the lines move farther apart. In every case, the spacing exactly equals the PRF (Fig. 12).

Incidentally, it may be instructive to note that since we maintained a constant pulse width, as we increased the PRF the number of lines decreased. Had we continued to increase the PRF, a point would ultimately have been reached where all of the power was concentrated into a single line. But then we would be transmitting a CW signal.

The important conclusion to be drawn from this experiment, though, is that the spectrum of a coherent pulsed signal consists of a series of lines that (1) occur at intervals equal to the PRF on either side of f_o and (2) fit within an envelope having a sin x/x shape with nulls at multiples of $1/\tau$ above and below f_o.

In one significant respect, however, this experiment was not realistic insofar as the operation of a great many radars is concerned. For each dial setting, the train of received pulses was at least several seconds long. In fact, for each new setting of the tuning dial, we had to wait several seconds for the receiver output meter to reach its final reading. By contrast, the train of pulses a search radar receives each time its beam sweeps across a target may be only a small fraction of a second long.

As we shall see, unless a pulse train is infinitely long—which no pulse train we will ever encounter could possibly be—the spectral lines have a finite width. This width is a function of the duration of the pulse train.

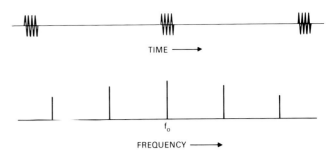

10. Coherent pulses produce output from receiver at evenly spaced intervals of frequency.

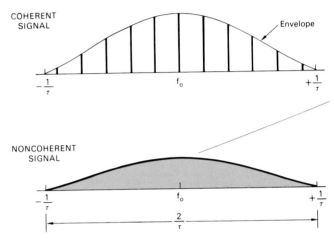

11. Spectral lines of coherent signal fit within envelope having same shape (sin x / x) as spectrum of noncoherent pulse train having equal pulsewidth, τ.

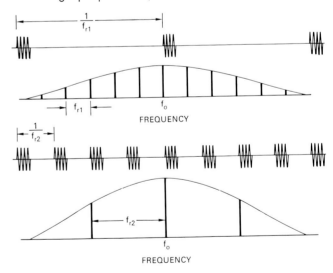

12. Spacing of spectral lines for a coherent pulse train equals PRF, f_r.

EARLIER METHODS OF ACHIEVING COHERENCE

Largely because the master oscillator—power amplifier was expensive to implement with the components then available, in the early airborne doppler radars various other techniques were used to achieve coherence. Some of these are still in use today.

INJECTION LOCKED MAGNETRON

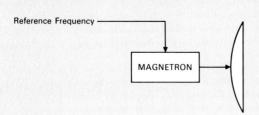

In one, called injection locking, the starting phase of a simple transmitter such as a magnetron is "locked" to the phase of a highly stable, continuously generated low power signal that is injected into the magnetron cavity.

In another, called coherent-on-receive, or COR, one measures the phase of each transmitted pulse relative to a continuously generated reference signal. An appropriate phase correction is then applied to the return received during the immediately following interpulse period.

Unfortunately, with injection locking, the degree of coherence generally leaves something to be desired. And with coherent-on-receive, only the first-time-around return is coherent, since the

phase correction is only valid for return from the immediately preceding transmitted pulse.

COHERENT ON RECEIVE (COR)

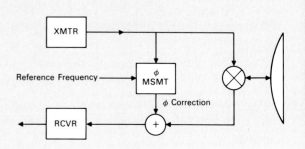

In still another approach, called noncoherent or clutter-referenced moving target indication, the equivalent of coherence is achieved by detecting the "beat" between the target echoes and the simultaneously received ground return. But as explained in Chapter 29, this technique has serious limitations.

CLUTTER REFERENCED MTI

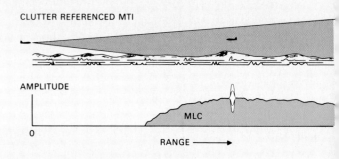

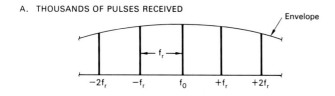

A. THOUSANDS OF PULSES RECEIVED

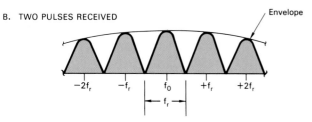

B. TWO PULSES RECEIVED

13. Whereas when thousands of pulses are received, spectral lines are narrow and sharply defined, when only two pulses are received, spectral lines broaden until they are contiguous.

Line Width versus Duration of Pulse Train

To find the relationship between line width and the length of a pulse train, we perform two more experiments.

Experiment No. 4: Two-Pulse Train. For this experiment, we use the same receiver and coherent transmitter as before. But, holding the PRF constant, we transmit only two pulses for each setting of the tuning dial.

The results are shown in Fig. 13, along with a repeat of the results of Experiment No. 3, for the same PRF and pulse width.

Whereas, when the pulse train was a thousand or more pulses long, the receiver output peaked up sharply at each multiple of the PRF; when it is only two pulses long, the plot of receiver output versus frequency is almost continuous. The output still reaches its maximum values at multiples of the PRF and falls off on either side of each peak, but it only reaches zero halfway between peaks. The null-to-null "line width" is f_r hertz!

Experiment No. 5: Eight-Pulse Train. We repeat the experiment, using the same PRF and pulse width, but this time transmitting four times as many pulses for each dial setting—eight as opposed to two. Although the signal still produces an output over a fairly broad band around each multiple of the PRF (Fig. 14) the spectral lines are now only one-fourth as wide—$f_r/4$ hertz as opposed to f_r hertz.

General Relationships. From the results of these two experiments, we conclude that the width of the spectral lines is inversely proportional to the number of pulses in the pulse train. Since for two pulses the line width equals the PRF, we further conclude that for N pulses it equals $(2/N)$ times the PRF.

$$LW_{nn} = \left(\frac{2}{N}\right) f_r$$

where

$$LW_{nn} = \text{null-to-null line width}$$
$$f_r = \text{pulse repetition frequency}$$
$$N = \text{number of pulses in the train}$$

If, for example, a pulse train contains 32 pulses, the line width is 2/32, or one-sixteenth, of the PRF.

The primary factor upon which line width depends however is not just the number of pulses, but the duration of the pulse train—its length in seconds. This becomes clear if we replace f_r in the expression for LW_{nn} with $1/T$, where T is the interpulse period. The expression then becomes $LW_{nn} = 2/(NT)$. Since N is the number of interpulse periods in the train, NT is the train's total length. Accordingly,

$$LW_{nn} = \frac{2}{\text{Length of train (seconds)}} \text{ Hz}$$

Equivalence of Pulse Train to Long Pulse. Interestingly, the results of this experiment are consistent with those of Experiment No. 2. They indicated that the null-to-null bandwidth for a single pulse is two divided by the length of the pulse in seconds: $BW_{nn} = 2/\tau$. If we were to transmit a single pulse, the length of a train of N pulses, its null-to-null bandwidth would be exactly the same as the null-to-null line width of the pulse train (Fig. 15). Thus, there is only one difference between the spectrum of a train of coherent pulses and the spectrum of a single pulse the same length as the train: the spectrum of the pulse train is repeated at intervals equal to the PRF.

The parallel between the spectra of a coherent pulse train and a single long pulse is noted here for two reasons. First, it makes remembering the spectrum of a pulsed signal a bit easier. Second, it will prove illuminating when we take up the explanation of the pulsed spectrum, in the next chapter.

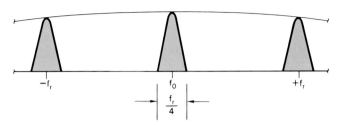

EIGHT PULSES RECEIVED

14. When eight pulses (instead of two) are received, null-to-null width of spectral lines is only one-fourth as great.

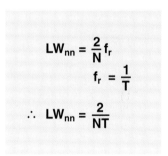

$$LW_{nn} = \frac{2}{N} f_r$$
$$f_r = \frac{1}{T}$$
$$\therefore LW_{nn} = \frac{2}{NT}$$

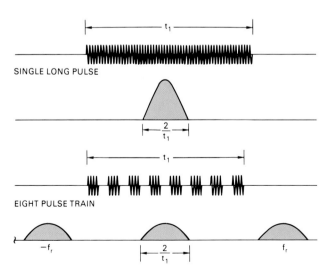

SINGLE LONG PULSE

EIGHT PULSE TRAIN

15. Individual spectral lines for a coherent pulse train differ from the spectrum of a single pulse of the same length only in being repeated at intervals equal to the PRF (f_r).

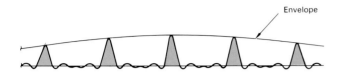

Envelope

16. Just as sidelobes flank the central spectral lobe of a single pulse, they also flank each line in the spectrum of coherent pulse train.

2. If ranges are extremely long it is possible to transmit perfectly coherent pulses and have some loss of coherence in the medium through which the waves propagate.

Spectral Sidelobes

What about spectral sidelobes? Just as they flank the mainlobe of the spectrum of a single pulse, sidelobes of half the null-to-null line width flank each "line" of the spectrum of a pulse train (Fig. 16). The line itself has a sin x/x shape.

Since the length of the train of pulses received from a target during any one scan of the radar antenna is invariably limited, the sidelobes are an important concern to the radar designer. For they tend to fill in the gaps between the spectral lines. Fortunately, as will be explained in Chap. 19, by suitably designing the doppler filters of the radar's signal processor, the sidelobes can generally be reduced to an acceptable level.

Conclusions Drawn from the Experiments. The conclusions we have drawn from our five simple experiments are summarized graphically in the panel on facing page. By way of underscoring their significance, let us return to the question raised earlier in this chapter: If the spectral width of a radar pulse may be many times the highest doppler frequency, how can a pulsed radar discern small doppler shifts in what may be extremely weak target return buried in strong ground clutter?

In light of the illustrations in the panel, the answer should be abundantly clear. A pulsed radar can readily discern these shifts if the following conditions are satisfied:

- The radar is coherent.[2]

- The PRF is high enough to spread the lines of the spectrum reasonably far apart.

- The duration of the pulse train is long enough to make the lines reasonably narrow.

- The doppler filters are suitably designed to reduce the spectral sidelobes.

As will be borne out in subsequent chapters, detecting doppler shifts under typical operating conditions may be a bit more involved than implied here. But the crux of the matter is satisfying these basic requirements.

Summary

Transmitting a radio frequency signal in pulses markedly changes in the signal's spectrum, as observed by the tuned circuit of a receiver.

Whereas the spectrum of a continuous wave of constant wavelength consists of a single line, the spectrum of a single pulse of the same wavelength covers a band of frequencies and has a sin x/x shape. The width of the central lobe of

RESULTS OF THE EXPERIMENTS

Continuous Wave (CW)
Infinite Length

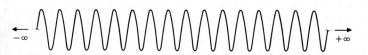

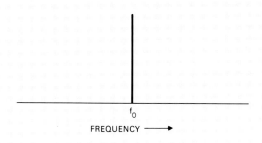

Single Pulse

Train of Noncoherent Pulses
(Random starting phase)

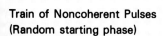

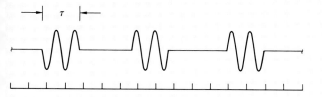

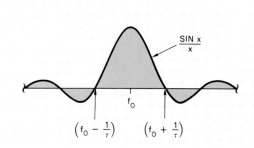

Train of Coherent Pulses
Infinite Length

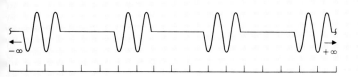

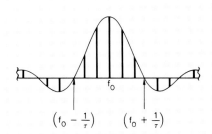

Train of Coherent Pulses
Limited Length

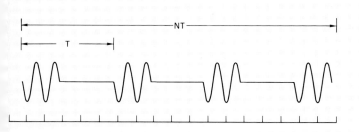

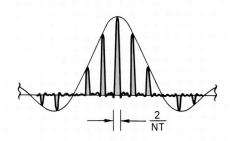

Some Relationships To Keep In Mind

• For a single pulse:

Null-to-null bandwidth $= \dfrac{2}{\tau}$

Where τ = pulse width

• For a coherent pulse train

Line spacing $= f_r$

Null-to-null line width $= \dfrac{2}{N} f_r = \dfrac{2}{NT}$

Where f_r = pulse repetition frequency

N = number of pulses in train

T = interpulse period

---

this spectrum varies inversely with pulse width. If the pulses are as narrow as those used in many radars, the central lobe may be several megahertz wide.

A train of pulses of random starting phase is said to be noncoherent. Its spectrum has the same shape as that of a single pulse.

Coherence is a consistence or continuity in the phases of successive pulses. Commonly, it is achieved by using a master oscillator–power amplifier transmitter. The pulses are then essentially cut out of a continuous wave.

The spectrum of a coherent pulse train of infinite length consists of lines at intervals equal to the PRF, within an envelope having the same shape as the spectrum of a single pulse. If the coherent pulse train is *not* infinitely long, the individual lines have a finite width and the same shape as the spectrum of a single pulse the length of the train. Line width is thus inversely proportional to the length of the train.

Mysteries of the Pulsed Spectrum Unveiled

17

In the preceding chapter, we became acquainted with the dramatic effect that pulsed transmission has on the spectrum of a radio wave. While it would suffice merely to memorize the relationships summarized at the end of that chapter, you will not only remember them better but gain a deeper insight into the operation of a radar if you understand the reasons for them.

This chapter gives the reasons. It begins by raising the fundamental question of exactly what is meant by the spectrum of a signal. This, as you'll see, is actually the crux of the matter. The chapter then explains the spectrum of a pulsed signal in two quite different ways: first, in terms of a conceptually simple but powerful analytical tool, called the Fourier series,[1] and second, in terms of what physically takes place when a radio frequency signal passes through a lossless narrowband filter. For those readers who have some familiarity with calculus, the essence of both explanations is presented in more precise, mathematical terms—the Fourier transform—at the end of the chapter.

1. The series is named for Jean Fourier, the 19th century mathematician and physicist who developed the concept.

Crux of the Matter

Much of the "mystery" which in many people's minds surrounds the spectrum of a pulsed signal stems from not having a clear picture of what is meant by spectrum.

Spectrum Defined. Broadly speaking, the spectrum of a signal is the distribution of the signal's energy over the range of possible frequencies. It is commonly portrayed as a plot of amplitude versus frequency (Fig. 1).

In the last chapter, we gained a rough physical feel for the spectrum of a pulsed signal by measuring the output

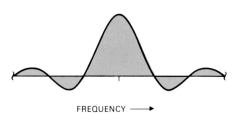

1. The spectrum of a signal is commonly portrayed as plot of amplitude versus frequency.

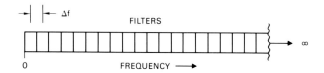

2. To explain its spectrum, a signal may be envisioned as being applied simultaneously to myriad lossless narrowband filters whose frequencies are infinitesimally closely spaced.

2. To be completely rigorous, besides plotting the amplitude of each filter's output, one must also indicate its phase.

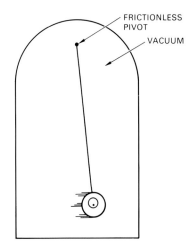

3. A lossless narrowband filter is analogous to a pendulum suspended from a frictionless pivot in a vacuum. Amplitude of swing corresponds to filter's output.

3. A more exact analogy is a mass suspended on a spring, since its restoring force is *directly* proportional to the displacement. But for small displacements, the analogy to a pendulum is very close.

that the signal produced when it was applied to a highly selective receiver, tuned a hertz at a time through a broad band of frequencies. Actually, you can define spectrum quite rigorously in these terms, provided you refine them as follows. Instead of envisioning the signal as being applied to a receiver whose frequency is periodically changed, think of it as being applied simultaneously to myriad lossless narrowband filters whose frequencies are infinitesimally closely spaced and cover the entire range from zero to infinity (Fig. 2). A signal's spectrum, then, is a plot of the amplitudes of the filter outputs versus the frequencies of the filters.[2]

By itself, this definition doesn't clear up any of the "mysteries." To explain them, we must in addition have a clear picture of what a lossless narrowband filter actually does.

What a Lossless Narrowband Filter Does. The most easily visualized mechanical analogy to a lossless narrowband filter is a pendulum suspended from a frictionless pivot in a vacuum (Fig. 3).

The frequency of the filter is the pendulum's natural frequency—the number of cycles per second that it would complete if deflected and allowed to swing freely.

The input signal is applied to the pendulum by a tiny electric motor at the center of the pendulum mass. On the shaft of this motor is an eccentric flywheel. The speed of the motor is such that for every cycle of the input signal the flywheel makes one complete revolution. Because of the flywheel's imbalance, a sinusoidally varying reactive force is exerted on the pendulum.[3] This force tends to make the pendulum swing alternately right and left. The effect is similar to that of a child "pumping" a swing (Fig. 4).

The filter's output is the amplitude to which the swing builds up over the duration of the input signal.

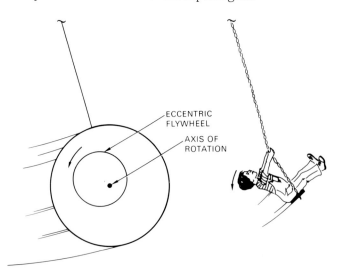

4. Motor driven eccentric flywheel makes one revolution for each cycle of the input signal. Reactive force is similar to that produced by a child "pumping" a swing.

By means of this analogy, it's not too difficult to explain why even the simplest ac signal has a broad frequency spectrum. Consider a signal that turns the flywheel at a rate of 1000 revolutions per second and has a duration of 1/10th second. To see what its spectrum is like, we apply the signal simultaneously to our myriad filters. As the flywheels start turning, all of the pendulums begin to swing. The extent to which each pendulum's swing builds up, however, depends upon the pendulum's natural frequency.

In the case of the pendulum whose frequency is exactly 1000 hertz (Fig. 5), with every turn of the flywheel, the amplitude of the swing increases by the same amount. The swing remains in phase with the sinusoidally varying forces exerted by the flywheel. After 1/10th of a second has elapsed and the flywheel has made 100 turns, the pendulum is swinging with an amplitude 100 times as great as when the input completed its first cycle.

In the case of a pendulum whose frequency is, say, 5 hertz less than 1000 (Fig. 6), the swing starts building up in the same way. But because of the pendulum's lower natural frequency, the phase of the swing gradually falls behind that of the flywheel's rotation. Consequently, the momentum of the pendulum and the reactive forces of the flywheel work against each other over a correspondingly increasing fraction of each cycle. When the input stops, the amplitude of this pendulum's swing is considerably less than that of the pendulum whose frequency is 1000 hertz. Nevertheless, the swing is substantial.

But in the case of the pendulum whose frequency is 10 hertz less than 1000 (i.e., the frequency of the input signal's first spectral null), the phase of the swing falls behind at a high enough rate that the swing is completely damped out by the time the input ends (Fig. 7).

However, for the pendulum whose frequency is 15 hertz less than 1000 (i.e., in the middle of the first sidelobe), the phase of the swing falls behind at a sufficiently high rate that the swing builds up and damps out and builds up once again before the input ends. Though the final amplitude of the swing is only a fraction of that of the pendulum whose frequency is 1000 hertz, this fraction is considerable—roughly 21 percent (Fig. 8).

Moving on to pendulums whose frequencies are farther and farther below 1000 hertz, we observe the familiar pattern of lobes and nulls. At corresponding points within successive lobes, the farther the lobe is from 1000 hertz, the more nearly the total time during which pendulum and flywheel work against each other equals the total time during which they work together. Hence, the less the final amplitude of the pendulum's swing is. But no matter how far

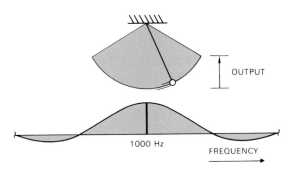

5. Swing of pendulum whose frequency is 1000 hertz stays in phase with reactive force of flywheel and builds up.

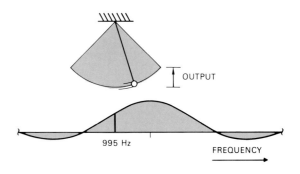

6. Momentum of pendulum whose frequency is 995 hertz works against flywheel part of the time, so buildup is not as great.

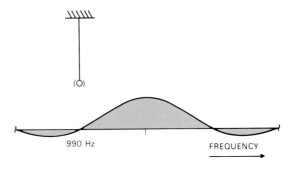

7. Swing of pendulum whose frequency is 990 hertz builds up initially, but is completely damped out when signal ends.

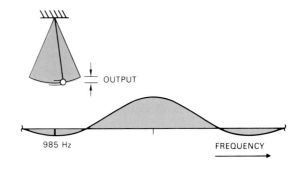

8. Swing of pendulum whose frequency is 985 hertz falls behind sufficiently fast that it builds up again before input ends.

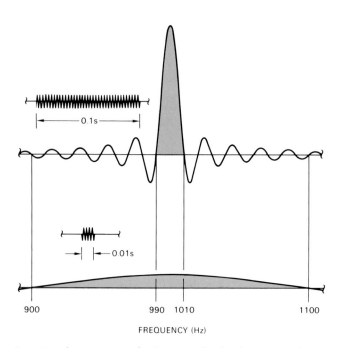

900   990   1010   1100

FREQUENCY (Hz)

9. Complete spectrum of 1/10 second pulse discussed in the text (top). Although most of the energy is centered around the carrier frequency (1000 hertz), if the pulse's duration is decreased to 1/100 second (bottom), the central spectral lobe alone spreads over a band 200 hertz wide.

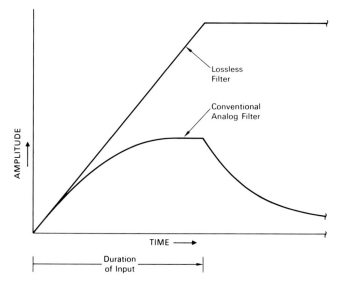

10. Difference between outputs of a lossless filter and a conventional analog filter. Lossless filter is a perfect integrator.

down in frequency we go, *only* at those frequencies for which the periods of buildup and damping are *exactly* equal is a pendulum completely at rest when the input ends.

For the pendulums whose frequencies are greater than 1000 hertz, the responses are similar.

Thus, the spectrum of every signal we encounter covers an immensely broad band of frequencies. True, the energy at most of these frequencies is minuscule. But the shorter the signal, the more widely its energy is spread. For example, if we reduce the duration of the signal we have been considering by a factor of 10, when the input ends, the shift in phase of each pendulum's swing will be only 1/10th as great as before.

So the nulls on either side of the signal's central spectral lobe will be 10 times farther apart than they were (Fig. 9) and the distribution of the signal's energy will be proportionately broader.

Following this general line of reasoning, a far less cumbersome graphic model of a lossless filter may be used. With it, later in this chapter, we will analytically deduce all of the results of the experiments of Chap. 16, practically in our heads.

At this point, though, one thing more should be said about narrowband filters. A lossless filter differs from most of the filters with which we are familiar in two important respects.

First, whereas the output of a *conventional* analog filter builds up fairly quickly to a "steady-state" value when a constant-amplitude input of the filter's frequency is applied, the output of a *lossless* filter continues to build up as long as the input continues (Fig. 10).

Second, whereas the output of a conventional filter decays after the input stops, the output of a lossless filter retains its last value for an unlimited time, unless the output is dumped in some way.

In short, a lossless filter can be thought of as efficiently integrating the energy of that component of the input signal which has the same frequency as the filter.

That's all very well, you may say. But how can a purely sinusoidal signal which completes a given number of cycles per second *really* have a component of energy at any other frequency? The fact is, it does. To see why, though, we must look a little more closely at the definition of frequency.

Definition of "Frequency." As defined in Chap. 4, the frequency of a sinusoidal signal is indeed the number of cycles the signal completes per second. But that definition, you may recall, was qualified as applying strictly to a continuous, unmodulated signal.

Although not generally thought of in this way, a pulsed radio wave, such as is transmitted by a radar, is actually a continuous wave (carrier) whose amplitude is modulated by a pulsed video signal. The latter has an amplitude of one during each pulse and zero during the periods between pulses (Fig. 11).

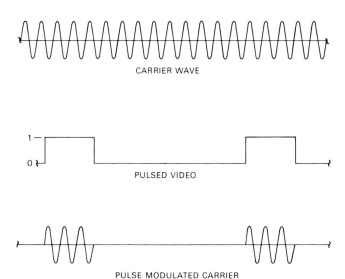

CARRIER WAVE

PULSED VIDEO

PULSE MODULATED CARRIER

11. A coherent pulsed radio-frequency signal is actually a continuous wave (carrier) whose amplitude is modulated by a pulsed video signal.

As we learned in Chap. 5, any wave whose amplitude is modulated invariably has sidebands, and a portion of the wave's energy is contained in each of these. (The signal of an AM broadcast station is an example.)

So one way of explaining how the energy of a pulsed radio wave is distributed in frequency is to visualize the spectrum in terms of the sidebands produced by a pulsed modulating signal. Fortunately, we can determine the nature of the sidebands quite easily, with the help of the Fourier series.

Fourier Series

It can be demonstrated both graphically and mathematically that any continuous, periodically repeated wave shape, such as the pulsed modulating signal just referred to, can be created by adding together a series of sine waves of specific amplitudes and phases whose frequencies are integer multiples of the repetition frequency of the wave shape. The repetition frequency is called the *fundamental*; the multiples of it are called *harmonics*. The mathematical expression for this collection of waves is the Fourier series. (See panel at top of next page.)

THE FOURIER SERIES

Any well-behaved periodic function of time, f(t), that can be assumed to continue from the beginning to the end of time can be represented by the sum of a constant, a_0, plus a series of sine terms whose frequencies are integer multiples of the repetition frequency, f_r.

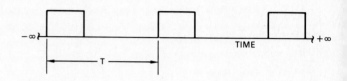

$$f(t) = A_0 + A_1 \sin(\omega_0 t + \phi_1) + A_2 \sin(2\omega_0 t + \phi_2) + A_3 \sin(3\omega_0 t + \phi_3) + A_4 \sin(4\omega_0 t + \phi_4) \ldots$$

(First Harmonic)   (Second Harmonic)   (Third Harmonic)   (Fourth Harmonic)

where $\omega_0 = 2\pi f_r$, $f_r = 1/T$, and $\phi_1, \phi_2, \phi_3, \phi_4 \ldots$ are the phases of the harmonics.

The phase angles can be eliminated by resolving the terms into in-phase and quadrature components.

$$f(t) = a_0 + \underbrace{a_1 \cos(\omega_0 t) + b_1 \sin(\omega_0 t)}_{} + \underbrace{a_2 \cos(2\omega_0 t) + b_2 \sin(2\omega_0 t)}_{} + \underbrace{a_3 \cos(3\omega_0 t) + b_3 \sin(3\omega_0 t)}_{} \ldots$$

(First Harmonic)   (Second Harmonic)   (Third Harmonic)

The complete series can be written compactly as the summation of n terms for which n has values of 1, 2, 3 . . . out to infinity.

$$f(t) = a_0 + \sum_{n=1}^{\infty} a_n \cos n\omega_0 t + b_n \sin n\omega_0 t$$

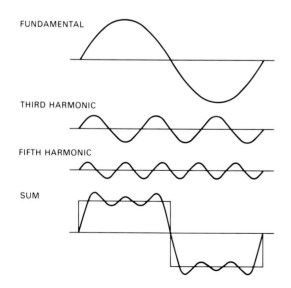

12. Square wave produced by adding two harmonics to the fundamental. (Because positive and negative excursions are of equal duration, amplitude of even harmonics is zero.)

Application to Rectangular Waves. The concept is illustrated graphically for a square wave in Fig. 12.

As you can see, the shape of the composite wave depends as much upon the phases of the harmonics as upon their amplitudes. To produce a rectangular wave, the phases must be such that all harmonics go through a positive or negative maximum at the same time as the fundamental.

A wave of a more general rectangular shape is illustrated in Fig. 13 at the top of the facing page.

Theoretically, to produce a perfectly rectangular wave an infinite number of harmonics would be required. Actually, the amplitudes of the higher order harmonics are relatively small so reasonably rectangular wave shapes can be produced with a limited number of harmonics. For example, a recognizably rectangular wave has been produced in Fig. 12 by adding only two harmonics to the fundamental.

214

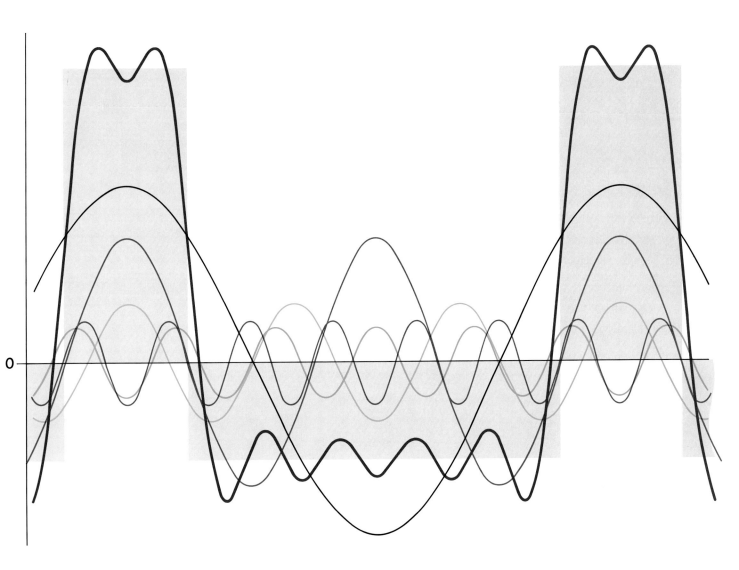

13. Rectangular wave produced by adding four harmonics to the fundamental. Shape of composite wave is determined by relative amplitudes and phases of harmonics. For shape to be rectangular, all harmonics must go through a positive or negative maximum at the same time as the fundamental.

And a still more rectangular wave has been produced in Fig. 13 (above) by adding only four harmonics.

The more harmonics included, the more rectangular the wave will be and the less pronounced the ripple.

In Fig. 14, the ripple has been reduced to negligible proportions—except at the sharp corners—by including 100 harmonics.

If the corners were rounded, as they commonly are in practice—e.g. the shape of a radar's transmitted pulses and the shape of the received echoes or the shape of the pulses in a digital computer—even this ripple would be negligible.

To create a train of *pulses*—i.e., a waveform whose amplitude alternates between zero and, say, one—with a series of sine waves, a zero-frequency, or dc, component must be added.

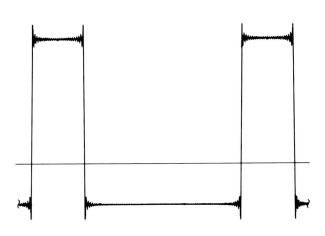

14. Rectangular waveshape produced by combining 100 harmonics. Note reduction in ripple.

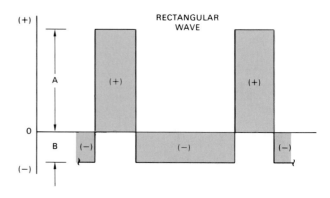

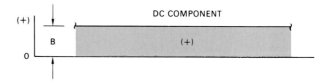

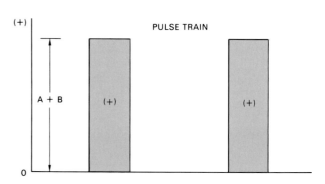

15. To produce a train of rectangular pulses from a rectangular wave, a dc component must be added.

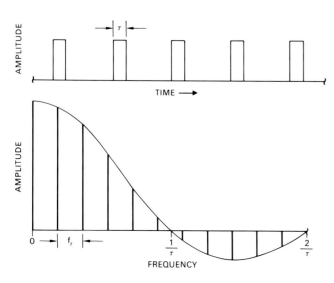

16. Portion of an infinitely long rectangular pulse train and the spectrum of the train.

Its value equals the amplitude of the negative loops of the rectangular wave, with sign reversed (Fig. 15, above).

Spectrum of a Train of Pulses. A portion of an infinitely long train of rectangular pulses is plotted in Fig. 16. Beneath it is a plot of amplitude versus frequency for the individual waves which would have to be added together to produce the waveform—the wave's spectrum. The equation relating the spectrum to the waveform is called the *Fourier transform*—see panel on facing page.

Each line of this spectrum, except the zero-frequency line, represents a sine wave which goes through a maximum at the same time as the fundamental. The phases of the waves are thus implicit in the plot. In alternate lobes of the envelope, the phase of the harmonics is shifted by 180°, a fact indicated by plotting the amplitudes of these harmonics as negative.

HE FOURIER TRANSFORM

Time and Frequency Domains. A graph (or equation) ·lating the amplitude of a signal to time represents the signal in hat is called the time domain.

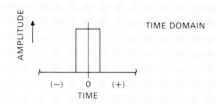

graph (or equation) relating the amplitude and phase of the gnal to frequency (the signal's spectrum) represents the signal what is called the frequency domain.

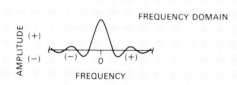

signal can be represented completely in either domain. onsequently, what one does to the signal shows up in both presentations.

witching Between Domains. The representation of a signal in ne domain can readily be transformed into the equivalent presentation in the other domain.

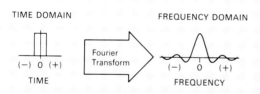

e mathematical expression for transforming from the time main to the frequency domain is called the Fourier transform.

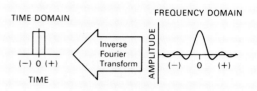

The mathematical expression for transforming from the frequency domain to the time domain is called the *inverse* Fourier transform. Together, the two transforms are called a transform pair.

Deriving the Transforms. To derive the Fourier transform, you write an expression for the signal as a function of time, substitute it for f(t) in the following equation

$$F(\omega) = \int_{-\infty}^{+\infty} f(t)\, e^{-j\omega t}\, dt$$

and perform the indicated integration.

Similarly, to derive the inverse transform, you write an expression for the signal as a function of ω, substitute it for F(ω) in this next equation

$$f(t) = \int_{-\infty}^{+\infty} \frac{1}{2\pi} F(\omega)\, e^{+j\omega t}\, d\omega$$

and perform the indicated integration. The variable ω is frequency in radians per second ($\omega = 2\pi f$), and $e^{-j\omega t}$ is the exponential form of the expression, $\cos \omega t - j \sin \omega t$.

Value of the Concept. The concept of the two domains and transformation between them is immensely useful. In radar work, in fact, it is indispensable. The crux of modern signal processing design is translating from one representation to the other. Range resolution and range measurement (except for chirp) may be readily perceived only in the time domain. Doppler resolution, doppler range-rate measurement, and certain aspects of high resolution ground mapping, on the other hand, may be readily perceived only in the frequency domain.

Since all of the waves are integer multiples of the fundamental, the frequency of which is f_r, the spacing between lines is f_r.

The first null in the envelope within which the lines fit occurs at a frequency equal to one divided by the pulsewidth, $1/\tau$. Subsequent nulls occur at multiples of $1/\tau$.

Spectrum of a Pulse Modulated Radio Wave. As explained in Chap. 5, when the amplitude of a carrier wave of frequency f_c is modulated by a single sine wave of frequency f_m, two sidebands are produced—one f_m above f_c, the other f_m below f_c.

Therefore, when the carrier of a coherent transmitter is modulated by a pulsed video signal, such as that illustrated in Fig. 17 (below), the sine wave represented by each line in the spectrum of the modulating signal produces two sidebands.

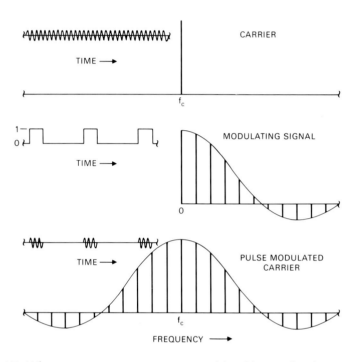

17. When a continuous carrier wave is modulated by an infinitely long pulsed video signal, each harmonic of the video signal produces a sideband above and below the carrier frequency.

The fundamental produces sidebands f_r hertz above and below the carrier. The second harmonic produces sidebands $2f_r$ above and below the carrier, and so on. The zero frequency line, of course, produces an output at the carrier frequency.

The spectrum of the envelope is thus mirrored above and below the carrier frequency. The resulting radio frequency spectrum is exactly the same as the spectrum we obtained for a continuous train of coherent pulses in Experiment 3 of the preceding chapter.

What the Spectral Lines Represent. One aspect of the spectrum of a pulsed carrier wave which some people have difficulty seeing is that each of the individual spectral lines represents a continuous wave. That is, a wave of constant amplitude and constant frequency, which continues uninterruptedly in time from the beginning to the end of the pulse train (Fig. 18). How can that be, when the transmitter is "on" for only a fraction of each interpulse period?

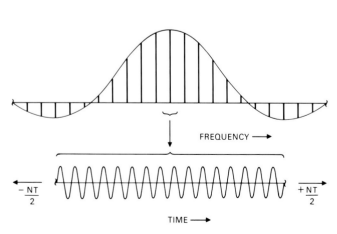

18. Each line in the spectrum of a pulse modulated carrier represents a single sine wave the length of the pulse train.

Briefly, the explanation is this. The amplitudes and phases of the fundamental and its harmonics are such that they completely cancel the carrier, as well as each other, during the periods between pulses. Yet they combine to produce a signal having the carrier's wavelength and the full power of the transmitter, during the brief period of each pulse.

This was illustrated for a rectangular wave in Fig. 14 and is illustrated in a cursory way for a pulsed radio wave by the phasor diagrams of Fig. 19.

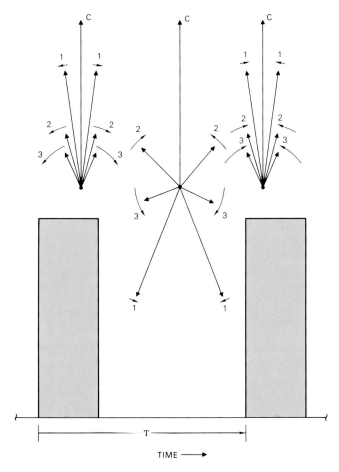

TIME ⟶

19. Phasors representing carrier (c) and first several sidebands of an infinitely long pulse train. During the pulses, they combine constructively. Between pulses, they cancel.

They show how the carrier and the first three sidebands above and below the carrier combine to produce the transmitted pulses. The phasor representing the carrier is synchronized with the strobe that provides the phase reference for the phasors (see Chap. 5).

Therefore, the phasors representing the upper sidebands rotate counterclockwise and the phasors representing the lower sidebands rotate clockwise. The higher the order of the individual sidebands, the more rapidly the phasors rotate.

4. Except the phasors representing harmonics in the odd numbered sidelobes, not shown in the figure. They are 180° out of phase with the others.

Since the harmonics are all integer multiples of the fundamental frequency and it equals the pulse repetition frequency, once every repetition period, all of the phasors line up. At this point, the phasors add constuctively.[4] Thereafter, the counter-rotating phasors rapidly fan out. Pointing essentially in opposite directions, they cancel the carrier and each other for the balance of the period, only to come together once again at the beginning of the next period.

That the waves represented by the phasors are continuous is borne out by the fact that when a pulsed signal is applied to a narrowband *analog* filter tuned to the frequency of one of the spectral lines, the filter's output is a continuous signal.

A pulsed signal has a true line spectrum, though, only if the signal is infinitely long. Otherwise, the spectral lines have a finite width. And how does the Fourier series tell us what the width is? This question can be answered most simply in terms of the spectrum of a single pulse. So let us first see what the Fourier series tells us about that.

Spectrum of a Single Pulse. Strictly speaking, the Fourier series applies to a signal only if the signal has a repetitive waveform that can be assumed to continue uninterruptedly from the beginning to the end of time. In some cases, though, we can safely make this assumption, even though the waveform may not be repetitive at all.

This is true in the case of a single rectangular pulse. We start with a continuously repetitive form of the pulse (Fig. 20a).

Keeping the pulse width constant, we gradually decrease the repetition frequency. As we do so, the lines of the pulsed signal's spectrum move closer and closer together (Fig. 20b). The envelope within which they fit, however, retains its original shape, since that is determined solely by the pulse width.

If we continue this process, stretching the time between pulses to weeks, to years, to eons, to an infinite number of eons, the separation between spectral lines ultimately disappears.

We end up with a single pulse and a continuous spectrum that has exactly the same shape as the envelope of the line spectrum of the continuously repetitive waveform (Fig. 20c). This, you may recall, is what we found the spectrum of a single pulse to be in Experiment 2 of the last chapter.

Incidentally, if we pursue the above logic a step further, we are led to an interesting conclusion. Since the pulse train of which this single pulse is actually a part is infinitely long, every point in the pulse's spectrum represents a con-

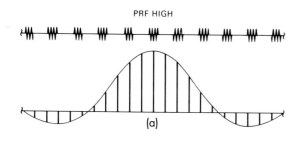

PRF HIGH

(a)

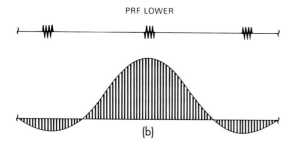

PRF LOWER

(b)

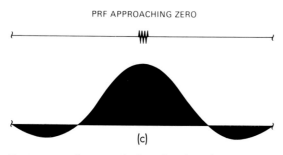

PRF APPROACHING ZERO

(c)

20. Continuous pulse train of infinite length and its spectrum. As PRF is reduced, spectral lines move closer together. As PRF approaches zero, spectrum becomes continuous.

tinuous wave of infinite duration. How can that be?

Of course, no wave we shall ever see will have extended to the end of time—and a lucky thing too. However, the spectra of the signals we are considering here are exactly the same *as if* the signals were comprised of infinitely long waves.

So, in modeling spectral characteristics, it matters little whether such long waves actually exist. With that point settled, let us return to the question of what the Fourier series tells us about spectral line width.

Line Width. Knowing the spectrum of a single pulse, we can easily find the spectrum of a pulse train of limited length, such as a radar would receive from a target. Suppose we wish to find the spectrum of a train of N pulses, having an interpulse period T, hence a total length NT.

We start by imagining an infinitely long pulse train. Each line of the spectrum of this train, remember, represents a continuous wave having a single frequency and an infinite duration—a true CW signal. Holding the PRF and pulse width constant, we gradually reduce the length of the train (Fig. 21).

Since the constituent CW signals are the same length as the train, each of them now becomes a single long pulse. As the length of this pulse decreases, the spectral line representing it gradually broadens into a sin x/x shape. When we finally reach the length, NT, of the pulse train in question, the null-to-null width of the central lobe of this "line" equals 2/(NT).

Thus, the Fourier series indirectly tells us that the spectrum of a pulse train of limited length differs from the spectrum of a train of infinite length only in that each spectral line has a sin x/x shape. The null-to-null width of the line is inversely proportional to the length of the pulse train.

$$\text{Line width} \quad = \quad \frac{2}{\text{Length of pulse train}}$$

That, you'll recall, is exactly what we found the line width to be in Experiments 4 and 5 of the preceding chapter.

Some people find the foregoing explanation a bit unsatisfying. Somehow, it makes them a little uneasy to assume—even though only for the sake of illustration—that, when you key a transmitter "on" for a millionth of a second, you are in effect transmitting myriad radio waves, waves that began ages ago and will continue for ages to come, waves that cancel one another completely throughout all of this vast time, except for the glorious fraction of a second when the transmitter is actually "on."

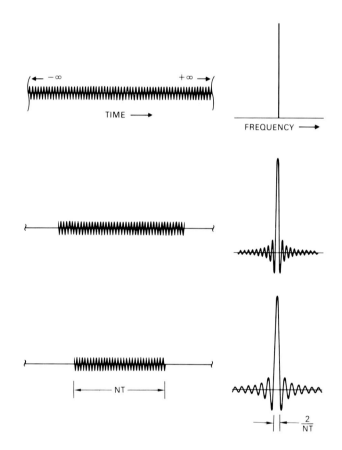

21. CW wave represented by a single line in the spectrum of an infinitely long pulse train. As the length of train is reduced, this wave becomes a single pulse and its spectrum broadens into a sin x/x shape.

If you are like these people, you may find it is helpful to consider the spectrum of a pulsed signal from the point of view of a lossless narrowband filter.

Spectrum Explained from a Filter's Point of View

As was explained on page 212, a lossless narrowband filter integrates the energy of a signal (wave) in such a way that the filter's output builds up to a large amplitude only if the frequency of the signal is the same as that to which the filter is tuned.

In essence, the filter determines how close the frequencies are to being the same by sensing the shift, if any, in the phases of successive cycles of the input signal, relative to a signal whose frequency is that of the filter.

Analogy of a Filter to a Ruler. If the wave crests of the input signal are represented graphically by a series of vertical lines spaced at intervals equal to a wavelength, the filter can be thought of as measuring the spacing between wave crests with an imaginary ruler. On this ruler, marks are inscribed at intervals of one wavelength for the frequency to which the filter is tuned.

Quite obviously, if the filter is tuned to the exact frequency of the wave, when the first mark is lined up with a wave crest, all subsequent marks will similarly line up (Fig. 22).

If the filter is tuned to a slightly different frequency, however, the first mark beyond the initial one will be displaced slightly from the next wave crest; the second mark will be displaced twice as much from the following wave crest; the third mark, three time as much, and so on (Fig. 23). The displacements correspond to the phases of the individual cycles of the signal as seen by the filter; the progressive increase in displacement corresponds to the progressive shift in phase from once cycle to the next.

Now, the amplitude of each cycle of the wave, as well as the phase of that cycle relative to the corresponding mark on the ruler, can be represented by a phasor (Fig. 24). What the narrowband filter does is add up—integrate—the phasors for successive cycles (Fig 25). If n phasors point in

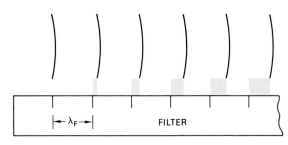

22. A lossless narrowband filter can be thought of as measuring the spacing of a signal's wave crests with an imaginary ruler.

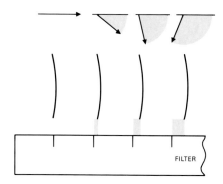

23. If the filter's frequency is higher than the signal's, the phase shift between the wave crests and the marks on the ruler builds up.

24. The amplitude and phase of each cycle of the wave can be represented by a phasor.

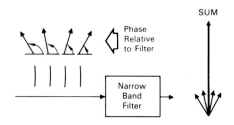

25. In essence, the filter adds up the phasors for successive cycles.

the same direction—i.e., the cycles they represent have the same phase—the sum will be n times the length of the phasors. If they point in slightly different directions, the sum will be less. And if they point in opposite directions—i.e., the cycles are 180° out of phase—they will cancel.

With this simple analogy in mind, let us analyze the results of some of the experiments performed in the preceding chapter.

Spectrum of a Single Pulse. To see why the spectrum of a single pulse is continuous over a band of frequencies 2/τ hertz wide, we measure a pulse τ seconds long with four different rulers. Each ruler represents a narrowband filter tuned to a different frequency.

In Fig. 26(a), the filter has the same frequency as the pulse's carrier (f_c). Consequently, the phases of the wave crests relative to the marks on the ruler are all the same. The phasors representing the individual cycles of the wave all point in the same direction. The pulse is eight cycles long. Assuming that the length of each phasor is one, their sum is eight.

In Fig. 26(b) the same pulse is applied to a filter having a higher frequency ($f_c + \Delta f$); the wavelength marks are closer together. As a result, there is a progressive shift in the phases of the wave crests relative to the marks. Over the length of the pulse, the shift builds up to a quarter of a wavelength. The phasors, therefore, fan out over 90°. Even so, their sum is nearly seven.

In Fig. 26(c), the filter has a considerably higher frequency ($f_c + 2\Delta f$). The total accumulated phase shift over the length of the pulse now is half a wavelength (180°). Still, the sum is nearly half what it was for the filter tuned to f_c.

In Fig. 26(d), the filter has a sufficiently high frequency ($f_c + 4\Delta f$) that the phase shift over the length of the pulse is one whole wavelength. As a result, the phasors are uniformly spread over 360°. Pointing in opposite directions, the phasors for cycles No. 1 and No. 5 cancel. So do the phasors for cycles No. 2 and No. 6, No. 3 and No. 7, and No. 4 and No. 8. The pulse produces no output from the filter; we have reached a frequency where there is a null in the pulse's spectrum.

What is this frequency? Over the duration of the pulse the oscillation of the filter that was tuned to the null frequency completed one more cycle than the pulse's carrier (Fig. 27). The duration of the pulse was τ seconds. So, the filter's frequency was 1/τ cycles per second (hertz) higher than the carrier frequency, f_c. The null frequency, therefore, is ($f_c + 1/\tau$).

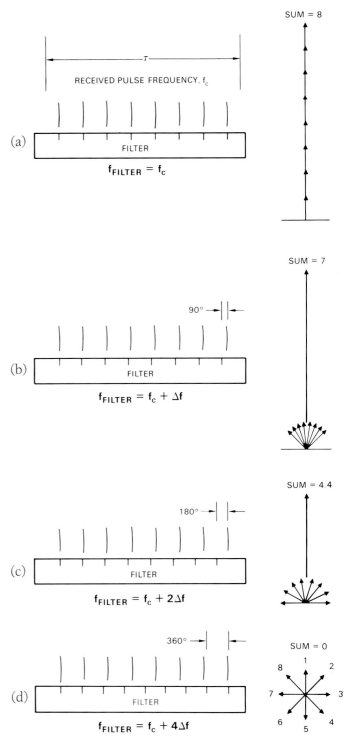

26. If a filter is tuned to progressively higher frequencies, the cumulative phase shift over the length of a pulse increases.

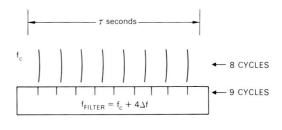

27. At null, in τ seconds filter completes one more cycle than signal.

223

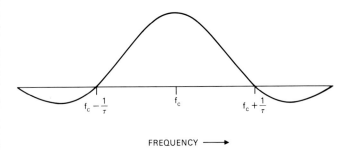

FREQUENCY ⟶

28. A plot of the phasor sums has a sin x/x shape with nulls $1/\tau$ hertz above and below the carrier frequency, f_c.

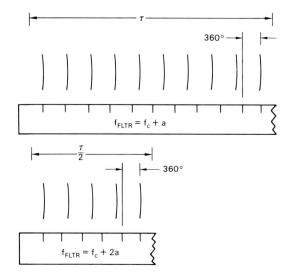

29. The shorter a pulse is, the greater the frequency difference must be to produce a 360° phase shift over the length of the pulse.

Following the same line of reasoning, we similarly find a null $1/\tau$ cycles per second below f_c (Fig. 28). The null-to-null bandwidth of a single pulse, therefore, is $2/\tau$ hertz—exactly as was observed in Experiment No. 2 of Chap. 16.

Upon a little reflection, the fundamental reason for the filter's response become clear. A difference in frequency is actually a continuous linear shift in phase. When a pulsed signal is applied to a filter, the rate of this shift in cycles per second equals the difference between the signal's carrier frequency and the frequency of the filter. In the case of a single pulse, only when this difference is large enough to make the total phase shift over the duration of the pulse equal one whole wavelength, do the individual cycles of the received wave entirely cancel. The shorter the pulse (Fig. 29), the greater the frequency difference must be to satisfy this condition. Conversely, the longer the pulse, the less the frequency difference must be. Thus, for a pulse whose duration is 1 microsecond, the null-to-null width of the central spectral lobe is $2 \div 10^{-6} = 2$ MHz. For a pulse whose duration is 1 second, the spectral line width is 2 hertz. And for a pulse whose duration is 1 hour, the spectral line width is only $2 \div (60 \times 60) \cong 0.00056$ Hz.

Spectrum of a Coherent Pulse Train. To see why the null-to-null width of the central spectral lobe is drastically reduced when the filter integrates a train of coherent pulses, we represent each pulse with a single phasor (Fig. 30).

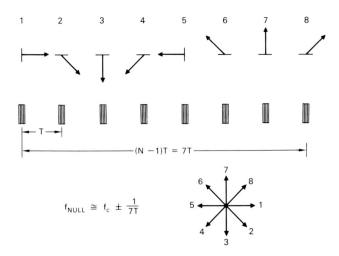

30. The amplitude and phase of each pulse in a train may be represented by a phasor. Spectral nulls occur when phase shift over length of train is 360°.

The nulls now occur when the total phase shift over the length of the pulse train is one wavelength. Since the train is many times longer than a single pulse, the frequency difference that produces a phase shift of one wavelength is many times smaller for the train than for a single pulse.

Take a train of 32 pulses, for example (Fig. 31). Suppose the interpulse period is 100 times the pulse width. The duration of the train, then, will be roughly 31 x 100 = 3100 times the duration of a single pulse, making the null-to-null bandwidth of the line only 1/3100 that of a single pulse.[5]

It is instructive to consider the effect of deleting every other pulse in this train. Because the length of the pulse train is essentially the same, the phasors for the remaining 16 pulses would still cancel at almost the same frequency; so the null-to-null bandwidth would be about the same. Since there would be only half as many pulses, however, the amplitude of the filter output would be only half as great. Finally, since the PRF would be only half as great, the pulses would produce an output from the filter at twice as many points within the envelope established by the pulse width. How do we explain why the pulse train produces an output at intervals equal to the PRF in the first place?

Repetition of Spectral Lines. As we have seen, when a train of pulses is applied to a filter, what causes the output of the filter to fall off as the filter is tuned away from the carrier frequency of the pulses is the pulse-to-pulse difference in the phase of the carrier, as seen by the filter. But, since phase angles are repeated every 360°, there is no way of telling whether the phase of any one pulse is the same as that of the preceding pulse or has been shifted by some multiple of 360°.

A pulse-to-pulse shift of 360° amounts to one cycle per interpulse period, corresponding to an increment of frequency equal to the PRF (Fig. 32). Consequently, there may be very little difference between a filter's response to a pulse train whose carrier frequency is the same as the filter frequency and its response to a pulse train whose carrier frequency is some integer multiple of the PRF above or below the filter frequency. In fact, the only difference is that due to the phase shift occurring from cycle to cycle over the duration of each pulse. Unless the multiple of the PRF is very high or the pulse width is a fairly large fraction of the interpulse period—i.e., unless the carrier frequency is near one end or the other of the envelope established by the pulse width—the difference is slight.

Mathematical Explanation of the Pulsed Spectrum

For those having at least a nodding acquaintance with calculus, the spectrum of a pulsed signal is derived mathematically in the following panel. If your interest is not so mathematical, then skip ahead to *"Results"* on page 230.

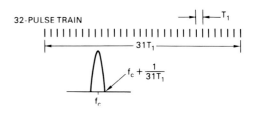

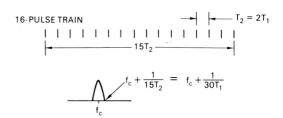

31. If every other pulse in a train is deleted, the output is reduced, but the bandwidth remains essentially unchanged.

5. Although the train contains 32 pulses, it is only one pulse width longer than 31 interpulse periods.

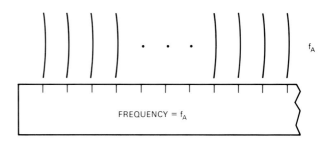

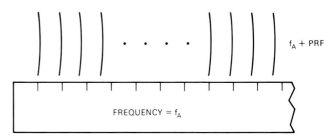

32. Pulse trains whose carrier frequencies equal the filter frequency, f_A, (top) and f_A plus the PRF (bottom). Only difference in the outputs produced by the two trains is that due to the cycle-to-cycle phase shift within each pulse of the second train.

MATHEMATICAL EXPLANATION OF THE PULSED SPECTRUM

In the text, the spectrum of a pulsed signal is explained in several quite different nonmathematical ways, at least one of which is a bit unconventional. While hopefully these explanations have provided some helpful insights, the spectrum can be explained much more rigorously *and* succinctly in purely mathematical terms.

Accordingly, a mathematical derivation of the spectrum of a simple, perfectly rectangular pulsed signal is presented on the third and fourth pages of this panel.

In case your math is a little rusty, a brief preliminary explanation of the derivation is given on this and the facing page.

General Approach.   Basically, the panel shows two things: first, the derivation of a mathematical expression for a pulse modulated carrier signal as a function of time, f(t), and second the transformation of this expression from the time domain to the frequency domain —in other words, the derivation of the Fourier transform for the signal.

The expression for the pulse modulated signal is derived by writing separate expressions for each of the following.

- An infinitely long pulsed video signal, $f_1(t)$, having an amplitude of 1, a pulse width τ, an interpulse period T, and a pulse repetition frequency (expressed in radians per second) of ω_0 where ($\omega_0 = 2\pi/T = 2\pi f_r$).

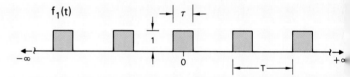

- A signal, $f_2(t)$, having an amplitude of 1 and a duration equal to the length of a train of N pulses whose interpulse period is T.

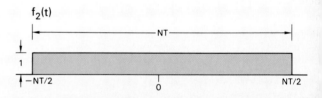

- An infinitely long carrier wave, $f_3(t)$, having an amplitude A and a frequency expressed in radians per second of ω_c, where ($\omega_c = 2\pi f_c$).

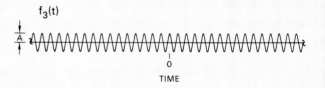

By positioning zero on the time axis in the center of one of the pulses, the coefficients of the sine terms are reduced to zero. (The signal has even symmetry.)

The expression for the pulsed video signal, $f_1(t)$, is obtained b evaluating the coefficients (a_0, a_1, a_2 . . .)* of the Fourier series in terms of the pulse width, τ, and interpulse period, T, of the signal and substituting these values into the series.

Multiplying the first two functions, $f_1(t)$ and $f_2(t)$, together gives an equation for the pulsed *modulating* signal. Multiplying the function for the carrier wave, $f_3(t)$, by this product yields th desired equation for the pulse *modulated* carrier, $f_4(t)$.

The Fourier transform of this function is then derived, yieldin the spectrum of the pulse modulated signal. The essence of bo derivations is briefly outlined on the next page.

Essence of the Derivations. In evaluating the coefficients of the Fourier series for the pulsed video signal, the key operation is multiplying the equation for the signal by cos ωt for the frequency of the harmonic whose coefficient we wish to obtain.

As illustrated in the figure below, when the instantaneous amplitudes of two sine waves of the same phase and frequency are multiplied together, their product is positive for both halves of every cycle. (The same is true of cosine waves.)

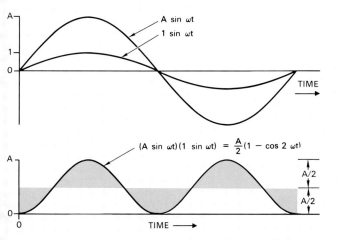

Consequently, if the product is integrated over a complete cycle, the result divided by the period of the cycle is one-half the product of the peak amplitudes of the two waves. If the amplitude of one of the waves is one, then the product is one-half the peak amplitude of the other wave.

$$\frac{1}{T}\int_0^T \frac{A}{2}(1-\cos 2\omega t)\,dt = \frac{1}{T}\left(\frac{At}{2}\right)\Big|_0^T = \frac{A}{2}$$

Yet, if the frequencies of the two waves are not the same, the sign of the product will alternate between (+) and (−). If the frequency of one wave is an integer multiple of the frequency of the other, when the product is integrated over the period of the lower frequency wave, the result will be zero.

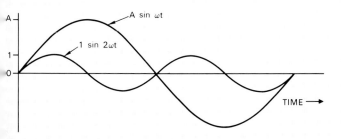

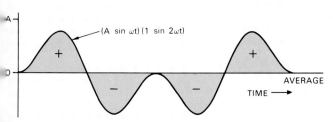

Thus, the coefficients of the Fourier series for our continuously repeating pulsed video signal can be found by multiplying the

mathematical expression for the waveform by the cosines of $\omega_0 t$, $2\omega_0 t$, $3\omega_0 t$, ... $n\omega_0 t$, in turn, integrating each product over the waveform's repetition period, T, and dividing by T/2. The dc coefficient (average amplitude) is found by integrating the expression for the wave alone over the period T and dividing by T.

Similarly, in deriving the Fourier transform, that component of the pulse modulated wave having a particular frequency, ω, can be found by multiplying the equation for the wave (as a function of time) by cos ωt − j sin ωt and integrating the product. In this case, since ω is not necessarily an integer multiple of the fundamental of the modulated wave, the product must be integrated over the entire duration of the pulse train, i.e., from −NT/2 to +NT/2. As with the Fourier series, the dc component is found by integrating the expression for the wave alone over the same period and dividing by its duration.

In deriving the Fourier transform, the sinusoidal functions are expressed most conveniently in exponential form. The relationships between the two forms were explained with phasors in Chapter 5 and are summarized in the diagram below.

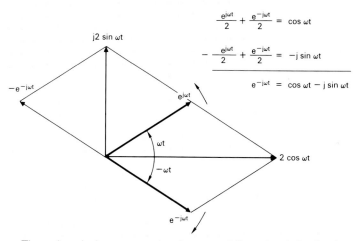

$$\frac{e^{j\omega t}}{2} + \frac{e^{-j\omega t}}{2} = \cos \omega t$$

$$-\frac{e^{j\omega t}}{2} + \frac{e^{-j\omega t}}{2} = -j \sin \omega t$$

$$e^{-j\omega t} = \cos \omega t - j \sin \omega t$$

The only calculus you need to know to follow the derivation is that the integral of the cosine of ωt is 1/ω times the sine of ωt and the integral of $e^{j\omega t}$ is −1/jω times $e^{-j\omega t}$.

$$\int \cos(\omega t)\,dt = \frac{1}{\omega}\sin \omega t$$

$$\int e^{-j\omega t}\,dt = \frac{1}{-j\omega}e^{-j\omega t}$$

One other reminder. If a quantity raised to a given power is multiplied by the same quantity raised to another power, the product is the quantity raised to the sum of the two powers. Thus,

$$e^{j\omega t} \times e^{j\omega_0 t} = e^{(j\omega t + j\omega_0 t)} = e^{j(\omega+\omega_0)t}$$

With the above relationships in mind, let us proceed with the derivations. The expression for the pulse modulated carrier is derived on the next page; the Fourier transform of this expression, on the facing page.

MATHEMATICAL EXPLANATION OF THE PULSED SPECTRUM (Con't.)

1. Continuous Pulsed Modulation Signal (Expressed as a Fourier Series)

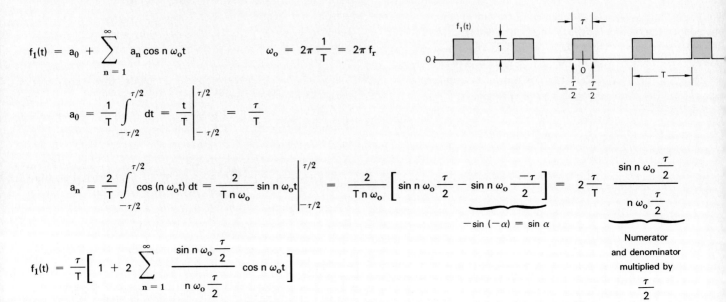

$$f_1(t) = a_0 + \sum_{n=1}^{\infty} a_n \cos n \omega_o t \qquad \omega_o = 2\pi \frac{1}{T} = 2\pi f_r$$

$$a_0 = \frac{1}{T} \int_{-\tau/2}^{\tau/2} dt = \frac{t}{T} \Big|_{-\tau/2}^{\tau/2} = \frac{\tau}{T}$$

$$a_n = \frac{2}{T} \int_{-\tau/2}^{\tau/2} \cos(n \omega_o t)\, dt = \frac{2}{T n \omega_o} \sin n \omega_o t \Big|_{-\tau/2}^{\tau/2} = \frac{2}{T n \omega_o}\left[\sin n \omega_o \frac{\tau}{2} - \underbrace{\sin n \omega_o \frac{-\tau}{2}}_{-\sin(-\alpha)\,=\,\sin \alpha}\right] = 2\frac{\tau}{T}\underbrace{\frac{\sin n \omega_o \frac{\tau}{2}}{n \omega_o \frac{\tau}{2}}}_{\substack{\text{Numerator} \\ \text{and denominator} \\ \text{multiplied by} \\ \frac{\tau}{2}}}$$

$$f_1(t) = \frac{\tau}{T}\left[1 + 2 \sum_{n=1}^{\infty} \frac{\sin n \omega_o \frac{\tau}{2}}{n \omega_o \frac{\tau}{2}} \cos n \omega_o t\right]$$

2. Duration of Pulse Modulation

$$f_2(t) = 1 \qquad \frac{-NT}{2} \le t \le \frac{NT}{2}$$

$$= 0 \qquad t < \frac{-NT}{2} \quad\text{and}\quad t > \frac{NT}{2}$$

3. Unmodulated Carrier

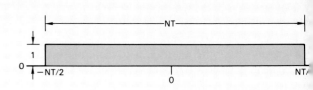

$$f_3(t) = A \cos \omega_c t$$

4. Pulse Modulated Carrier (Product of expressions 1, 2, and 3)

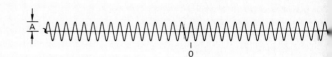

$$f_4(t) = f_1(t) \cdot f_2(t) \cdot f_3(t)$$

$$= \frac{A\tau}{T}\left\{1 + 2 \sum_{n=1}^{\infty} \frac{\sin n \omega_o \frac{\tau}{2}}{n \omega_o \frac{\tau}{2}} (\cos n \omega_o t)\right\} \cos \omega_c t \qquad \frac{-NT}{2} \le t \le \frac{NT}{2}$$

$$= \frac{A\tau}{T}\left\{\cos \omega_c t + \sum_{n=1}^{\infty} \frac{\sin n \omega_o \frac{\tau}{2}}{n \omega_o \frac{\tau}{2}}\left[\cos(\omega_c + n \omega_o)t + \cos(\omega_c - n \omega_o)t\right]\right\}$$

228

. Fourier Transform of Pulse Modulated Carrier

$$F(j\omega) = \int_{-\infty}^{\infty} e^{-j\omega t} f_4(t)\, dt$$

$$F(j\omega) = \frac{A\tau}{T}\left[\underbrace{\int_{-NT/2}^{NT/2} e^{-j\omega t}\cos\omega_c t\, dt}_{①} + \sum_{n=1}^{\infty}\frac{\sin n\omega_o \frac{\tau}{2}}{n\omega_o \frac{\tau}{2}}\left\{ \underbrace{\int_{-NT/2}^{NT/2} e^{-j\omega t}\cos(\omega_c + n\omega_o)t\, dt}_{②} + \underbrace{\int_{-NT/2}^{NT/2} e^{-j\omega t}\cos(\omega_c - n\omega_o)t\, dt}_{③} \right\} \right]$$

$$① = \frac{1}{2}\int_{-NT/2}^{NT/2} e^{-j\omega t}(e^{j\omega_c t} + e^{-j\omega_c t})\, dt = \frac{1}{2}\int_{-NT/2}^{NT/2} e^{-j(\omega + \omega_c)t}\, dt + \frac{1}{2}\int_{-NT/2}^{NT/2} e^{-j(\omega - \omega_c)t}\, dt$$

$$= \frac{e^{-j(\omega + \omega_c)t}}{-2j(\omega + \omega_c)}\Bigg|_{-NT/2}^{NT/2} + \frac{e^{-j(\omega - \omega_c)t}}{-2j(\omega - \omega_c)}\Bigg|_{-NT/2}^{NT/2}$$

$$= \frac{e^{-j(\omega + \omega_c)NT/2} - e^{j(\omega + \omega_c)NT/2}}{-2j(\omega + \omega_c)} + \frac{e^{-j(\omega - \omega_c)NT/2} - e^{j(\omega - \omega_c)NT/2}}{-2j(\omega - \omega_c)}$$

$$= \frac{NT}{2}\left\{ \frac{\sin\left[(\omega + \omega_c)\frac{NT}{2}\right]}{(\omega + \omega_c)\frac{NT}{2}} + \frac{\sin\left[(\omega - \omega_c)\frac{NT}{2}\right]}{(\omega - \omega_c)\frac{NT}{2}} \right\}$$

$$② = \frac{NT}{2}\left\{ \frac{\sin\left[(\omega + \omega_c + n\omega_o)\frac{NT}{2}\right]}{(\omega + \omega_c + n\omega_o)\frac{NT}{2}} + \frac{\sin\left[(\omega - \omega_c - n\omega_o)\frac{NT}{2}\right]}{(\omega - \omega_c - n\omega_o)\frac{NT}{2}} \right\}$$

$$③ = \frac{NT}{2}\left\{ \frac{\sin\left[(\omega + \omega_c - n\omega_o)\frac{NT}{2}\right]}{(\omega + \omega_c - n\omega_o)\frac{NT}{2}} + \frac{\sin\left[(\omega - \omega_c + n\omega_o)\frac{NT}{2}\right]}{(\omega - \omega_c + n\omega_o)\frac{NT}{2}} \right\}$$

$$F(j\omega) = \frac{A\tau N}{2}\left[\overset{\text{Carrier}}{\frac{\sin\left[(\omega + \omega_c)\frac{NT}{2}\right]}{(\omega + \omega_c)\frac{NT}{2}}} + \overset{\text{Envelope}}{\sum_{n=1}^{\infty}\frac{\sin n\omega_o \frac{\tau}{2}}{n\omega_o \frac{\tau}{2}}}\left\{ \overset{\text{Lower Sidebands}}{\frac{\sin\left[(\omega + \omega_c + n\omega_o)\frac{NT}{2}\right]}{(\omega + \omega_c + n\omega_o)\frac{NT}{2}}} + \overset{\text{Upper Sidebands}}{\frac{\sin\left[(\omega + \omega_c - n\omega_o)\frac{NT}{2}\right]}{(\omega + \omega_c - n\omega_o)\frac{NT}{2}}} \right\} \right.$$

$$\left. + \frac{\sin\left[(\omega - \omega_c)\frac{NT}{2}\right]}{(\omega - \omega_c)\frac{NT}{2}} + \sum_{n=1}^{\infty}\frac{\sin n\omega_o \frac{\tau}{2}}{n\omega_o \frac{\tau}{2}}\left\{ \frac{\sin\left[(\omega - \omega_c + n\omega_o)\frac{NT}{2}\right]}{(\omega - \omega_c + n\omega_o)\frac{NT}{2}} + \frac{\sin\left[(\omega - \omega_c - n\omega_o)\frac{NT}{2}\right]}{(\omega - \omega_c - n\omega_o)\frac{NT}{2}} \right\} \right]$$

229

Results. The final equation obtained in the panel on preceeding page is the Fourier transform for a train of N perfectly rectangular pulses having these characteristics:

- Carrier frequency, $\omega_c = 2\pi f_c$

- Pulse width, τ

- PRF, $\omega_o = 2\pi f_r$

- Interpulse period, T

- Duration, NT

The transform consists of two similar sets of terms. The first set applies to frequencies having negative values; the second set, to frequencies having positive values.

So we can examine it more easily, the positive-frequency portion of the transform is repeated in Fig. 33. The first term inside the braces represents the spectrum of the central spectral line—the carrier. Immediately following the summation sign is the sin x/x term giving the envelope within which the other spectral lines fit. The remaining terms represent the lines above and below the carrier.

By substituting appropriate values for N (number of pulses), one can apply this same equation to pulse trains of virtually any length.

Beneath the equation is a plot of the spectrum—amplitude versus frequency in radians per second—obtained by

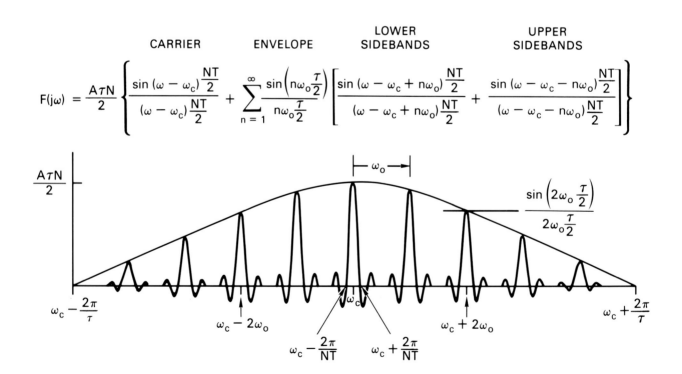

33. Positive-frequency portion of the Fourier transform for a rectangular train of N pulses. The pulses have a width τ, a carrier frequency of ω_c, a PRF of ω_0, and an interpulse period of T.

evaluating the equation for values of ω covering a wide enough range of positive frequencies to include the entire central lobe of the envelope. The first pair of nulls in the envelope occurs $2\pi/\tau$ radians per second above and below the carrier frequency ω_c. Within the envelope, spectral lines occur above and below the carrier frequency, at intervals equal to the PRF, ω_0. Each line has a sin x/x shape, with nulls $2\pi/NT$ above and below the line's central frequency.

Needless to say, these results are identical with those deduced earlier in the chapter, first, with the aid of the Fourier series and, second, simply with phasors. Hopefully, one or another of these explanations has removed the veil of mystery (if indeed there was one) from the spectrum of the pulsed signal.

Significance of the Negative-Frequency Terms. Many people are puzzled by the negative-frequency components of the Fourier transform. Yet they needn't be.

It so happens that these components are what reflect the difference between the transform for a signal whose carrier is a cosine wave and the transform for a signal whose carrier is a sine wave. In the case of the transform for a cosine wave (Fig. 34A), such as the one we just derived (carrier is A cos ω_ct), the algebraic signs of the negative-frequency terms are the same as the signs of the corresponding positive-frequency terms. Whereas, in the case of the transform for a sine wave (carrier is A sin ω_ct), the signs of the negative-frequency terms (Fig. 34B) are the opposites of the signs of the corresponding positive-frequency terms.

If the signal is considered alone, the negative-frequency terms have no significance—i.e., contribute no additional information regarding what frequencies are present. Since cos ($-\omega$t) = cos ωt, the energy represented by the negative-frequency terms of the transform for a cosine wave merely adds to the energy represented by the corresponding positive-frequency terms. And since sin ($-\omega$t) = $-$sin ωt, the energy represented by the negative-frequency terms of the transform for a sine wave likewise merely adds to the energy represented by the positive frequency terms.

However, in the case of a signal that has been resolved into I and Q components (see Chap. 5, page 67), the negative-frequency terms do contribute additional information. As we shall see in Chap. 19, when the signal is translated to the video range, the signal's Fourier transform will have only negative-frequency terms if the frequency of the original signal was lower than that of the reference signal used in the frequency translation. And the transform will have only positive-frequency terms if the frequency of the original signal was higher.

TRANSFORM FOR COSINE WAVE
(Carrier = A cos ω_ct)

A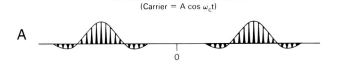

TRANSFORM FOR SINE WAVE
(Carrier = A sin ω_ct)

B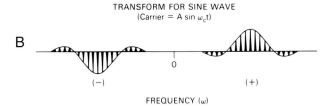

($-$)   ($+$)

FREQUENCY (ω)

34. Comparison of frequency spectra for a signal whose carrier is a cosine wave and a signal whose carrier is a sine wave. Note that the algebraic signs of the negative frequency terms are reversed for the sine wave.

What the Amplitudes Represent. One important question, though, remains to be answered. All of the spectra shown thus far have been plots of amplitude versus frequency. But nothing has been said about how this amplitude relates to the amplitude of the wave in the time domain, or even what units it is expressed in.

We can clear up this deficiency by examining the first term of the Fourier transform—the term which establishes the peak amplitude of the envelope

$$\frac{A\tau N}{2}$$

The factor, A, you may recall, was defined in the panel as the peak amplitude of the carrier wave. Assuming that A is a voltage, then the spectrum is a plot of voltage versus frequency.

Going a step further, since power is proportional to voltage squared, by squaring the values of amplitude given by the Fourier transform, we can obtain the *power* spectrum of the pulsed signal (Fig. 35).

Energy, of course, is power times time. It can be shown mathematically[6] that the total area under the power spectrum equals the total *energy* of the pulsed signal. The power spectrum thus illustrates how the energy of the signal is distributed in frequency. For example, by measuring the area encompassed by the central line of the power spectrum and dividing it by the total area encompassed by the spectrum, we can tell what fraction of the signal's energy is contained in that line.

6. Parseval's formula.

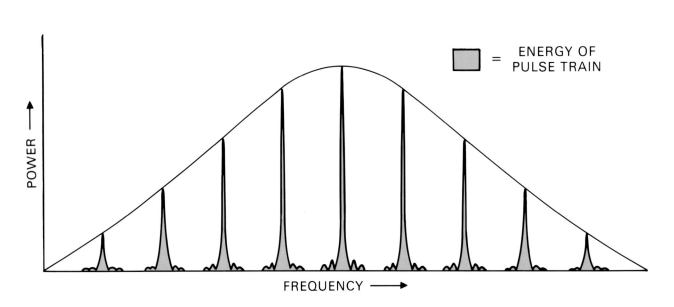

35. If the amplitude represented by a signal's Fourier transform is a voltage, then a plot of the amplitude squared versus frequency is the signal's power spectrum, and the area under this plot corresponds to the signal's energy.

Summary

The spectrum of a signal is the distribution of the signal's energy over the range of possible frequencies. One way of explaining this distribution is to envision the signal as being applied simultaneously to myriad lossless narrowband filters whose frequencies are infinitesimally closely spaced and cover the complete range of frequencies. Each filter can be envisioned as a pendulum suspended from a frictionless pivot in a vacuum and driven by the reactive force of an eccentric flywheel rotating at the frequency of the input signal.

A pulsed radio frequency signal such as a radar transmits is actually a continuous wave (carrier) whose amplitude is modulated by a video signal having an amplitude of one during each pulse and zero between pulses. So another way of explaining how the energy of a pulsed radio wave is distributed in frequency is in terms of the sidebands produced by the video modulating signal.

A continuous rectangular wave, such as the modulating signal, can be constructed by adding together a series of sine waves of appropriate amplitudes and phases, whose frequencies are multiples of the wave's repetition frequency, plus a dc signal of appropriate amplitude—the Fourier series. When the amplitude of the carrier wave is modulated by the pulsed wave, each of these sine waves produces sidebands on either side of the carrier frequency.

The spectrum of a single pulse may be found by starting with a pulse modulated wave that is endlessly repetitive and decreasing the repetition frequency to zero. The spectrum of a pulse train of limited length can then be found by treating each of the sine waves comprising the pulse modulated wave as a single pulse the length of the train.

The spectrum of a pulsed carrier may also be explained in terms of the progressive phase shift of the carrier relative to the frequency to which a narrowband filter is tuned. For a single pulse, nulls in the filter output occur at frequencies for which the phase shift over the length of the pulse is 360° or a multiple thereof.

NORTHROP F–89J SCORPION (1956)

This was the final version of the Scorpion series which entered service in 1951 as the first USAF jet fighter designed specifically as a radar interceptor. The F–89J was a very large aircraft carrying a heavy armament load which included Hughes Falcon radar-guided missiles, Douglas Genie nuclear-tipped rockets and 2.75 inch folding-fin rockets. The Hughes E–9 Radar Fire Control System guided the Scorpion to firing position and provided radar illumination for the Falcon missiles.

Sensing Doppler Frequencies

18

There are two basic reasons for sensing doppler frequencies. One is to separate—resolve—returns received simultaneously from different objects. The other is to determine range rates.

In this chapter we will concern ourselves only with sensing doppler frequencies and detecting differences between them. We will see how this may be done with a bank of doppler filters; then, in principle, how the filtering is handled in both analog and digital mechanizations. Finally, we will see why adequate dynamic range is so essential in a doppler radar.

Doppler Filter Bank

How can a radar detect the echoes from many different sources simultaneously and, in the process, sort them out on the basis of differences in doppler frequency?

Conceptually, it is quite simple. The received signals are applied to a bank of filters, commonly referred to as doppler filters (Fig. 1).

Each filter is designed to pass a narrow band of frequencies (Fig. 2). Ideally it produces an output only if the frequency of a received signal falls within this band. Actually, because of filter sidelobes, it may produce some output for signals whose carrier frequencies lie outside the band. If the return is to be sorted by range as well as doppler frequency, a separate filter bank must be provided for each range increment.

Moving up the bank from the lower end, each filter is tuned to a progressively higher frequency. To minimize the loss in signal-to-noise ratio occurring when adjacent filters

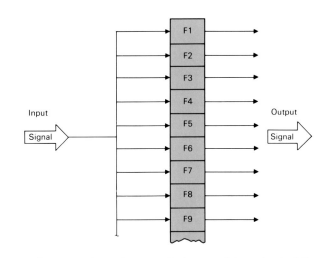

1. The received signals are applied in parallel to a bank of filters.

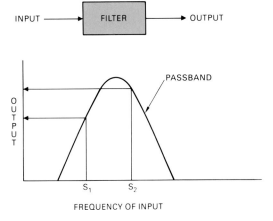

2. Neglecting sidelobes, each filter passes only a narrow band of frequencies. The closer a signal is to the center frequency, the greater the output.

235

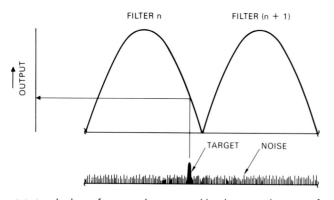

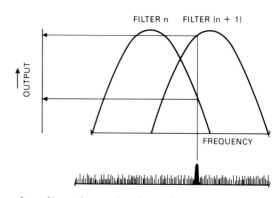

3.   To minimize the loss of output when a signal lies between the center frequencies of two filters, the passbands overlap.

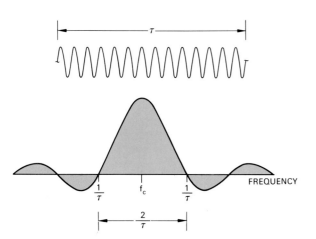

4.   Spectrum of sinusoidal signal of duration, τ.

straddle a target's frequency, the center frequencies of the filters are spaced so the passbands overlap (Fig. 3, above). Thus, if a target's doppler frequency gradually increases, an output is produced, first, primarily from one filter; next, more or less equally from that filter and the next filter up the line; then, primarily from that second filter, and so on.

Bandwidth of the Filters. As we learned in Chap. 10, a narrowband filter achieves its selectivity by integrating the signals applied to it over a period of time. The width of the band of frequencies passed by the filter depends primarily upon the length of the integration time, t_{int}.

Though you may not have realized it at the time, the relationship between the bandwidth and t_{int} was demonstrated indirectly by the experiments of Chap. 16. There we learned that the spectrum of a sinusoidal signal of duration τ (single pulse) has a sin x/x shape such as that shown in Fig. 4. Each point on this plot corresponds to the output the signal would produce from a narrowband filter that integrates the signal throughout its entire duration. The plot was in fact obtained by progressively tuning the filter to each of a great many different frequencies.

We can find the relationship between the filter's bandwidth and t_{int} simply by repeating the experiment as follows:

- Hold the tuning of the filter constant and progressively change the frequency of the applied signal.

- Limit the filter's integration time (t_{int}) and make the signal at least as long as t_{int}.

Now, instead of representing the spectrum of the applied signal, the plot represents the output characteristic of the narrowband filter.

The central lobe of this characteristic is the filter's passband, and the center frequency of the central lobe is the filter's resonant frequency. Since t_{int} in this last experiment corresponds directly to τ in the earlier one, the filter's null-to-null bandwidth is $2/t_{int}$ (Fig. 5). For ease of comparison, the horizontal scale factor used in this figure was adjusted to make the positions of the nulls the same as in Fig. 4. Bear in mind, though, that integration times are generally on the order of milliseconds, whereas pulse widths are on the order of microseconds.

As with the mainlobe of an antenna radiation pattern, a more useful measure of filter bandwidth than the null-to-null width is the width of the central lobe at the points where the power of the output is reduced to half its maximum value—the 3-dB bandwidth. Just as with a uniformly illuminated antenna, that width is approximately half the null-to-null width.

$$BW_{3\,dB} \cong \frac{1}{t_{int}}$$

To realize this bandwidth, of course, the duration of the applied signal must at least equal t_{int}. In fact, filter bandwidth is often selected on the basis of the maximum available integration time.

If the radar is pulsed, the number of pulses that must be integrated to achieve a given bandwidth is equal to t_{int} times the PRF. A useful rule of thumb derived from this relationship is, *the 3-dB bandwidth of a filter equals the PRF divided by the number of pulses integrated.*

The bandwidth given by the above equation, it should be noted, is the *minimum* achievable bandwidth. Depending on the mechanization, a practical filter may have a substantially broader passband as a result of losses or, in digital mechanizations, deliberately introduced "weighting."

Passband of the Filter Bank. If the PRF is greater than the spread between the maximum positive and negative doppler frequencies for all significant targets—or if the radar is not pulsed—enough filters must be included in the bank to bracket the anticipated doppler frequencies. For example, if the PRF were 180 kilohertz, the maximum anticipated positive doppler frequency 100 kilohertz, and the maximum anticipated negative doppler frequency −30 kilohertz (Fig. 6), then the passband of the filter bank would have to be at least 100 + 30 = 130 kHz wide to pass the return from all targets.

On the other hand, if the PRF is less than the anticipated spread in doppler frequencies (as it often must be made to reduce range ambiguities), the passband of the bank should be made no greater than the PRF. The reason, of course, is that the spectral lines of a pulsed signal occur at intervals equal to the PRF, and it is desirable that any one target

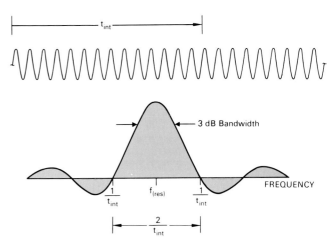

5. Output characteristic of a narrowband filter to which a signal at least as long as the filter integration time, t_{int} is applied.

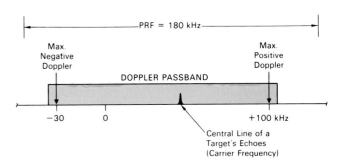

6. If the PRF is greater than the spread between the maximum positive and negative doppler frequencies, the doppler passband should be made wide enough to encompass these frequencies.

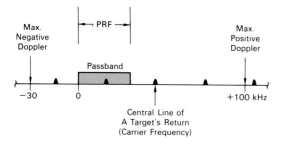

7. If the PRF is less than the spread of doppler frequencies, the passband should be made no wider than the PRF so that a target will appear at only one point within the band.

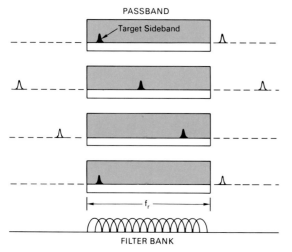

8. If the width of the filter bank's passband equals f_r or less, only one line of the target's spectrum will fall within it, regardless of the target's doppler frequency.

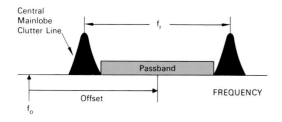

9. Passband may be offset from f_0 to avoid mainlobe clutter.

10. For analog filtering, the doppler spectrum is shifted to a low intermediate frequency.

appear at only one point in the filter bank's passband.

Depending on the target's doppler frequency, the spectral line falling within the passband in this case may not be the target's central one (carrier frequency). It may be one of the lines (sideband frequencies) above or below it (Fig. 7). But since the lines are harmonically related, which one it is doesn't matter. What is important is that for each target one and only one line falls within the passband.

That this requirement is satisfied when the width of the passband equals f_r is illustrated in Fig. 8. It shows a portion of the spectrum of a target's echoes for each of several progressively higher doppler frequencies. These frequencies all happen to be such that the target's central line (carrier frequency) lies outside the figure. Superimposed over the spectrum is a mask with a window in it, representing the passband of a filter bank, f_r hertz wide.

In the first plot of Fig. 8, one of the target's spectral lines falls in the lower end of the passband. With the progressive increase in doppler frequency, in subsequent plots this same line appears progressively farther up in the band. In the last plot, the doppler frequency is sufficiently high that the line we have been observing is actually above the passband; the next lower frequency line now appears in the lower end of the passband.

It can similarly be shown that the target will always appear somewhere within the passband regardless of where we position it. Without causing any problems, therefore, we can shift the passband up or down relative to the transmitter frequency, f_0. It is, in fact, often advantageous to do so. In low and medium PRF radars, for example, the passband is generally made somewhat less than f_r hertz wide and shifted up in frequency so it conveniently lies between the central and next higher lines of the ground return that is received through the antenna's mainlobe (Fig. 9). (Actually, to simplify mechanization, the frequencies of the doppler filters are not changed. Instead the spectrum of the radar return is shifted relative to the filter bank. The net result, however, is the same.)

The doppler filters making up the bank may be either analog or digital. While both types perform essentially the same function, they differ radically in implementation.

Analog Filters

These are essentially tuned electrical circuits. Since with them it is easier to obtain the desired selectivity at comparatively low radio frequencies, the spectrum of the radar return is generally translated to an intermediate frequency on the order of 50 megahertz or less (Fig. 10). In the process, track is kept of the position of the doppler spectrum relative to the transmitter (or in some cases mainlobe clutter) frequency.

Filters' Basic Function. What the filters actually are sensitive to is not frequency, per se, but phase shift—a doppler frequency being, in fact, a continuous phase shift. To see how an analog filter would detect this shift, it is necessary to know a little more about the filter.

In its simplest form, a tuned electrical circuit consists of a capacitor and an inductor (Fig. 11). If a charge is placed on the capacitor, a current surges back and forth between the plates of the capacitor through the inductor, alternately discharging the capacitor and charging it back up again with the opposite polarity. The number of these cycles completed per second depends upon the capacitance of the capacitor and the inductance of the inductor and is called the *resonant frequency of the circuit.*[1] The inductor and capacitor naturally have some losses (resistance). Consequently, the passband is invariably wider than $1/t_{int}$. The lower the losses, the closer the passband approaches this limit.

Analogy to a Pendulum. As with the lossless narrowband filter of the previous chapter, the response of a tuned electrical circuit to an alternating current signal is analogous to the more readily visualized response of a pendulum to a series of impulses (Fig. 12). The first impulse starts the pendulum swinging. Subsequent impulses increase the swing. If the pendulum is allowed to swing freely for a time and another series of impulses is applied, they will do one of three things. If they are in phase with the swing, they will increase it. It they are not quite in phase with it, they will not increase it as much. And if they are out of phase with it, they will tend to damp it out. When the process is repeated many times, the amplitude of the swing builds up to a large value if, and only if, there is a continuity of phase from one series of impulses to the next (i.e., the impulses are coherent) and the frequency of the impulses is the same as the pendulum's natural frequency. Because of friction with the air and in the pivot, some of the energy imparted to the pendulum is lost, so the oscillation builds up somewhat more slowly than might otherwise be expected and dies out after the impulses stop.

The impulses, of course, correspond to the individual cycles of the signal applied to the electrical circuit. Each series of impulses corresponds to a received pulse. The amplitude to which the swing builds up corresponds to the amplitude of the filter's output.[2]

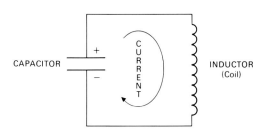

11. An analog filter is a tuned electrical circuit—in simplest form, a capacitor and an inductor.

1. The resonant frequency is $1/2 \pi \sqrt{LC}$, where L is the inductance and C is the capacitance.

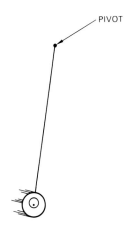

12. Response of a tuned electrical circuit to an alternating current signal is analogous to the response of a pendulum to a series of impulses applied by an eccentric flywheel driven by an electric motor.

2. The pendulum's motion corresponds to the current; the restoring force on the pendulum, to the charge on the capacitor; the mass, to the inductance of the inductor; and the friction, to the resistance of the tuned circuit.

Why is the tuned circuit called an *analog* filter in the first place? Because the electrical characteristics of the circuit elements are analogous to the mathematical operation necessary to isolate a given band of frequencies—integration of the current by the capacitor, differentiation of the current by the inductor, and weighting by the resistance.

Practical Filters. To achieve the desired sharpness of tuning, a quartz crystal, which has the same electrical characteristics as an exceptionally sharply tuned combination of capacitance and inductance, is substituted for the capacitor and inductor (Fig. 13). When wider bandwidths are required, two or more crystals having slightly different resonant frequencies may be used (Fig. 14).

13. Portion of a bank of analog doppler filters used in a representative airborne radar. (The black units are filters; the smaller units, threshold detectors.)

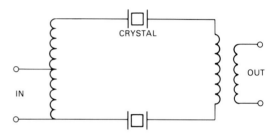

14. Circuit of an analog filter in which two crystals tuned to slightly different frequencies are used to provide a slightly broader passband than that of a single crystal.

To detect the presence of a target, the filter output may be applied to a threshold detector.

Digital Filtering

As we just saw, an analog filter is implemented with circuit elements whose electrical characteristics are *analogous* to mathematical operations. By contrast, a digital filter is implemented with the logic of a digital computer, which performs these same operations numerically—a process called "forming" the filter digitally. Why do the filtering this way?

There are several reasons. Perhaps the most compelling is accuracy. Once the radar return has been accurately converted to digital numbers, all subsequent signal processing is essentially error-free. (There are, of course, quantization and round-off errors, but these can be kept within acceptable bounds through proper system design.) All results are repeatable, no adjustments are required, and performance doesn't degrade with the passage of time.

Also, where a great many doppler filters are required and a variety of operating modes is desired, the size and weight of the equipment needed to implement the radar can be substantially reduced through digital filtering. In fact, it is only through digital filtering that many of today's advanced multimode airborne radars are even feasible.

Converting the radar return into digital form for input to the computer requires some additional operations. Generally, a radar receiver's intermediate frequency is too high to make analog-to-digital conversion convenient, so at the outset (Fig. 15) the receiver output is translated downward to the video frequency range—zero (dc) to several megahertz. The resulting video signal, it might be noted, is similar to the signal which controls the intensity of the cathode ray beam that "paints" the pictures on a TV screen. Since this signal is continuously varying and the numbers into which it will be converted are discrete,[3] the signal must be sampled at short intervals. Finally, each sample must be converted to an equivalent binary digital number. The numbers are applied as inputs to the computer that forms the filters. In the following paragraphs, each of these steps will be explained briefly.

Translation to Video Frequencies. The radar receiver's IF output signal is translated to video frequencies by comparing it with a reference signal whose frequency corresponds to the transmitter frequency, f_0, translated to the receiver's IF. (In some cases, an offset is added to the reference frequency, but we will assume no offset here.)

Before considering how the comparison is made, it will help to have a clear picture in mind of the relationship between the reference signal and the IF output produced by a target. This relationship is illustrated for three representative situations by the phasor diagrams of Fig. 16. In each diagram, the imaginary strobe light that illuminates the phasors is synchronized with the reference signal, so the phasor representing it remains fixed.

In the first diagram, the frequency of the target signal equals f_0—no doppler shift. Consequently, the phasor representing the target return also remains fixed. The angle, ϕ, corresponds to the phase of the target signal relative to the reference signal.

In the second diagram, the target has a positive doppler frequency. The target phasor, therefore, rotates counterclockwise, with ϕ increasing at a rate proportional to the doppler frequency, f_d.

$$\dot{\phi} = 2\pi f_d \text{ radians per second}$$

In the third diagram, the target's doppler frequency is negative; so the target's phasor rotates clockwise. Again, the phase angle ϕ changes at a rate proportional to the doppler frequency.

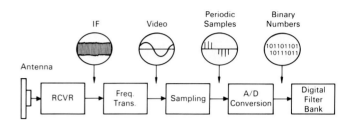

15. For digital filtering, the IF output of the receiver must be translated to video frequencies, sampled, and converted to binary numbers.

3. Discontinuous in time—i.e., the value of each number is separate and distinct from that of the preceding number.

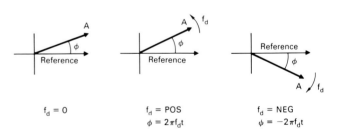

16. Three possible relationships between the reference signal supplied to the synchronous detector and the IF output (A) produced by the return from a target.

HOW THE SYNCHRONOUS DETECTOR WORKS

BASIC FUNCTION. The synchronous detector discussed in the text compares a doppler-shifted input signal with an unshifted reference signal and produces an output whose amplitude is proportional to the amplitude (A) of the input signal times the cosine of the phase (ϕ) of the input signal relative to the reference signal.

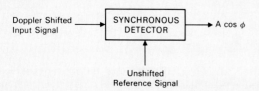

For purposes of explanation, we'll assume here that the reference signal has an amplitude k and a frequency of ω_o radians per second.

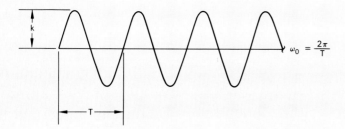

Reference Signal = k sin ω_o t.

Since a doppler frequency shift is actually a continuous phase shift, at any one instant of time the doppler-shifted input signal can be thought of as having a frequency equal to the reference frequency (ω_o) but being shifted in phase relative to the reference signal by ϕ radians.

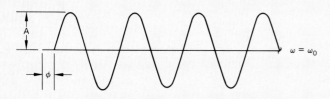

Input Signal = A sin (ω_ot + ϕ)

WHAT THE DETECTOR DOES. In essence, the detector does two things: (1) multiplies the instantaneous value of the input signal by the instantaneous value of the reference signal and (2) applies the resulting signal to a lowpass filter.

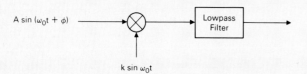

The filter's passband is wide enough to pass the highest doppler frequency that may be encountered but narrow enough to reject completely any signal whose frequency is as high as or higher than ω_o.

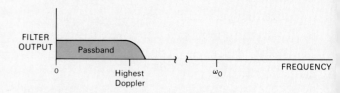

PRODUCTS OF THE MULTIPLICATION. By means of a simp trigonometric identity, the input signal can be shown to consis of two components:

$$\underset{(1)}{} \qquad \underset{(2)}{}$$
$$A \sin (\omega_o t + \phi) = A (\sin \phi)(\cos \omega_o t) + A (\cos \phi)(\sin \omega_o t)$$

When term 1 is multiplied by the expression for the reference signal (k sin ω_ot), the product

$$= kA \ (\sin \phi)(\cos \omega_o t)(\sin \omega_o t)$$

$$= \frac{kA}{2} (\sin \phi)(\sin 2\omega_o t)$$

Because of its high frequency ($2\omega_o$), the signal represented by this product is rejected by the lowpass filter.

However, when term 2 is multiplied by k sin ω_ot, the produ expands mathematically into two terms.

$$= kA (\cos \phi)(\sin^2 \omega_o t)$$

$$= kA (\cos \phi)\left[\frac{1}{2} + \frac{1}{2} \cos 2 \omega_o t\right]$$

Because of its high frequency ($2\omega_o$), the signal represented by the second of these terms is also rejected by the lowpass filte The sole output of the filter, then

$$= \frac{kA}{2} (\cos \phi)$$

If k is taken as being equal to two

$$V_{output} = A \cos \phi$$

where A is proportional to the amplitude of the input signal a ϕ is the signal's phase relative to the reference signal. Since is a cosine function, it is called the in-phase or I output.

REFERENCE SHIFTED 90°. If we shift the phase of the reference signal, i.e., insert a delay which makes the signal applied to the detector equal k sin (ω_ot − 90°), the same inp signal will produce an output equal to A cos (ϕ − 90°). Since the cosine of any angle minus 90° equals the sine of the ang the output voltage is proportional to the *sine* of ϕ.

$$V_{output} = A \sin \phi$$

Again, "A" is proportional to the amplitude of the input signal and ϕ is the signal's phase relative to the unshifted reference. Since this is a sine function, it is called the quadrature or Q output.

Now, the IF output signal is compared to the reference signal by a circuit called a *synchronous detector* (Fig. 17). As explained in detail on the facing page, it produces an output voltage proportional to the amplitude of the received signal times the cosine of the phase angle, ϕ, relative to the reference signal.

$$V_{output} = A \cos \phi$$

where A is proportional to the amplitude of the received signal and ϕ is its phase (Fig. 18).

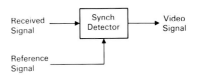

17. For digital filtering, a synchronous detector translates the received signal to the video frequency range.

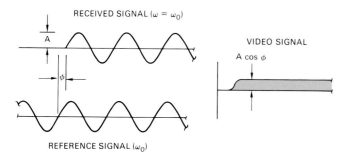

18. Amplitude of output pulse is proportional to cosine of received pulse's phase relative to the reference signal.

We can conveniently visualize the detector's output, therefore, as the projection of the phasor representation of the received signal on the x axis, Fig. 19. If the target's doppler frequency is zero, the output voltage (x) will be constant.

Its exact value may lie anywhere between zero and A, depending upon the signal's phase. If the target's doppler frequency is *not* zero, the output (x) will be a cosine wave having an amplitude, A, and a frequency equal to the target's apparent doppler frequency.

If the radar is pulsed, unless the duty factor is very high, the output pulse produced by each target echo will represent only a fraction of a cycle of the target's apparent doppler frequency. Nevertheless, by observing successive pulses, we can get an idea of the amplitude of the target return and tell its doppler frequency (Fig. 20).

But we won't be getting everything out of the echoes that we might. Since x varies cyclically as the phasor rotates, we will on average throw away half the received energy—the component A sin ϕ in Fig. 19. Also, in certain applications where the time-on-target is short compared to the period of the doppler frequency, the echoes may all be received when cos ϕ is so small that they cannot be detected.

More importantly, we will not be able to tell in which direction the phasor is rotating. For a given rate of rotation, the projections of the phasor on the x axis are the same,

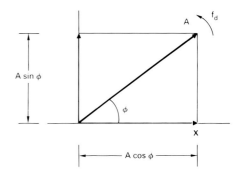

19. The output which a received signal produces from a single synchronous detector may be visualized as the projection of the phasor representation of the received signal on the x axis.

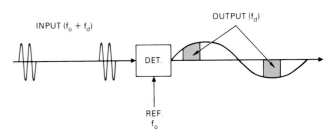

20. Output of the synchronous detector for a pulsed input signal having a duty factor of 25 percent and an apparent doppler frequency equal to half the PRF.

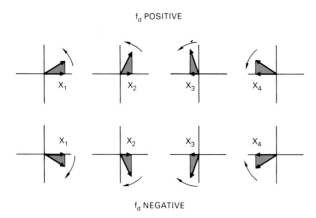

f_d POSITIVE

f_d NEGATIVE

21. Detector output for successive echoes. In-phase or quadrature component alone is the same for both positive and negative doppler frequencies.

whether the phasor rotates clockwise or counterclockwise (Fig. 21). On the basis of these projections alone, we have no way of telling whether the target's doppler frequency is positive or negative. Indeed, in simple MTI radars, which process only one component of the return, all doppler frequencies are indicated as positive. The negative half of the doppler spectrum is said to be "folded over" onto the positive half. Thus, a target whose doppler frequency is, say, $-\frac{1}{4}f_r$ will also appear to have a doppler frequency of $+\frac{1}{4}f_r$ (Fig. 22).

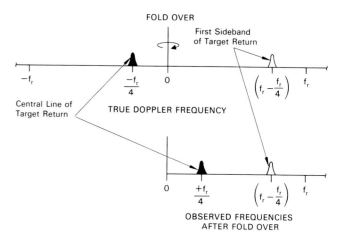

FOLD OVER

First Sideband
of Target Return

$-f_r$ $\dfrac{-f_r}{4}$ 0 $\left(f_r - \dfrac{f_r}{4}\right)$ f_r

Central Line of
Target Return

TRUE DOPPLER FREQUENCY

0 $\dfrac{+f_r}{4}$ $\left(f_r - \dfrac{f_r}{4}\right)$ f_r

OBSERVED FREQUENCIES
AFTER FOLD OVER

22. If only one component of the return is processed, the negative portion of the doppler spectrum will be folded over onto the positive portion.

The above limitations can be eliminated by simultaneously applying the IF output to a second synchronous detector to which the same reference signal is applied, but with a 90° phase lag.[4] Since the cosine of $(\phi - 90°)$ equals the sine of ϕ, the output voltage of this detector is proportional to the amplitude of the target return times the *sine* of the phase angle, ϕ, relative to the unshifted reference.

$$V_{\text{output 2}} = A \sin \phi$$

We can conveniently visualize this second output as the projection of the phasor representation of the target return onto the y axis, as in Fig. 23. Now, if the second detector's output (y) lags behind the first detector's output (x), we know that the phasor is rotating counterclockwise: the target's doppler frequency is positive. On the other hand, if y leads x, we know that the phasor is rotating clockwise: the target's doppler frequency is negative.

The x projection is called the *in-phase* or *I component*; the y projection, the *quadrature* or *Q component*. Together, the two projections describe the phasor completely. Their vector sum equals the length of the phasor (A).

4. To avoid imbalances, a single detector and A/D converter may instead be alternately used for I and Q channels. The sampling rate must then be doubled.

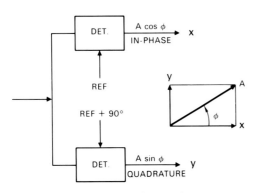

DET. A cos φ x
IN-PHASE

REF

REF + 90°

DET. A sin φ y
QUADRATURE

23. Two channel detector system. Reference frequency for quadrature channel has 90° phase lag.

244

Their ratio,[5] together with their algebraic signs, unambiguously indicate the phase angle, ϕ; hence, both the rate and the direction of the phasor's rotation.

Sampling the Video Signals. So that the continuously varying outputs of the I and Q detectors can be converted to digital numbers, they must be sampled at short intervals of time. Because the outputs may be rapidly varying, the sampling rate must be precisely controlled to avoid introducing errors. Depending upon the design of the radar, the rate may range anywhere from a few hundred thousand to hundreds of millions of samples per second.

In a CW radar, the rate must at least equal the width of the band of frequencies to be passed by the doppler filter bank. If the rate is less than this, the sampling will introduce frequency ambiguities. The reason is that sampling converts the CW signal into a pulsed signal whose repetition frequency is the sampling rate (Fig. 24). The spectral lines of a pulsed signal, of course, recur at intervals equal to the repetition rate. Consequently, if two signals are received whose true doppler frequencies differ by more than the sampling rate, the observed difference in their frequencies will be the true difference *minus* the sampling rate (Fig. 25).

In a pulsed radar, sampling corresponds to the range gating performed in analog mechanizations. If only one sample is taken during each interpulse period—as might be done if the PRF were high and the duty factor close to 50 percent—the radar is said to have a single range gate (Fig. 26).

5. Actually, the ratio is the tangent of the phase angle; the arctangent of the ratio is the phase angle.

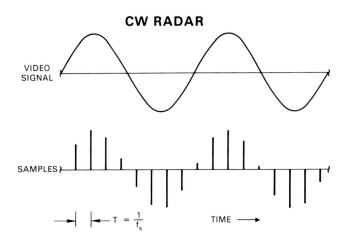

24. When a CW video signal is sampled, it is converted to a pulsed signal whose PRF is the sampling rate, f_s.

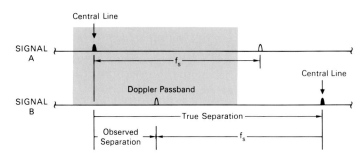

25. Doppler spectra of two CW signals after sampling when sampling rate f_s is less than signals' true frequency separation.

PULSED RADAR—Single Range Gate

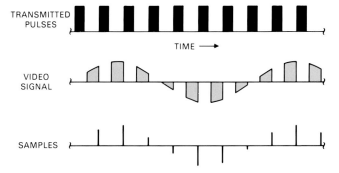

26. If only one sample is taken between transmitted pulses, the radar has the equivalent of a single range gate. Note that the samples are taken at the end of the sampling intervals.

Obviously, the sampling rate in this case would be the PRF.

Where the equivalent of more than one range gate is required, the sampling rate must equal the PRF times the number of range gates. Each sample then represents the return of a single pulse from a given range increment—or, if range is ambiguous, from a number of range increments separated by the unambiguous range, R_u (Fig. 27).

PULSED RADAR—Multiple Range Gates

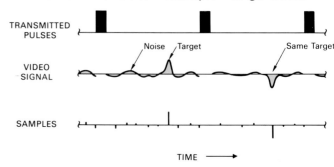

27. If more than one sample is taken between transmitted pulses, successive samples represent returns from different ranges (or sets of ranges if range is ambiguous).

In either case, the sampling is generally performed with "sample-and-hold" circuits. Separate circuits are provided for the outputs of both the I and the Q detector (Fig. 28). At precisely the required intervals, these circuits sense the instantaneous values of the output voltages. They hold the samples long enough for them to be converted to digital numbers, then dump them, and the process repeats.

Analog-to-Digital Conversion. Exactly how this conversion is done depends upon the length of time between samples, the maximum rate at which the voltage being sampled may change, the required conversion accuracy, cost considerations, and so on.

In essence, though, most mechanizations are much the same. The A/D converter compares the voltage of each sample it receives with a succession of progressively higher voltages of precisely known value. When the closest of these is found, the converter outputs a binary number equal to the known voltage. The panel on the facing page illustrates in general how this process might be performed.

Separate converters (or converter channels) are provided for the I and Q samples. The continuous stream of binary numbers emerging from the converter(s) is supplied to the computer which forms the doppler filters.

Forming the Filters. During each successive integration time, t_{int}, the computer mathematically forms a separate bank of doppler filters for every range gate.

Each filter in the bank for a given range gate receives as inputs the same set of numbers (x_n, y_n) from the A/D converter (Fig. 29). If return is being received from a target, each pair of numbers constitutes the x and y components of one sample of a signal whose amplitude corresponds to the

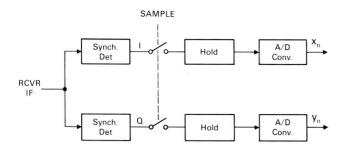

28. At regular intervals, I and Q video signals are momentarily sampled. Samples are held long enough to be converted to numbers.

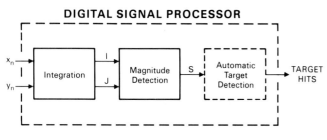

29. Functions performed by a digital filter during each successive integration time, t_{int}.

OW AN A/D CONVERTER WORKS

HAT IT DOES. In essence, an analog-to-digital converter mpares each sample of the signal that is to be digitized with a own scale of incrementally increasing voltage.

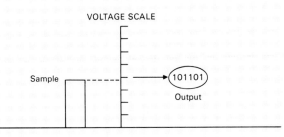

then outputs a binary number corresponding to the voltage ep the sample comes closest to equaling. To give an idea of w these functions might be performed, the operation of a dimentary two-digit-plus-sign converter is outlined below.

HE VOLTAGE SCALE. This may be obtained by applying an tremely stable voltage equal to the maximum possible peak-to-ak excursion of the samples across a chain of precision sistors (voltage divider).

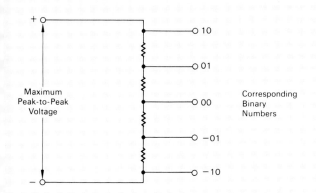

voltage at each tap of this divider corresponds to one of the cession of binary numbers to which the samples may be verted. The difference in voltage from tap to tap, ΔV, responds to the value of the least significant digit of these nbers—in this example, binary 1.

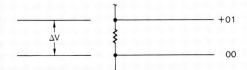

the numbers output by the converter will be correctly nded off, the taps are offset toward the negative end of the ider by half the intertap voltage, ΔV.

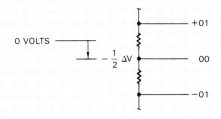

As a result, the voltages of all of the taps are half an increment less positive than the positions of the taps would imply. The voltage of the central tap, for instance, is not O but $-\frac{1}{2}\Delta V$.

COMPARING THE VOLTAGES. A bank of comparator circuits compares the voltage of the sample that is to be digitized with the voltage of each tap.

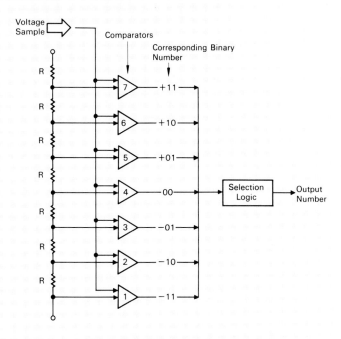

Any one comparator produces an output only if the sample is more positive than the tap to which the comparator is connected. For example, if the voltage of the sample were $+\frac{1}{4}\Delta V$, an output would be produced by Comparator 4 but not by Comparator 5.

SELECTION LOGIC. Each time a voltage sample is applied to the converter, a logic circuit identifies the highest numbered comparator from which an output is produced and outputs the binary number corresponding to this tap. For instance, in the above example where the sample produced an output from Comparator 4

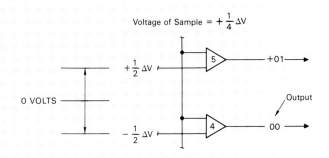

but not from Comparator 5, the sample would be assumed to lie somewhere in the range from $-\frac{1}{2}\Delta V$ to $+\frac{1}{2}\Delta V$, and a zero would be output.

power of the target return and whose frequency is the target's doppler frequency. The job of the filter is to integrate these numbers in such a way that if the doppler frequency is the same as the filter's frequency, the sum will be large, but otherwise it will not.

As will be explained in detail in the next chapter, filter projects successive x and y components onto a coordinate system (i, j) that rotates at the frequency the filter is tuned to and sums the i and j projections separately. At the end of the integration period, the magnitude of the integrated signal is computed by vectorially adding the two sums. To detect targets automatically, the vector sum may be applied to a threshold detector.

The series of computations (algorithm) which must be performed to accomplish the integration and compute the magnitude of the vector sum is called the *discrete Fourier transform* (DFT).[6] Although the arithmetic is simple, the required volume of computing can be enormous. Moreover, to keep up with the flood of incoming data, the computing must be done at exceptionally high speeds. By suitably organizing the computations, however, the volume can be slashed. The procedure commonly employed is called the *fast Fourier transform*.

Providing Adequate Dynamic Range

As noted in earlier chapters, the relative amplitudes of the returns received from different sources and different ranges may vary enormously. Echoes from large short-range targets may be as much as a billion times stronger than the echoes of small distant targets, and ground return may be many times stronger than the strongest target echoes. While the problem may be alleviated by suitably varying the system's gain with time, it cannot be avoided because often both strong and weak returns are received simultaneously. The system must not only be able to handle signals of maximum strength, but provide a wide enough range of output levels at any one time so that small differences in output due to simultaneously received echoes from small distant targets can be detected. The solution, of course, is to provide adequate dynamic range.

By dynamic range is meant the spread between (1) the minimum incremental change in the amplitude of the input to a circuit or system which will produce a discernible change in output and (2) the maximum peak-to-peak amplitude which the input can have without saturating the output. That is, without reaching a point where the output no longer responds to a further increase in input. Beyond

6. It's called "discrete" because it applies to samples of a continuous function taken at discrete intervals of time.

this point, the output becomes a distorted representation of the input.

Providing adequate dynamic range is an important consideration in the design of the receiving and signal processing system of any radar. But in the case of a radar which must sense doppler frequencies it is crucial. For if the dynamic range is inadequate, not only may weak signals be masked by strong signals, but spurious signals will be created. These signals, whose frequencies may be quite different from those of the received signals, may appear falsely as echoes from other targets or interfere with the detection of true targets.

Source of the Spurious Signals. The spurious signals are of two types: *harmonics* and *cross modulation* products.

Harmonics are signals whose frequencies are multiples of another signal's frequency. That harmonics are created when a system's output is limited by saturation can be demonstrated simply by lopping off the top and bottom of a sine wave (Fig. 30).

The result is a nearly square wave. As we saw in Chap. 17, such a wave is made up of a series of sine waves whose frequencies are multiples of the frequency of the square wave.

If a system's passband is narrow enough, the harmonics may lie outside the passband and so be rejected (Fig. 31). Otherwise, they may cause problems.

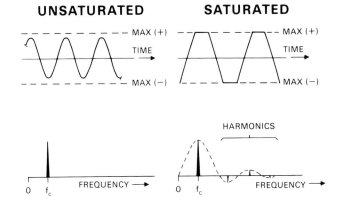

30. Harmonics are created when a signal's peak amplitude is limited by saturation.

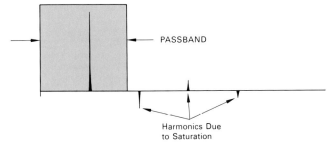

31. If passband is narrow enough, harmonics due to saturation may be eliminated.

Cross modulation is the modulation of one signal by another. It is produced if the sum of two or more signals of different frequency is limited by saturation. The products of cross modulation are sidebands. They, of course, occur both above and below the frequencies of the modulated signals.

Consequently, if the frequencies of the saturating signals are closely spaced, a great many cross modulation products will be passed by a system regardless of the width of its passband.

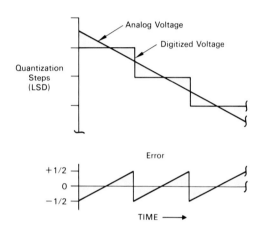

32. If a gradually changing voltage is represented by digital numbers, the error due to quantization has a triangular shape and a peak amplitude equal to half the value of the least significant digit (LSD).

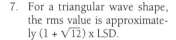

7. For a triangular wave shape, the rms value is approximately $(1 + \sqrt{12})$ x LSD.

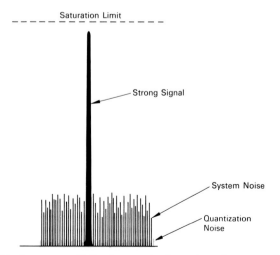

33. Ideally, you would like the quantization noise to be one-tenth or less of the system noise, and the saturation limit to be sufficiently far above the system noise to accommodate the strong signals.

Avoiding Saturation. The creation of harmonics and cross modulation products may be avoided simply by avoiding saturation.

Toward this end, in designing a signal processing system, the average signal level is usually kept as low as possible without risking the loss of weak signals in the locally generated noise. Enough dynamic range is then provided to prevent strong signals from saturating the system. Generally, this approach leads to a tradeoff between saturation, on the one hand, and low-level noise on the other.

Dealing with Quantization Noise. If the signal processor is digital, the problem of low-level noise is exacerbated by the presence of so-called quantization noise. It is the inevitable result of representing signal amplitudes which are continuously variable, with digital numbers which are graduated in finite steps—quanta.

The effect is illustrated for a linearly changing signal in Fig. 32. After being digitized, the signal actually consists of the sum of two signals: (1) a quantized replica of the original analog signal and (2) a triangular error wave having a peak amplitude equal to half the value of the least significant digit (LSD).

If the original signal is comprised of periodic samples of the return from a given range, a simple triangular error wave is generally not produced. For successive samples are about as likely to fall at one point as another between the steps of the A/D converter's reference voltage. Consequently, this undesirable byproduct of digitization is more or less random and so is customarily categorized as noise.

Quantization noise in both the A/D converter and the processor puts a lower limit on the signal levels that can be handled by a system. A common figure of merit for an A/D's dynamic range is the ratio of (1) the maximum peak signal voltage the A/D can handle to (2) the rms value of the quantization error voltage.[7]

To avoid degrading signal-to-noise ratios, you would like the quantization noise to contribute only negligibly to the overall system noise. For that, the level of the incoming signals must be set high enough that the level of the noise accompanying the signals is substantially higher than the quantization noise—ideally, on the order of 10 times higher (Fig. 33).

To prevent saturation by strong signals, then, the dynamic range must be correspondingly increased. This may require increasing the number of digits in the numbers used to represent the signals or handling the processing in a more clever way or both.

Summary

To sort out the radar return from various objects according to doppler frequency, the receiver output is applied to a bank of narrowband filters. If sorting by range is also desired, a separate bank is provided for each range increment. The width of the passband of a narrowband filter is primarily determined by the filter's integration time but is increased by losses. So that return will not be lost when a target straddles two filters, the passbands are made to overlap. So that only one line of a target's spectrum will fall within the band of frequencies bracketed by the bank, the passband of the bank is made no greater than the PRF.

The filters may be either analog or digital. For analog filtering, the radar return is translated to a relatively low intermediate radio frequency. Each filter typically uses one or more quartz crystals. Its response to an incoming signal is analogous to that of a pendulum's response to a succession of impulses.

For digital filtering, the IF output of the receiver is translated to video frequencies by applying it to a pair of synchronous detectors, along with a reference signal whose frequency corresponds to that of the transmitter. The outputs of the detectors represent the in-phase (I) and quadrature (Q) components of the return. Quadrature components are needed to preserve the sense of the doppler frequencies.

The outputs of the synchronous detectors are sampled at a precisely controlled rate. In a pulsed radar, sampling corresponds to the range gating of an analog processor. The sampling rate equals the PRF times the number of range gates desired.

Each sample is converted to a binary number by comparing its voltage with a succession of progressively higher voltages of precisely known value. The numbers are then supplied to a special purpose computer, which implements the filters.

In the case of both analog and digital filters, targets may be detected automatically by applying the filter outputs to threshold detectors.

CONVAIR F–102A DELTA DAGGER (1957)

This milestone aircraft was the first supersonic all-weather interceptor in the world. Its unique features included a delta wing design, six Hughes Falcon missiles carried internally, and the Hughes MG–10 Radar/Weapon Control System. Late in its service life the Dagger flew escort missions with B52's during the Vietnam conflict.

How Digital Filters Work

19

I n the preceding chapter we saw how, for digital filtering, the radar returns are translated to video frequencies by a pair of synchronous detectors and sampled at precisely timed intervals. And we learned how the samples are converted to digital numbers. We were told that the numbers are then supplied to a computer (signal processor), which "forms" a separate bank of doppler filters for each sampling interval (range gate).[1] But little was said about the way in which the filters are formed.

In this chapter, we will learn how that is done. After briefly reviewing what the stream of numbers supplied to the computer represents, we will derive the simple set of equations (algorithm) which the filter must repeatedly compute to form a filter—the discrete Fourier transform—and see how the required mathematical operations may be organized. Finally, we will briefly consider what can be done to reduce the sidelobes which invariably occur on either side of a filter's passband.

The organization of a complete filter bank and the ingenious approach taken to minimizing the otherwise staggering computing load (the fast Fourier transform) are covered in the next chapter.

Inputs to the Filter

Before delving into the details of the digital filter's operation, it will be well to have a good physical picture of what the digitized samples that are the inputs to the filter actually represent. We can gain a picture of this sort quite easily by viewing the output of one of a pulsed radar's two synchronous detectors on a range trace.

1. Depending on the radar's PRF, the returns passed by any one range gate may all come from the same range increment or they may come from many ambiguous range increments separated by R_u.

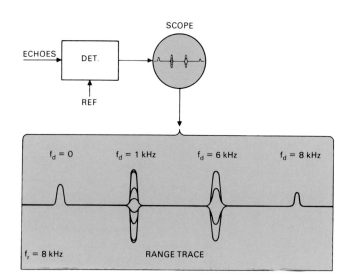

1. Output of a single-channel synchronous detector displayed on a range trace. Echoes of equal amplitude are being received from four targets having different doppler frequencies.

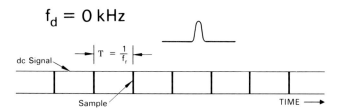

2. Detector output for target with zero doppler frequency has constant amplitude.

Detector Output Displayed on a Range Trace.

Let us suppose that the output of the in-phase (I) detector is supplied to the vertical deflection circuit of an oscilloscope on which is displayed a horizontal range trace. The I output, you'll recall, equals A cos ϕ, where A is the amplitude and ϕ is the radio-frequency phase of the target return relative to the reference signal supplied to the detector. The radar's PRF, we'll say, is 8 kilohertz. Echoes are being received from four targets. Their doppler frequencies are 0, 1, 6, and 8 kilohertz. So that we can isolate the echoes of each target on the range trace, the targets have been positioned at progressively greater ranges (Fig. 1). To isolate the effects of the frequency differences, we will assume that the amplitudes of the received echoes are all the same.

Despite this similarity, the "pips" which the four targets produce are all quite different. Not only do they vary in height, but some fluctuate and others do not. To appreciate why this is so, you must bear in mind that the height of a target pip drawn on the oscilloscope during any one range sweep (interpulse period) corresponds to the detector output produced by a single target echo. Since in this case the period of even the highest doppler frequency is a great deal longer than the width of a radar pulse, each pip is essentially a sample taken at a single point in a cycle of the target's doppler frequency.

For the target having zero doppler frequency, the height of successive pips is constant (Fig. 2). The reason is fairly obvious. Since the target echoes have the same frequency as the reference signal, their phase, relative to it, does not change from one echo to the next. The detector output for the range increment in which this target resides is a pulsed dc voltage. As was explained in the last chapter, the amplitude of this voltage may lie anywhere between zero and plus or minus A, depending upon the phase (ϕ) of the target echoes (Fig. 3).

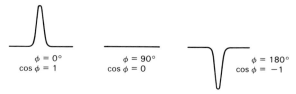

3. Depending on the echo's radio frequency phase, the amplitude of detector output for zero doppler frequency may be anywhere between +1 and –1 times echo's amplitude.

One of the reasons for providing both I and Q channels, of course, is to eliminate this variability. Since the output of the I detector equals A cos ϕ and the output of the Q detector equals A sin ϕ, the magnitude of the vector sum of the two outputs for all values of ϕ equals A.

From the standpoint of filtering, though, the important characteristic of the I and Q samples when the doppler shift is zero is that their individual amplitudes do not fluctuate.

Moving on to the target whose doppler frequency is 1 kilohertz (Fig. 4), the amplitude of its pips fluctuates widely from pulse to pulse. The reason for this, too, can be readily seen. Because the echoes do not have the same radio frequency as the reference signal, their phase relative to it changes from pulse to pulse. The amount of change is 360° times the ratio of the target's doppler frequency to the PRF. In this case (doppler frequency, 1 kilohertz; PRF, 8 kilohertz) the ratio is 1/8. The doppler frequency "wave" is, in effect, being sampled at intervals of 360° x ¹⁄₈ = 45°. (Again, the magnitude of the vector sum of the I and Q samples equals A, but the phase of the sum cycles through 360° at a rate equal to the doppler frequency.)

For the target whose doppler frequency is 6 kilohertz (Fig. 5), the story is the same. The only difference is that, in this case, the samples are taken at intervals of 360° x ⁶⁄₈ = 270°.

As a result, not only does the amplitude of the detector's output fluctuate widely, but as the target return slides into and out of phase with the reference frequency, the samples alternate between positive and negative signs.

For the target whose doppler frequency is 8 kilohertz, though, the pips once again have a constant amplitude (Fig. 6). The reason, of course, is that the doppler frequency and the PRF are equal. The samples are all taken at the same point in the doppler frequency cycle. There is, in fact, no way of telling whether the doppler frequency is zero, or f_r, or some integer multiple of f_r.

Similarly, if echoes are received from a target having a doppler frequency of 9 kilohertz (Fig. 7, below), the pips it produces will fluctuate at exactly the same rate as the pips produced by the target having a doppler frequency of 1 kilohertz. The observed frequency is ambiguous.

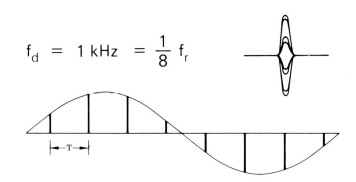

$$f_d \ = \ 1 \text{ kHz} \ = \ \frac{1}{8} \, f_r$$

4. Detector output varies sinusoidally from pulse to pulse.

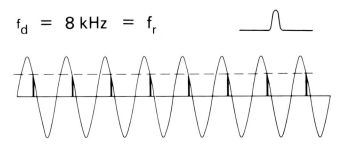

$$f_d \ = \ 6 \text{ kHz} = \frac{6}{8} \, f_r$$

5. Detector output alternates between positive and negative values.

$$f_d \ = \ 8 \text{ kHz} \ = \ f_r$$

6. Detector output is similar to that for target with zero doppler frequency.

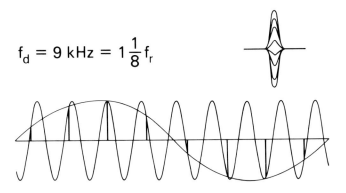

$$f_d = 9 \text{ kHz} = 1\frac{1}{8} f_r$$

7. Detector output fluctuates at exactly the same rate as for a target whose doppler frequency is 1 kilohertz (9 kHz – 8 kHz).

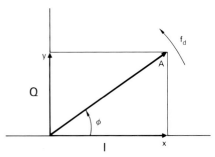

8. If the sine wave is represented by a phasor (A), the I component is the projection of the phasor on the x axis and the Q component is the projection of the phasor on the y axis.

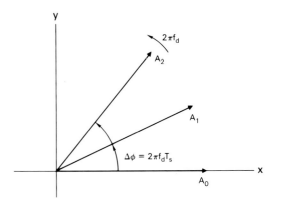

9. The amount that the phasor steps ahead from sample to sample (Δφ) is proportional to the target's doppler frequency.

2. In certain applications digital doppler filtering may be preceded by some analog filtering (for clutter reduction), which converts the return to a CW signal. The sampling rate then is generally not the PRF.

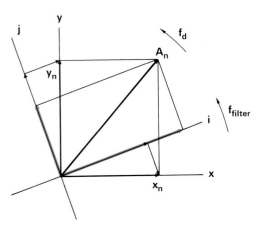

10. The filter projects the x and y components of the phasor, A_n, onto a coordinate system (i, j) that rotates at the frequency to which the filter is "tuned."

Phasor Representation of the Samples. Everything we have seen by viewing the detector output at a point on a range trace corresponding to a particular target's range can be presented much more conveniently in a phasor diagram, such as that shown in Fig. 8. Moreover, a phasor diagram presents the outputs of both the I and Q detectors simultaneously. The length of the phasor corresponds to the amplitude (A) of the target's echoes. The angle (φ) which the phasor makes with the x axis, corresponds to the radio-frequency phase of the echoes relative to the reference signal. The length, x, of the projection of the phasor on the x axis corresponds to the output of the in-phase (I) detector; the length, y, of the projection on the y axis corresponds to the output of the quadrature (Q) detector.

The phasor rotates at the target's apparent doppler frequency, i.e., its true doppler frequency or true doppler frequency plus or minus an integer multiple of the sampling rate. If this frequency is positive (greater than the reference frequency), rotation is counterclockwise (Fig. 9); if negative, rotation is clockwise. The amount that the phasor steps ahead from sample to sample (Δφ) is 2π radians (360°) times the doppler frequency times the length of the sampling interval:

$$\Delta\phi = 2\pi f_d T_s$$

where f_d is the apparent doppler frequency and T_s is the sampling interval. If the sampling rate is the PRF (as it generally is in all-digital signal processors),[2] T_s is the interpulse period, T.

What the Filter Does

Digital filtering is simply a clever way of adding up (integrating) successive samples of a continuous wave so that they produce an appreciable sum *only* if the wave's frequency lies within a given narrow band—i.e., produce a sum equivalent to the output which would be produced if the continuous wave were applied to a narrow band analog filter. If the variation in amplitude from sample to sample corresponds closely to the resonant frequency of the equivalent analog filter, the sum builds up; otherwise, it does not.

What the filter does, in effect, is project the x and y components of the phasor representation of the samples onto a rotating coordinate system (i and j in Fig. 10). The rate at which the coordinates rotate—number of revolutions per second—is made equal to the center frequency of the band the filter is intended to pass. This rate, f_f, can be thought of as the filter's resonant frequency: the frequency to which it is "tuned."

If the frequency of the sampled wave is the same as that of the filter, the angle between the phasor (A) and the rotating coordinate system will be the same for every sample (Fig. 11).

Consequently, after N samples have been received, the sum of the projections of the x and y components of A on the i axis will be N times as great as after a single sample has been received. The same will be true of the sum of the projections on the j axis.

On the other hand, if the frequencies of the sampled wave and the filter differ sufficiently, the angle between the phasor and the rotating coordinate system (i, j) will vary cyclically and the projections will tend to cancel.

At the end of the integration time, the sum of the projections on the i axis (I) is added vectorially to the sum of the projections on the j axis (J). The magnitude of the overall vector sum is the output of the filter. The quantities I and J are then dumped and the integration is repeated for the next N samples.

We can visualize the process most easily if we imagine that we are riding on the rotating coordinate system (Fig. 12). We then see only the rotation of the phasor relative to the i and j axes—that is, the rotation ($\dot{\Phi}$) due to the difference (Δf) between the frequency of the sampled wave and the frequency of the filter. As we just saw, if Δf is zero, the phasor will be in the same relative position each time a sample is taken. But if the frequencies are different, the phasor will be in progressively different positions (Fig. 13).

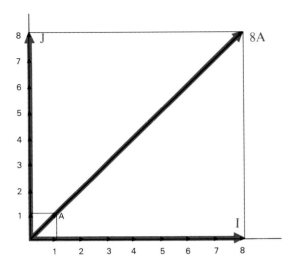

11. If the target's doppler frequency and the filter frequency are the same, after eight pulses have been integrated, the magnitude of the vector sum of the phasor's projections on i and j will be eight times the phasor's amplitude.

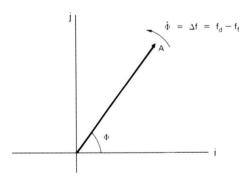

12. If we were to ride on the rotating coordinate system (i, j), we would see only the rotation, Φ, of the phasor (A) due to the difference between the frequencies of the sampled wave and the filter.

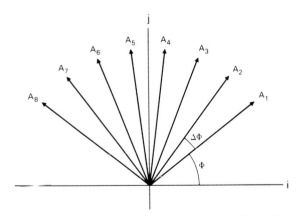

13. If the frequencies of the sampled wave and the filter differ, the phases of successive samples will differ by $\Delta\Phi$.

The phase difference, $\Delta\Phi$, between successive positions is directly proportional to the frequency difference. In radians

$$\Delta\Phi = (2\pi\, T_s)\, \Delta f$$

where 2π is the number of radians in one revolution, T_s is the sampling interval, and Δf is the difference between the frequencies of the sampled wave and the filter.

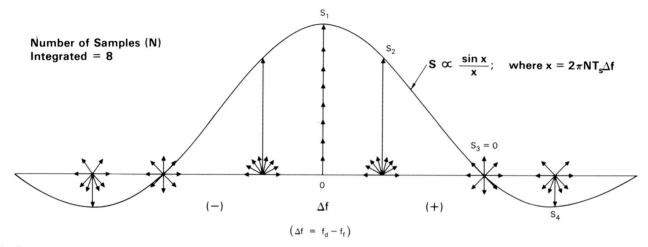

Number of Samples (N)
Integrated = 8

$$S \propto \frac{\sin x}{x}; \quad \text{where } x = 2\pi NT_s \Delta f$$

$(\Delta f = f_d - f_f)$

14. If difference (Δf) between frequencies of sampled signal and filter is increased, phasors representing successive samples will fan out increasingly and the magnitude of their sum (S) will vary as sin x/x.

As you can see from the phasor diagrams in Fig. 14, if the frequency difference Δf (hence also $\Delta\Phi$) is gradually increased, the phasor positions fan out increasingly, and the extent to which the samples cancel correspondingly increases.

A point is soon reached where the phasor positions are fanned out over a full 360°. The sum of the samples is then zero; a null in the filter characteristic has been reached. Beyond this null, the filter output goes through a succession of sidelobes.

For a given amplitude of the sampled signal, a plot of the amplitude of the filter output versus Δf has a sin x/x shape. The band of frequencies between the first pair of nulls is the filter's passband.

Since 360° is 2π radians, after N samples have been integrated, the value of $\Delta\Phi$ for which the first nulls occur is 2π divided by N.

$$\Delta\Phi_N = \frac{2\pi}{N}$$

By substituting the expression we derived earlier for $\Delta\Phi_N$ ($2\pi T_s \Delta f$), we can find the difference, Δf, between the frequencies of the wave and the filter at the nulls.

$$2\pi T_s \Delta f = \frac{2\pi}{N}$$

$$\Delta f = \frac{1}{NT_s}$$

The number of samples integrated (N) times the sampling interval (T_s) is of course the filter integration time (t_{int}). Therefore, the null-to-null bandwidth is

$$BW_{nn} = \frac{2}{NT_s} = \frac{2}{t_{int}}$$

258

The 3-dB bandwidth of a sin x/x curve is roughly half the null-to-null bandwidth, so

$$BW_{3\,dB} \cong \frac{1}{t_{int}}$$

Thus, the more samples integrated and the longer the sampling interval—i.e., the greater t_{int}—the narrower the passband will be. Interestingly, if integration is performed serially, the passband is in a sense dynamic; throughout the integration period it narrows, reaching its final width only at the end of the period.

As noted previously, if the samples received as inputs by the filter are due to the return from a target, the amplitude (A) of the phasor representation of the sampled signal will be proportional to the power of the target echoes. The filter output then will be proportional to the power of the echoes times the integration time and so will be proportional to the total *energy* received from the target during t_{int}. The constant of proportionality will have its maximum value if the target's doppler frequency is centered in the filter passband. The constant will be zero if the doppler frequency is the same as one of the null frequencies. Otherwise, the constant will have some intermediate value determined by the filter's sin x/x output characteristic.

Discrete Fourier Transform

The algorithms which must be repeatedly computed to project the in-phase and quadrature components of successive samples of a wave (x_n and y_n) onto the rotating coordinates (i and j) can be derived on sight from the geometry of Fig. 15. They are

$$i_n = x_n \cos \theta_n + y_n \sin \theta_n$$

$$j_n = y_n \cos \theta_n - x_n \sin \theta_n$$

where i_n and j_n are the projections of x_n and y_n on the rotating coordinates and θ_n is the angle between the coordinate systems.

The terms $\cos \theta_n$ and $\sin \theta_n$ are called the filter coefficients. The amount that θ changes from one sampling interval to the next—$\Delta\theta$—is 2π times the product of the filter frequency and the sampling interval, T_s. If the sampling rate is the PRF, then $T_s = 1/f_r$, and

$$\Delta\theta = \frac{f_f}{f_r}\, 360°$$

where f_f is the filter's frequency and f_r is the PRF.

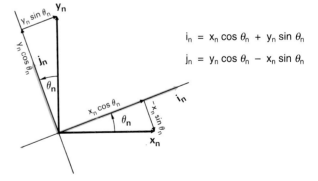

$$i_n = x_n \cos \theta_n + y_n \sin \theta_n$$

$$j_n = y_n \cos \theta_n - x_n \sin \theta_n$$

15. The algorithms which must be repeatedly computed to perform the integration may be derived on sight from the relationships illustrated here.

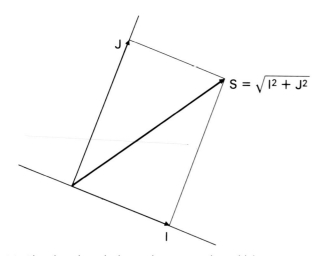

$$S = \sqrt{I^2 + J^2}$$

16. The algorithm which must be computed to add the sums vectorially at the end of the integration period may be derived on sight from the relationships illustrated here.

3. The Fourier transform for a continuous function of time (pulse-modulated sine wave) was derived in Chap. 17.

As the values of I and J are computed, they are separately summed. At the end of the integration period, the sums are added vectorially (Fig. 16). For this, the equation which must be solved is

$$S = \sqrt{I^2 + J^2}$$

where I and J are the sums of the values of i and j, respectively, for the samples taken during the integration period. The output of the filter is the quantity, S. (Often, to avoid taking the square root, which is comparatively time consuming, an algorithm approximating $\sqrt{I^2 + J^2}$ is used to compute S.)

Taken together, the iterations of the algorithms for I and J for a single integration period plus the algorithm for S constitute the discrete Fourier transform, DFT. "Discrete" connotes that the transform applies to discrete-time *samples* of a time-varying function—the sampled wave—rather than to the continuous function of time, itself.[3]

Implementing the DFT

The sequence of operations which a computer must carry out to implement the discrete Fourier transform may be most easily seen if we consider first a simple filter processing one component (say the in-phase component, x) of the return from a single range increment. We can then move on to the processing of return from a contiguous series of range increments and, finally, to the processing of both in-phase and quadrature components.

Single-Channel Filter. To project the component x_n, of successive pulses onto a single rotating coordinate, i, only one term of the DFT must be repeatedly computed.

$$i_n = x_n \cos \theta_n$$

Other than summing successive values of i, no other computation is required.

For any one range increment (Fig. 17), the input to the computer consists of a succession of digital numbers—x_1, x_2, x_3, etc. Since all of these numbers represent echoes reflected from the same range, successive numbers are received by the computer at the same point in successive interpulse periods. Their arrivals are thus separated by T seconds, T being the interpulse period.

Now, the computer's job is to multiply each number by the coefficient, $\cos \theta_n$; add up the products for some prespecified number of inputs, N; and output the accumulated

17. Inputs to a single-channel filter for a single-range increment. The filter's job is to multiply successive values of the in-phase component of the target signal (x_1, x_2, etc.) by prestored coefficients and sum the products.

sum. The computer repeats this process over and over again. For purposes of explanation, we will assume here that N is 16.

If we represent the pulse-to-pulse change in θ for the particular frequency to which the filter is tuned by $\Delta\theta$ and for the sake of simplicity assume that the initial value of θ is $\Delta\theta$, the values of the coefficients for each set of 16 inputs will be cos $\Delta\theta$, cos $2\Delta\theta$, cos $3\Delta\theta$, cos $4\Delta\theta$, . . . cos $16\Delta\theta$ (Fig. 18).

When the first input, x_1, is received, the computer multiplies it by cos $\Delta\theta$ and adds the product into a register, where it is held for T seconds. When the second number, x_2, is received, the computer multiplies it by cos $2\Delta\theta$, retrieves the stored product from the register, adds the two products together, and returns the sum to the register (Fig. 19).[4]

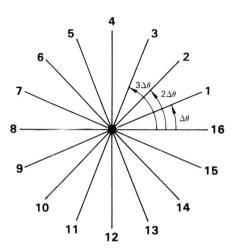

18. Values of θ for a filter designed to integrate 16 pulses.

4. "Register" here connotes a memory position which only an entire digital number may be entered into or retrieved from.

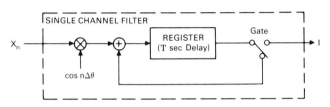

19. Functions performed by a single-channel filter. After N inputs have been integrated, gate closes and sum (I) is output.

The process is repeated for x_3 through x_{16} without change. However, before x_{17} is received, a switch (gate) is temporarily thrown, allowing the sum which has accumulated in the register to pass on to the output. The switch is then returned to its original position, and the integration process is repeated for x_{17} through x_{32}. In general, for every value of x, the filter must perform one multiplication and one addition.

Processing Returns from Successive Ranges. If returns from more than one range increment are to be processed, instead of receiving an input of only one number in every interpulse period, the computer receives a continuous stream of numbers. During the first interpulse period, it successively multiplies the number (x_1) for each range increment by cos $\Delta\theta$, storing the individual products in separate registers. During the next interpulse period, the computer multiplies the number (x_2) for each range increment by cos $2\Delta\theta$, adds this product to the previously stored product, and so on.

Thus, over the same integration period, the computer forms a separate filter—tuned to the same frequency—for the returns from every range increment. If, for example, returns from 100 range increments are processed, the computer forms 100 filters during every integration time for just this one doppler frequency.

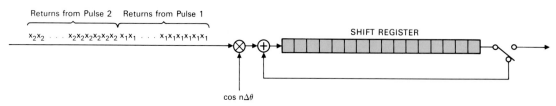

Returns from Pulse 2   Returns from Pulse 1

$x_2x_2 \ldots x_2x_2x_2x_2x_2x_2 \; x_1x_1 \ldots x_1x_1x_1x_1x_1x_1$

SHIFT REGISTER

$\cos n\Delta\theta$

20. When returns from successive range increments are processed, the sums for the individual increments may be stored in a shift register. As each new number is received, sums shift one position to the right.

In simple processors, the products may be stored in what is called a shift register (Fig. 20). It has as many storage positions as there are range increments. Each time a new product is produced, the stored sums are all shifted one position (to the right in the figure). The new product is added to the sum that has spilled out of the last position, and the resulting sum is stored in the memory position which has just been vacated (at the left end of the register).

Since the simple filters we have considered so far process only one component of the return, they cannot, of course, discriminate between positive and negative doppler frequencies. If the frequency to which a filter is tuned is 10 kilohertz, for example, the filter will pass return whose frequency is $(f_0 - 10 \text{ kHz})$, just as well as return whose frequency if $(f_0 + 10 \text{ kHz})$. To differentiate between positive and negative doppler frequencies, both in-phase and quadrature components must be processed. The computing then is done essentially in two parallel channels.

Two-Channel Processing. A two-channel filter is similar to a single-channel filter, but is more complex. Two registers are required—one for the in-phase channel and one for the quadrature channel (Fig. 21). For each echo from the same range increment, the filter receives two inputs, x_n and y_n.

For the in-phase channel, x_n is multiplied by $\cos n\Delta\theta$, just as in the single-channel filter. However, y_n must simultaneously be multiplied by $\sin n\Delta\theta$. This second product must then be added to the first. The sum $(x_n \cos n\Delta\theta + y_n \sin n\Delta\theta)$ is stored in the register.

Similarly, for the quadrature channel, y_n is multiplied by $\cos n\Delta\theta$ and x_n is multiplied by $\sin n\Delta\theta$. In this case, though, the second product is subtracted, and the difference is stored in the register.

At the end of each integration period, the magnitude of the integrated target signal is calculated by vectorially adding the outputs of the two channels. For this, each output is individually squared. The squares are then added together, and the square root is taken of the sum. This root, or an approximation of it such as illustrated at the top of the facing page, is the output of the filter.

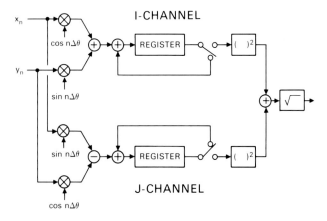

I-CHANNEL

x_n

$\cos n\Delta\theta$

y_n

$\sin n\Delta\theta$

REGISTER

$(\;\;)^2$

$\sqrt{\;}$

$\sin n\Delta\theta$

REGISTER

$(\;\;)^2$

$\cos n\Delta\theta$

J-CHANNEL

21. Mechanization of a two-channel filter. Inputs x_n and y_n are applied to both channels. At end of every integration period, I and Q outputs are squared and summed, and the square root is taken of the sum.

ALGORITHM FOR APPROXIMATING $\sqrt{I^2 + J^2}$

Although it looks simple enough, taking the vector sum, $\sqrt{I^2 + J^2}$ is a comparatively long process. For the square root can be found only through an iterative series of trials.

To save computing time, therefore, the value of $\sqrt{I^2 + J^2}$ is commonly approximated. The simplest of several possible algorithms for making this approximation is this:

- Subtract I from J (or vice versa) to find which is smaller.
- Divide the smaller quantity by two. (Doing this in binary arithmetic is easy; you just shift the number right one binary place.)
- Add the result to the larger quantity. The sum is $\sim \sqrt{I^2 + J^2}$

The error of approximation varies with the value of the phase angle Φ. But at most it is only a fraction of decibel.

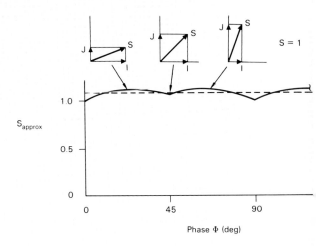

All of the foregoing computations—a total of eight for each pair of numbers that is input plus four for the filter output—must be repeated for every filter that is formed. If 32 pairs of numbers are integrated, this amounts to a total of $(8 \times 32) + 4 = 260$ computations per filter. If, as in the earlier example, returns from 100 range intervals are processed, a total of $260 \times 100 = 26{,}000$ computations must be performed for just this *one* frequency during every integration time.

Sidelobe Reduction

As was illustrated in Fig. 14, the passband of a digital filter has sidelobes similar to the sidelobes of a linear array antenna. Unless something is done to reduce these, a target's echoes may be detected in the outputs of several adjacent doppler filters, or, if the echoes are especially strong, in the outputs of a considerable portion of the filter bank.

Fortunately, filter sidelobes yield to the same reduction technique as antenna sidelobes. (This technique was described in Chap. 18.) Just as antenna sidelobes are due to the radiation from the radiators at the ends of the array, filter sidelobes are due to the pulses at the beginning and end of the pulse train. By progressively reducing the amplitudes of these pulses, the spectral sidelobes can be substantially reduced.

This process, called *amplitude weighting*, is carried out "wholesale," before the digitized video is supplied to the doppler filters (Fig. 22). Following every transmission, the numbers representing the I and Q components of the return from each range increment are multiplied by a weighting coefficient. The coefficient is changed from one pulse to the next according to a prescribed pattern, which

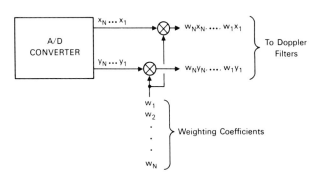

22. Outputs of A/D converter are multiplied by weighting coefficients before being supplied to doppler filters.

is repeated for each train of pulses that is to be integrated. If this pattern has been suitably selected, the sidelobes can be reduced to an acceptable level. In the process, the passband is widened somewhat just as the mainlobe of an antenna is widened by illumination tapering. But this is generally a small price to pay for the reduction achieved in the sidelobes.

What is an acceptable sidelobe level? Naturally this depends upon the application. The characteristic of a weighted filter for a representative fighter application is shown in Fig. 23.

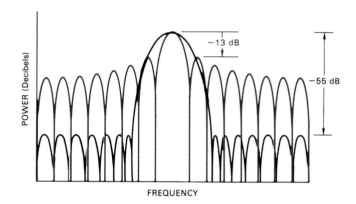

23. Reduction in sidelobe level achieved by weighting the inputs to a representative doppler filter. Note broadening of passband.

Incidentally, even when a doppler filter's sidelobes have been acceptably reduced, some return will invariably get through the sidelobes. If strong enough, therefore, return that is outside a filter's passband can still be detected in the filter's output.

For this reason, it is essential that strong ground return be filtered out *before* the radar return is applied to a bank of doppler filters.

Filtering Actual Signals

In the foregoing discussion, we considered a somewhat artificial situation in which the numbers supplied to the filter represented a continuous train of echoes from a single target and nothing more. How, you may ask, does the filter respond in the real world, where echoes may be received simultaneously from more than one target, where a target's echoes may be accompanied by strong ground return, and where sometimes there may be no echoes at all, only noise?

Let us assume that the radar receiver and signal processor are reasonable "linear" and (as discussed in Chap. 18) saturation has been avoided in all stages of receiving and signal processing up to the filter. The input to the filter then will simply represent the algebraic sum of the instantaneous

264

values of all of the simultaneously received signals, multiplied of course by the system gain up to that point. If the filter itself doesn't saturate—i.e., if the integrated signal doesn't "overflow" the filter's registers—the output will be the same as if each of these signals had been integrated individually and the individual outputs had then been superimposed.

Suppose, for example, that the doppler frequency of a given signal, S_1, lies in the center of the filter's passband and the doppler frequency of a stronger signal, S_2, lies outside (Fig. 24). The ratio of the outputs produced by the two signals will equal the ratio of the powers of the two signals times the ratio of the filter's gain at its center frequency to its gain at the frequency of S_2. If, say, S_2 is 30 dB stronger than S_1 but the filter's gain at S_2's frequency is down 55 dB from the gain at the center of the passband, the output produced by S_1 will be 55 dB – 30 dB + 25 dB stronger than the output produced by S_2. The outputs would be exactly the same as if the two signals had been received at different times.

And noise? As explained in Chap. 10, depending upon its relative phase, noise falling in the filter passband may combine with a target signal either destructively or constructively or somewhere in between. As a result, the filter output produced by an otherwise detectable target may sometimes fail to cross the detection threshold, and vice versa. At times, too, the integrated noise alone may exceed the threshold.

What if the filter saturates? That, too, is possible. To avoid making errors when this happens, the signal processor is usually designed to sense overflows in a filter's registers and discard the filter output when they occur (Fig. 25).

One more question: What if the reception of a train of target echoes is not synchronized with the filter's integration period? Suppose the first echo of the train is received half way through t_{int}.

Synchronization between the radar antenna's time on-target and the integration time of the doppler filters is, of course, entirely random. The first pulse in a train of target echoes is as likely to arrive in the middle of the integration period as at the beginning (Fig. 26). Consequently, on an average, the integrated signal in the output of the filter generally falls short of the maximum possible value. In calculating detection probabilities, this difference is normally accounted for by including a loss term in the range equation (see Chap. 11).

All told, though, performance of a digital filter in a real life situation is very much as has been described here. If saturation has been avoided and strong ground return has

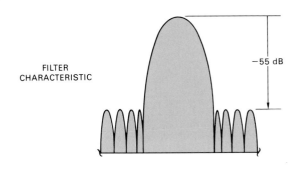

FILTER CHARACTERISTIC

−55 dB

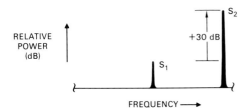

RELATIVE POWER (dB)

+30 dB

S_2

S_1

FREQUENCY →

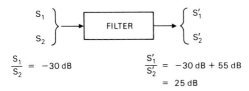

S_1 } → FILTER → { S_1'

S_2 { S_2'

$$\frac{S_1}{S_2} = -30\ dB \qquad \frac{S_1'}{S_2'} = -30\ dB + 55\ dB$$
$$= 25\ dB$$

24. What happens when two signals (S_1 and S_2) of different frequency are simultaneously applied to a filter tuned to the frequency of S_1. Although S_1 is only $1/1000$th as strong as S_2, the output produced by S_1 will be 25 dB stronger.

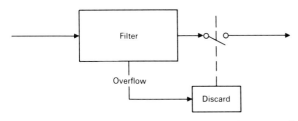

Filter

Overflow

Discard

25. If an overflow occurs in a filter's registers, it will be sensed and filter output will be discarded.

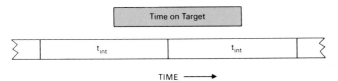

Time on Target

t_{int} t_{int}

TIME →

26. Synchronization of antenna's time on any one target and filter's integration time (t_{int}) is completely random.

largely been rejected in advance, a well designed digital filter will separate target echoes from clutter and noise on the basis of their differences in doppler frequency, quite as effectively as a well designed analog filter.

Summary

A digital doppler filter receives as inputs a succession of pairs of digital numbers. If echoes from a target are being received, each pair constitutes the x and y components of a phasor representing one sample of a signal whose amplitude corresponds to the power of the target echoes and whose frequency is the target's doppler frequency. The job of the filter is to integrate these numbers in such a way that if the doppler frequency is the same as the filter's frequency, the sum will be large but otherwise it will not.

In essence, the filter projects successive x and y components onto a coordinated system that rotates at the frequency the filter is tuned to and sums the components separately. At the end of the integration period, the magnitude of the integrated signal is computed by vectorially adding the two sums. The simple algorithm which must be repeatedly computed to perform the integration and obtain the magnitude of the vector sum is called the discrete Fourier transform (DFT).

A plot of the filter output versus doppler frequency for a pulse train of given length and power has a sin x/x shape. Its peak value is proportional to the total energy of the pulse train. Its nulls occur at intervals equal to $1/t_{int}$ on either side of the central frequency.

To reduce the sidelobes of this pattern, the numbers representing the pulses at the beginning and end of the train are progressively scaled down—a process called amplitude weighting.

Barring nonlinearities and saturation, when several signals are received simultaneously, the filter output is the same as if the signals had been integrated separately and the results had been superimposed. Since receipt of a train of pulses cannot be synchronized with the filter's integration time, the filter output is on an average less than the potential maximum value.

Some Relationships To Keep In Mind

- Filter passbands:

    $$\text{Null-to-null} = \frac{2}{t_{int}}$$

    $$\text{Between half power points} = \frac{1}{t_{int}}$$

    where t_{int} = filter integration time)

- Operations per sample required to form a filter with the DFT

    Multiplications = 4

    Additions = 4

- Operations required to approximate $\sqrt[2]{I^2 + J^2}$

    Subtractions = 1

    Division by 2 = 1

    Additions = 1

The Digital Filter Bank and the FFT

20

In Chaps. 18 and 19, we learned that the simultaneously received returns from several different targets may be sorted in accordance with the targets' doppler frequencies by applying the I and Q components of successive samples of the receiver output from the same range to a bank of digital filters (Fig. 1). With the aid of phasor diagrams, we saw that a digital filter achieves its selectivity by rotating the phases of successive samples through progressively larger angles proportional to the desired filter frequency and summing the rotated samples (Fig. 2). This iterative process, we learned, is performed with an algorithm called the discrete Fourier transform (DFT).

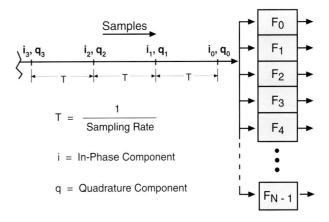

1. Digital filter bank. I and Q components of successive samples of the receiver output are applied in parallel to N digital filters. Filter F_0 passes dc . Filters F_1, F_2, F_3, etc. are tuned to progressively higher frequencies.

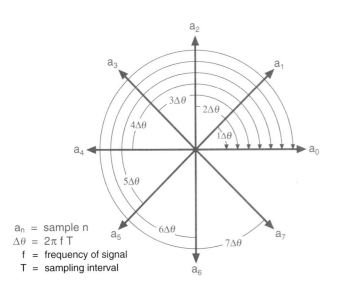

a_n = sample n
$\Delta\theta = 2\pi f T$
  f = frequency of signal
  T = sampling interval

2. Phase rotations which a digital filter must perform to bring successive samples—a_1, a_2, a_3, etc.—of a signal having a given doppler frequency into phase with the first sample, a_0. Sample-to-sample phase advance due to a positive doppler shift (closing target) is counterclockwise. The phase rotations needed to remove these advances, are clockwise.

DFT

$$I = \sum_{n=0}^{N-1} i_n \cos n\,\Delta\theta \; + \; q_n \sin n\,\Delta\theta$$

$$Q = \sum_{n=0}^{N-1} q_n \cos n\,\Delta\theta \; - \; i_n \sin n\,\Delta\theta$$

i_n = in-phase component of sample n

q_n = quadrature component of sample n

n = sample number (0, 1, 2, . . . (N -1)

$\Delta\theta = 2\pi f T$

f = desired frequency of filter

T = time between samples

3. The discrete Fourier transform (DFT). The iterative equations for forming a single digital filter from the in-phase and quadrature components of a sequence of discrete samples of a continuous wave require performing eight simple computations per sample.

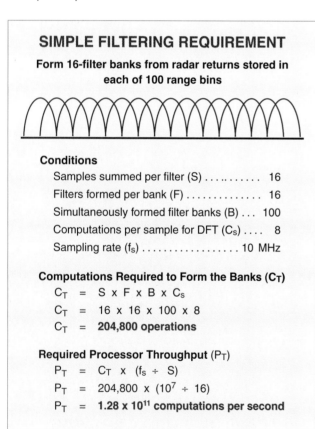

SIMPLE FILTERING REQUIREMENT

Form 16-filter banks from radar returns stored in each of 100 range bins

Conditions

Samples summed per filter (S) 16

Filters formed per bank (F) 16

Simultaneously formed filter banks (B) . . . 100

Computations per sample for DFT (C_s) 8

Sampling rate (f_s) . . . . 10 MHz

Computations Required to Form the Banks (C_T)

C_T = S x F x B x C_s

C_T = 16 x 16 x 100 x 8

C_T = **204,800 operations**

Required Processor Throughput (P_T)

P_T = C_T x ($f_s \div$ S)

P_T = 204,800 x ($10^7 \div$ 16)

P_T = **1.28 x 10^{11} computations per second**

4. Digital computation which would be needed to satisfy even a simple radar signal processing requirement if the filters are formed individually with the DFT. To do the processing in real time, all of the computations must be completed by the time the radar has taken the next set of samples.

For forming a single filter, the DFT is quite efficient. Only eight simple arithmetic operations must be performed per sample (Fig. 3). But if the number of samples and the number of filters per bank are large, and, if filter banks are needed for many successive range increments, forming the filters with the DFT requires an immense amount of processing. Moreover, if the filters must be formed in real time—that is, if all of the computations performed on one set must be completed before the processor receives the next set of samples from the radar—an exceptionally high processing throughput is required (Fig. 4).

In this chapter, we will see how the required throughput may be dramatically reduced by forming each filter bank with an algorithm called the fast Fourier transform (FFT). Following a brief overview of the basic concept, we'll examine the FFT for a small filter bank; then, see how its design may be extended to banks of virtually any size. Finally, we'll derive a simple rule of thumb for quickly estimating the amount of processing needed to form a filter bank with the FFT—as opposed to the DFT— and get a feel for the rapid increase in the FFT's processing savings with the size of the bank.

Basic Concept

The efficiency of the FFT is achieved in two basic ways. First, the key parameters of the bank are selected so that the phase rotations applied to successive samples to form successive filters are harmonically related. Second, the phase rotation and summation of samples for all of the filters are consolidated into a single, multi-step process which exploits these harmonic relationships to eliminate duplications that occur when the filters are formed individually. To illustrate, let us consider a representative FFT.

A Representative FFT

Over the years, many different versions of the FFT have evolved. The basic principles underlying them all, however, are most simply illustrated by the original Cooley-Tukey algorithm.

For it, the parameters of the filter bank are selected in accordance with these rules.

- Make the number of filters (N) equal to a power of two—e.g., 2, 4, 8, 16, 32, 64, 128, 256, 512, etc.

- Form each filter by summing N successive samples.

- Make the incremental phase rotation, $\Delta\theta$, for the lowest-frequency filter (F_1) equal to 360°÷ N.

- Make the incremental phase rotations for successive filters whole multiples of $\Delta\theta$ (see Fig. 5, next page).

INCREMENTAL PHASE ROTATIONS

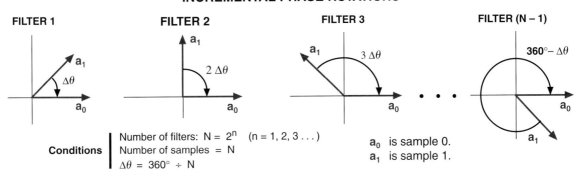

5. Incremental phase rotations for an eight-filter bank whose parameters are selected for implementation with the Cooley-Tukey FFT. This selection results in all phase rotations being harmonically related. (Filter 0, which sums samples of constant phase (dc) requires no phase rotations.)

This choice of parameters leads to a high degree of symmetry throughout the bank and results in all of the phase rotations applied by the filters being multiples of $\Delta\theta$. Consequently, the samples can be partially summed for more than one filter at a time, and the required phase rotations can be incrementally applied to both the individual samples and the partial sums.

Just how this is done may be most easily seen by considering the FFT for a four-filter bank. While such a bank may seem trivially small, the FFT for it is simple to describe and suffices to illustrate virtually all of the fundamental features of the algorithm for any sized bank.

Required Phase Rotations. The phase rotations needed to form our four filter bank are illustrated in Fig. 6. The phasors in the diagram for each filter represent successive samples of a continuous wave whose frequency is to be passed by that particular filter. The curved arrows indicate the phase rotations necessary to bring all of the samples into phase, so they will produce the maximum possible output when summed.

The incremental phase rotation for Filter 1 is 90 degrees.

$$\Delta\theta = \frac{360°}{N} = \frac{360°}{4} = 90°$$

The incremental phase rotations for Filter 2 and Filter 3 are integral multiples of $\Delta\theta$: $2\Delta\theta$ and $3\Delta\theta$.

For Filter 0, of course, none of the samples are rotated. For Filter 2, sample a_2 also is not rotated, and samples a_1 and a_3 are both rotated by $2\Delta\theta$. For Filter 1 and Filter 3, sample a_2 also is rotated by $2\Delta\theta$.

The way in which the FFT takes advantage both of these duplications and of the harmonic nature of the phase rotations is most clearly illustrated by the processing flow diagram for the algorithm.

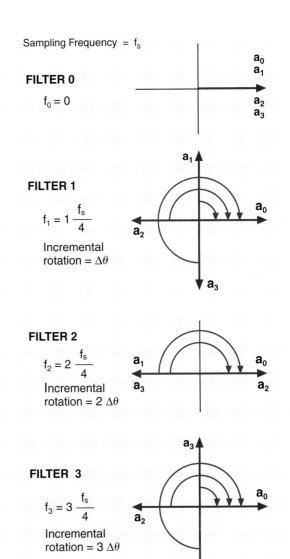

6. Phase rotation and summation requirements for a four-filter bank. The phasor diagram for each filter illustrates the required phase rotations of successive samples of a wave whose frequency is to be passed by the filter.

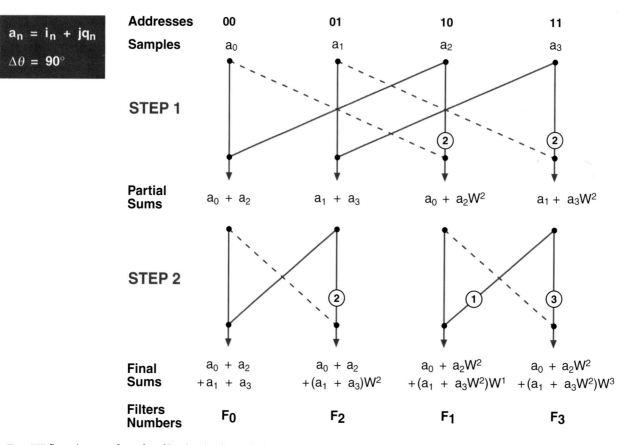

$$a_n = i_n + jq_n$$

$$\Delta\theta = 90°$$

Addresses	00	01	10	11
Samples	a_0	a_1	a_2	a_3

STEP 1

| Partial Sums | $a_0 + a_2$ | $a_1 + a_3$ | $a_0 + a_2W^2$ | $a_1 + a_3W^2$ |

STEP 2

| Final Sums | $a_0 + a_2$ $+a_1 + a_3$ | $a_0 + a_2$ $+(a_1 + a_3)W^2$ | $a_0 + a_2W^2$ $+(a_1 + a_3W^2)W^1$ | $a_0 + a_2W^2$ $+(a_1 + a_3W^2)W^3$ |
| Filters Numbers | F_0 | F_2 | F_1 | F_3 |

7. FFT flow diagram for a four-filter bank. The circled numbers represent multiples of the basic phase increment, $\Delta\theta$. By convention, in the lists of samples included in the sums, these numbers are shown as powers of the complex operator, W.

Processing Flow Diagram. The flow diagram for the bank is shown in Fig. 7. Before discussing it, however, the conventions used in plotting the flow may warrant some explanation.

- The numbers at the top of the diagram identify memory locations in the signal processor. For reasons which will become clear later on, these "addresses" are written in binary form.

- The lowercase letters—a_0, a_1, a_2, and a_3—represent the samples to be summed. Each stands for a complex number specifying the amplitudes of the sample's i and q components (see inset).

- The vertical and slanted lines indicate the flow of these numbers. To avoid confusion where the lines cross, those lines slanting to the right are dashed.

- At the points where the lines meet, the I and Q components of the complex quantities they represent are separately summed.

- Each circled number represents a phase rotation. The number indicates what multiple of the incremental rotation, $\Delta\theta$, is applied. A circled 3, for example, indicates a rotation of $3\Delta\theta$.

When all four samples needed to form the filter bank have been received and placed in the assigned memory locations, two major processing steps are performed.

In Step 1 (see repeat of Fig. 7, right), each sample has added to it the sample that is two memory locations away and is returned to its original location. Before this addition is made to the samples in memory locations 10 and 11, their phases are rotated by $2\Delta\theta$. For your convenience, beneath each summation point, the samples included in the sum are listed. By convention, in these lists the phase rotations are indicated as powers of the operator W. For example, W^2 stands for a phase rotation of $2\Delta\theta$. (See the panel below.)

In Step 2, each of the partial sums produced in Step 1 similarly has one of the other partial sums added to it. In this case, though, the sums that are added are taken from adjacent memory locations. Again, below the summation points, the compositions of the sums are listed. Each of these sums includes all four samples.

Referring back to Fig. 6, you will see that these sums meet the requirements of each of the four filters. The sum

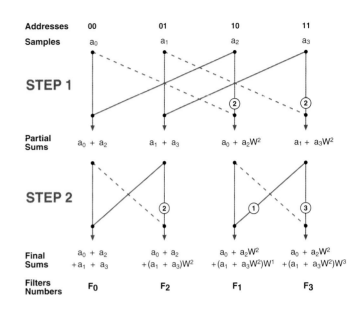

THE COMPLEX OPERATOR W

In this chapter, the DFT is conveniently expressed in terms of the complex operator, W, rather than sine and cosine functions. If you are unfamiliar with complex notation, you may find correlating the two methods of expression illuminating.

The panel on page 69 explained how the amplitude and phase of a sinusoidally varying signal represented by a phasor may be mathematically expressed as an exponential function.

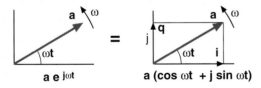

When either of the above functions is used to represent one of N *discrete* samples of a signal, t is the period between the time sample a_n was taken and the time the first sample in the series, a_0, was taken. The product, ωt, then is the phase angle, θ_n, of a_n relative to a_0.

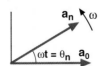

As explained in the text, in a digital filter bank, θ_n is equal to the basic phase increment for the bank, $\Delta\theta$, times the filter number, f, times the

sample number, n. For the FFT, $\Delta\theta = 2\pi/N$. Thus,

$$\omega t = \frac{2\pi}{N} f n$$

By (a) substituting this term for ωt in the exponential expression for the sample and (b) reversing the algebraic sign of j to indicate a clockwise phase rotation that will bring a_n into phase with a_0, we get:

$$a_n e^{-j\omega t} = a_n e^{-j\frac{2\pi}{N} f n}$$

Now, the operator W is a shorthand representation of the non-varying portion of the exponential.

$$W = e^{-j\frac{2\pi}{N}}$$

Consequently,

$$a_n W^{f r} = a_n e^{-j\frac{2\pi}{N} f r}$$

Expressed in terms of W, the exponential function can be mathematically manipulated, like any other constant raised to a given power.

$$W^{(a+b)} = W^a W^b$$
$$W^0 = 1 \quad \text{and} \quad a W^0 = a$$

The exponent applied to W indicates phase rotation in multiples of $\Delta\theta$. For example,

$$W^4 = \text{a phase rotation of } 4\Delta\theta$$

Since $\Delta\theta = 2\pi/N$, a rotation of $N\Delta\theta$ is 2π radians, or 360°. Thus,

$$W^{(mN)} = W^0 \qquad m = 1, 2, \ldots$$
$$W^{(N+m)} = W^m$$

stored in memory location 00 satisfies the requirement for Filter 0 (which passes dc): none of the samples are rotated. The sum in memory location 01, satisfies the phase rotation requirements of Filter 2. The sum in memory location 10 satisfies the requirements of Filter 1. And the sum in memory location 11, satisfies the requirements of Filter 3.

Reduction in Computations. A count of computations indicated in the flow diagram of Fig. 7 reveals that, for this small filter bank, the FFT has reduced the number of complex additions from 16—which would have been required if the filters were formed individually with the DFT—to 8. And it has reduced the number of phase rotations from 16 to only 5. Even so, the harmonic relationship of the phase rotations can be further exploited in two basic ways to reduce the required number of phase rotations still more.

First, a phase rotation of $(N/2) \times \Delta\theta = 180°$. The equivalent of that rotation can be achieved simply by giving the quantity whose phase is to be rotated a negative sign (Fig. 8). For this four-filter bank, $N/2 = 2$; hence, a rotation of $2\Delta\theta = 180°$. Consequently, in Step 1 the phase rotations of $2\Delta\theta$ applied to samples a_2 and a_3 can be eliminated by changing their algebraic signs. So can the phase rotation of $2\Delta\theta$ applied in Step 2 to the partial sum in memory location 01. The number of phase rotations may thus be reduced from 5 to 2.

Second, in Step 2, the partial sum stored in memory location 11 is rotated by $1\Delta\theta$ before being added to the sum in memory location 10 and by $3\Delta\theta$ before the sum in memory location 10 is added to it. A rotation of $3\Delta\theta$ equals a rotation of $1\Delta\theta + 2\Delta\theta$. Consequently, by applying the $1\Delta\theta$ rotation to the sum residing in memory location 11 before it is included in the two following summations, the $3\Delta\theta$ rotation can be replaced with a $2\Delta\theta$ rotation, and it in turn can be replaced with a change in algebraic sign (Fig. 9, below).

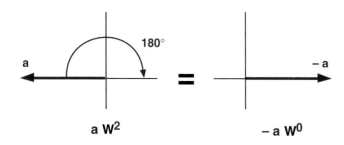

8. A phase rotation of 180°—represented by W^2 in this chapter's four-filter example—can be eliminated simply by changing the algebraic sign of the complex number representing the sample (or partial sum) whose phase is to be rotated.

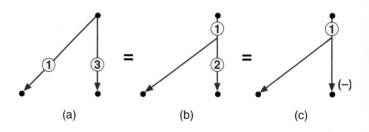

(a) (b) (c)

9. Eliminating a phase rotation by taking advantage of the harmonic nature of the rotations. In this example, $2\Delta\theta = 180°$. Consequently, by moving the rotation of $1\Delta\theta$, shown in (a), ahead of the branch in the flow diagram, the rotation of $3\Delta\theta$ in the right-hand branch can be reduced to $2\Delta\theta$, as in (b), and replaced with a simple change in algebraic sign, as in (c).

With these changes, the required number of rotations is reduced to only 1 (Fig. 10).

From the flow diagram for this small filter bank, four significant conclusions may be drawn regarding the FFT for any sized bank.

- In each step of processing, N summations are performed, in which two complex numbers are algebraically added.

- In general, before each addition is performed, the phase of one of these numbers must be rotated. But half of these rotations can be eliminated through a change in algebraic sign. And no rotations are required for the first step.

- In each successive step, the number of samples included in every sum is doubled. Since N is a power of 2, all N samples can be summed for each of the N filters in $\log_2 N$ steps, thereby substantially reducing the number of summations as well as the number of phase rotations required.

- No intermediate results need to be saved; the final sums for the N filters end up residing in the same memory locations as were initially occupied by the N samples from which the filters were formed.

Rotating the Phases. While the flow diagram specifies the required amount of each phase rotation, it doesn't illustrate how the rotation is produced. Actually, there is no reason to do so. For all of the rotations are produced with the same two simple equations as are used to rotate the phase of a single sample in the DFT:

$$i_2 = i_1 \cos n\Delta\theta + q_1 \sin n\Delta\theta$$

$$q_2 = q_1 \cos n\Delta\theta - i_1 \sin n\Delta\theta$$

where i_1 and q_1 are the in-phase and quadrature components of the complex number whose phase is to be rotated, $n\Delta\theta$ is the desired amount of rotation, and i_2 and q_2 are the in-phase and quadrature components of the number after rotation. Figure 11 presents the processing flow diagram for this single complex multiplication.

Identifying the Filter Outputs. As you may have noticed in Figs. 7 and 10, although the filter outputs do indeed occupy the same memory locations as were initially occupied by the samples from which the filters were formed, the outputs are not all in the same numeric order. The output of filter F_1 ends up in the location initially occupied by

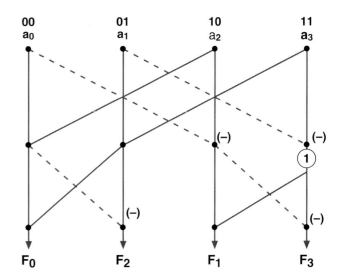

10. FFT flow diagram for the four-filter bank after simplification. By (1) replacing phase rotations of $2\Delta\theta$ with changes in algebraic sign, (2) replacing the phase rotation of $3\Delta\theta$ with the already required rotation of $1\Delta\theta$ (for filter F_1), and (3) changing the algebraic sign at the bottom of the far right leg of the diagram, the required number of rotations has been reduced to only one.

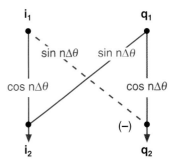

11. Phases are rotated with the same equations (one complex multiply) as used to rotate the phase of a single sample with the DFT.

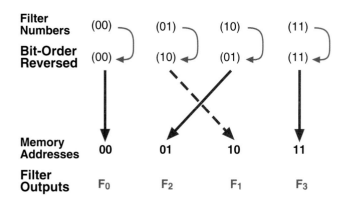

Filter Numbers (00) (01) (10) (11)

Bit-Order Reversed (00) (10) (01) (11)

Memory Addresses 00 01 10 11

Filter Outputs F₀ F₂ F₁ F₃

12. Simple procedure for identifying the memory locations of the filter outputs. Write each filter number in binary form and reverse the order of the bits—last bit first, first bit last. The result is the address of the filter's output.

sample a₂; and the output of filter F₂, in the location initially occupied by sample a₁. We identified the outputs by correlating them with the phasor diagrams for the filters. While that was easily done for our small four-filter bank, it is not convenient for larger banks.

It turns out that, because of the binary nature of the algorithm, you can tell which memory locations the outputs occupy simply by writing the filter numbers in binary form and reversing the order of the digits—last digit first, first digit last—as illustrated in Fig. 12.

Forming Magnitudes. As with the DFT, the final step in forming a filter bank with the FFT is combining the I and Q components of the final sum for each filter. As we saw in Chap. 19, the components may be combined either by taking the square root of the sum of their squares or by executing a more easily computed algorithm, such as that presented on page 263.

In some cases, it is desired to further process a filter bank's outputs before the phase information implicit in the I and Q components is lost. This last step may then be postponed.

FFTs for Filter Banks of Any Size

In practice, filter banks containing many more than four filters are generally required. Following the same basic approach as outlined above, FFTs may be designed for filter banks of virtually any size. The first step, of course, is determining what phase rotations must be performed.

Determining the Required Phase Rotations. For banks of four, or even eight filters, the required phase rotations can be determined on sight, as we just have done, from phasor diagrams for the individual filters. But for larger banks, the rotations are best determined mathematically.

The panel on the next page presents a simple mathematical derivation of the original Cooley–Tukey FFT for an eight-filter bank. The derivation begins with the equation for forming the eight filters directly with the DFT— expressed in terms of the complex variable, W. The filter and sample numbers in this equation are then expanded in binary form, phase rotations of 360° and multiples thereof are eliminated, and the summation is carried out separately for each binary digit. A series of recursive equations is thus produced which specifies the phase rotations and summations to be performed in each of the FFT's $\log_2 N$ processing steps.

Following that general procedure, computer programs can readily be written with which the FFT for any sized fil-

DERIVATION OF THE COOLEY-TUKEY FFT

For An Eight-Filter Bank

The derivation begins with the equation for the DFT, expressed in terms of the complex operator, W.

$$\chi(f) = \sum_{n=0}^{N-1} a(n)\, W^{fn}$$

Output. Filter f — Sample n — Phase rotation, Filter f, Sample n

f = filter number = $0, 1, 2 \ldots (N-1)$
n = sample number = $0, 1, 2 \ldots (N-1)$
$N = 8$

When f and n are written in binary form and expanded, they become:

$$f = 4f_2 + 2f_1 + f_0$$
$$n = 4n_2 + 2n_1 + f_0$$

The digits, $f_2, f_1, f_0,$ and n_2, n_1, n_0 can each take on a value of 1 or 0.

$$\therefore fn = (4f_2 + 2f_1 + f_0)\,4n_2 + (4f_2 + 2f_1 + f_0)\,2n_1 + (4f_2 + 2f_1 + f_0)\,n_0$$

Carrying out those multiplications which will produce factors of 16 and 8, and bracketing the terms containing them,

$$fn = \left[16 f_2 n_2 + 8 f_1 n_2\right] + 4 f_0 n_2 + \left[8 f_2 n_1\right] + (2 f_1 + f_0)\,2 n_1 + (4 f_2 + 2 f_1 + f_0)\,n_0$$

Writing the above expression as a power of W, and taking advantage of the fact that $W^{(m+n)} = W^m W^n$, we have

$$W^{fn} = \left[W^{(16 f_2 n_2 + 8 f_1 n_2)}\right] W^{4 f_0 n_2} \left[W^{8 f_2 n_1}\right] W^{(2 f_1 + f_0)\,2 n_1}\; W^{(4 f_2 + 2 f_1 + f_0)\,n_0}$$

Since N = 8, phase rotations of W^8 and W^{16} equal W^0. Since $W^0 = 1$ (no rotation), the bracketed terms for these rotations drop out of the equation, leaving:

$$W^{fn} = W^{4 f_0 n_2}\; W^{(2 f_1 + f_0)\,2 n_1}\; W^{(4 f_2 + 2 f_1 + f_0)\,n_0}$$

Substituting this expression for W^{fn} in the equation for $\chi(f)$, writing the arguments of $\chi(f)$ and a(n) in binary form, and replacing the single summation sign with separate signs for each binary digit of n, we have:

$$\chi(f_2, f_1, f_0) = \sum_{n_0=0}^{1}\; \sum_{n_1=0}^{1}\; \sum_{n_2=0}^{1} a(n_2, n_1, n_0)\, W^{4 f_0 n_2}\; W^{(2 f_1 + f_0)\,2 n_1}\; W^{(4 f_2 + 2 f_1 n + f_0)\,n_0}$$

$A_1 (f_0, n_1, n_0)$
$A_2 (f_0, f_1, n_0)$
$A_3 (f_0, f_1, f_2)$

As indicated by the horizontal bracketing, the equation can now be broken into three recursive equations, which specify the phase rotations and partial summations to be performed in each of the three successives processing steps of the FFT for the eight filter bank.

$$A_1(f_0, n_1, n_0) = \sum_{n_2=0}^{1} a(n_2, n_1, n_0)\; W^{4 f_0 n_2}$$

$$A_2(f_0, f_1, n_0) = \sum_{n_1=0}^{1} a(n_2, n_1, n_0)\; W^{4 f_0 n_2}\; W^{(2 f_1 + f_0)}$$

$$A_3(f_0, f_1, f_2) = \sum_{n_0=0}^{1} a(n_2, n_1, n_0)\; W^{4 f_0 n_2}\; W^{(2 f_1 + f_0)\,2 n_1}\; W^{(4 f_2 + 2 f_1 n + f_0)\,n_0}$$

The last sum [$A_3(f_0, f_1, f_2)$] consists of the outputs of the bank's 8 filters. Note the bit reversal in the argument of A_3.

$$\chi(f_2, f_1, f_0) = A_3(f_0, f_1, f_2)$$

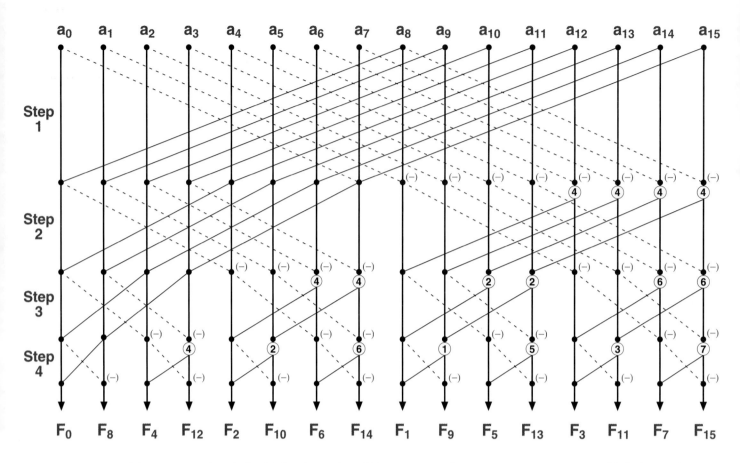

13. Basic structure of the FFT. For each multiple of 2 that the number of filters is increased, one more processing step is added. The FFT shown here is for a 16-filter bank; hence, is a four-step procedure.

ter bank can be generated in virtually no time at all. The flow diagram for a 16-filter bank produced in that way and subsequently simplified with the techniques that were illustrated in Fig. 9 is presented in Fig. 13, above.

From it, we can conclude that in the general case of a filter bank of any size, for every multiple of 2 that the number of filters (N) is increased, one more processing step must be performed.

In the first step each of the samples has added to it and subtracted from it the sample stored N/2 memory locations away.

In the second step, each of the resulting partial sums has added to it and subtracted from it the partial sum N/4 memory locations away, thereby doubling the number of samples included in each partial sum.

In the third step, each of these partial sums has added to it and subtracted from it, the partial sum N/8 memory locations away, again doubling the number of samples included in each sum.

This process is continued until all N samples are included in the final sum for each filter. A total of $\log_2 N$ processing steps are thus carried out in forming the filter. Since the flow diagram for even a relatively small filter bank is unwieldy, in practice the information the diagram contains is generally printed out in tabular form.

The FFT Butterfly. When examining Fig. 13, you may have noticed a striking similarity in the flow patterns for the individual partial summations. In every case, the phase rotation (when required) is made to one of the two quantities that is to be algebraically summed—the one on the right, in that figure—before the addition and subtraction are performed.

The pattern for this generic operation is shown in Fig. 14, along with the corresponding processing instruction. Because of its wing-like shape when diagramed, this instruction is called the *FFT butterfly*.

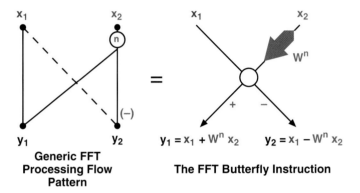

Generic FFT Processing Flow Pattern

The FFT Butterfly Instruction

$$y_1 = x_1 + W^n x_2 \qquad y_2 = x_1 - W^n x_2$$

14. Derivation of the FFT butterfly instruction from the generic flow pattern of the individual partial summations which are the building blocks of the FFT algorithm.

Having determined the appropriate pairings of memory positions for each processing step, the entire filter bank can be formed by iteratively carrying out this basic instruction on the contents of each successive pair of memory positions.

Rules of Thumb for Estimating Number of Computations

While the tremendous reduction in computing load which may be realized through the use of the FFT should by now be apparent, it is interesting to consider quantitatively what it amounts to in the more general case of large filter banks

If we were to use the DFT to form an N-filter bank, we would expect to have to perform one phase rotation and one complex addition for each sample per filter. Since N samples must be summed to form a filter and the bank contains N filters, forming the bank would require a total of N^2 phase rotations and N^2 complex additions.

On the other hand, to form the bank with the FFT, we would expect to have to perform N phase rotations and N

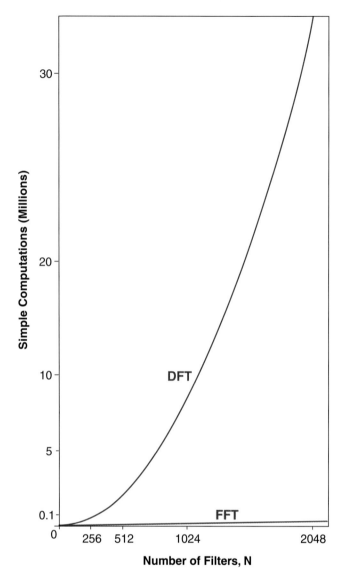

15. Reduction in number of computations achieved by forming a filter bank with the FFT rather than with the DFT. As the number of filters in the bank increases, the reduction becomes immense.

complex additions in each of only $\log_2 N$ processing steps, for a total of $N \log_2 N$ rotations and $N \log_2 N$ additions. However, as the FFT butterfly diagram makes clear, half of the rotations are eliminated (as was illustrated in Fig. 9) by substituting for them changes in algebraic sign. Thus, the total processing requirements for the FFT are:

$$\text{Phase rotations} \ldots 0.5 \, N \log_2 N$$
$$\text{Complex additions} \ldots N \log_2 N$$

As we have seen, for small filter banks the FFT reduces the number of phase rotations substantially more than the above rule implies. But for the general case of larger banks, the required number of rotations closely approaches that given by the rule.

To illustrate the FFT's efficiency, it will be instructive to apply these rules to a reasonably large bank, say, one containing 2,048 filters.

Since $\log_2 2,048 = 11$, the bank can be formed with $0.5 \times 2,048 \times 11 = 11,264$ phase rotations. Each of these requires 6 simple mathematical operations—4 multiplications, 1 addition, and 1 subtraction—for a total of 67,584. Adding to this figure two simple operations for each of $2,048 \times 11$ complex additions, brings the total number of simple operations needed to form the bank with the FFT to 112,640.

By contrast, to form the same bank directly with the DFT, it would take $2,048 \times 2,048 \times 8 = 33,554,432$ simple operations (Fig. 15). That is roughly 300 times as much computing!

Moreover, because the number of processing steps increases only as the logarithm of the number of filters, as the size of the filter increases, the reduction in computations achievable with the FFT increases dramatically.

Summary

The FFT is an algorithm which vastly reduces the amount of processing necessary to form a bank of digital filters with the DFT. Its efficiency is achieved primarily by choosing the parameters of the bank so that they are harmonically related and consolidating the formation of the filters into a single multiple-step process.

By making the number of filters, N, equal to a power of two and the number of samples summed equal to N, the processing is accomplished in $\log_2 N$ steps. In the first step, each sample is algebraically summed with one of the other samples. In each succeeding step, certain phase rotations are performed, and each partial sum is algebraically summed with one of the other partial sums.

The required phase rotations and pairing of the quantities to be summed in each step can readily be determined mathematically. The basic processing instruction for performing the individual partial summations—consisting of a phase rotation, a complex addition, and a complex subtraction—is called the FFT butterfly. The phase rotations themselves are performed the same way as in the DFT.

The partial sums are returned to the same memory locations as held the quantities that were summed. The final sums are read out in the order of the filter numbers they apply to, by labeling the memory locations with the numbers—in binary form—of the samples they originally held and reversing the order of the bits in these numbers.

Whereas N^2 complex multiplies and N^2 complex additions must be performed to form an N-filter bank with the DFT, only $0.5\,N\,Log_2 N$ complex multiplies and $N\,Log_2 N$ complex additions must be performed to form the same bank with the FFT.

A POINT OF NOMENCLATURE

In this chapter, we have described the filter banks formed with the FFT in terms of the number of filters they contain, N, or of the number of samples summed to form the filters—the two numbers being the same for any one bank.

Signal processing experts, however, commonly describe filter banks in terms of points, the word "point" meaning the number of data points (samples) used to form the bank. A 16-filter bank, for example, is referred to as a 16-point filter, or a 16-point FFT.

Accordingly, in a functional block diagram, a filter bank may be conveniently represented by a box with the number of points written beneath it:

16 point

McDONNELL F–101B VOODOO (1959)

This two-place interceptor was adapted from the single-seat F–101A fighter bomber. The Hughes MG–13 Radar Weapon Control System directed the firing of three Hughes Falcon missiles and two Douglas MB–1 nuclear-tipped rockets carried on a unique rotating missile bay. This aircraft enjoyed a long service life of over 20 years with the USAF and the Canadian Forces.

Measuring Range Rate

21

In many radar applications, knowing a target's present position (angle and range) relative to the radar is not enough. Often one must be able to predict the target's position at some future time. For that, we must also know the target's angular rate and its range rate.

Range rate may be determined by one of two general methods. In the first, called *range differentiation*, the rate is computed on the basis of the change in the measured range with time. In the second and generally superior method, the radar measures the target's doppler frequency—which is directly proportional to the range rate.

In this chapter, we will look at both methods briefly. We will then take stock of potential doppler ambiguities and see how they may be resolved.

Range Differentiation

If we plot the range of a target versus time, the slope of the plot is the range rate (Fig. 1). A downward slope corresponds to a negative rate; an upward slope, to a positive rate.

Determining the slope—hence the range rate—is easy. You select two points on the plot which are separated by a small difference in time and pick off the difference in their ranges. Dividing the range difference by the time difference yields the range rate

$$\dot{R} = \frac{\Delta R}{\Delta t}$$

where

$$\dot{R} = \text{range rate}$$
$$\Delta R = \text{difference in range}$$
$$\Delta t = \text{difference in time}$$

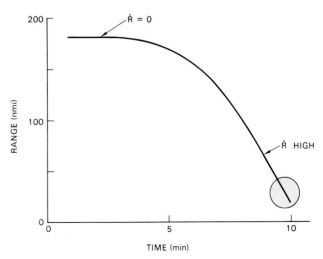

1. Range rate, $\dot{R}$, corresponds to the slope of a plot of range versus time.

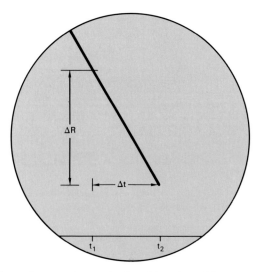

2. Slope of range plot can be found by taking difference in range (ΔR) at points separated by short increment of time (Δt).

1. With differentiation, the time difference (Δt) is infinitesimal.

If ΔR is taken as the difference between the current range and the range Δt seconds earlier, Ṙ corresponds to the current range rate. This process approximates differentiation (Fig. 2).[1]

In essence, range rate is measured in this way both by non-doppler radars and by doppler radars when operating under conditions where doppler ambiguities are too severe for range rate to be measured directly by sensing doppler frequency.

If the range rate is changing, the shorter Δt is made, the more closely the measured rate, Ṙ, will follow the changes in the actual rate; i.e., the less the measured rate will lag behind the actual rate (Fig. 3)—a quality referred to as *good dynamic response*.

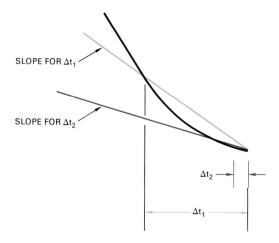

3. The shorter Δt is made, the closer the measured slope will follow changes in the actual range rate.

Unfortunately, a certain amount of random error, or "noise" is invariably present in the measured range. Though small in comparison to the range itself, the noise can be appreciable in comparison to ΔR. In fact, the shorter Δt is made, the smaller ΔR will be and therefore the greater the extent to which the noise will degrade the rate measurement (Fig. 4).

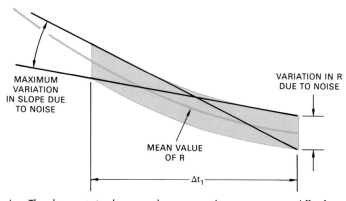

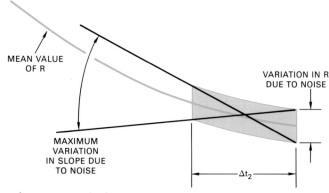

4. The shorter Δt is, the more the measured range rate may differ from the actual rate as a result of noise in the measured range.

The noise can be reduced by subsequently "smoothing" the measured range rate, but the effect of smoothing on response time is essentially the same as that of increasing Δt. The performance achieved with this method of range rate measurement is thus a compromise between smooth tracking and good dynamic response.

When a target is tracked without stopping the antenna's search scan (track-while-scan), the dynamic response is still further limited by the fact that Δt is stretched to a full scan frame time. If the radar's range measurement sensitivity is sufficiently high, though, a useful estimate of the range rate can usually be provided during the dwell time by extrapolation.

Doppler Method

A doppler radar can not only measure range rates with greater precision than is possible with range differentiation, but make the measurement directly.

In the absence of doppler ambiguities, a target's doppler frequency may be determined simply by noting in which filter of the doppler filter bank the target appears (Fig. 5), or, if it appears in two adjacent filters, by interpolating between the center frequencies of the filters on the basis of the difference in their outputs. In translating the doppler spectrum to the frequency of the filter bank, accurate track must have been kept of the relative position of the transmitter frequency, f_0. To measure the doppler frequency, one need only count down the bank from the target's position to the frequency corresponding to f_0—or, if the doppler spectrum has been offset from f_0 to put the mainlobe clutter at zero frequency, count to the bottom of the filter and add the offset.

The doppler frequency is determined with greater precision during single-target tracking. In this mode, the receiver output is usually applied in parallel to two adjacent doppler filters, whose passbands overlap near their −3 dB points (Fig. 6).

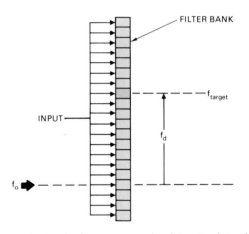

5. A target's doppler frequency may be determined simply by noting the target's position in the filter bank.

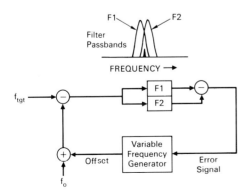

6. In single-target tracking, an offset is added to the target's frequency to center it exactly between two filters. The doppler shift is then determined by precisely measuring the offset.

An automatic tracking circuit (velocity servo) then shifts the doppler spectrum so it is offset from f_0 by just enough to cause the target to produce equal outputs from the two filters. The offset is thus maintained equal to the target's doppler frequency, and the doppler frequency is measured by measuring the offset.

The range rate is computed from the doppler frequency with the inverse of the expression for doppler frequency derived in Chap. 15:

$$\dot{R} = -\frac{f_d \lambda}{2}$$

where

$\dot{R}$ = range rate

f_d = doppler frequency

λ = wavelength

The rules of thumb we learned in that chapter can similarly be inverted. For an X-band radar, range rate in feet per second is nearly equal to the doppler frequency in hertz divided by 20.

And in knots, the range rate is nearly equal to the doppler frequency in hertz divided by 35.

But how do we know that it is the target echoes' carrier frequency we have observed and not a sideband frequency some multiple of f_r above or below the carrier? Isn't any doppler frequency we measure, hence the range rate we compute from it, inevitably ambiguous?

Yes. But whether the ambiguity is significant depends on the PRF and the magnitudes of the closing rates that may be encountered.

Potential Doppler Ambiguities

To get a feel for the significance of doppler ambiguities at different PRFs, let us consider a hypothetical operational situation.

Hypothetical Situation. We'll assume that a radar is operating against targets that may be detected anywhere within a 120°-wide sector, dead ahead (Fig. 7). The targets may be flying in any direction.

Their speeds, too, may vary but are not expected to exceed 1000 knots. The maximum speed of the aircraft carrying the radar, we'll say, is also 1000 knots.

Under these conditions, the maximum closing rate which the radar might encounter—i.e., the rate when both the radar-bearing aircraft and the target are flying at maxi-

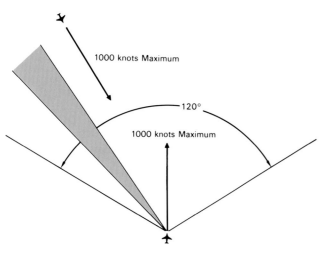

1 FPS ≈ 20 Hz

1 KNOT ≈ 35 Hz

1000 knots Maximum

120°

1000 knots Maximum

7. Hypothetical situation used to illustrate conditions under which doppler measurements may be significantly ambiguous.

mum speed, nose-on (Ṙ is negative for such approaches) — would be –1000 – 1000 = –2000 knots (Fig. 8). At X-band, this rate would produce a doppler shift of roughly 2000 x 35 = 70 kHz.[2]

The maximum opening rate (Ṙ is positive) would occur if a target were at the largest azimuth angle (60°) and flying at maximum speed away from the radar, while the radar-bearing aircraft was flying at its minimum cruising speed. That speed, we'll say, is 400 knots. The range rate then would be +1000 – (0.5 x 400) = +800 knots. This rate produces a doppler shift of –800 x 35 Hz = –28 kHz.

Thus, provided the radar does not encounter a significant target whose speed exceeds 1000 knots or whose azimuth exceeds 60°, the spread between maximum positive and negative doppler frequencies would be 70 – (–28) = 98 kHz (Fig. 9).

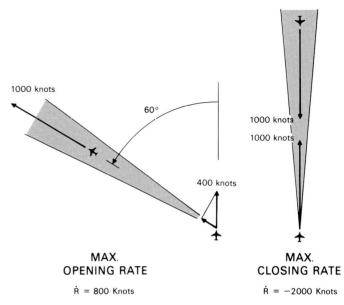

MAX.
OPENING RATE

Ṙ = 800 Knots

MAX.
CLOSING RATE

Ṙ = –2000 Knots

8. Flight geometry producing maximum negative doppler frequency, left; maximum positive doppler frequency, right.

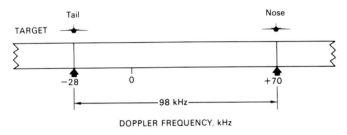

DOPPLER FREQUENCY, kHz

9. Spread between maximum positive and negative doppler frequencies for hypothetical situation.

2. Remember that negative range rates (range decreasing, –Ṙ) result in positive doppler frequencies and vice versa according to the equation

$$f_d = -\frac{2\dot{R}}{\lambda}$$

PRF Greater Than Spread of Doppler Frequencies. Suppose, now, that in the above described situation, the radar's PRF is 120 kilohertz. To cover the band of anticipated doppler frequencies (–28 kilohertz to +70 kilohertz) with a little room to spare, let's say we provide a doppler filter bank having a bandwidth extending from a little below –28 kilohertz to a little above +70 kilohertz (Fig. 10).

If we encounter a target having the maximum anticipated closing rate—doppler frequency of +70 kilohertz—the carrier frequency (central spectral line) of its echoes will fall just inside the high frequency end of the passband. Since the first pair of sidebands are separated from the carrier by the PRF (120 kilohertz), the sideband nearest the passband will have a frequency of 70 – 120 = –50 kHz, well below the lower end of the passband.

Similarly, if we encounter a target having the maximum anticipated negative doppler frequency (–28 kilohertz), the carrier frequency of its echoes will fall just inside the lower end of the passband (Fig. 11). The nearest sideband in this case will have a frequency of –28 + 120 = 92 kHz, well above the upper end of the passband.

MAXIMUM CLOSING RATE

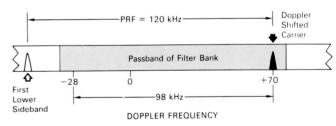

DOPPLER FREQUENCY

10. If PRF *exceeds* spread between the maximum positive and negative doppler frequencies, carrier of most rapidly closing target will fall in passband and nearest sideband will lie below it.

MAXIMUM OPENING RATE

DOPPLER FREQUENCY

11. Carrier of maximum opening-rate target will similarly fall in passband and nearest sideband will lie above it.

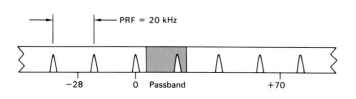

DOPPLER FREQUENCY

12. If PRF is less than spread of maximum closing rates, radar has no direct way of telling which repetition of carrier frequency it is observing.

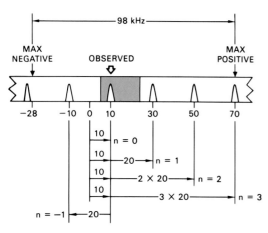

DOPPLER FREQUENCY, kHz

13. If PRF is 20 kilohertz and observed doppler frequency is 10 kilohertz, true doppler could have any of these values: –10, 10, 30, 50, and 70 kilohertz.

Thus, if the PRF is greater than the spread between the maximum anticipated positive and negative doppler frequencies, the only spectral line of the target echoes producing an output from the filter bank will be the central line—the carrier. The difference between its frequency and the transmitter's carrier frequency is the target's true doppler frequency. Hence, no significant ambiguities will exist.

This would *not* be so, however, if the PRF were less than the spread between the maximum positive and negative doppler frequencies—as it often must be to satisfy other operational requirements.

PRF Less Than Spread of Doppler Frequencies. Suppose that in this same hypothetical situation—difference between maximum anticipated positive and negative doppler frequencies equals 98 kilohertz—we reduce the PRF to only 20 kilohertz (Fig. 12). The separation between a target echoes' carrier frequency and first pair of sidebands, as well as between successive sidebands above and below them, is now only one-sixth of what is was before.

So that the return from any one target will appear at only one point within the passband, we must make it somewhat less than 20 kilohertz wide. But if the sidebands are only 20 kilohertz apart, no matter where we position the passband, there is no direct way of telling whether the target return that appears in the bank is the echoes' carrier or a sideband, or, which sideband it might be. To determine the target's true doppler frequency—hence its range rate—we must resolve the ambiguity.

Resolving Doppler Ambiguities

To resolve doppler ambiguities, we must have some way of telling what whole multiple of the PRF, if any, separates the observed frequency of the target echoes from the carrier frequency. If not too great, this multiple—n—may readily be determined. There are two common ways: range differentiation and PRF switching.

Range Differentiation. Generally, the simplest way to determine the value of n is to make an approximate initial measurement of the range rate by the differentiation method. From this rate, we compute the approximate value of the true doppler frequency. Subtracting the observed frequency from the computed value of the true frequency and dividing by the PRF yields the factor n.

Suppose, for example, that the PRF is 20 kilohertz and the observed doppler frequency is 10 kilohertz (Fig. 13). The true doppler frequency then could be –10 kilohertz plus any whole multiple of 20 kilohertz up to 70 kilohertz. The approximate value of the true doppler frequency computed from the initial range rate measurement, let's say, turns out to be 50 kilohertz. The difference between this frequency and the observed doppler frequency is 50 – 10 = 40 kHz. Dividing the difference by the PRF, we get n = 40 ÷ 20 = 2 (Fig. 14). The echoes' carrier is separated from the observed doppler frequency by two times the PRF.

Although in this simple example we assumed that the initial range-rate measurement was fairly precise, it need not be particularly accurate. As long as any error in the doppler frequency computed from the initial rate measurement is less than half the PRF, we can still tell in which PRF interval the carrier lies and so tell what n is. The initially computed "true" doppler frequency, for example, might have been only 42 kilohertz, almost half way between the two nearest possible exact values (30 and 50 kilohertz) (Fig. 15).

Nevertheless, this rough initially computed value (42 kilohertz) would still be accurate enough to enable us to find the correct value of n. The difference between the initially computed value of the doppler frequency and the observed value is 42 – 10 = 32 kHz. Dividing the difference by the PRF, we get 32 ÷ 20 = 1.6. Rounding off to the nearest whole number, we still come up with n = 2.

After having determined the value of n just this once, we can, by tracking the target continuously, determine the true doppler frequency, hence compute $\dot{R}$ with considerable precision, solely on the basis of the observed frequency.

PRF Switching. The value of n can also be determined with a PRF switching technique similar to that used to resolve range ambiguities (see Chap. 12). In essence, this technique involves alternately switching the PRF between two relatively closely spaced values and noting the change, if any, in the target's observed frequency.

Naturally, switching the PRF will have no effect on the target echoes' carrier frequency f_c. It, of course, equals the carrier frequency of the transmitted pulses plus the target's doppler frequency and is completely independent of the PRF. But not the sideband frequencies above and below f_c. Because these frequencies are separated from f_c by multiples of the PRF, when we change the PRF, the sideband frequencies correspondingly change (Fig. 16).

Which direction a particular sideband frequency moves—up or down—depends upon two things: (1) whether the sideband frequency is above or below f_c and (2) whether the PRF has been increased or decreased. An upper sideband will move up if the PRF is increased and down if it is decreased. A lower sideband, on the other hand, will move down if the PRF is increased and up if it is decreased.

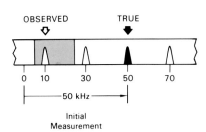

14. By making initial measurement of $\dot{R}$ with differentiation method, true doppler frequency, hence value of n, can be immediately determined.

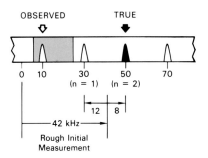

15. Initial measurement of true doppler frequency need not be particularly accurate. If error is less than half the PRF, value of n can still be found.

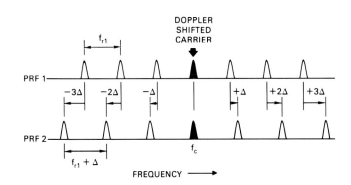

16. If PRF is changed, each sideband frequency shifts by amount, nΔ, proportional to multiple of f_r separating it from carrier.

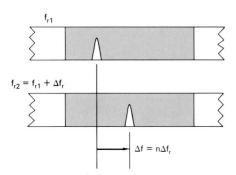

17. By noting the change in observed frequency when PRF is switched, multiple (n) of f_r contained in true frequency can be determined.

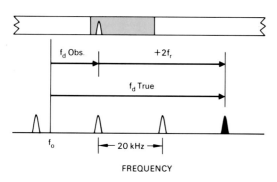

FREQUENCY

18. True doppler frequency is computed by adding n times f_r to observed doppler frequencies. (Here n = 2.)

How much the observed doppler frequency moves also depends upon two things: (1) how much the PRF has been changed and (2) what multiple of the PRF separates the observed frequency from f_c. If the PRF is changed by 1 kilohertz, the first set of sidebands on either side of f_c will move 1 kilohertz; the second, 2 kilohertz; the third, 3 kilohertz, and so on. If the PRF is changed by 2 kilohertz, each set of sidebands will move twice as far, and so on.

By noting the change, if any, in the target's observed doppler frequency, we can immediately tell where f_c is relative to the observed frequency (Fig. 17). If the observed frequency does not change, we know that it is f_c. If it does change, we can tell from the direction of the change whether f_c is above or below the observed frequency. And we can tell from the amount of the change by what multiple of the PRF f_c is removed from the observed frequency.

Thus, the factor n by which the PRF must be multiplied to obtain the difference between the echoes' carrier frequency f_c and the observed frequency is

$$n = \frac{\Delta f_{obs}}{\Delta f_r}$$

where

Δf_{obs} = change in target's observed frequency when PRF is switched

Δf_r = amount PRF is changed

If, for example, an increase in PRF (Δf_r) of 2 kilohertz caused a target's observed doppler frequency to increase by 4 kilohertz, the value of n would be 4 ÷ 2 = 2.

In order to avoid the possibility of "ghosts" when returns are simultaneously received from more than one target, the PRF must generally be switched from one to another of three values, instead of two, just as when resolving range ambiguities. As explained in Chap. 12, switching the PRF has the disadvantage of reducing the maximum detection range.

Calculating the Doppler Frequency. Having determined the value of n by either of the methods just outlined, we can compute the target's true doppler frequency, f_d, simply by multiplying the PRF by n and adding the product to the observed frequency (Fig. 18)

$$f_d = n f_r + f_{obs}$$

where

f_r = PRF before the switch

f_{obs} = target's observed doppler frequency

Summary

A target's range rate may be determined either by continuously measuring its range and calculating the rate at which the range changes—a process that approximates differentiation—or by measuring the target's doppler frequency. Because of inevitable random errors in the measured range, the differentiation method tends to be less accurate and provides poorer dynamic response.

The doppler method not only can be extremely precise, but can be nearly instantaneous. The observed doppler frequencies, however, are inherently ambiguous. Unless the spread between the maximum anticipated positive and negative doppler frequencies is less than the PRF—and the consequence of occasionally mistaking a very high-speed target for a lower speed one is negligible—the ambiguities must be resolved.

To resolve them, the number of times, n, that the PRF is contained in the difference between the observed frequency and the true frequency must be determined. If n is not too large, it can readily be found either by measuring the range rate initially with the differentiation method or by switching the PRF and observing the direction and amount that the observed doppler frequency changes.

LTV F–8E CRUSADER (1960)

This truly outstanding design has been in service for over 25 years. It fir
flew as a day fighter in 1955. It was later equipped with a Magnavox AP
125 radar and transformed into a successful all-weather fighter armed wi
Sidewinder missiles and four 20 millimeter cannons.

PART V

Return from the Ground

CONVAIR F-106A DELTA DART (1960)

This mach 2 interceptor was the first operational aircraft to be equipped with an airborne digital computer. The Hughes MA-1 Weapon Control System is virtually an entire avionics suite providing fully automatic flight control after initial radar lockon. This classic combination of airframe, engine, avionics, and weaponry was in service with the USAF for an incredible 25+ years. The aircraft is shown here firing an MB-1 Genie at Tyndall Air Force Base in 1961.

Sources and Spectra of Ground Return

22

G round return falls into three categories: mainlobe return, sidelobe return, and altitude return, which is sidelobe return received from directly beneath the radar. Mainlobe return is signal for many applications—ground mapping, altimetry, doppler navigation, etc. But both mainlobe and sidelobe return are clutter for radars which must detect airborne targets or moving targets on the ground (Fig. 1).

The principal means of discerning target echoes from ground clutter is doppler resolution. In ground based applications, separating targets from clutter is straightforward. Since the radar is stationary, all of the clutter has essentially one doppler frequency—zero. In airborne applications, however, this is far from true. Consequently, the way in which the clutter is distributed over the band of possible frequencies—its doppler spectrum—and the relationship of this spectrum to the doppler frequencies of anticipated targets critically influence a radar's design.

In this chapter, we'll consider what determines the amplitude of the ground return. We will then examine the doppler spectrum of each of the three categories of ground return, and the relationship of the composite spectrum to the doppler frequencies of target aircraft in representative situations. Finally, we'll consider the problem of exceptionally strong sidelobe return reflected by certain objects on the ground.

For simplicity, we will assume that the radar is transmitting at a sufficiently high PRF that doppler ambiguities are avoided. Their effect—which can make ground clutter much more difficult to deal with—will be covered in the next chapter.

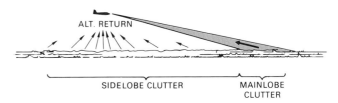

1. The three categories of ground clutter.

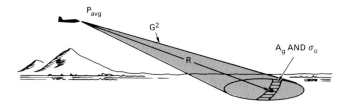

2. Factors which determine the power of the return from a patch of ground: two-way gain of the radar antenna, range to the patch, area of the patch, and backscattering coefficient, σ_o.

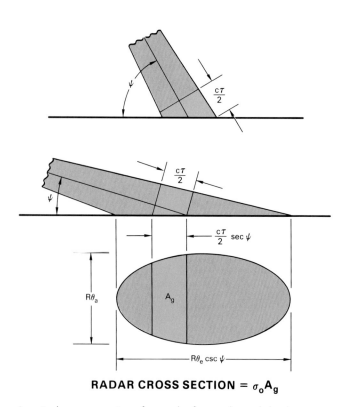

RADAR CROSS SECTION = $\sigma_o A_g$

3. Radar cross section of a patch of ground equals backscattering coefficient, σ_o, times resolvable ground area, A_g. Depending on pulse width, τ, at steep grazing angles (ψ) A_g may be determined solely by a radar's doppler and angular resolution and ψ. Generally, at shallow angles A_g also is limited by τ.

What Determines the Amplitude of the Ground Return

In general, ground return is governed by the same basic factors as return from an aircraft. For a given transmitter frequency, the power of the return received from a small patch of ground (Fig. 2) is

$$P_r \propto \frac{P_{avg} G^2 \sigma_o A_g}{R^4}$$

where

$$P_{avg} = \text{average transmitted power}$$

$$G = \text{gain of radar antenna in the direction of the patch } (G^2 = \text{two-way gain})$$

$$\sigma_o = \text{factor called the incremental backscattering coefficient}$$

$$A_g = \text{resolvable area of ground (ground patch)}$$

$$R = \text{range of ground patch}$$

The backscattering coefficient, σ_o, is the radar cross section of a small increment of ground area, ΔA.

There are three reasons for using an incremental coefficient. First, the ground viewed by a radar is more or less continuous. Second, the extent to which a given radar isolates the return from any one portion of the total ground area depends upon the radar's design—antenna beamwidth, etc. (In contrast, the radar cross section of a discrete target, such as an aircraft, is independent of the radar's design.) Third, the backscattering coefficient may vary considerably from one increment of ground to the next. As a rule, though, statistical averages of the coefficient are used for different types of terrain.

When the appropriate values of σ_o is multiplied by the area of a particular patch of ground—say the area at a specific azimuth and elevation that is delineated by a radar's range, angle, and doppler resolution—the product is the radar cross section (σ) of the patch (Fig. 3).

Like the radar cross section of a discrete target, σ_o is the product of three factors:

• Geometric area

• Reflectivity

• Directivity

4. Factors of which the incremental backscattering coefficient, σ, is a product.

Geometric area is the projection of the incremental area, ΔA, onto a plane perpendicular to the line of sight from the radar (Fig. 5). This projection determines how much transmitted energy will be intercepted by ΔA. (The power of the intercepted radiation equals the power density of the incident waves times the projected area.) The depth (vertical dimension) of the projected area is foreshortened in proportion to the cosine of the angle of incidence, $\boldsymbol{\theta}$. Consequently, σ_0 decreases as the angle of incidence increases. Put another way, as the grazing angle (Fig. 6) approaches zero, so does the value of σ_0.

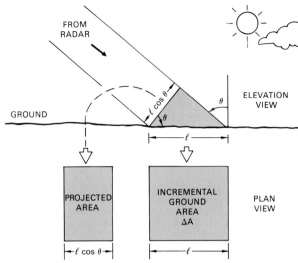

5. The term for area included in the backscattering coefficient σ_0, is the projection of the incremental area, ΔA, onto a plane perpendicular to the line of sight to the radar.

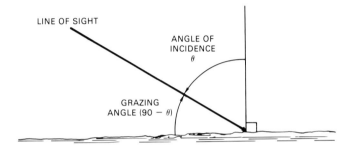

6. As the angle of incidence, θ, approaches 90° (the grazing angle approaches 0°), the projection of ΔA onto a plane normal to the line of sight to radar goes to zero.

Reflectivity is the ratio of scattered energy to intercepted energy. It varies widely with the conductivity and dielectric constant of the ground, as well as with the nature of the objects on it.

The *directivity* of an incremental ground area, like the directivity of a discrete target, is the ratio of (a) the energy scattered back toward the radar to (b) the energy that would have been backscattered if the scattering had been isotropic. This ratio depends in a complex way upon the angle of incidence ($\boldsymbol{\theta}$, in Fig. 5), the roughness of the surface relative to the wavelength of the incident radio waves, the polarization of the waves, the presence of man-made objects, etc.

Since the purely geometrical relationship between angle of incidence and σ_0 has nothing to do with the nature of the ground, analysis may sometimes be simplified by mathematically cancelling it out. To do so, one merely divides σ_0 by cos $\boldsymbol{\theta}$. The resulting coefficient, called the normalized backscattering coefficient, is represented by the Greek letter, gamma, γ.[1]

1. The normalized coefficient is also represented by the Greek letter eta (η).

$$\gamma = \frac{\sigma_0}{\cos \theta}$$

Any variation in the value of γ with the angle of incidence is due solely to variations in reflectivity and directivity. Over most angles of incidence, though, γ is more or less constant, and that is why it is used.

TYPICAL BACKSCATTERING COEFFICIENTS

Terrain	Coefficients (dB)*	
	σ_o	γ
Smooth Water	− 53	− 45.4
Desert	− 20	− 12.4
Wooded Area	− 15	− 7.4
Cities	− 7	0.6

*Values for a 10° grazing angle and 10 GHz frequency.

Backscattering coefficients are normally expressed in dB. The greater the fraction of the incident energy scattered back in the direction of the radar is, the greater (less negative) the coefficient. Where the gain due to directivity is high—as, for example, at small angles of incidence over water—the decibel value of σ_o becomes positive.

Table 1 lists the values of both σ_o and γ for common types of terrain at comparatively shallow grazing angles (large angles of incidence). When operating over smooth water, little or no energy is reflected back to the radar from such angles. Over desert, the return is substantially greater. Over wooded areas with a liberal sprinkling of man-made structures, it is greater still. Over cities, it is even greater.

Mainlobe Return

Mainlobe return—or *mainlobe clutter* (MLC) as it is called when it is not desired—is produced whenever the mainlobe intercepts the ground, as when looking down or flying at low altitudes and not looking up. It may be received from long ranges, even when flying at high altitudes and looking straight ahead.

Because the ground area intercepted by the mainlobe can be extensive and the gain of the mainlobe is high, mainlobe return is generally quite strong—far stronger than the return from any aircraft.

Frequency Versus Angle. The spectral characteristics of mainlobe return are best understood by visualizing the ground area illuminated by the mainlobe as consisting of a large number of small, individual patches (Fig. 7). The doppler frequency of each patch is proportional to the cosine of the angle, L, between the radar velocity and the line of sight to the patch.

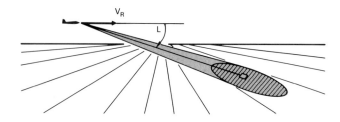

7. Area illuminated by mainlobe may be thought of consisting of many small ground patches, each at a different look angle.

$$f_d = \frac{2 V_R \cos L}{\lambda}$$

where

V_R = velocity of radar

L = angle between V_R and line of sight to ground patch

λ = wavelength

The angle L is not the same for every patch. As a result, the collective return occupies a band of frequencies.

When the antenna is looking straight ahead (Fig. 8), the doppler frequency of the return from patches near the center of the illuminated area (L ≈ 0) very nearly equals its maximum possible value: $f_{d\max} = 2 V_R / \lambda$.

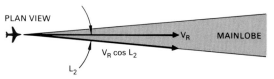

PLAN VIEW

V_R

MAINLOBE

$V_R \cos L_2$

L_2

8. When looking straight ahead, closing rate for all look angles within the beam is about the same, and $f_d \approx 2V_R/\lambda$.

The frequencies of those patches farther from the center are somewhat lower. But, since the angles to these patches are small and the cosine of a small angle is very nearly one, the band of frequencies covered by the mainlobe return when looking straight ahead is quite narrow.

As the azimuth and depression angles of the antenna increase, the cosine of L for patches at the center of the illuminated ground area decreases (Fig. 9). Consequently, the frequency of these patches decreases. At the same time, the spread between the values of cos L for patches at the two edges of the area increases, causing the band of frequencies covered by the mainlobe clutter to become wider.

To give you a quantitative feel for these relationships, the cosine of the angle L off the center of the illuminated area is plotted in Fig. 10 for values of L between ± 90°. The vertical scale gives the corresponding doppler frequencies for a radar velocity of 800 feet per second (approximately 480 knots) and a wavelength of 0.1 foot (3 centimeters).

Superimposed over the graph are two vertical bands. Each brackets those angles encompassed by a mainlobe having a beamwidth of 4°. The band in the center is for an antenna azimuth angle of zero. The other band is for an antenna azimuth angle of 60°. (In both cases, the antenna depression angle is zero, and the aircraft is assumed to be at very low altitude.)

When the azimuth is 0°, the central doppler frequency of the return is 16 kilohertz. Yet, when the azimuth has increased to 60°, this frequency is only 8 kilohertz—a decrease of 50 percent (cos 60° = 0.5).

The width of the band of frequencies spanned by the return, on the other hand, is vastly greater at the larger antenna angle. When the azimuth is 0°, the doppler frequency of a patch at the edge of the illuminated ground area ($f_{d_{max}}$ cos 2°) is so close to that of a patch at the center ($f_{d_{max}}$) that the difference cannot be read from the graph. Actually, it is about 10 hertz. Yet (since the cosine changes much more rapidly with angle at large angles), when the azimuth is 60°, the return spans a band of frequencies nearly 1 kilohertz wide—$f_{d_{max}}$ (cos 58° − cos 62°) = 16 (0.53 − 0.47) = 0.96 kilohertz.

Influence of Beamwidth, Speed, and Wavelength. For any one antenna azimuth (and/or depression) angle, the wider the mainlobe, the wider the band of mainlobe frequencies will be (Fig. 11). If the beamwidth were increased from 4° to 8°, the width of the band for an antenna angle of 60° would be 1.9 kilohertz—nearly twice that for the 4° beam.

Both the center frequency and the width of the band vary directly with the speed of the radar ($f_{d_{max}} \propto V_R$). If it

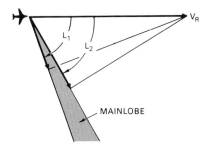

9. As angle L increases, value of cos L for patches at center of beam decreases, and spread between values for patches at edges of beam increases.

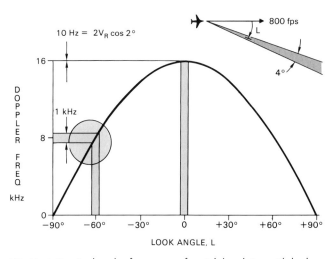

10. Variation in doppler frequency of mainlobe clutter with look angle. Vertical bands represent width of lobe (λ = 0.1 foot).

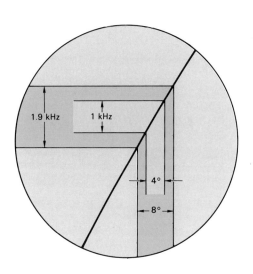

11. The wider the mainlobe, the wider the band of mainlobe clutter frequencies.

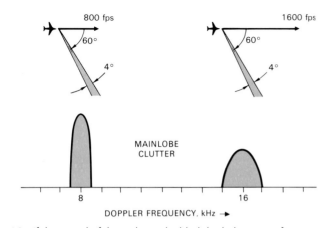

12. If the speed of the radar is doubled, both the center frequency and the width of the spectrum will double.

decreases, they decrease; if it increases, they increase. Suppose the center frequency is, say, 8 kilohertz. If the speed were doubled, this frequency as well as the frequencies at the edges of the band would double. Not only would the entire band shift up by 8 kilohertz, but its width would double (Fig. 12).

The width and center frequency vary inversely with wavelength ($f_{d_{max}} \propto 1/\lambda$). The longer the wavelength, the narrower the band will be, and vice versa. Other conditions being the same, at S-band wavelengths (10 centimeters), the band is only three-tenths as wide as at X-band wavelengths (3 centimeters) (Fig. 13).

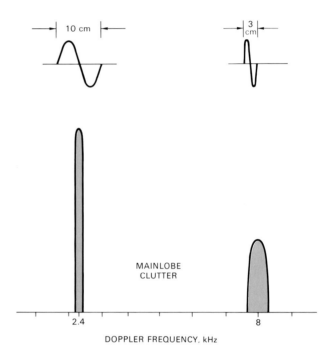

13. If the wavelength of the radar transmitter is decreased, both the center frequency and the spectral width of the mainlobe clutter will increase proportionately.

Effect of Antenna Scan. During search, the antenna scans back and forth through an azimuth angle which may be ±70 or more degrees. As it sweeps from one extreme to straight ahead (Fig. 14, top of next page), the mainlobe clutter band moves up in frequency and simultaneously squeezes into a narrow line. As the sweep continues to the other extreme, the clutter moves down in frequency and spreads to its original width. Thus the band appears to "breathe."

Significance. Because of its strength, spectral width, and variability, mainlobe return can be difficult to contend with when searching for aircraft. On the other hand, the strength and spectral width are advantageous when ground mapping. Then, the stronger the mainlobe return, the better;

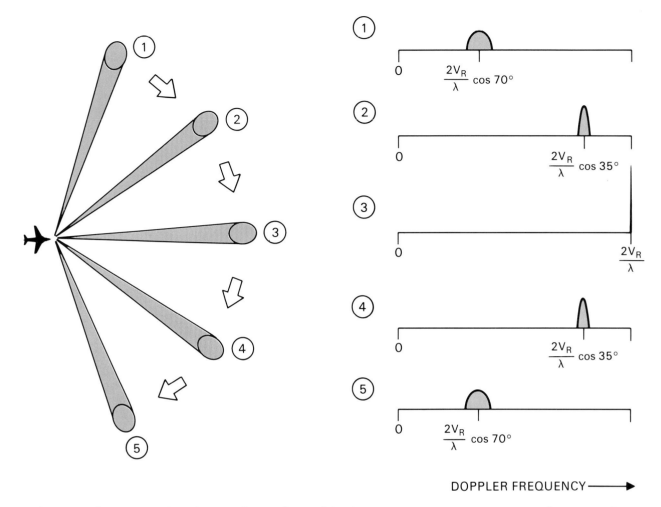

14. As the antenna beam sweeps through its search scan, the mainlobe clutter spectrum moves out to its maximum frequency and squeezes into a narrow line, then returns again.

and the wider the band of frequencies it occupies, the higher the angular resolution that can be obtained through doppler processing.

Sidelobe Clutter

The radar return received through the antenna's sidelobes is always undesirable and so is called *sidelobe clutter* (SLC). Excluding the altitude return, sidelobe clutter is not nearly as concentrated (less power per unit of doppler frequency) as mainlobe clutter. But it covers a much wider band of frequencies.

Frequency and Power. Sidelobes extend in virtually all directions, even to the rear. Therefore, regardless of the antenna look angle, there are always sidelobes pointing ahead, behind, and at virtually every angle in between. As a result, the band of frequencies covered by the sidelobe clutter extends from a positive frequency corresponding to the radar's velocity ($f_d = 2V_R/\lambda$) to an equal negative (less than the transmitter's) frequency (Fig. 15).

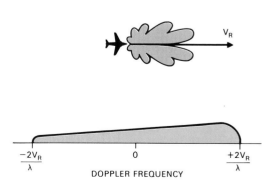

15. Because sidelobes are radiated in all directions, sidelobe clutter extends from a positive frequency corresponding to the radar's velocity to an equal negative frequency.

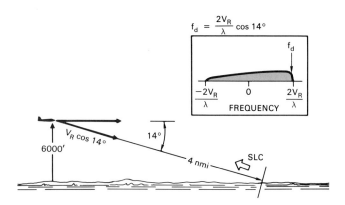

16. At an altitude of 6,000 feet, sidelobe return from a range of only 4 nautical miles will have a doppler frequency almost equal to the maximum sidelobe clutter frequency.

While the power radiated in any one direction through the sidelobes is relatively slight, the area illuminated by the sidelobes is extremely large. Moreover, much of the clutter comes from relatively short ranges. This is so, even out to the ends of the sidelobe clutter spectrum. As illustrated in Fig. 16, since the cosine of a small angle is nearly equal to one, if a radar is at an altitude of 6000 feet, return from a range of only 4 nautical miles (depression angle = 14°) will have a doppler frequency only 3 percent less than the maximum ($2V_R/\lambda$).

In aggregate, therefore, not only can the sidelobe clutter power be substantial, but the clutter may be spread more or less uniformly over a broad band of doppler frequencies.

Impact on Target Detection. The extent to which the clutter interferes with target detection depends on the frequency discrimination the radar provides. This is illustrated by the map in Fig. 17 (below).

Plotted on it are lines of constant doppler frequency. They are called isodoppler contours. Each line represents the intersection between the ground and the surface of a cone around the radar's velocity vector. Since the angle between this vector and every point on the cone is the same, the return from every point along the contour has the same doppler frequency. Just as the distances between contour lines on a relief map correspond to a fixed interval of elevation, so the distances between isodoppler lines correspond to a fixed interval of doppler frequency.

Let us suppose now that the doppler interval corresponds to the minimum difference in doppler frequency which can be discerned by a particular radar—i.e., its doppler resolution. If the radar differentiates between tar-

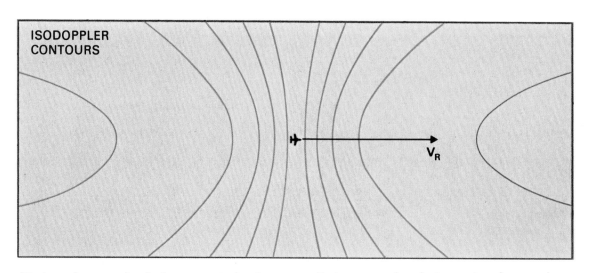

17. Lines of constant doppler frequency—isodoppler contours. Each corresponds to the intersection of a cone about the radar's velocity vector with the ground.

gets and clutter solely on the basis of doppler frequency, a target falling amid the sidelobe clutter must compete with the return from the entire strip of ground between the contours bracketing the target's doppler frequency.

And, as is made clear in Fig. 18, depending upon the target's range rate, much of this ground may be at substantially closer range than the target.

To appreciate this point, bear in mind that the strength of radar return is inversely proportional to the fourth power of the range from which it is received. For given values of antenna gain and backscattering coefficient, the power of the return from a ground patch at a range of, say, 1 mile is $(10/1)^4 = 10,000$ times (40 dB greater than) that of the return from a ground patch of the same size at a range of 10 miles.

If the radar also provides range resolution (as by range gating), the target must compete only with that portion of the clutter passed by the same range gate as the target's echoes.

Other Factors Governing Sidelobe Clutter Strength. The strength of the sidelobe return from any one patch of ground depends upon several factors besides range. One is the gain of the particular sidelobe within which the patch lies. In a representative fighter radar, the two-way gain of the first sidelobe beyond the mainlobe is on the order of 100 times (20 dB) stronger than that of the weaker sidelobes.

The strength of the return also varies widely with the nature of the terrain included in the patch—its backscattering coefficient.

As explained previously, as the grazing angle increases, the backscattering coefficient also increases. Consequently, even though a radar is closer to the ground at low altitudes, sidelobe clutter may be most severe when flying at moderate rather than low altitudes.

Significance. Clearly, the extent to which sidelobe clutter is a problem depends upon many things.

- Frequency resolution provided by the radar
- Range resolution provided by the radar
- Gain of the sidelobes
- Altitude of the radar
- Backscattering coefficient and angle of incidence

Also, as already noted, certain man-made objects can be immensely important sources of sidelobe clutter. (They will be discussed separately at the end of this chapter.)

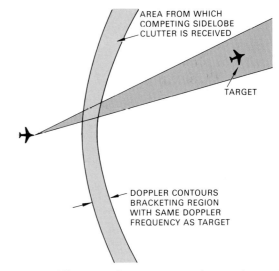

18. If a radar differentiates between target echoes and ground return solely on basis of doppler frequency, sidelobe clutter can present a serious problem.

Altitude Return

Beneath an aircraft, there is usually a region of considerable extent within which the ground is so close to being at a single range that the sidelobe return from it appears as a spike on a plot of amplitude versus range (Fig. 19).

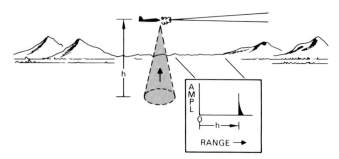

19. Altitude return comes from a large area, often at very close range.

This return is called the *altitude return,* since generally its range equals the radar's absolute altitude.

Relative Strength. The altitude return is not only much stronger than the surrounding sidelobe clutter but may be as strong or stronger than the mainlobe clutter.[2] For, the area from which it comes not only may be very large but is often at extremely close range.

This point is illustrated in Fig. 20. A radar is at an altitude (h) of 6000 feet over flat terrain. As you can see, the slant range to the ground at an angle of incidence θ equals h/cos θ. The cosine of a small angle being only slightly less than one, even when θ is as much as 22° the slant range to the ground is only 500 feet greater than the altitude (vertical range).

If the slant range at an angle of incidence of 22° is rotated about the vertical axis, it traces a circle on the ground having a diameter of roughly 5000 feet (Fig. 21).

A circle this size contains nearly 20 million square feet. Thus, the radar receives all of the return from a 20-million-square-foot area, at a range of just 1 nautical mile, in the round-trip transit time for a range increment of 500 feet. That time is only 1 microsecond.

Furthermore, at near vertical incidence, the backscattering coefficient tends to be very large. Over water, the coefficient is enormous. Little wonder, then, that the altitude return appears as a sharp spike in a plot of amplitude versus range.

2. In the case of an altimeter, the mainlobe return is the altitude return.

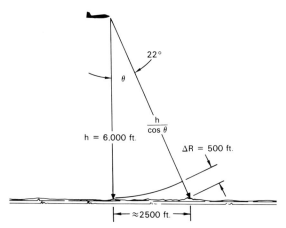

20. At an altitude of 6,000 feet, even at an angle of incidence θ of 22°, the slant range to the ground is only 500 feet greater than the altitude.

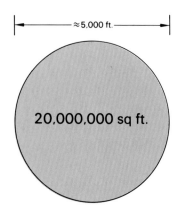

21. Yet, at an altitude of 6,000 feet, an angle of incidence of 22° encompasses a circle having an area of 20 million square feet.

Doppler Frequency. The altitude return peaks up in a plot of amplitude versus frequency, too, but not sharply. The reason can be seen from Fig. 22 at the top of the facing page.

The projection of the radar velocity, V_R, on the slant range to the ground equals $V_R \sin \theta$. Unlike the cosine, the sine of an angle changes most rapidly as the angle goes through zero. While the doppler frequency of the clutter is zero when θ is zero, it increases to nearly 40 percent of its maximum value ($2 V_R/\lambda$) at an angle θ of only 22°. Return from a circle of ground that produces a sharp spike in a plot of amplitude versus range, therefore, produces only a broad hump in a plot of amplitude versus doppler frequency (Fig. 23).

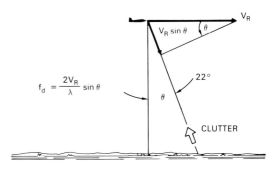

22. Doppler frequency of return from a given angle of incidence, θ, is proportional to projection of radar velocity onto slant range at that angle, hence to the sine of θ.

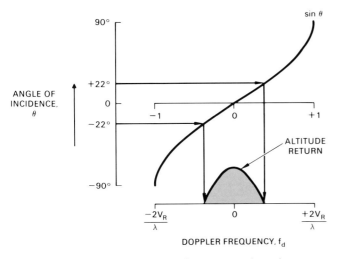

23. Since sine θ changes most rapidly as θ goes through zero, altitude return is spread over a comparatively broad band of doppler frequencies.

Normally, the doppler frequency of the altitude return is centered at zero. However, if the altitude of the radar is changing—as when the aircraft is climbing, diving, or flying over sloping terrain—this will not be so. For a dive, the doppler frequency will be positive (Fig. 24); for a climb, it will be negative. Even though the frequency is generally fairly low, it can be considerable. In a 30° dive, for instance, the altitude would be changing at a rate equal to half the radar velocity.

Significance. Despite its strength, the altitude return is usually less difficult to deal with than the other ground returns. Not only does it come from a single range, but its range is predictable; and, as we have seen, its frequency is generally close to zero—though that unfortunately is the doppler frequency of a target pursued at constant range (e.g., a tail chase with zero closing rate, $\dot{R} = 0$).

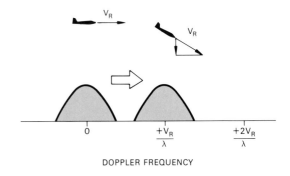

24. Doppler frequency of altitude return is normally low, but may be quite high in a dive.

Relation of Clutter Spectrum to Target Frequencies

Having become familiar with the characteristics of mainlobe clutter, sidelobe clutter, and altitude return individually, let us look briefly at the composite clutter spectrum and its relationship to the frequencies of the echoes from repre-

NOSE ASPECT APPROACH

ALTITUDE RETURN

MLC

TARGET

SIDELOBE CLUTTER

(−)   0   (+)

DOPPLER FREQUENCY

25. Doppler frequency of target is greater than that of any ground return.

sentative airborne targets in typical operational situations. Again, we will assume that the PRF is high enough to avoid doppler ambiguities.

Figure 25 illustrates the relationship between target and clutter frequencies for a nose-aspect approach. Because the target's closing rate is greater than the radar's velocity, the target's doppler frequency is greater than that of any of the ground return.

Figure 26 (below) shows the relationship for a tail chase. Because the target's range rate is less than the radar's velocity, the target's doppler frequency falls within the band of sidelobe clutter. Just where, depends upon the range rate.

TAIL CHASE, LOW CLOSING APPROACH

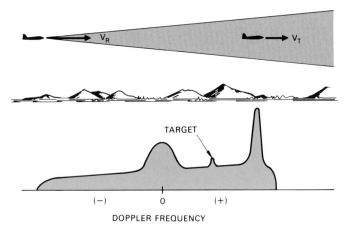

TARGET

(−)   0   (+)

DOPPLER FREQUENCY

26. Doppler frequency of target falls within sidelobe clutter.

In Fig. 27, the target's velocity is perpendicular to the line of sight from the radar; the target has the same doppler frequency as the mainlobe clutter. Fortunately, a target will attain such a relationship only occasionally and usually will remain in it fleetingly.

In Fig. 28, the target's closing rate is zero; the target has the same doppler frequency as the altitude return.

TARGET VELOCITY PERPENDICULAR TO L.O.S.

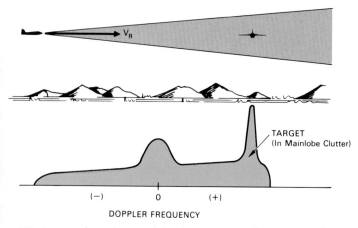

TARGET (In Mainlobe Clutter)

(−)   0   (+)

DOPPLER FREQUENCY

27. Target is buried in mainlobe clutter. Fortunately, a target will attain such a relationship only occasionally and usually will remain in it fleetingly.

TAIL CHASE, ZERO CLOSING RATE ($V_R = V_T$)

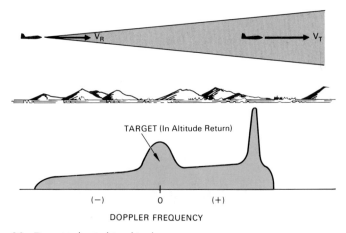

TARGET (In Altitude Return)

(−)   0   (+)

DOPPLER FREQUENCY

28. Target is buried in altitude return.

Figure 29 shows two opening targets. Target A has an opening rate that is greater than the radar's ground speed (V_R), so this target appears in the clear beyond the negative-frequency end of the sidelobe clutter spectrum. On the other hand, Target B has an opening rate less than V_R. So this target appears within the negative frequency portion of the sidelobe clutter spectrum.

With these situations as a guide, the relationship of the doppler frequencies of target return and ground return for virtually any situation can easily be pictured (Fig. 30). Bear in mind, though, that at lower PRFs doppler ambiguities can occur which may cause a target and a ground patch having quite different range rates to appear to have the same doppler frequency.

The consequences of such ambiguities will be discussed in detail in the next chapter. The signal processing commonly performed to separate target echoes from ground clutter in the various operational situations presented here, as well as when operating at PRFs which make doppler frequencies ambiguous, will be described in Chaps. 26, 27, and 28.

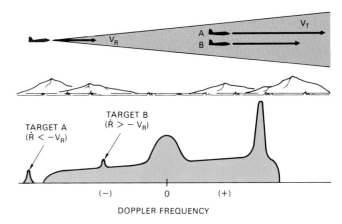

29. If range rate is more negative than $-VR$, target appears in clear (A) below sidelobe clutter spectrum; otherwise, it appears in negative frequency half of the spectrum (B).

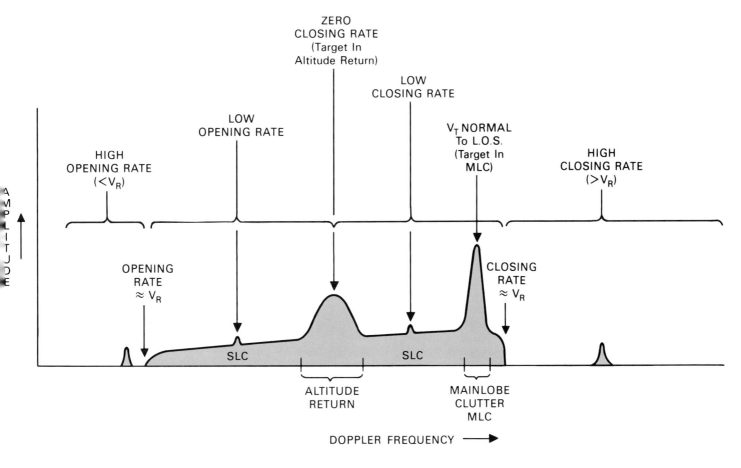

30. Relationships between target doppler frequency and ground-clutter spectrum for various target closing rates. (Doppler frequencies assumed to be unambiguous.)

31. Viewed straight-on, a smooth[3] flat plate can have an immense radar cross section.

σ = 320,000 sq. ft.

3. The surface variations are small in comparison to a wavelength.

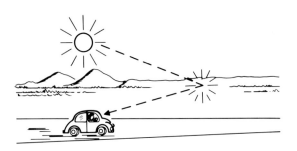

32. Radar echoes from certain man-made objects on the ground are comparable to reflection from a window struck from just the right angle by the sun.

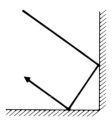 

33. Corner formed by two flat surfaces is retroreflective over wide range of angles; that formed by three surfaces, over wider range still.

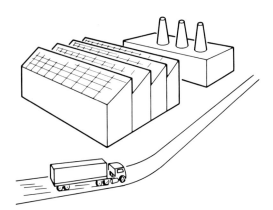

34. Portions of a building may act like corner reflectors; a truck like several corner reflectors.

Return from Objects on the Terrain

The return from certain man-made structures can be very strong. Viewed straight on by an X-band radar, a smooth, flat metal sign only 4 feet on a side, for example, has a radar cross section on the order of 320,000 square feet (Fig. 31) compared to 10 square feet or less for a small aircraft in some aspects.

This may sound absurd, but not if you stop to think. Most of the power intercepted by the sign when viewed straight-on will be reflected back in the direction of the radar. The sign is a specular (mirrorlike) reflector. It acts, in fact, just like an antenna that is trained on the radar and reradiates all of the transmitted power it intercepts, back in the radar's direction. At X-band frequencies, for example, an antenna with a 16-square foot aperture has a gain of around 20,000. Multiply the area of the sign by this gain (16 x 20,000) and you get a radar cross section of 320,000 square feet.

In principle, the radar return from a flat reflecting surface, such as a sign, is directly comparable to the intense reflections one frequently gets from the windshield of a car or the window of a hillside house when it is struck from just the right angle by the early morning or late afternoon sun (Fig. 32).

Whereas a single flat surface such as a sign must be viewed from nearly straight-on to reflect the incident energy back to the radar, two surfaces forming a 90° corner will do so over a wide range of angles in a plane normal to the intersection of the surfaces. They are what is called *retroreflective*. If a third surface is added at right angles to the other two (forming a corner reflector), the range of angles over which the surfaces will be retroreflective may be increased to nearly a quarter of a hemisphere (Fig. 33). This, incidentally, is the way bicycle reflectors work. Portions of a large building may act like corner reflectors, and a vehicle such as a truck may look like a group of corner reflectors (Fig. 34).

Because of their enormous radar cross sections, retroreflective objects on the ground can produce sidelobe return as strong or stronger than the echoes from distant aircraft received through the mainlobe. Furthermore, because the objects are of limited geographic extent—they are discrete as opposed to distributed reflectors—all of the return from one of them has very nearly the same doppler frequency and comes from very nearly the same range. The return may appear to the radar, therefore, exactly as if it came from an aircraft in the mainlobe.

Naturally, since these objects are virtually all man-made, they are much more numerous in urban than in rural areas.

Nevertheless, they may be encountered almost anywhere. In the little populated region of Southern California's Antelope Valley, for example, the long, low corrugated metal sheds of the turkey ranchers (Fig. 35) return tremendously strong echoes.

Depending on the use of the radar, special measures may be required to reduce or eliminate sidelobe return of this sort. Mainlobe return from such objects is usually not a problem, since its doppler frequency is generally different from that of targets of interest. But if the objects are moving or have extraordinarily large radar cross sections, the mainlobe return, too, can be a problem.

Summary

The same factors govern the power of ground return as that from an aircraft. Backscattering from the ground, however, is expressed in terms of an incremental coefficient, σ_0, which must be multiplied by the area of a ground patch to obtain its radar cross section, σ. The coefficient σ_0 varies with angle of incidence, frequency, polarization, electrical characteristics of the ground, roughness of the terrain, and nature of the objects on it. Generally, the variation with angle is due primarily to foreshortening of the ground area as viewed from the radar.

The most important ground return—and the only return of interest for ground mapping etc.—is that received through the antenna's mainlobe. When the antenna is looking straight ahead, the doppler frequency of this return corresponds to the radar's full ground speed. As the look angle increases, the frequency decreases and spreads over an increasingly broad band. Both the center frequency and the width of this band increase directly with the radar's velocity and are inversely proportional to wavelength.

Ground return received through the sidelobes, though comparatively weak at any one frequency, extends from a positive frequency corresponding to the radar's full velocity $(2V_R/\lambda)$ to an equal negative frequency. The portion received from directly below—the altitude return—is especially strong, particularly over water. It appears as a spike on a plot of amplitude versus range and as a broad hump on the doppler spectrum. Its center doppler frequency is normally zero.

Man-made objects on the ground may be highly retroreflective and can produce sidelobe return as strong as target echoes received through the mainlobe.

If the PRF is high enough to eliminate ambiguities, whether a target's echoes and the ground return have different doppler frequencies—as is desirable for ground clutter rejection—depends upon the target's closing rate. As long

35. Even in little populated regions, there may be numerous structures having tremendous radar cross sections.

as it is greater than the radar's velocity, the echoes will lie outside the ground return. Otherwise, they must compete with sidelobe return. Only if the target is flying at right angles to the line of sight from the radar, will its echoes have the same frequency as the mainlobe return. Only if the closing rate is zero, will the target echoes have the same frequency as the altitude return.

However, as we shall see in the next chapter, doppler ambiguities can cause a target and ground patches, having quite different range rates, to appear to have the same doppler frequency, thereby greatly compounding the problem of separating target echoes from clutter.

Effect of Range and Doppler Ambiguities on Ground Clutter

23

I n the last chapter, we surveyed the sources of ground return and became acquainted with its doppler spectrum. But we did not consider the profound effects of range and doppler ambiguities on ground return. Although we discussed both types of ambiguities in detail in earlier chapters, the discussions there involved only target return. If a radar is searching for or tracking a target in the presence of ground clutter, however, the consequences of ambiguities in the clutter are quite different from the consequences of the same ambiguities in the target return.

In the case of a target, we are interested in the target itself and in the *value* of its range or doppler frequency. Since a target such as an aircraft is essentially a point source, ambiguities simply give the observed range or doppler frequency more than one possible value. If the ambiguities are not too severe, we can resolve them through such techniques as PRF switching.

In the case of ground clutter, on the other hand, we are interested only in *differences* in range and doppler frequency which will enable us to separate the clutter from the target echoes. Since the sources of the clutter generally are widely dispersed, ambiguities tend to wash out these differences. About all that may be accomplished through resolution techniques such as PRF switching is to shift blocks of clutter about.

In this chapter, after briefly considering the dispersed nature of ground clutter, we will examine the effects of ambiguities on the range and doppler profiles for a representative flight situation and see how they compound the problem of separating target echoes from clutter.

Dispersed Nature of the Clutter

As we saw in Chap. 22, when the antenna beam strikes the ground, it usually illuminates an area which is extensive in both range and angle. Furthermore, the antenna invariably has sidelobes through which it radiates an appreciable amount of energy (Fig. 1).

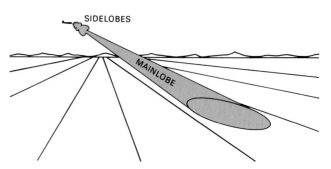

1. When the antenna beam strikes the ground, it illuminates an area extensive in both azimuth and elevation. Additionally, sidelobes illuminate ground in virtually all directions.

Ground return of various amplitudes is thus received from a great many different ranges and directions. Since the direction to a point on the ground in large measure determines the point's range rate, the return also covers a broad band of doppler frequencies.

Naturally, any dispersion in range and frequency of the ground return makes the problem of separating target echoes from it more difficult. Ambiguities in range and doppler frequency compound the problem by causing clutter from more than one block of ranges and more than one portion of the doppler spectrum to be superimposed. The effect of this can be visualized most clearly by examining separately the range and doppler-frequency profiles for a representative flight situation.

We will assume that a radar-equipped aircraft is flying at low altitude over terrain from which a considerable amount of ground return is received. The radar antenna is looking down at a slightly negative elevation angle and to the right at an angle of about 30°.

Two targets, A and B, are in the antenna's mainlobe. Target A is being overtaken from the rear and so has a low closing rate. Target B is approaching the radar head-on and so has a high closing rate (Fig. 2).

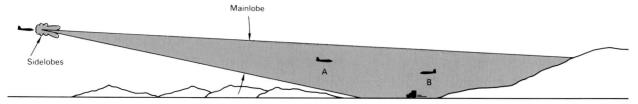

2. Representative flight situation. Targets include both low and high closing rate aircraft plus a truck.

For purposes of illustration, Target A has been placed at a range from which only sidelobe return is being received; Target B, at a range from which both mainlobe and sidelobe return are being received (Fig. 3).

Within the ground patch illuminated by the mainlobe is a truck. It is heading toward the radar and so has a slightly higher closing rate than the ground it is traveling over.

The flight situation diagram is repeated in Fig. 4, with the corresponding "true" range profile beneath it. This profile is simply a plot of the amplitude of the radar return versus the range of its sources relative to the radar—i.e., slant range, as opposed to horizontal range on the ground. Sidelobe clutter, you will notice, extends outward from a range equal to the radar's altitude. Notice how rapidly it decreases in amplitude as the range increases.

The echoes from Target A stand out clearly above the sidelobe clutter. By contrast, the echoes from Target B and the truck are completely obscured by the much stronger mainlobe clutter. Even though we know exactly where to look for these echoes, we cannot distinguish them from the clutter.

Toward the left end of the range profile, you will notice two strong spikes. The one at zero range is due to what is called transmitter spillover—energy from the transmitter that leaks into the receiver during transmission, despite all efforts to block it. The second spike is the altitude return.

Range Ambiguities

Range ambiguities arise when all of the echoes from one pulse are not received before the next pulse is transmitted. As explained in detail in Chap. 12, when return is received from beyond the unambiguous range, R_u, it is impossible to tell which pulse a particular echo belongs to (Fig. 5).

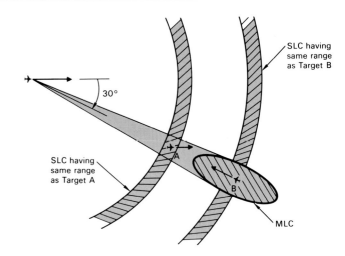

3. Target A is at a range from which only sidelobe return is received; target B, at a range from which both mainlobe and sidelobe return are received.

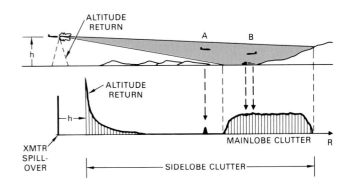

4. True range profile of representative flight situation. Target A can be seen clearly above sidelobe clutter, but target B and truck are obscured by mainlobe clutter.

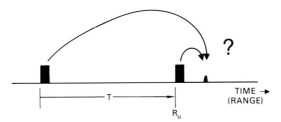

5. Range ambiguities occur when return from one pulse is being received after the next pulse is transmitted. Range corresponding to interpulse period, T, is R_u.

But, far more important from the standpoint of clutter rejection, the returns from ranges separated by R_u are received simultaneously. The echoes from a target, therefore, must compete with ground return not only from the target's own range but from every range that is separated from it by a whole multiple of R_u.

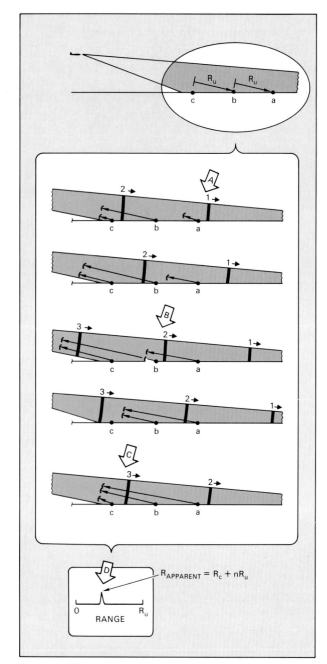

6. Return of echoes of three successive transmitted pulses from points on the ground which are separated in range by the unambiguous range, R_u.

Returns from Ranges Separated by R_u. To illustrate the effect of range ambiguities on the range profile observed at the output of the receiver, Fig. 6 traces the paths of the echoes of three successive transmitted pulses from three points on the ground—"a," "b," and "c"—which are separated in range by R_u. The key points in the figure (called out by the large lettered arrows) are the following.

(A) An echo of pulse No. 1 is reflected from the most distant point, "a."

(B) This echo reaches the next most distant point, "b," just as it is reflecting an echo of pulse No. 2.

(C) Traveling together, these two echoes similarly reach the near point, "c," just as it is reflecting an echo of pulse No. 3. The three echoes travel the remaining distance to the radar together.

(D) They arrive simultaneously and appear on the radar display as though received from a single range. All three points have the same apparent range.

An instant later, the echoes of the same pulses from points just beyond "a," "b," and "c" arrive simultaneously. An instant after that, so do the echoes from points just beyond these, and so on.

Thus, the range profile is, in effect, broken into segments, R_u wide, which are superimposed, one over the other (Fig. 7).

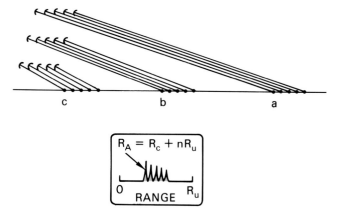

7. Echoes from points just beyond "c," "b," and "a" are likewise received simultaneously. So are echoes from points just beyond these, and on and on.

Range Zones. Now the range of point "a" in Figure 6 was selected more or less at random. It could have been any range within the region from which return is received.

Let us assume now that the range of "a" is such as to place point "c" at zero range (right at the radar antenna). Then the echoes of all ranges between "c" and "b" would be first-time-around echoes; i.e., they would be echoes of the immediately preceding (last) transmitted pulse.

The echoes from all ranges between "b" and "a" would be second-time-around echoes; they would be echoes of the pulse before the immediately preceding one.

Likewise, the echoes from all points between "a" and a point R_u beyond "a" would be third-time-around echoes, and so on.

The particular segments into which the range profile would in this case be divided (Fig. 8) are called *range zones* (or *ambiguity zones*).

Although the true range profile could similarly be divided into any number of different sets of contiguous zones R_u wide, this particular set was chosen for two reasons. First, it conveniently starts at zero range. Second, the true range of every point within any one zone is the point's apparent range plus the same whole multiple of R_u.

Now, the higher the PRF is, the shorter R_u will be, hence the narrower the range zones. The narrower the zones, the greater the number of segments into which the true range profile will be divided and from which returns will be received simultaneously.

As shown in Chap. 12, R_u very nearly equals 80 nautical miles divided by the PRF in kilohertz.[1]

$$\text{Width of range zones } = \frac{80}{f_r} \text{ nmi}$$

If, for example, the PRF were 4 kilohertz, the range zones would be 20 nautical miles wide. If returns were received from ranges out to 60 miles, the true range profile would be broken into 3 zones.

Range Zones Superimposed. The effect of breaking the true range profile for our representative flight situation into three range zones is illustrated in Fig. 9. There the returns from Zones 2 and 3 are placed beneath the return from Zone 1, and the corresponding ranges within the zones are lined up. Beneath these plots is the composite profile that would appear at the output of the receiver.

As you can see, superimposed over the echoes from Target A are not only the sidelobe clutter from the target's own range but the much stronger close-in sidelobe clutter from the corresponding range in Zone 1 and the still stronger mainlobe clutter from the corresponding range in Zone 3.

Similarly, superimposed over the echoes from Target B and the truck are not only the mainlobe clutter from their own ranges but the sidelobe clutter from the corresponding ranges in Zones 1 and 2.

As the PRF is increased and the range zones narrow, the amount of clutter that may be superimposed over any one

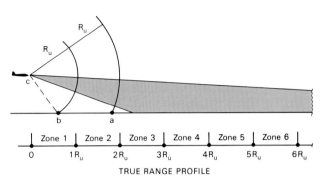

8. If point "c" in the preceding example had been moved in to zero range and points "b" and "a" had been moved equally, the true range profile would have been broken into segments called range zones.

1. In kilometers, R_u equals 150 divided by the PRF.

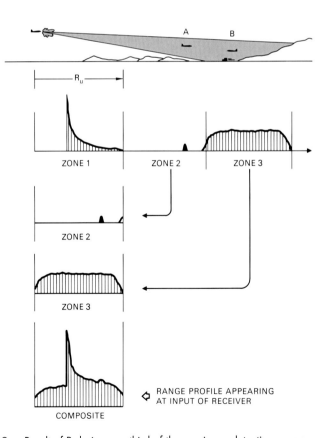

9. Result of R_u being one-third of the maximum detection range. Range profile for representative flight situation is broken into three range zones. Mainlobe clutter blankets composite profile seen by radar.

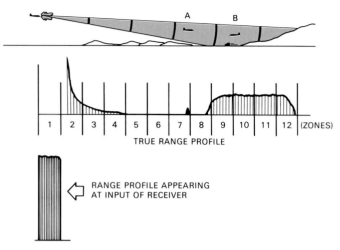

10. The higher the PRF, the narrower the range zones and the more deeply the clutter is piled up.

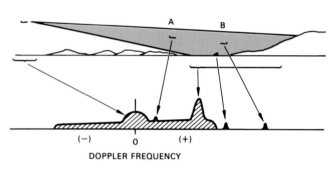

11. True doppler frequency profile of representative flight situation. Target B and truck, having higher closing rates than the ground, appear in clear.

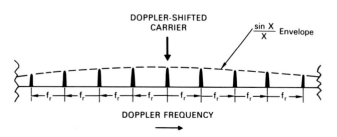

12. Each element of radar return has sideband frequencies separated from the doppler-shifted carrier by multiples of the PRF, f_r.

target's echoes increases (Fig. 10). If the PRF is increased without limit, a point is ultimately reached where the radar transmits continuously.

A target's echoes must then compete with the ground clutter received from all ranges.

Clearly, the more deeply the clutter is piled up, the less able the radar will be to isolate target echoes from the clutter on the basis of differences in range, and the greater the extent to which the radar must depend on other means, such as differences in doppler frequency, for clutter rejection.

Doppler Profile

The flight profile is shown with the true doppler frequency profile beneath it in Fig. 11. This profile is a graph of the amplitude of the radar return versus doppler frequency. In plotting it, no attempt was made to differentiate the return received at one point in the interpulse period from the return received at another. The profile represents the return from all ranges.

Sidelobe clutter extends from zero doppler frequency out in both positive and negative directions to frequencies corresponding to the radar's full velocity ($f_d = \pm 2V_R/\lambda$). The spike at zero frequency is the transmitter spillover. The broad hump under it is the altitude return. The narrower hump near the maximum positive sidelobe-clutter frequency is mainlobe clutter.

Target A is being overtaken, so its doppler frequency falls below that of the mainlobe clutter, in the band of frequencies blanketed by sidelobe clutter. Because a good deal of this clutter comes from shorter ranges than the target's, the target echoes barely protrude above the clutter. If the target were smaller or at much greater range, its echoes might not even be discernible.

Since Target B and the truck are approaching the radar and are nearly dead ahead, they have higher doppler frequencies than any of the clutter.

Doppler Ambiguities

As we learned in Chap. 16, when transmission is pulsed, each element of the radar return has sideband frequencies separated from the doppler shifted carrier frequency by multiples of the pulse repetition frequency, f_r (Fig. 12). Thus, the portion of the altitude return which has zero doppler frequency also appears to have doppler frequencies of $\pm f_r$, $\pm 2f_r$, $\pm 3f_r$, $\pm 4f_r$ etc.

Similarly, the portion of the return having a true doppler frequency of, say, +100 hertz also appears to have doppler frequencies of $(100 \pm f_r)$, $(100 \pm 2 f_r)$, $(100 \pm 3 f_r)$, $(100 \pm 4 f_r)$, etc.

314

The same is true of the return received from every other point in the true doppler frequency profile (Fig. 13). The entire profile is, therefore, repeated at intervals equal to f_r above and below the carrier frequency of the transmitted pulses. Because it is made up of doppler-shifted return from the central spectral line of the transmitted signal, the true profile is commonly referred to as the *central line return*. Repetitions of the spectrum, or of portions of it such as the mainlobe clutter, are then referred to as *PRF lines*.

If f_r is sufficiently high, the repetitions of the doppler spectrum will in no way affect the ability of the radar to discriminate between target echoes and ground clutter. In fact, if the radar's doppler passband is no more than f_r hertz wide, the repetitions will not even be seen by the doppler filter bank (Fig. 14).

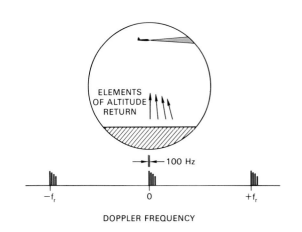

13. Spectrum of a portion of the altitude return. Elements shown here are separated in doppler frequency by 100 hertz. Each element has sidelobes separated from the doppler shifted carrier frequency by multiples of the PRF.

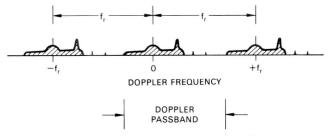

14. If f_r is high enough, the nearest sidebands will be entirely outside the passband.

However, if f_r is less than the width of the true doppler profile (as it often must be made to reduce or eliminate range ambiguities), the repetitions will overlap, and the observed doppler frequencies will be ambiguous. This condition is illustrated in Fig. 15 for a value of f_r that is only one-half the width of the true profile. In this case, the true profile is overlapped by the repetitions immediately above and below it. For clarity, each repetition is plotted on a separate baseline. In actuality, they would all merge into a single composite profile. The central portion of the composite profile is shown at the bottom of the figure.

Any overlapping of the repetitions of the profile, such as that just illustrated, can result in a target's echoes and ground clutter passing through the same doppler filter(s), even though the true doppler frequencies of the target and the clutter may be quite different. Examples of this are Target B and the truck, in Fig. 15. Whereas the true doppler frequencies of these targets are actually higher than those of any of the clutter, in the composite profile, both targets are nearly obscured by sidelobe clutter.

Because the sideband frequencies are separated from the central-line frequencies by multiples of the PRF, any one segment of the composite spectrum is identical to every

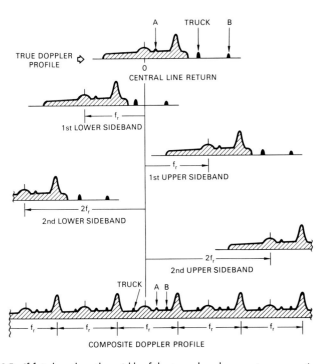

15. If f_r is less than the width of the true doppler spectrum, repetitions of the spectrum due to sideband frequencies will overlap and actually merge to from the single composite profile shown at bottom of the figure.

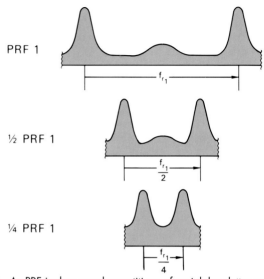

PRF 1

½ PRF 1

¼ PRF 1

16. As PRF is decreased, repetitions of mainlobe clutter spectrum move closer together, leaving less room in which to detect targets.

other segment of the same width on either side. As noted in previous chapters, it is for this reason that the passband of the doppler filter bank need be no more than f_r hertz wide.

The repetitions of the true doppler profile overlap increasingly as the PRF is reduced (Fig. 16). From the standpoint of clutter rejection, reducing the PRF has two main effects. First, more and more sidelobe clutter piles up in the space between successive mainlobe clutter lines. Second, and more important, the mainlobe clutter lines move closer together. Since the width of these lines is independent of the PRF, reducing the PRF causes the mainlobe clutter to occupy an increasingly larger percentage of the receiver passband and causes the altitude return and other close-in sidelobe clutter to pile up increasingly in the space between.

As the percentage of the passband occupied by mainlobe clutter increases, it becomes increasingly difficult to reject even the mainlobe clutter on the basis of its doppler frequency without at the same time rejecting a large percentage of the target echoes. If carried to its extreme, the overlap would ultimately reach a point where the mainlobe clutter completely blanketed the receiver passband.

Clearly, the lower the PRF, the more severe the effect of doppler ambiguities on ground clutter.

Summary

Since ground clutter is widely dispersed, range and doppler ambiguities greatly compound the problem of isolating target echoes from the clutter.

In effect, range ambiguities break the range profile into zones, which are superimposed on one another. Because of this superposition, a target's echoes may be received simultaneously with clutter not only from the target's own range but from the corresponding range in every other range zone. Increasing the PRF narrows the range zones and increases the number of zones that are superimposed, thereby making it increasingly difficult to isolate the target echoes.

Doppler ambiguities cause successive repetitions of the doppler profile to overlap. Because of this, a target's echoes may have to compete with clutter whose true doppler frequency is quite different from the target's. Increasing the PRF moves successive repetitions of the mainlobe clutter spectrum farther apart, thereby making it easier to isolate the target echoes. Thus the relationship between PRF and doppler ambiguities is just the opposite of that between PRF and range ambiguities. The lower the PRF, the more severe the effect of doppler ambiguities on ground clutter; and the higher the PRF, the more severe the effect of range ambiguities.

Separating Ground-Moving Targets from Clutter

24

I n earlier chapters, very little was said about the detection of moving targets on the ground—cars, trucks, tanks. Except that low PRFs are generally required for air-to-ground operation, it was tacitly assumed that separating *ground moving targets* (GMTs) from competing ground return on the basis of differences in doppler frequency is no different than separating airborne moving targets from ground return.

This assumption, however, is only partially true. For the radial component of the velocity of many ground moving targets is so low that the returns from them are embedded in mainlobe clutter and cannot be separated from it by conventional moving target indication (MTI) techniques.

In this chapter, after briefly examining the problem, we will be introduced to two highly effective techniques for detecting such targets. One is called *Classical DPCA*, for *displaced phase center antenna*. The other and newer technique is called *notching* or *clutter nulling*. We'll take up Classical DPCA first; then, notching; and, finally, a combination of the two. In closing we'll touch on the adaptation of these techniques to precise angle measurement.

Problem of Detecting "Slow" Moving Targets

The chief problem in detecting moving targets in low-PRF modes—whether on the ground or in flight—is separating the target returns from mainlobe clutter. As explained in detail in Chap. 26, by employing a reasonably long antenna and flying at comparatively low speeds, we can reduce the width of the mainlobe clutter spectrum and spread its repetitions far enough apart to provide a fairly

WHAT AN ANTENNA PHASE CENTER IS

Every antenna has a phase center. It is that point in space where, if a hypothetical omnidirectional point-source radiator were placed . . .

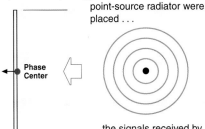

Phase Center

. . . the signals received by it from any source within the field of regard of the real antenna

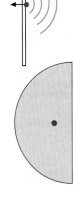

would have the same radio frequency phase as signals from the same source received by the real antenna.

If weighting of the antenna (for sidelobe reduction) is symmetrical, the position of the phase center will be same for all look angles. But if the weighting is nonsymmetrical—as in a half aperture of a mono-pulse antenna—the position of the phase center will be a function of the look angle.

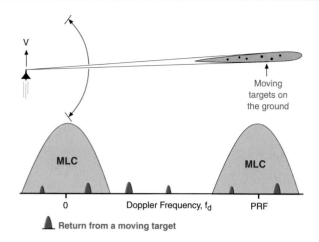

Return from a moving target

1. Doppler spectrum of ground moving targets. Special techniques are needed to detect targets whose *true* doppler frequencies fall within mainlobe clutter (MLC).

> 1. Some shift may also be due to so-called "internal motion" of the clutter scatterers, e.g., wind-blown trees.

DPCA TRANSMISSION

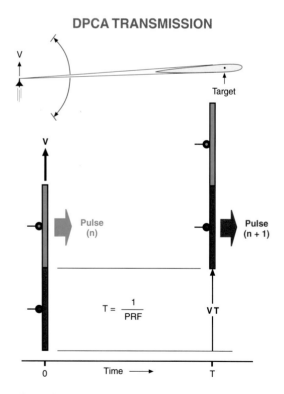

2. In Classical DPCA, radar transmits alternate pulses with forward and aft segments. By adjusting velocity, V, and PRF so radar advances distance between segments' phase centers during interpulse period, pulse n and (n + 1) are transmitted from the same point in space.

wide clutter-free region in which to detect moving targets (Fig. 1). Moreover, any target whose *apparent* doppler frequency falls within the mainlobe clutter can be periodically moved into this region by switching the PRF among several different widely separated values.

However, if the radial component of a target's velocity is so low that its *true* doppler frequency lies within the mainlobe clutter, no amount of PRF switching will move the target's returns out of the clutter. Consequently, in many applications a special "slow-moving-target" indication capability is needed. Conceptually, the simplest is Classical DPCA.

Classical DPCA

This technique takes advantage of the fact that the doppler shift in the frequency of the returns received from the ground is due entirely[1] to the aircraft's velocity. Specifically, this shift—which is manifest as a progressive pulse-to-pulse shift in the phase of the returns from any one range—is the result of the forward displacement of the radar antenna's phase center (defined in the panel on the preceeding page) from one interpulse period to the next.

For any two successive pulses, therefore, the shift can be eliminated by displacing the antenna phase center by an equal amount in the opposite direction before the second pulse of the pair is transmitted. The second pulse will then be transmitted from the same point in space as the first.

And how does one displace an antenna's phase center? Generally, the radar is provided with a two-segment side-looking electronically steered antenna. The aircraft's velocity and the radar's PRF are adjusted so that during each interpulse period the aircraft will advance a distance precisely equal to that between the phase centers of the two antenna segments (Fig. 2).

Successive pulses then are alternately transmitted by the two segments:

- Pulse (n) by the forward segment, pulse (n + 1) by the aft segment;

- Pulse (n + 2) by the forward segment, pulse (n + 3) by the aft segment, and so on.

As a result, the pulses of every pair—e.g., (n) and (n + 1)— are transmitted from exactly the same point in space.

The returns of each pulse are received by the antenna segment which transmitted the pulse. When the return from any one range, R, is received, of course, the phase center of that segment will have advanced a distance equal to the aircraft velocity, V, times the round-trip transit time, t_R, for the range R.

But if V is constant, this advance will be the same for both pulses. Therefore, the round-trip distance traveled by the pulses to any one point on the ground will be the same, making the phases of the returns the same (Fig. 3).

So, for each resolvable range interval the radar returns received from the ground may be canceled simply by passing the digitized video outputs of the radar receiver through a single-pulse-delay clutter canceler. As illustrated below, it delays the return of pulse (n) by the interpulse period, T, and subtracts it from the return of pulse (n + 1).

CLUTTER CANCELER

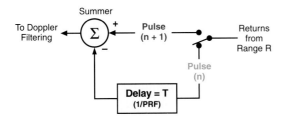

Successive returns from a *moving* target, however, will differ in phase as a result of the radial component of the target's velocity. Consequently, they will not cancel but will produce a useful output.

Effective as this technique is, it has four limitations:

1. The PRF is tied to the aircraft's velocity

2. Very tight constraints are placed on aircraft and antenna motion

3. The phase and amplitude characteristics of the two antenna segments and of the receive channels for both segments must be precisely matched

4. Only half of the aperture is used at any one time

The fourth limitation may be partially removed by adjusting the velocity and PRF so that during the interpulse period the phase centers advance by only half the distance between them (Fig. 4), the entire aperture may be used for transmission. But still only half the aperture may be used for reception. Returns of pulse (n) must be received by the forward segment; returns of pulse (n + 1), by the aft segment.

Although both pulses are not transmitted from the same point in space and returns from the same ranges are not received at the same points in space, the result is the same as if they were. For as indicated in the table at right, the phase centers' total displacement for transmission and reception is the same for both pulses. Therefore, the round-trip distance traveled to any one point on the ground is the same for both pulses—just as when the pulses are alternately transmitted by the fore and aft antenna segments.

COMPLETE DPCA CYCLE

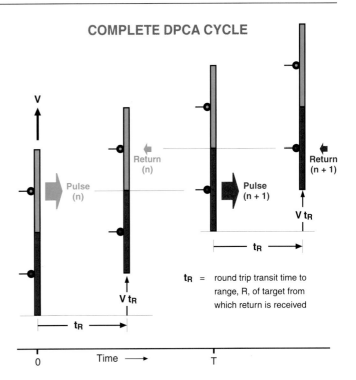

3. Returns of each pulse are received by the same segment that transmitted the pulse. Consequently, round-trip distances traveled by both pulses n and (n + 1) will be equal.

MODIFIED DCPA

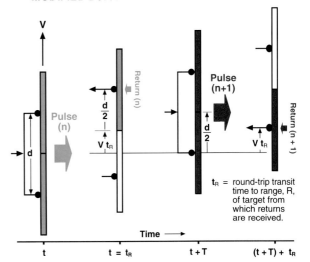

4. To transmit with full aperture, velocity, V, and PRF are adjusted so the radar travels only half the distance between phase centers during interpulse period. Returns of pulse **n** are received by the forward antenna segment; returns of pulse **(n + 1)**, by the aft segment. So the round trip distance traveled by both pulses is the same.

Pulse	Displacement of Phase Centers		
	Transmit	Receive	Total
(n)	0	$V t_R + \dfrac{d}{2}$	$V t_R + \dfrac{d}{2}$
(n + 1)	$\dfrac{d}{2}$	$V t_R$	$V t_R + \dfrac{d}{2}$

319

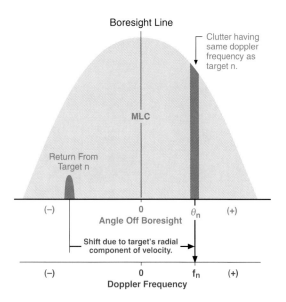

5. Relationship between return received from a slowly moving target on the ground and mainlobe clutter. Because of target's radial component of velocity, clutter having the same doppler frequency as the target, f_n, is received from a different angle off the boresight line.

HOW A NOTCH IS MADE

In a Two-Segment Antenna's Receive Pattern At an Angle θ_n Off the Boresight Line

1. Calculate the difference in distance, Δ_d, traveled to the two phase centers, A and B, by returns from a distant point at the angle, θ_n, off the boresight line.

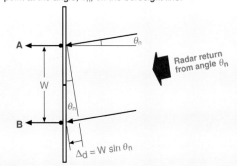

2. Convert Δ_d to phase, ϕ.

$$\phi = \frac{2\pi}{\lambda}\, \Delta_d = \frac{2\pi}{\lambda}\, W \sin\theta_n$$

3. Subtract ϕ from π radians (180°). Divide by 2. Result is the phase rotation, $\Delta\phi$, which—when made in opposite directions to antenna outputs A and B—will increase the phase difference, ϕ, between returns received from θ_n to 180°.

$$\Delta\phi = \frac{\pi - \phi}{2}$$

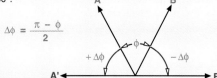

5. Sum the phase-rotated outputs, A' and B'. The returns received from θ_n will then cancel, producing the equivalent of a notch in the antenna receive pattern at angle θ_n

6. To produce a notch on the opposite side of the boresight line, reverse the directions of the two phase rotations.

Notching Technique

Notching has the advantage over Classical DPCA of not requiring that the PRF be tied to aircraft velocity and of relaxing the constraints on aircraft and antenna motion. Mainlobe clutter is rejected without rejecting target returns by taking advantage of the doppler shift due to the radial component of a target's velocity, small as it may be. Because of this shift, clutter having the same doppler frequency as target n comes from a slightly different angle, θ_n, off the boresight line (Fig. 5). Therefore, by placing a notch in the antenna receive pattern at θ_n, the clutter can be rejected without rejecting the return from target n (Fig. 6).

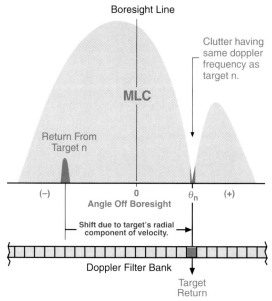

6. Placing a notch in the antenna receive pattern at angle, θ_n, from which the radar receives mainlobe clutter having the same doppler frequency, f_n, as target, n, prevents clutter from interfering with the target's detection. Doppler filtering isolates the target return from other clutter.

Moreover, the return from target n is isolated from clutter received from other directions—and therefore having other doppler frequencies—by doppler filtering.

Because a target's angular position and radial component of velocity generally are not known in advance, and because returns from targets in different directions may be received simultaneously, a separate notch must be formed for each of the N resolvable doppler frequencies. To avoid rejecting target returns along with the clutter, notching must be performed after, not before, doppler filtering.

The notches are produced with an interferometric technique similar to that used in phase-comparison monopulse angle tracking (see panel, left). As with Classical DPCA, a two-segment electronically steered antenna is typically used. So that very small differences in doppler frequency may be resolved, dwell times are increased to allow predetection integrated over long periods, t_{int}. So that the notching can be done in the signal processor, a separate receive and signal processing channel is generally provided for each antenna segment.

Implementation of the notching process is illustrated in abbreviated form in Fig. 7. For every range bin in both Channel A and Channel B, a separate doppler filter bank is formed. The outputs of each pair of filters, n, passing returns of the same frequency, f_n, from the same range, m, are then rotated in opposite directions through the angle, $\Delta\phi_n$. This rotation causes the returns received by the two antenna segments from the ground at an angle θn off the boresight line to be 180° out of phase.

The phase-rotated returns are summed, with the result that the ground returns from θ_n cancel, while returns from targets at any other angle off the boresight line whose doppler frequency is f_n do not.

For radars in which monopulse sum and difference signals for angle tracking are produced ahead of the receiver (i.e., at microwave frequencies), notching is performed similarly with the outputs of the sum and difference channels. In that case, though, rather than being phase rotated and summed, the outputs of corresponding doppler filters, f_n, are weighted and summed to shift the null of the difference output to the angle θ_n off boresight.

In view of the fact that a good many targets on the ground will have high enough radial velocities to fall in the clutter-free portion of the doppler spectrum, notching is generally time shared with conventional moving-target-indication processing.

Combined Notching and Classical DPCA

Generally, notching provides very good mainlobe clutter cancellation. But, under some conditions—such as when frame-time requirements limit dwell times hence achievable doppler resolution—clutter rejection performance can be substantially improved by combining notching and Classical DPCA. This improvement may at any time be traded to various degrees for an easing of Classical DPCA's strict constraints on aircraft and antenna motion and/or for uncoupling PRF from aircraft velocity.

Implementation differs from that just described primarily in that for each range bin, two doppler filter banks are formed from the outputs of each receive channel, and the inputs to one of these banks are delayed by the interpulse period, T (Fig. 8).

Although further improvements in clutter rejection performance can be expected, it should be borne in mind that, since the cancellation techniques rely on clutter scatterers being stationary, cancellation ultimately will be limited by the "internal motion" of the clutter.

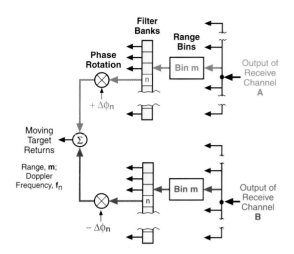

7. Implementation of notching technique. Video outputs of receive channels A and B are collected in range bins. For each range bin, m, a doppler filter bank is formed. Output of each filter, n, is rotated in phase in Channel A by $+\Delta\phi_n$ and in Channel B by $-\Delta\phi_n$. Rotated outputs are then summed, creating the equivalent of a notch in the antenna receive pattern at θ_n, while passing returns from targets at other angles whose doppler frequency is f_n.

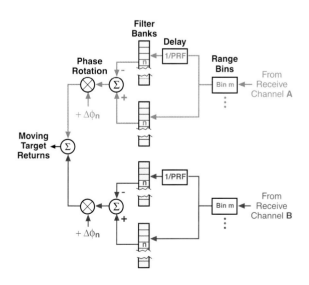

8. Combination of classical DPCA and notching. Technique can be used to ease constraints DPCA places on aircraft and antenna motion or improve clutter rejection performance of notching in applications that limit dwell time.

321

Precise Angle Measurement

While the DPCA and notching techniques enable a target to be detected which would otherwise be hopelessly embedded in mainlobe clutter, they don't tell at what angle within the antenna beam the target is located. For although the direction of the interfering clutter can be determined from the frequency of the doppler filter that passes the target return, without knowing the target's radial velocity, hence its doppler frequency, it is impossible to tell directly how far removed from that angle the target actually is.

In conventional operation of a two-segment antenna and a two-channel receiving system, a target's precise direction may be obtained by comparing the phases, or amplitudes, of the outputs the target produces from the two channels. But the slow-moving-target detection techniques fully utilize the outputs of both channels for clutter rejection and target detection.

Accordingly, where precise angle measurement is required, a three-segment antenna and three receive channels are generally provided. As illustrated in Fig. 9, the outputs of receive channels A and B are used to provide clutter rejection and target detection for an effective phase center half way between the phase centers of antenna segments A and B. The outputs of channels B and C are similarly used to provide clutter rejection and target detection for an equivalent phase center half way between those of antenna segments B and C. The target's precise direction is then estimated on the basis of the distance between the two effective phase centers and the difference in phase of the two output signals.

Summary

Targets whose true doppler frequencies fall in mainlobe clutter are commonly separated from the clutter with either Classical DPCA or notching, both of which employ a two-segment side-looking antenna.

For DPCA, aircraft velocity and PRF are adjusted so the radar advances the distance between the segments' phase centers during the interpulse period. By transmitting and receiving alternate pulses with fore and aft segments, both pulses travel the same round-trip distance to any one point on the ground; so MLC can be eliminated by a clutter canceller.

For notching, radar returns are sorted with a doppler filter bank, and a notch is placed in the antenna receive pattern for each filter. Because of the doppler shift due to a target's velocity, MLC having the same doppler frequency as the target will come from a different direction than target return. So, the clutter is "notched out" without rejecting target returns.

By combining DPCA with notching, greater flexibility may be obtained than with either technique alone.

To pinpoint a detected target's position within the antenna's mainlobe, a third antenna segment must be provided.

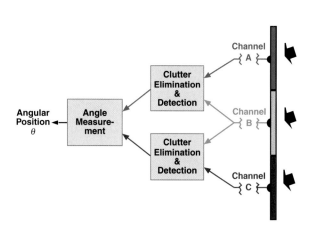

9.   Approach to precise angle measurement. Since returns received by two antenna segments are required for clutter elimination and target detection, a third segment must be provided to determine target's angle, θ, within the radar's beam.

PART VI

Air-to-Air Operation

McDONNELL DOUGLAS F—4 PHANTOM (1960)

The very successful F—4 can be called the ubiquitous jet fighter of the supersonic era. Over 5000 were built for the USN, the USAF, the USMC and our allies around the world. All versions are equipped with Westinghouse radars beginning with the APQ—72 in early versions and the AWG—10 pulse doppler radar in later versions.

The Crucial Choice of PRF 25

Few parameters of a pulsed radar are more important than the PRF. This is particularly true of doppler radars. Other conditions remaining the same, the PRF determines to what extent the observed ranges and doppler frequencies will be ambiguous. That, in turn, determines the ability of the radar not only to measure range and closing rate directly, but to reject ground clutter. In situations where substantial amounts of clutter are encountered, the ability to reject clutter crucially affects the radar's detection capability.

In this chapter we will survey the wide range of pulse repetition frequencies employed by airborne radars and see in what regions significant range and doppler ambiguities may occur. We will then take up the three basic categories of pulsed operation — low, medium, and high PRF— and learn what their relative merits are.

Primary Consideration: Ambiguities

The pulse repetition frequencies used by airborne radars vary from a few hundred hertz to several hundred kilohertz (Fig. 1). Exactly where, within this broad spectrum, a radar will perform best under a given set of conditions depends upon a number of considerations. The most important of these are range and doppler ambiguities.

PRFs, kHz

10	40	100	200	300

1. PRFs used by airborne radars range all the way from a few hundred hertz to several hundred kilohertz.

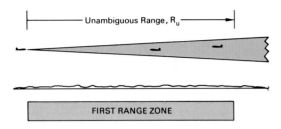

2. Assuming all return from beyond first range zone is rejected, this zone is a region of unambiguous range.

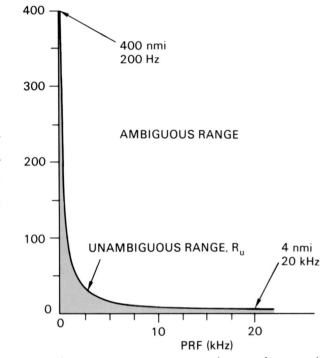

3. Area under curve encompasses every combination of range and PRF for which a target's observed range will be unambiguous, assuming all return from beyond first range zone is either negligible or rejected.

Range Ambiguities. As we have seen, for range to be unambiguous, the echoes from the most distant detectible targets must be first-time-around echoes. In other words, all sources of detectable return must lie in the first range zone: their ranges must be less than the unambiguous range, R_u (Fig. 2). Because targets of large radar cross section, as well as return from the ground, may be detected at exceptionally long ranges, range is almost always ambiguous.

However, if the PRF is sufficiently low so that the maximum required operating range falls within the first range zone, ambiguities can be eliminated by rejecting any return received from beyond R_u. (Techniques for rejecting it are described in Chap. 12.) Under this condition, the first range zone is a region of unambiguous range.

The first range zone extends to a range very nearly equal to 80 nautical miles divided by the PRF in kilohertz. In Fig. 3, this range is plotted versus PRF. The area under the curve encompasses every combination of PRF and true range for which range is unambiguous. The area above the curve encompasses every combination for which range is invariably ambiguous.

Notice how rapidly the curve plunges as the PRF is increased. From a range of 400 miles at a PRF of 200 hertz, it drops to 10 miles at a PRF of 8 kilohertz and to only 4 miles at a PRF of 20 kilohertz.

Doppler Ambiguities. Like range, doppler frequency is inherently ambiguous. Whether the ambiguities are significant, however, depends not only upon the PRF but upon the wavelength and the spread between the maximum opening and closing rates likely to be encountered. The maximum closing rate is usually the rate of the most rapidly approaching target. The maximum opening rate may be either that of the most rapidly opening target or that of the ground from which sidelobe clutter is received behind the radar (Fig. 4). In fighter applications, it is generally the latter. This rate is very nearly equal to the maximum velocity of the aircraft carrying the radar.

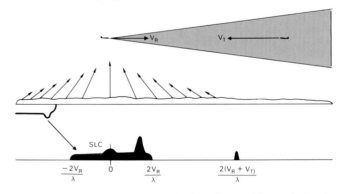

4. Maximum opening rate is usually that of ground from which sidelobe clutter is received behind radar. Maximum closing rate is that of fastest approaching target.

The relationship between the PRF and the doppler frequency at which ambiguities arise in a clutter environment is illustrated in Fig. 5.

It shows the doppler profile for the flight situation presented in Chap. 23 and includes both the true profile (central-line frequencies) and the next-higher-frequency repetition of it (first upper sideband frequencies). Corresponding points in the two profiles are, of course, separated by the pulse repetition frequency. A high closing rate target (B) appears in the clear region above the highest true clutter frequency. If this target's closing rate were progressively increased, the target would move up the doppler frequency scale and into the repetition of the sidelobe clutter spectrum. On the basis of doppler frequency, alone, the radar would have no way of separating the target echoes from the sidelobe clutter, even though their true doppler frequencies are quite different.

From the standpoint of clutter rejection, therefore, the highest unambiguous doppler frequency a target can have—i.e., the highest frequency at which the target will not have to compete with clutter whose true doppler frequency is different from the target's—equals the pulse repetition frequency minus the maximum sidelobe clutter frequency. The latter frequency, as we just noted, corresponds to the radar's velocity.

$$\text{Maximum unambiguous doppler} \; = \; \text{PRF} - \left(\frac{2V_R}{\lambda}\right)$$

The maximum closing rate for which the doppler frequency will be unambiguous in a clutter environment is plotted versus PRF in Fig. 6. A wavelength of 3 centimeters and a radar velocity of 1000 knots are assumed. The plot decreases linearly from a closing rate of about 8000 knots at a PRF of 300 kilohertz to a closing rate of 1000 knots (radar's ground speed) at a PRF of 70 kilohertz. (It is terminated at this point since at lower PRFs the maximum positive and negative sidelobe clutter frequencies overlap.)

The area beneath the curve encompasses every combination of PRF and closing rate for which the observed doppler frequencies are unambiguous. Conversely, the area above the curve encompasses every combination for which the doppler frequencies are ambiguous. For example, at a PRF of 250 kilohertz, and a closing rate of 5000 knots the doppler frequency is unambiguous, whereas at a PRF of 150 kilohertz and the same closing rate, the doppler frequency is ambiguous.

If the radar-carrying aircraft's ground speed is greater than 1000 knots, the area beneath the curve will be correspondingly reduced, and vice versa.

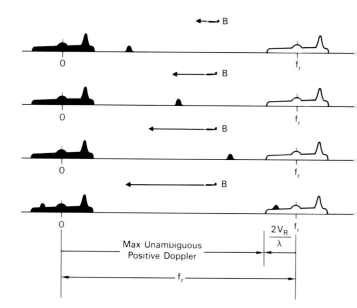

5. True doppler profile and next higher repetition of it. As component of target velocity along line of sight to radar increases, target moves through doppler clear region and ultimately enters repetition of negative frequency sidelobe clutter.

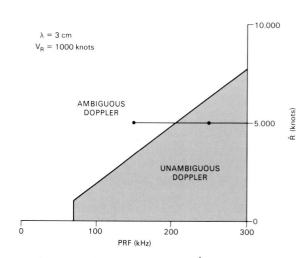

6. Combinations of PRF and closing rate, $\dot{R}$, for which observed doppler frequencies are unambiguous.

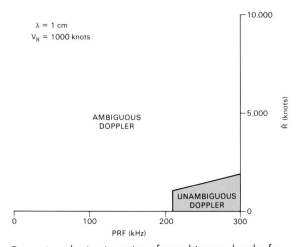

7. Dramatic reduction in region of unambiguous doppler frequencies resulting from decrease in wavelength λ from 3 to 1 centimeter.

Since doppler frequency is inversely related to wavelength, the shorter the wavelength, the more limited the region of unambiguous doppler frequencies will be. To illustrate the profound effect of wavelength on doppler ambiguities, Fig. 7 plots the maximum closing rate at which the doppler frequency will be unambiguous for a one centimeter wavelength. Not only is the area under the curve comparatively small, but even at a PRF of 300 kilohertz, the maximum closing rate at which the PRF is unambiguous is less than 2000 knots.

Yet, at a wavelength of 10 centimeters the area under the curve extends from 1000 knots at 21 kilohertz to about 27,000 knots (off the scale) at 300 kilohertz (Fig. 8).

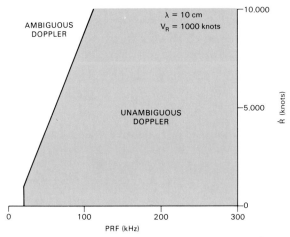

8. Dramatic *increase* in region of unambiguous doppler frequency resulting from increase in wavelength to 10 centimeters.

Putting the Plots in Perspective. The regions of unambiguous range and unambiguous doppler frequency (for λ = 3 cm and V_R = 1000 knots) are shown together in Fig. 9. Drawn to the scale of this diagram, the region of unambiguous range (first range zone) is quite narrow. To the right of it is the comparatively broad region of unambiguous doppler frequencies. In between is a region of considerable extent within which both range and doppler frequency are ambiguous.

Clearly, the choice of PRF is a compromise. If the PRF is increased beyond a relatively small value, the observed ranges will be ambiguous. And unless the PRF is raised to a much higher value than that, the observed doppler frequencies will be ambiguous.

While both range and doppler ambiguities make clutter rejection difficult, as we shall see their effects on a radar's operation are fortuitously quite different. It turns out that by designing a radar to operate over a wide range of PRFs and by judiciously selecting the PRF to suit the operational requirements at the time, the difficulties can be almost completely obviated.

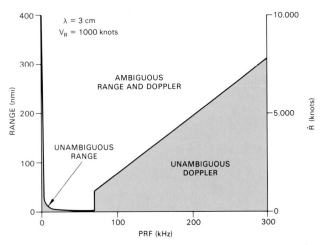

9. When region of unambiguous range is plotted to same scale as PRF, it becomes apparent that choice of PRF is, at best, a compromise.

The Three Basic Categories of PRF

Because of the tremendous impact the choice of PRF has on performance, it is customary to classify airborne radars in terms of their PRFs. Recognizing that the regions of unambiguous range and unambiguous doppler frequency are very nearly mutually exclusive, three basic categories of PRF have been established: "low," "medium," and "high."

These are defined in terms not of the numerical value of the PRF per se, but of whether the PRF is such that the observed ranges and/or doppler frequencies are ambiguous. While exact definitions vary, all are similar. The following is a widely used, consistent set of definitions.

- A low PRF is one for which the maximum range the radar is designed to handle lies in the first range zone. In the absence of return from beyond this zone, *range* is *un*ambiguous.

- A high PRF is one for which the observed *doppler frequencies* of all significant targets are *un*ambiguous.

- A medium PRF is one for which neither of these conditions is satisfied. Both *range* and *doppler* frequency are *ambiguous*.

Which category a particular PRF falls in depends to a considerable extent upon the operating conditions. A PRF of 4 kilohertz—first range zone extending to 20 nautical miles—would be "low" if the maximum target range were less than 20 miles (Fig. 10). Yet the same PRF, 4 kilohertz, would be "medium" if the maximum range were greater than 20 miles and the spread between maximum positive and negative doppler frequencies exceeded 4 kilohertz.

Similarly, a PRF of 20 kilohertz might be "medium" for a 3-centimeter radar (X-band), yet "high" for a 10-centimeter radar (S-band) if, say, the radar's velocity were 200 knots and the velocity of the fastest target, 1000 knots—maximum closing rate 1200 knots (Fig. 11).

CATEGORIES OF PRF

PRF	RANGE	DOPPLER
HIGH	Ambiguous	Unambiguous
MED	Ambiguous	Ambiguous
LOW	Unambiguous	Ambiguous

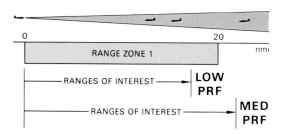

PRF = 4 kHz

10. A PRF of 4 kilohertz would be LOW if maximum range of interest was less than 20 miles; yet MEDIUM if maximum range of interest was greater than 20 miles.

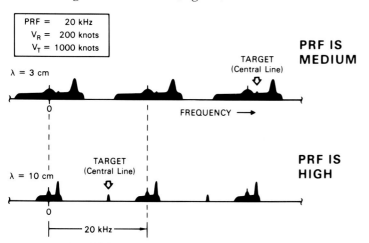

11. A PRF of 20 kilohertz might be medium at a wavelength of 3 centimeters, yet high at a wavelength of 10 centimeters.

PRFs USED AT X-BAND

12. In practice, not all of the PRFs within each category are used. Reason will be made clear in subsequent chapters.

In practice, not all of the possible PRFs within each category are used for any one radar band (Fig. 12, above). At X-band, for example, PRFs in the low category typically run from 250 to 4000 hertz; PRFs in the medium category are on the order of 10 to 20 kilohertz; PRFs in the high category may range anywhere from 100 to 300 kilohertz.

One should not get the idea that the classifications are technicalities of little practical importance. To the contrary, in the everyday world of radar development and application, they have proved to be immensely useful—especially regarding fighter and airborne early warning radars. As we shall see in subsequent chapters, whereas changing the PRF within any one category does not alter the radar's design in any fundamental way, changing the PRF from one category to another radically affects both the radar's signal processing requirements and its performance.

In fighter radars, the three categories of PRF complement one another nicely from the standpoint of performance.

Low PRF Operation. Because range is unambiguous at low PRFs, this mode of operation has two important advantages. First, range may be measured directly by simple precise pulse delay ranging. Second, as will be explained in the next chapter, virtually all sidelobe return can be rejected through range resolution.[1]

However, unless the mainlobe clutter is separated in range from the targets the radar encounters, it can be rejected only on the basis of differences in doppler frequency. And because of the overlapping of successive repetitions of the doppler spectrum at low PRFs, the clutter cannot be rejected without also rejecting the returns from a considerable portion of the spectrum in which targets may appear. If the wavelength is long enough, the ground speed low enough ($f_d \propto V_R/\lambda$), and the antenna large enough ($\theta_{3dB} \propto \lambda/d$) the mainlobe clutter spectrum will be sufficiently narrow that the possible loss of target return is quite tolerable.

But for the conditions under which most fighter radars must operate—short wavelength, small antenna, and potentially high ground speed—if the PRF is low enough to extend the first range zone out to reasonably long ranges—30 or 40 miles—the unambiguous doppler spectrum is col-

LOW PRFs

ADVANTAGES	LIMITATIONS
1. Good for air-to-air look-up and ground mapping.	1. Poor for air-to-air look-down—much target return may be rejected along with mainlobe clutter.
2. Good for precise range measurement and fine range resolution.	2. Ground moving targets can be a problem.
3. Simple pulse delay ranging possible.	3. Doppler ambiguities generally too severe to be resolved.
4. Normal sidelobe return can be rejected through range resolution.	

1. Except for return from point targets of exceptionally large radar cross section.

330

lapsed (telescoped) to the point that mainlobe clutter occupies most of the doppler passband (Fig. 13). Consequently, when the clutter is rejected the return from most of the target region will be rejected. Also, since target echoes of widely different true doppler frequencies are indistinguishably intermixed, not only is it impossible to resolve doppler ambiguities, but the radar is susceptible to interference from ground moving targets, GMTs.

Because of the severity of the mainlobe clutter problem, the use of low PRFs for air-to-air operation in fighters, which employ short wavelengths and comparatively small antennas, is today restricted largely to situations where mainlobe clutter can be avoided:

- When flying over water, which (because of its more nearly mirrorlike surface) has a relatively low backscattering coefficient at moderate to low grazing angles

- When looking up in search of targets at higher altitudes

- When the mainlobe strikes the ground beyond the maximum range of interest (Fig. 14), and clutter from beyond the first range zone is rejected through other means than doppler resolution.

For ground mapping, low PRFs are ideal. Because mainlobe ground return is then the only return of interest, its overwhelming strength is an asset, not a liability. Moreover, the unambiguous observation of range which low PRFs provide is essential.

What about synthetic array ground mapping? For it (Fig. 15), the unambiguous observation of doppler frequencies, too, is essential. Happily, the PRF can generally be made high enough to prevent the repetitions of the mainlobe clutter spectrum from overlapping, while providing an adequately long maximum unambiguous range.

High PRF Operation. The problem of mainlobe clutter can be solved by operating at high PRFs. The width of the mainlobe clutter spectrum is generally only a small fraction of the width of the band of true target doppler frequencies, so that at high PRFs mainlobe clutter does not appreciably encroach on the region of the spectrum in which targets are expected to appear. Moreover, since all significant doppler ambiguities are eliminated at high PRFs, mainlobe clutter can be rejected on the basis of doppler frequency without at the same time rejecting echoes from targets. Only if a target is flying nearly at right angles to the line of sight from the radar—a condition which occurs rarely and is usually maintained for only a short time—will its echoes have the same doppler frequency as the clutter and so be rejected.

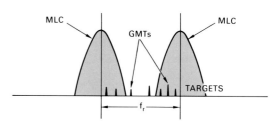

13. IF PRF is made low enough to provide reasonably long unambiguous ranges, most of target return will be rejected along with mainlobe clutter, MLC. Also, ground moving targets, GMTs, cannot be directly discerned from airborne targets.

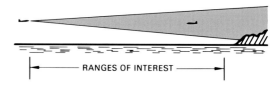

14. Use of low PRFs for air-to-air operations in fighter-type aircraft is limited to situations where mainlobe clutter is not a problem—e.g., over water or where mainlobe does not strike ground within ranges of interest.

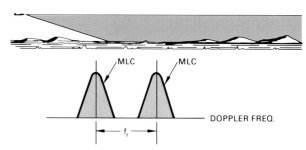

15. For ground mapping, unambiguous range provided by low PRFs is essential. However, for SAR mapping, PRF must also be high enough that repetitions of mainlobe clutter do not overlap.

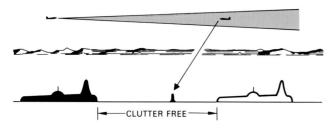

16. High PRFs provide clutter-free region in which to detect high closing-rate targets.

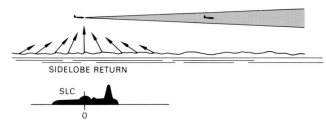

17. But, return from low closing-rate targets must compete with sidelobe return, much of it from short range.

HIGH PRFs

ADVANTAGES	LIMITATIONS
1. Good nose-aspect capability—high-closing-rate targets appear in clutter-free region of spectrum.	1. Detection range against low-closing-rate targets may be degraded by sidelobe clutter.
2. High average power can be provided by increasing PRF. (Only moderate amounts of pulse compression, if any, are needed to maximize average power.)	2. Precludes use of simple, accurate pulse delay ranging.
3. Mainlobe clutter can be rejected without also rejecting taget echoes.	3. Zero-closing rate targets may be rejected with altitude return and transmitter spillover.

Operation at high PRFs has other important advantages. First, between the band of central-line sidelobe clutter frequencies and the first repetition of this band, a region opens up in which there is absolutely no clutter (Fig. 16). It is here that the doppler frequencies of approaching targets lie—those for which long detection ranges normally are desired. Second, closing rates can be measured directly by sensing doppler frequencies. Third, for a given peak power, average transmitted power can be maximized simply by increasing the PRF until a duty factor of 50 percent is reached. High duty factors can also be obtained at low PRFs, but this requires increasing the pulse width and employing large amounts of pulse compression to provide the degree of range resolution which is essential in low PRF operation.

The principal limitation of high PRF operation is that detection performance may be degraded by sidelobe clutter when operating against low closing-rate tail-aspect targets. At the high PRFs typically employed in fighter radars, the return from virtually all ranges is collapsed (telescoped) into a range interval little wider than that occupied by a target's echoes. Consequently, the sidelobe clutter can only be rejected by resolving the return in doppler frequency.

Much of the sidelobe clutter falling within the same resolvable frequency increment as a target's echoes will have been reflected from very much shorter ranges and so will be quite strong (Fig. 17).

When flying at moderate to low altitudes over terrain that has a high backscattering coefficient, unless the target has a large radar cross section or is at short range, its echoes may be lost in the clutter, and there will be no way of extracting them from it.

Also, if little or no range discrimination is provided, zero closing-rate targets will be rejected along with the altitude return.

Another disadvantage of very high PRFs is that they make pulse delay ranging more difficult. As the PRF is increased, range ambiguities become more severe. To resolve them, the radar must switch among more PRFs.

Ultimately, a point is reached where range must be measured by more complex, less accurate techniques, such as frequency modulation ranging. In any event, because of losses incurred in resolving ambiguities, range can be measured only at the expense of a reduction in maximum detection range.

Nevertheless, when mainlobe clutter is a problem and long detection ranges are desired against approaching targets, the advantages of high PRFs far outweigh any of these disadvantages.

Medium PRF Operation. Medium PRFs were conceived as a solution to the problems of detecting tail-aspect targets in the presence of both mainlobe and strong sidelobe clutter, thereby providing good all-aspect coverage. If the maximum required operating range is not exceptionally long, the PRF can be set high enough to provide adequate separation between the periodic repetitions of the mainlobe clutter spectrum without incurring particularly severe range ambiguities.

Mainlobe clutter can then be isolated from the bulk of the target return on the basis of its doppler frequency. And individual targets can be isolated from the bulk of the sidelobe clutter through a combination of range and doppler resolution.

Also, ground moving targets—being close to the frequency of the mainlobe clutter—can be rejected along with it, without rejecting an unacceptable additional amount of possible target return (Fig. 18).

Range ambiguities are more easily resolved than at high PRFs so that pulse delay ranging is possible. While doppler frequencies are ambiguous, these ambiguities are also moderate enough to be resolved.

On the negative side, because of the range and doppler ambiguities, both nose- and tail-aspect targets may have to compete with close-in sidelobe clutter (Fig. 19). This problem, of course, can be avoided by switching among several different PRFs.

MEDIUM PRFs

ADVANTAGES	LIMITATIONS
1. Good all-aspect capability—copes satisfactorily with both mainlobe and sidelobe clutter.	1. Detection range against both low and high closing-rate targets can be limited by sidelobe clutter.
2. Ground-moving targets readily eliminated.	2. Must resolve both range and doppler ambiguities.
3. Pulse delay ranging possible.	3. Special measures needed to reject sidelobe return from strong ground targets.

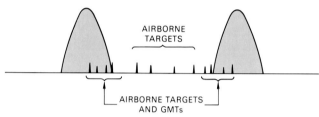

18. With medium PRFs, repetitions of mainlobe clutter are widely enough separated so that it and return from ground moving targets can be rejected without rejecting undue amount of target return.

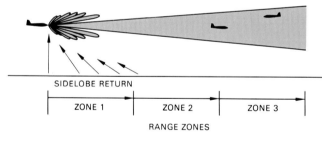

19. While targets may still have to compete with close-in sidelobe clutter, the clutter can be avoided by switching among several different PRFs.

But the resulting reduction in the integration time for each PRF limits the maximum detection range.

Nevertheless, where extremely long detection range is not required—as when operating at moderate to low altitudes, in lookdown situations, or in tail chases—adequate detection range can generally be achieved.

Another consequence of the range and doppler ambiguities encountered at medium PRFs is that sidelobe return from ground targets of large radar cross section can be a serious problem. Special measures must generally be taken to eliminate this return lest it be confused with return from airborne targets.

333

CATEGORIES OF PRF

PRF	RANGE	DOPPLER
HIGH	Ambiguous	Unambiguous
MED	Ambiguous	Ambiguous
LOW	Unambiguous	Ambiguous

Summary

Pulse repetition frequencies used by airborne radars range from a few hundred hertz to several hundred kilohertz at X-band. Generally, only at extremely low PRFs is range unambiguous—and then only if all return from beyond the first range zone is excluded or negligible. Conversely, only at considerably higher PRFs are doppler frequencies largely unambiguous. Thus, the choice of PRF is generally a compromise.

Three categories of PRF have been established: low, high, and medium. Which category a particular PRF falls in depends on the operational situation.

A low PRF is one for which the maximum required operating range falls within the first range zone. Simple pulse delay ranging can be used, and sidelobe clutter can be almost entirely removed through range resolution. But, in fighter radars, doppler ambiguities are generally so severe that mainlobe clutter cannot be rejected without rejecting much possible target return, and GMTs may be a problem.

A high PRF is one for which doppler frequencies of all significant targets are unambiguous. Mainlobe clutter can be rejected without rejecting target return, and a clutter-free region is provided in which approaching targets appear. Also, high average power can be obtained by increasing the PRF. While this mode is excellent for nose-aspect targets, because of range ambiguities sidelobe clutter may severely limit performance against tail-aspect targets. Range rates can be measured directly, but pulse delay ranging may be difficult or impractical because of severe range ambiguities.

A medium PRF is one for which both range and doppler frequency are ambiguous. But if the value of the PRF is judiciously selected, the ambiguities are comparatively easy to resolve. Consequently, good all-aspect performance can be provided despite both mainlobe and sidelobe clutter, as well as GMTs. Maximum detection range, however, is limited by close-in sidelobe clutter, and sidelobe return from large-RCS objects on the ground may be a problem.

Low PRF Operation

26

A low PRF is, by definition, one for which the first range zone—the zone from which first-time-around echoes are received—extends at least to the maximum range the radar is designed to handle (Fig. 1). In the absence of return from beyond this zone, the observed ranges are unambiguous. (The first range zone, you'll recall, extends out to the so-called maximum unambiguous range, R_u, which in nautical miles roughly equals 80 divided by the PRF in kilohertz.)

Typically, low PRFs range from around 250 hertz (R_u = 320 nmi) to 4000 hertz (R_u = 20 nmi). Unfortunately, at such PRFs unless the wavelength is relatively long and/or the closing rate relatively low, the observed doppler frequencies are highly ambiguous.

Low PRFs are essential for most air-to-ground uses. And they are superior to both medium and high PRFs for certain air-to-air applications, e.g. early warning. But for use in fighter aircraft, where target return generally must compete with mainlobe clutter, low PRFs have serious limitations.

In this chapter, we will take a closer look at low PRF operation. We will see how target echoes may be separated from ground clutter and how the signal processing may be performed; then, take stock of advantages and limitations and see how the limitations may be alleviated.

Differentiating Between Targets and Clutter

To see what must be done to separate target echoes from ground clutter, let us look at the range and doppler profiles that would be observed by a low PRF radar for a fighter aircraft in a representative flight situation.

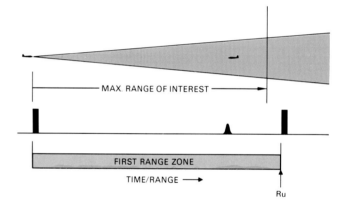

1. A low PRF is one for which the first range zone extends at least to the maximum range the radar is designed to handle.

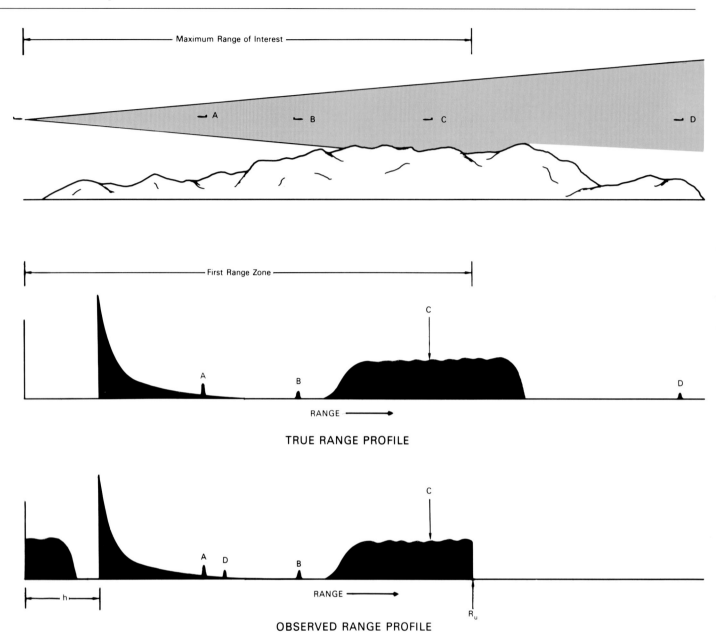

TRUE RANGE PROFILE

OBSERVED RANGE PROFILE

2.   Range profile for low PRF radar in representative operational situation. Out to maximum range of interest, ranges observed by radar directly correspond to true ranges.

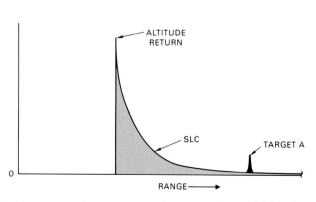

3.   A target's echoes are generally stronger than sidelobe clutter from target's own range.

Range Profile. As illustrated in Fig. 2, within the first range zone the observed ranges are true ranges. The altitude return, sidelobe clutter, and mainlobe clutter are all clearly identifiable—as is the return from targets A and B, which are outside the mainlobe clutter. Target C, however, is completely obscured by mainlobe clutter. Target D, which is beyond the first range zone, appears falsely at a much closer range, but we will defer considering it until later.

From even a cursory inspection of the portion of the profile in which sidelobe clutter alone appears (Fig. 3), one thing is immediately clear: the echoes from a target will generally be stronger than the sidelobe clutter received from the target's own range.

This is perhaps more clearly illustrated by the range profile observed at the output of the receiver (Fig. 4). In it, the amplitudes of both the target echoes and the sidelobe clutter are more or less independent of range over the portion of the profile in which strong sidelobe clutter is received. This characteristic is due to a feature called *sensitivity time control* (STC), which is generally employed when operating at low PRFs.

Now, if we slice the range profile into increments matching the width of the received pulses[1] and isolate the energy

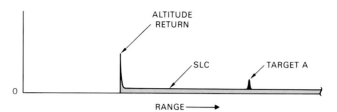

4. Range profile at output of receiver. Sensitivity time control (STC) makes amplitude of output independent of range to prevent saturation by close-in return.

1. Compressed width, if pulse compression is used.

SENSITIVITY TIME CONTROL

At low PRFs, saturation of the receiving system by strong return from short ranges is commonly avoided without loss of detection sensitivity at greater ranges through a feature called sensitivity time control, STC.

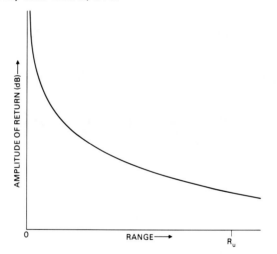

After each pulse has been transmitted, the system gain, which initially is greatly reduced, is increased with time to match the decrease in amplitude of the radar return with range. Maximum gain is usually reached well before the end of the interpulse period.

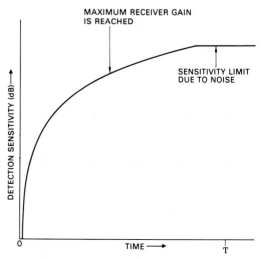

Thereafter, the increase in sensitivity is continued by lowering the detection threshold until the noise limit is reached—i.e., to the point where the threshold is just far enough above the mean noise level to limit the false-alarm probability to an acceptable value.

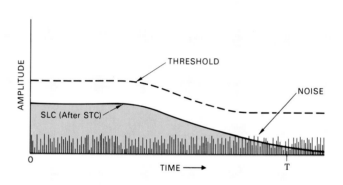

Thus, maximum sensitivity is provided at long ranges, where it is needed to detect the weak echoes of distant targets, while the strong return from short ranges is prevented from saturating the system.

STC may be applied at various points in a system. From whatever point it is applied, it helps prevent saturation in all following stages.

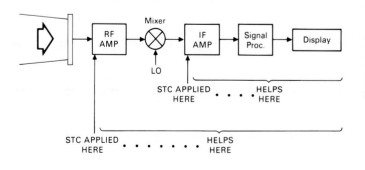

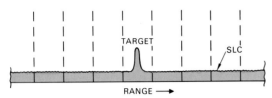

5. If range profile is sliced into narrow increments, a target's echoes can be discerned from sidelobe clutter on basis of amplitude.

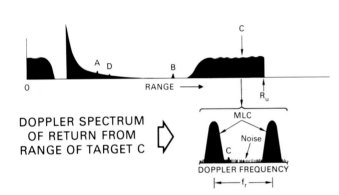

DOPPLER SPECTRUM OF RETURN FROM RANGE OF TARGET C

6. A target at a range from which mainlobe clutter is received can only be detected if it's doppler frequency is different from that of the clutter.

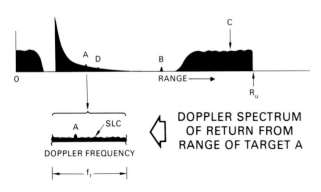

DOPPLER SPECTRUM OF RETURN FROM RANGE OF TARGET A

7. Target A, at short range, must compete with sidelobe clutter. But since the clutter comes from the target's own range, the target echoes are stronger than the clutter.

contained in each increment, we can differentiate between target echoes and sidelobe return by noting the differences in the amplitude from one increment to the next (Fig. 5).

But we are not so fortunate with mainlobe clutter. Isolating the returns from narrow range increments only moderates the problem. For the mainlobe clutter from a target's own range is generally much stronger than the target's echoes. Moreover, even when mainlobe clutter is received from a range that is some multiple of R_u beyond the target's range (multiple-time-around return), it will generally be stronger than the target echoes. To differentiate between target echoes and simultaneously received mainlobe clutter, we must look for differences in doppler frequency.

Doppler Profile. Since range is largely *un*ambiguous when low PRFs are used, the appearance of the doppler profile varies considerably with the point in the interpulse period at which the profile is observed. In other words, the returns from different range increments may have quite different doppler profiles.

The doppler profile for the range increment in which Target C resides is illustrated in Fig. 6. The most prominent features of this profile are the periodic repetitions of the mainlobe clutter spectrum. These occur at intervals equal to the pulse repetition frequency, f_r. Though they may be quite wide, they are commonly called "lines" or, in view of their spacing, PRF lines. (The central line is the carrier frequency; the others are sideband frequencies.)

Between successive PRF lines can be seen the thermal background noise and Target C. Sidelobe clutter from this range happens to be so weak that it is below the noise level. As you may have noticed, had Target C's doppler frequency been a little lower, it might have been obscured by the mainlobe clutter.

Target B is also at a sufficiently long range that the accompanying sidelobe clutter is below the noise level. Since in this particular flight situation mainlobe clutter is not received from Target B's range, this target will appear in the clear regardless of its doppler frequency. It must compete only with background noise.

At shorter ranges, such as that of Target A, sidelobe clutter is much stronger than noise. Nevertheless, because the accompanying sidelobe clutter comes from the same range as the target's echoes, the target appears above the clutter (see Fig. 7).

Target D is a second-time-around target and so is not wanted. Although it happens to be stronger than the accompanying first-time-around sidelobe clutter, it can be prevented from reaching the display by PRF jittering.

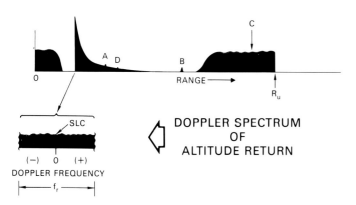

DOPPLER SPECTRUM
OF
ALTITUDE RETURN

8. Altitude return is spread over such a broad band of frequencies it is indistinguishable from the other sidelobe clutter.

Figure 8 shows the doppler profile at the range of the altitude return. As was explained in Chap. 22, the altitude return is generally spread over a band of frequencies whose width exceeds most low PRFs. Consequently, in the doppler profile for a range increment from which altitude return is received, it is indistinguishable from the other sidelobe clutter.

Because the altitude return is spread over such a broad band of doppler frequencies, if doppler filtering is employed, a target may be detected above the altitude return, provided the target's echoes are very strong, as they may well be at very short ranges. (For the altitude return to be at a short range, of course, the radar must be at low altitude.)

Figure 9 shows a fairly broad portion of the doppler spectrum for a range from which mainlobe clutter is received. To eliminate the clutter, we must reject not only the band of frequencies in which the central line lies but bands of equal width at intervals equal to f_r throughout the receiver's IF passband. (Along with the clutter, some target echoes may also be rejected.) Only sidelobe clutter and noise, plus the (unrejected) target echoes will remain. The target echoes may then be separated from the sidelobe clutter and noise on the basis of differences in amplitude and doppler frequency.

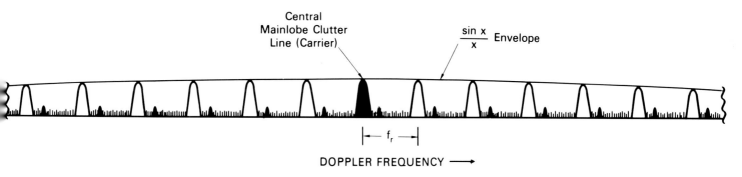

9. Doppler profile for range from which mainlobe clutter is received consists of periodic repetitions of mainlobe clutter spectrum, with sidelobe clutter and target echoes in between.

Signal Processing

One approach to mechanizing the signal processing functions outlined in the preceding paragraphs is illustrated by the block diagram of Fig. 10 (bottom of the page).

Basic Mechanization. As illustrated in Fig. 10, the IF output of the receiver is fed to a synchronous detector (such as described in Chap. 18), which converts it to I and Q video signals. The frequency of the reference signal supplied to the detector is such as to place the central line of mainlobe clutter at zero frequency (dc). The central line is picked, since its frequency doesn't change when the PRF is changed, whereas the frequencies of the other lines do. (The importance of this will be made clear later on.)

An analog-to-digital converter samples the video signals at intervals matching the width of the transmitted pulses.[2] The output of the converter therefore is a stream of numbers representing the I and Q components of the returns from successive range increments. The numbers are sorted by range increment into separate range bins.

To reduce the amount of mainlobe clutter, the numbers for each range increment are passed through a separate clutter canceller. As with the A/D converter, each clutter canceller has both I and Q channels.

To reduce the mainlobe clutter *residue* in the output of the canceller, as well as to minimize the amount of noise and simultaneously received sidelobe clutter with which a target must compete, the output of each clutter canceller is integrated in a bank of doppler filters (as described in Chaps. 19 and 20). So that the filter bank can be implemented with the fast Fourier transform, the passband of the bank is made equal to the PRF. Processing of the outputs for

2. Compressed width, if pulse compression is used.

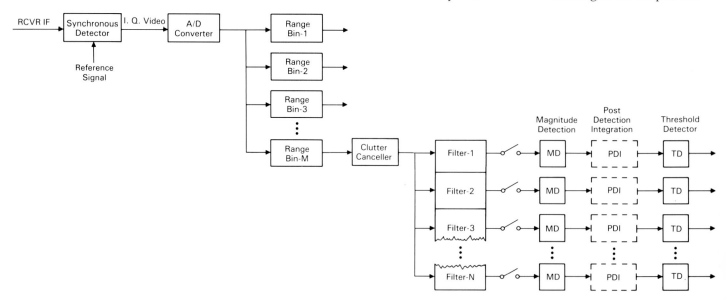

10. Signal processing functions in a low-PRF radar employing doppler filtering. In applications where mainlobe clutter is avoided, clutter cancellers may be eliminated.

WHAT A RANGE BIN IS

A range bin is a memory location in which are temporarily stored successive pairs of numbers (x_n, y_n) representing the I and Q samples of the radar return received at a given point in the interpulse period. A separate "bin" therefore must be provided for each sampling interval (range gate). To the extent that range is unambiguous, the numbers stored in any one bin represent successive returns from a single range increment, hence the name "range" bin.

Because of the correspondence of the range bins to the sampling intervals (when A/D conversion follows I and Q detection), "range bin" has come to be used synonymously with "sampling interval" as well as "range gate."

MEMORY LOCATIONS FOR SAMPLES

	BIN-1	BIN-2	BIN-3		BIN-M
PULSE-1	x_1, y_1	x_1, y_1	x_1, y_1	$\cdots$	x_1, y_1
PULSE-2	x_2, y_2	x_2, y_2	x_2, y_2	$\cdots$	x_2, y_2
PULSE-3	x_3, y_3	x_3, y_3	x_3, y_3	$\cdots$	x_3, y_3
$\vdots$	$\vdots$	$\vdots$	$\vdots$		$\vdots$
PULSE-N	x_N, y_N	x_N, y_N	x_N, y_N	$\cdots$	x_N, y_N

the filters covering the mainlobe clutter regions (filters at the ends of the bank) then is simply not completed.

At the end of every filter integration time, the magnitude of each desired filter's output is detected. If the integration time is less than the length of the time-on-target for the radar antenna, some postdetection integration (PDI) may subsequently be provided (see Chap. 10).

In either event, at intervals equal to the time-on-target, the integrated output of each doppler filter is applied to a threshold detector, which determines whether the sum represents a target.

Three aspects of this mechanization warrant elaboration: how the clutter canceller works, how the detection threshold is set, and how the central mainlobe clutter line is maintained at dc.

Clutter Canceller. In simplest form, each channel of a digital clutter canceller consists of a short-term memory and a summer (Fig. 11). The memory holds each of the numbers received from the analog-to-digital converter for one interpulse period $(1/f_r)$. The summer then subtracts the stored number from the currently received number and outputs the difference. Thus, each number output by the canceller corresponds to the *change* in amplitude of the return from a particular range during the preceding interpulse period.

Now, as explained in Chap. 18, the outputs which a synchronous detector produces for successive returns from any

CLUTTER CANCELLER

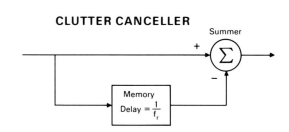

11. In simplest form, a clutter canceller consists of short-term memory and a summer. Memory holds signal for one interpulse period; summer subtracts delayed signal from undelayed signal.

THE CLASSIC DELAY–LINE CLUTTER CANCELLER

The original application of the clutter canceller was providing MTI in ground based radars. Their mechanization, of course, was analog.

Bipolar video from a phase sensitive detector in the receiver was passed through a delay line (e.g. a quartz crystal) which introduced a delay equal to the interpulse period. The delayed signal was then subtracted from the undelayed signal. Since ground return had no doppler shift, the video signal produced by the return from the ground at any one range was essentially constant from one interpulse period to the next. The video signal produced by a moving target, however, fluctuated at the target's doppler frequency. The clutter, therefore, cancelled whereas the target signal did not.

Analog cancellers were used to provide MTI in airborne early warning radars, as well as in early fighter radars. Digital cancellers, however, have the compelling advantages of avoiding problems of delay instability and of being adjustment free. Consequently, although most delay line cancellers are still analog, the cancellers used in all modern airborne radars are digital.

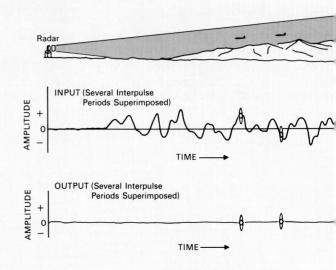

one range are in essence instantaneous samples of a signal whose amplitude corresponds to the amplitude of the returns and whose frequency is the doppler frequency of the returns. Illustrated in Figs. 12 through 14 are successive samples of three such signals. The frequencies of the signals are 0, f_r, and $f_r/2$. All of the samples are taken at intervals equal to the interpulse period ($1/f_r$).

Naturally, successive samples of a video signal having zero frequency—such as the central mainlobe clutter line—have the same magnitude and the same algebraic sign (Fig. 12).

Therefore, when one sample is subtracted from the other, they cancel.

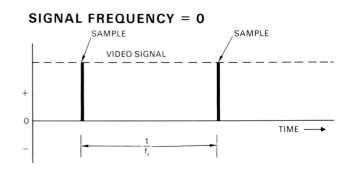

12. Periodic samples of video signal having frequency of zero (dc signal). Samples have the same amplitude and algebraic sign.

The same is true for a signal whose frequency is f_r—such as the first mainlobe clutter line above the central one. Since the sampling interval is equal to the period of the wave, the samples in this case are all taken at the same point in every cycle (Fig. 13).

SIGNAL FREQUENCY = f_r

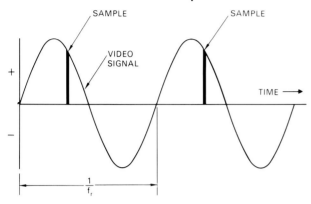

13. Samples of video signal having frequency equal to PRF (f_r). Again, samples have the same amplitude and sign.

But for a frequency of $f_r/2$, the result is just the opposite. Because the sampling interval is only half the period of the wave, the samples are alternately positive and negative (Fig. 14).

When one is subtracted from the other, the difference is twice the magnitude of the individual samples.

For frequencies above and below $f_r/2$, the differences become progressively smaller.

As a result, a plot of the canceller's output versus frequency for a constant-amplitude input has an inverted "U" shape (Fig. 15).

SIGNAL FREQUENCY = $\dfrac{f_r}{2}$

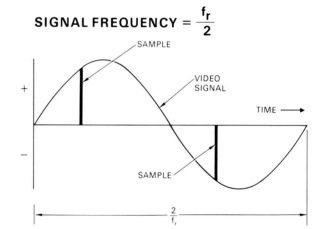

14. Samples of video signal having doppler frequency equal to $f_r/2$. While amplitudes are same, algebraic signs alternate.

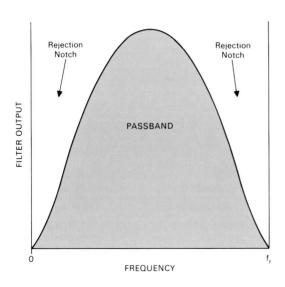

15. Output produced by simple single-delay clutter canceller for constant amplitude input.

343

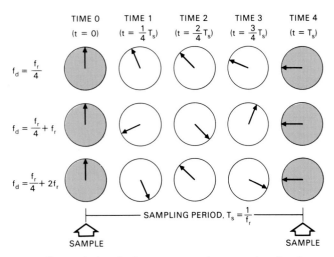

16. If target's doppler frequency equals some value, f_d, plus a whole multiple of the sampling rate, f_r, the output from the canceller will be the same as if the doppler frequency were f_d.

BEFORE CANCELLATION

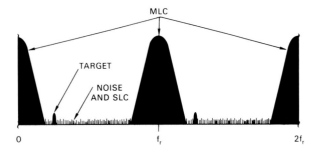

AFTER CANCELLATION

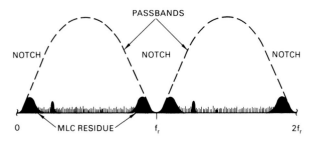

18. If clutter canceller's rejection notches are made wide enough, mainlobe clutter will largely cancel. Output will consist of mainlobe clutter residue, target echoes, sidelobe clutter, and noise.

What about frequencies higher than f_r? As illustrated with simple phasor diagrams in Fig. 16, when a signal is sampled at a given rate—in this case, f_r—the samples will be exactly the same if the signal's frequency is $f_d + f_r$ as they would be if their frequency were f_d.

The same is true if the signal's frequency is f_d plus any multiple of f_r.

The canceller's output characteristic, therefore, repeats identically at intervals of f_r from 0 (dc) on up (Fig. 17).

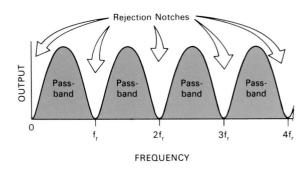

17. Clutter canceller's output characteristic repeats at intervals of f_r from dc on up.

The regions in which the output approaches zero—i.e., at dc and multiples of f_r—are called rejection notches. If any one of the mainlobe clutter lines is placed at dc (normally we place the central line there), every line will fall in a rejection notch, and the clutter will tend to cancel—hence the name, *clutter canceller.*

As you've probably already noticed, the notches of the simple canceller just described are much narrower than the clutter lines may sometimes be. But they can readily be widened. The simplest way is to connect more than one canceller together in series, i.e., cascade them.

If the rejection notches are made sufficiently wide and the mainlobe clutter is centered in them, it will largely cancel (Fig. 18).

The output then will represent target echoes, sidelobe clutter, and background noise—plus, of course, any mainlobe clutter residue.

The doppler filters following each clutter canceller not only eliminate most of the mainlobe clutter residue but substantially reduce both the amplitude of the competing sidelobe clutter and the mean level of the noise. By suitably setting the target detection threshold, we can still further reduce the possibility of clutter and noise producing false alarms.

Detection Threshold. As explained in Chap. 10, for each doppler filter, we set the threshold a predetermined amount higher than the average of the outputs of the corresponding filters for several range bins on either side (Fig. 19). Provided the threshold offset has been correctly chosen and the averaging has been properly done, the probability of clutter crossing the threshold may be reduced to an acceptably low value, while providing adequate sensitivity for the detection of target echoes.

If you examine the range profile of the receiver output with the mainlobe clutter removed, you will notice that, as the range increases, a point is eventually reached where the sidelobe clutter is submerged beneath the background noise (Fig. 20).

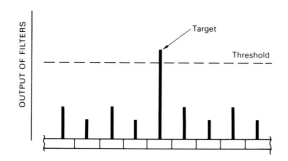

19. By setting the target detection threshold for each filter output far enough above the average of the outputs of the adjacent filters, the probability of clutter crossing the threshold can be reduced to an acceptable value.

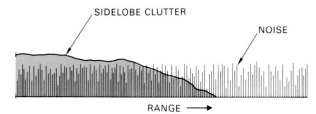

20. Long range end of range profile seen at receiver output, with mainlobe clutter removed. Sidelobe clutter ultimately becomes submerged in receiver noise.

Beyond this range, the noise determines the detection threshold. Thus, when low PRFs are being used, detection range is usually limited, not by sidelobe clutter, but only by background noise.

Tracking the Mainlobe Clutter. As we saw in Chap. 22, the mainlobe clutter spectrum varies continually. As the antenna look angle increases, the center frequency of the spectrum decreases, and the width increases from nearly a line to a broad hump (Fig. 21). As the speed of the radar increases, both the frequency and the width increase.

Consequently, to keep the mainlobe clutter lines in the clutter canceller's rejection notches, the frequency offset provided by the synchronous detector must track the changes in clutter frequency. From a knowledge of antenna look angle and ground speed, the frequency of the clutter lines can readily be predicted. Changes in frequency can then be tracked by appropriately adjusting the reference frequency supplied to the synchronous detector (Fig. 22).

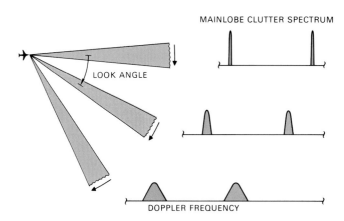

21. As look angle increases, mainlobe clutter spectrum broadens and shifts down in frequency.

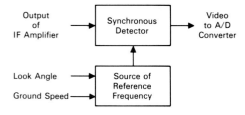

22. By continuously adjusting the reference frequency supplied to the synchronous detector to account for changes in look angle and ground speed, mainlobe clutter can be kept in rejection notches of clutter canceller.

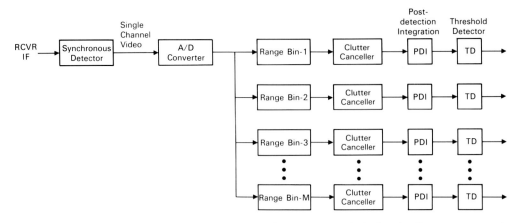

23. Block diagram of less sophisticated, single-channel signal processor in which doppler filtering is not employed.

Less Sophisticated Signal Processing

In many low-PRF pulse-doppler radars (Fig. 23, above), signal processing is much simpler than that just described. The clutter canceller is directly followed by postdetection integration and target detection. With the consequent elimination of doppler filtering, two channel processing is not essential. So, at a sacrifice of 3 dB in signal energy, single-channel processing can be employed. With it, the need for magnitude detection is eliminated. Naturally, these simplifications result in a considerable reduction in detection sensitivity.

Incidentally, in a single-channel processor, when the central line is placed at dc, the portion of the doppler spectrum lying below it folds over onto the positive-frequency portion. Consequently, return whose doppler frequency is lower than the central-line frequency appears in the bipolar video output just as it would if its frequency were an equal amount higher than the central-line frequency. There is no way of telling whether a target's range rate is positive or negative.

In still simpler non-doppler radars, the receiver output is applied to a simple envelope detector. It converts the output to a video signal, which is supplied directly to the display.

Advantages and Limitations

Low PRF operation has both compelling advantages and limitations. Among the advantages are these:

- Target ranges can be measured directly by the simple, highly precise pulse-delay method.

- Sidelobe clutter can largely be rejected through range resolution.

- Sensitivity time control (STC) can be used to provide wide dynamic range.

- Signal processing requirements can be met quite simply.

- Detection range is usually limited only by background noise.

Among the principal limitations associated with low PRFs are the following:

- If a target's doppler frequency is such that the target's echoes fall within one of the clutter filter's rejection notches, the radar will be "blind" to the target.

- Although first-time-around echoes are received from all ranges out to the maximum range the radar is designed to handle, there is little other than sensitivity time control and obstructions in the line of sight to prevent multiple-time-around echoes of strong targets beyond R_u from appearing falsely to be within the radar's range. (Mainlobe clutter from beyond R_u is, of course, rejected on the basis of doppler frequency, just as is the mainlobe clutter from ranges out to R_u.)

- In older radars and simpler modern radars employing magnetron transmitters, duty factors are typically low. So high peak powers are usually required to obtain reasonable detection ranges.

- In fighters, limitations on antenna size require use of wavelengths so short that doppler ambiguities are severe. Not only is direct measurement of closing rates impractical, but airborne targets are difficult to distinguish from moving targets on the ground.

Getting Around the Limitations

In the paragraphs that follow, we will examine the limitations of low PRF operation and see what can be done to alleviate them.

Doppler Blind Zones. Perhaps the most significant limitation of low PRF operation is that due to the so-called *"doppler blind zones."* The bands of doppler frequency falling within the clutter canceller's rejection notches and the passbands of the doppler filters whose outputs are not processed are blocked out on either side of the central line of mainlobe clutter (Fig. 24). If a target's doppler frequency lies within any of these "zones," the echoes' carrier and sideband frequencies will fall in the rejection notches, and the target will not be seen. Hence the name, blind zones.

At low PRFs a target's doppler frequency may be many times the PRF, so the target is about as likely to appear at any one point within a span of frequencies equal to the PRF as at any other. Therefore, the probability of a target being in the blind zones at any one time is roughly equal to the ratio of the width of the rejection notches to the PRF.

This probability can be reduced in several ways.

One is simply to increase the PRF, thereby spreading the blind zones farther apart. The extent to which the PRF can

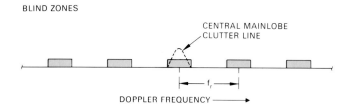

24. Bands of frequency falling within rejection notches of clutter canceller/doppler filter bank. Radar is blind to any target whose true doppler frequency lies within one of these bands.

347

be raised, however, is limited by the maximum range of interest. Generally, that is at least on the order of 20 miles, which puts an upper limit on PRF of about 4 kilohertz ($80 \div 20 = 4$ kHz).

Another way of reducing the severity of the blind zones is to reduce their width. The extent to which that can be done is, of course, limited by the width of the mainlobe clutter lines. They can be narrowed in one or more of the following ways.

- Increasing the size of the antenna, hence reducing the beamwidth or allowing use of longer wavelengths.

- Limiting the speed of the radar-bearing aircraft, hence reducing the spread of the mainlobe clutter frequencies.

- Limiting the maximum antenna look angle, hence further reducing the spread of the mainlobe clutter frequencies.

In radars for applications such as early warning and surveillance, where the speed of the radar-bearing aircraft is low, blind zones can be narrowed to the point that they are not a serious problem by employing large antennas. As illustrated in the upper half of Fig. 25 (below), for a radar having a 20-foot long antenna and a velocity of only 300 knots, even in the worst case (azimuth angle of 90°), the doppler-clear region will at least be as wide as the blind region at a PRF as low as 200 hertz.

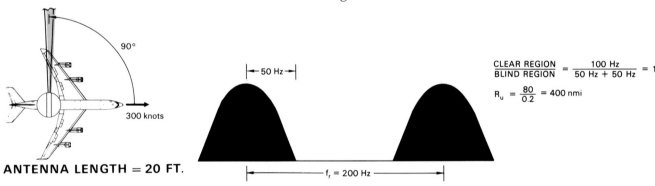

ANTENNA LENGTH = 20 FT.

$$\frac{\text{CLEAR REGION}}{\text{BLIND REGION}} = \frac{100 \text{ Hz}}{50 \text{ Hz} + 50 \text{ Hz}} = 1$$

$$R_u = \frac{80}{0.2} = 400 \text{ nmi}$$

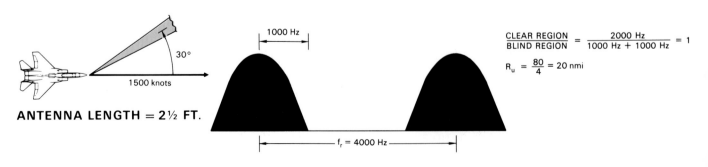

ANTENNA LENGTH = 2½ FT.

$$\frac{\text{CLEAR REGION}}{\text{BLIND REGION}} = \frac{2000 \text{ Hz}}{1000 \text{ Hz} + 1000 \text{ Hz}} = 1$$

$$R_u = \frac{80}{4} = 20 \text{ nmi}$$

25. Where long antennas are practical and radar velocities are low, the ratio of doppler-clear to blind region is sufficiently large that blind zones are not a serious problem. But in fighters, blind zones force the use of higher PRFs and/or limited look angles.

However, in radars for fighter applications, where antenna size is limited and radar speeds can be high, blind zones can occupy an excessive portion of the doppler spectrum. About the only recourse one has (beyond increasing the PRF) is to limit the maximum look angle.

As illustrated in the lower half of Fig. 25, the radar for a fighter, whose antenna is generally on the order of $2\,1/2$ feet in diameter and whose maximum velocity may well be on the order of 1500 knots or more, a one-to-one ratio of clear to blind regions can be achieved only by limiting the azimuth angle to a maximum of no more than 30° and raising the PRF to 4000 hertz. Usually neither such a severe restriction of azimuth angle nor such a high PRF is attractive. Consequently, in those fighter applications where mainlobe clutter is a problem, medium or high PRFs are commonly used.

Regardless of the severity of the blind zones, we can substantially reduce the probability of a target remaining in a blind zone throughout an entire time on target. Since the blind zones are all separated from the zone at zero frequency by multiples of the PRF, we can move them about by changing the PRF. The central line will of course remain at dc, since it is the carrier frequency. In principle, if we use enough different, widely separated PRFs we can periodically uncover every part of the spectrum (Fig. 26, below). However, since the time-on-target is divided among the different PRFs, PRF switching reduces detection sensitivity. The more PRFs that are used, the more the sensitivity will be reduced.

A common alternative is to "jitter" or sweep the PRF between two values. If these are suitably chosen, targets whose closing rates fall within a limited span of interest—e.g., rates corresponding to aircraft velocities around Mach 1—can be kept continuously in the clear.

If a target's doppler frequency is already known, the target may be kept out of the blind zones by adaptively selecting

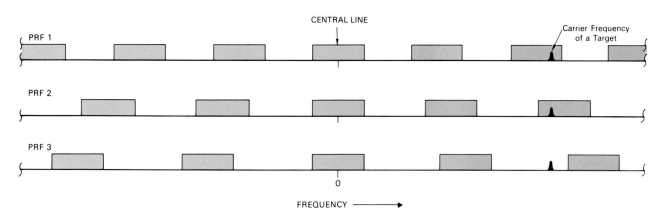

26. By periodically changing the PRF, blind zones can be shifted, reducing the possibility that any one target will remain in a blind zone for the entire time-on-target.

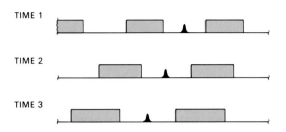

27. If a target's doppler frequency is known, it can be kept continually in the clear by adaptively changing the PRF.

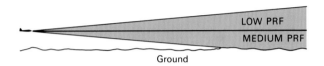

28. A possible mode in low altitude applications employs low PRFs on the upper bar of the search scan, where the beam does not strike the ground, and medium PRFs on lower bar.

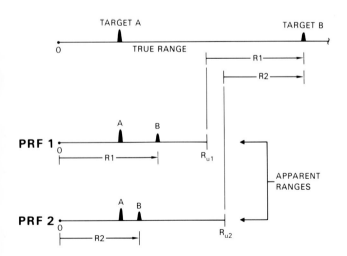

29. If the PRF is changed by a small amount, the observed range of a target beyond the unambiguous range (R_u) will change, but the observed range of a target in the first range zone will not.

the PRF. That is, the PRF may be selected so the zones will straddle the target's frequency (Fig. 27). The necessary *a priori* doppler information may be obtained by detecting the target in a high PRF search mode. Or, it may be available as a result of tracking the target in range.

If mainlobe clutter is not a problem—e.g., if the mainlobe intercepts the ground only at shorter or longer ranges than those of interest or if it does not intercept the ground at all, blind zones can be avoided simply by not discarding any return. That is, by eliminating the clutter cancellers and processing the outputs of all the doppler filters.

In low altitude applications (Fig. 28), a possible mode is one in which the radar employs low PRFs (for long range detection) on the upper bar of the antenna search scan, where mainlobe clutter is not encountered, and medium or high PRFs on the lower bars (for good performance in mainlobe clutter).

Multiple-Time-Around Echoes. The problems of multiple-time-around target echoes may be moderated to some extent by sensitivity time control (STC).

To illustrate, let us assume that the unambiguous range is 20 miles. If return is received from a target at 21 miles, it will appear to have a range of 1 mile. However, its echoes will be only $(1/21)^4 = 0.000005$ times as strong (-53 dB) as the echoes from a target of the same radar cross section and aspect at a range of 1 mile. With STC, because detection sensitivity is greatly reduced during the initial portion of the interpulse period, this unwanted target will likely not be detected.

On the other hand, if the target were at a range of, say, 39 miles, this would not necessarily be so. The target would then have an apparent range of 19 miles. Its echoes would be $(19/39)^4 = 0.0625$ times as strong (-12.5 dB) as those of an equivalent target at 19 miles, hence might be detected.

If multiple-time-around targets are a problem, they may be identified by changing the PRF (Fig. 29). As discussed in detail in Chap. 12, if the PRF is changed by a small amount, the observed ranges of these targets will correspondingly change, whereas the observed ranges of the first-time-around targets will not. Therefore, by periodically changing the PRF and looking for changes in the observed target ranges, the multiple-time-around targets can be spotted and prevented from reaching the display.

Low Duty Factor. Within the capabilities of the transmitter that is used, reasonably high duty factors can be achieved at low PRFs by transmitting very long pulses and employing large amounts of pulse compression to achieve the desired range resolution (Fig. 30). Surprisingly, if this is done in the absence of mainlobe clutter, a low PRF radar can actually obtain greater search detection ranges than a high PRF radar employing the same average power even against nose hemisphere targets.

What makes the difference are the losses incurred by the high PRF radar due to eclipsing—return being received while the radar is transmitting and the receiver is blanked out.

True, even at low PRFs, a considerable amount of return may be lost as a result of eclipsing. But eclipsing is much less of a problem at low PRFs than at medium and high PRFs.

For as long as a returned pulse is not received at exactly the same time as the transmitter is transmitting, some of the pulse will get through to the receiver, and with low PRFs only the return from zero range is so synchronized. As the range increases the portion of the return getting through increases. For targets at ranges greater than one pulse length, none of the return is lost.

This is so, of course, only up to the point where the trailing edge of the echo from a target at the maximum range of interest is received as the leading edge of the next pulse is being transmitted. In other words, the interpulse period must be at least one pulse width longer than the round-trip transit time for the most distant target of interest (Fig. 31).

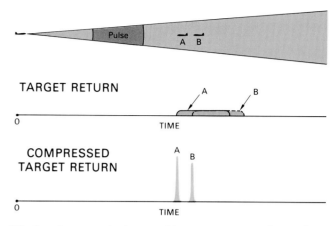

TARGET RETURN

COMPRESSED TARGET RETURN

30. Duty factor can be increased by transmitting very long pulses and using large amounts of pulse compression to obtain the desired range resolution.

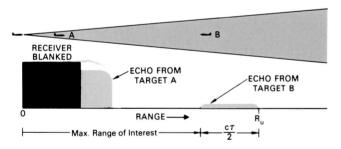

31. To avoid eclipsing of long-range targets, the interpulse period must be at least one pulsewidth longer than the transit time for the most distant target of interest.

Provided this requirement is met, duty factors of up to 20 percent may be used by a radar operating at low PRFs without incurring a significant eclipsing loss.

In contrast, at medium and high PRFs, because of range ambiguities, the severity of eclipsing is independent of the range from which the return is received. The eclipsing loss increases directly with the duty factor.

Moving Targets on the Ground. In air-to-ground operations detecting ground moving targets (GMTs) may be a primary objective, but in air-to-air operations *rejecting* GMTs

32. When searching for aircraft over areas where hundreds of vehicles may be moving on the ground, means must be provided to eliminate these targets from the radar display.

may be essential. When operating over territory where there are hundreds of moving vehicles on the ground—cars, trucks, trains, and so forth (Fig. 32, above)—a radar may detect a great many more GMTs than airborne targets. The GMTs may so clutter up the display that the operator cannot discern his own targets among them—even though the targets have been solidly detected and are clearly displayed.

If the separation between mainlobe clutter lines is sufficiently wide, the number of filters at the ends of the doppler filter bank whose outputs are not used can be increased enough to exclude GMTs without rejecting an unacceptable amount of possible target return.

If the lines are too closely spaced for this, the GMTs may be identified by observing the effect of PRF switching on their apparent doppler frequencies. Because of the greater ground speeds of airborne targets, at such low PRFs an aircraft's apparent doppler frequency will generally be its true doppler frequency minus some multiple of the PRF. Consequently, the apparent frequencies of these targets will usually change when the PRF is switched. On the other hand, the observed doppler frequencies of ground moving targets, whose speeds are much lower, will generally be true frequencies and so will not change (Fig. 33). By disregarding those threshold crossings which occur in the same doppler filter after the PRF has been switched, GMTs may be prevented from appearing on the display.

In air-to-ground applications where GMTs rather than airborne moving targets are of interest, the same procedure is commonly used in reverse: *airborne* moving targets are prevented from appearing on the display by discarding those threshold crossings which do *not* occur in the same doppler filter after the PRF has been switched.

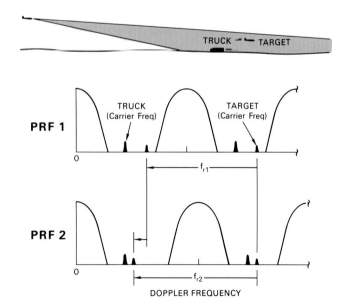

33. Ground moving targets may usually be distinguished from airborne targets because their apparent doppler frequencies do not change if the PRF is changed slightly.

Summary

At low PRF, mainlobe clutter may largely be eliminated by offsetting the doppler spectrum so the central line is at dc and passing the return through a clutter canceller and bank of doppler filters. To keep the clutter in the canceller's rejection notches, the "offset" must be varied with radar speed and antenna look angle.

Sidelobe clutter as well as mainlobe clutter residue and background noise are then minimized through a combination of range gating and doppler filtering. Maximum detection range is usually limited only by receiver noise.

The principal limitation of low PRFs is doppler blind zones—regions in the doppler spectrum for which a target's "observed" doppler frequency is the same as that of the mainlobe clutter. The zones are the same width as the mainlobe clutter lines and are spaced at intervals equal to the PRF. While not a serious problem where large antennas and low radar speeds are practical, in fighter radars blind zones can be acceptably reduced only by employing such a high PRF that R_u, hence the maximum operating range, is severely reduced or by limiting the maximum look angle.

The possibility of a target remaining in a blind zone for an entire time-on-target may be minimized by switching among widely separated PRFs. Alternatively, a limited span of doppler frequencies may be kept clear by jittering or sweeping the PRF. Or, the doppler frequency of a given target may be kept clear by adaptively selecting the PRF.

Multiple-time-around target echoes may be singled out by jittering the PRF and be blocked from reaching the display. Both PRF switching and PRF jittering, however, reduce detection sensitivity.

By transmitting very long pulses and employing pulse compression to provide adequate range resolution, duty factors of up to 20 percent may be achieved.

In fighter applications, severe doppler ambiguities make discrimination between airborne and ground moving targets difficult. The problem can be alleviated at the cost of wider blind zones by discarding the outputs of a larger number of filters at the ends of the doppler filter bank or by noting whether a target appears in the same doppler filter after the PRF has been changed slightly.

Because of the blind zone problem, low PRFs are generally used only where mainlobe clutter can be avoided or where large antennas and low radar speeds are practical.

LOW PRFs

ADVANTAGES	LIMITATIONS
1. Good for air-to-air look-up and ground mapping.	1. Poor for air-to-air look-down—much target return may be rejected along with mainlobe clutter.
2. Good for precise range measurement and fine range resolution.	2. Ground moving targets can be a problem.
3. Simple pulse delay ranging possible.	3. Doppler ambiguities generally too severe to be resolved.
4. Normal sidelobe return can be rejected through range resolution.	4. Higher peak powers or larger amounts of pulse compression generally required.

LOCKHEED YF–12 BLACKBIRD (1964)

This unique aircraft was the first interceptor design to be equipped with a pulse doppler radar, the Hughes ASG–18. Concepts proven with the ASG–18 were later refined and incorporated into the current line of modern Hughes radars for the F–14, F–15, and F–18 fighters.

Medium PRF Operation

27

A medium PRF is, by definition, one for which both range and doppler frequency are ambiguous (Fig.1). In practice, only the lower reaches of the relatively wide band of PRFs satisfying this definition are actually used. In general, the optimum value increases with the radar's radio frequency. For the X-band, medium PRFs typically range from about 8 to 16 kilohertz—slightly higher than the top of the low PRF range, which falls somewhere between 2 and 4 kilohertz.

Medium PRF operation was conceived as a means of getting around some of the limitations of low and high PRFs in fighter applications. The primary reason for operating above the low PRF region is to improve the radar's ability to contend with mainlobe clutter and GMTs. And the primary reason for operating below the high-PRF region is to improve the radar's ability to contend with sidelobe clutter in tail hemisphere (low-closing-rate) approaches.

In this chapter, we will take a closer look at medium PRF operation. We will see what must be done to separate targets from clutter and how the signal processing is performed. We will then take up the problems of rejecting ground moving targets, eliminating blind zones, minimizing sidelobe clutter, and rejecting sidelobe return from those targets on the ground which have exceptionally large radar cross sections.

Differentiating Between Targets and Clutter

As in the preceding chapter, to get a clear picture of the problem of rejecting ground clutter, let us look at the range and doppler profiles for a representative flight situation. We will assume that the radar has a PRF of 10 kilohertz and that the maximum range of interest is 24 nautical miles.

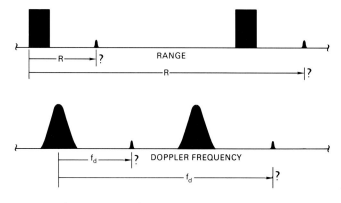

1. A medium PRF is one for which both range and doppler frequency are ambiguous.

Range Profile. This profile with the true range profile above it, and the flight situation from which the profiles were derived above that, is illustrated in Fig. 2.

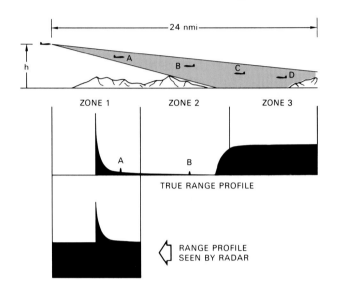

2. Range profile of a representative flight situation. PRF is such that the maximum range of interest is broken into three range zones.

The width of the range zones is about 8 miles ($80 \div 10 = 8$). The maximum range of interest, therefore, is in effect broken into three segments, 8 miles long.

As seen by the radar, however, these are indistinguishably superimposed. Ground clutter completely blankets the observed range interval. Mainlobe clutter extends from one end of it to the other. Strong sidelobe clutter received from short ranges covers a substantial portion of it. None of the targets are discernible.

Except in the case of very large targets in comparatively light clutter—such as are encountered in modes of operation provided for detecting and tracking ships—no amount of range discrimination alone is going to enable the radar to isolate the target echoes from the clutter. To reject both mainlobe and sidelobe clutter, we must rely heavily on doppler frequency discrimination.

Doppler Profile. As in the case of low PRFs, this profile consists of a series of mainlobe clutter lines separated by the pulse repetition frequency, f_r (Fig. 3). Between any two

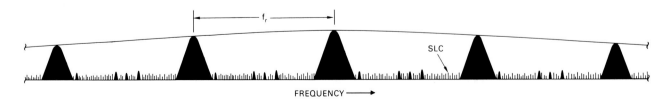

3. Doppler profile of the representative flight situation. Mainlobe clutter lines are much more widely spaced than for low PRF operation; other conditions remaining the same.

consecutive lines (Fig. 4) appears most, but not all, of the sidelobe clutter and the return from most but not necessarily all of the targets. The rest of the sidelobe clutter and target return is indistinguishably intermixed with the mainlobe clutter.

Rejecting Mainlobe Clutter. While the doppler profiles for low and medium PRF operation are similar, there is one important difference: other conditions remaining the same, at medium PRFs the mainlobe clutter lines are spread farther apart. Since the width of the line is independent of the PRF, there is considerably more "clear" room between them in which to detect targets. Even if the mainlobe clutter is reasonably broad, it can be rejected on the basis of its doppler frequency without at the same time, on an average, rejecting the return from an inordinately large fraction of the radar's targets.

Rejecting Sidelobe Clutter. Because of the more severe range ambiguities, this is not as simple as at low PRFs. To illustrate, in Fig. 5 the range profile as seen by the radar is repeated with the mainlobe clutter removed. You will notice two things in this plot. First, the sidelobe clutter has a sawtooth shape. Second, only the short-range target (A) can be discerned above the clutter. Targets B and C are still obscured.

The sawtooth shape is due to the strong sidelobe return from the first range zone being superimposed over the weaker return from subsequent range zones (Fig. 6).

As for the obscured targets, Target B, in the second range zone, must compete not only with sidelobe clutter from its own range but with the far stronger clutter from the corresponding range in the first range zone. Targets C and D, in the third range zone, must compete not only with sidelobe clutter from their own range but with the much stronger return from the corresponding ranges in the first and second zones.

The clutter can, of course, be reduced substantially. It comes not only from different ranges but from different angles. Since returns from different angles have different doppler frequencies, we can differentiate between the target echoes and a great deal of the competing sidelobe clutter if we sort the return by *both* range and doppler frequency.

Sorting by range may, of course, be done by range gating (sampling), just as in low PRF operation. The range gates will isolate the returns received from relatively narrow strips of ground at constant range. Because of range ambiguities, though, the return passed by each gate will come from not just one strip, but several. And, as already noted, one or more of these strips may lie at relatively short range.

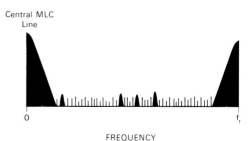

4. Portion of the doppler profile processed by the radar. The doppler spectrum is normally shifted to place the central line of mainlobe clutter at zero frequency (dc).

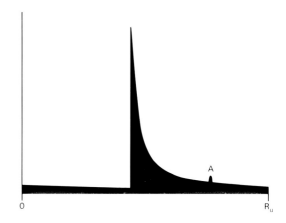

5. Range profile as seen by the signal processor, with mainlobe clutter removed. Only the short-range target can be discerned above the clutter.

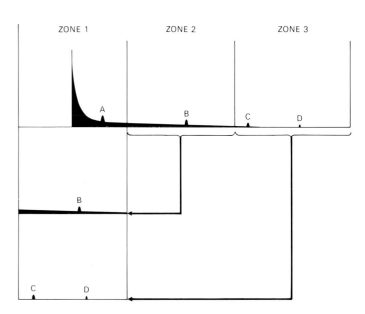

6. Sawtooth shape is due to strong sidelobe return from the first range zone being superimposed over the weaker return from the second and third zones.

Still, the reduction in clutter obtained through range gating will be substantial (Fig. 7).

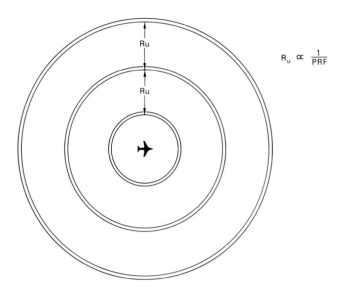

$$R_u \propto \frac{1}{PRF}$$

7. Each range gate passes the return from a series of circular strips (only three of which are shown here) separated from one another by R_u.

Sorting by doppler frequency may be accomplished by applying the output of each range gate to a bank of doppler filters. They will isolate the returns received from strips of ground lying between lines of constant angle relative to the radar's velocity (Fig. 8).

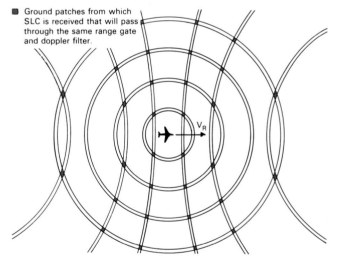

■ Ground patches from which SLC is received that will pass through the same range gate and doppler filter.

8. Each doppler filter receives only that portion of the total side-lobe return passed by a single range gate. The filter passes only that fraction of this return which comes from strips of ground whose angles relative to the radar's velocity are such that the return falls in the filter's passband.

Because of doppler ambiguities, though, any one filter will pass the return from not just one strip, but several. Nevertheless, the amount of clutter with which a target's echoes must compete will be only a fraction of that passed by the range gate.

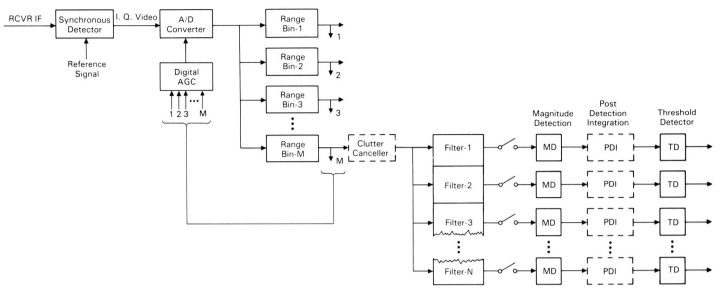

9. How signal processing for medium PRF operation may be handled. Clutter canceller is optionally included to reduce dynamic range required of doppler filters. Postdetection integration (PDI) may be provided if filter integration time is less than time-on-target.

Signal Processing

As illustrated in Fig. 9 (above) for medium PRFs, signal processing is quite similar to that for low PRFs. There are, however, three main differences. First, because range ambiguities preclude the use of sensitivity time control, additional automatic gain control is needed to avoid saturation of the A/D converter. Second, to further attenuate the sidelobe clutter (which because of range ambiguities piles up more deeply at medium PRFs), the passbands of the doppler filters may be made considerably narrower. Third, additional processing is required to resolve range and doppler ambiguities.

As at low PRFs, the first step in processing the IF output of the radar receiver is to shift the doppler spectrum so as to place the central mainlobe clutter line at dc. Again, the shift must be dynamically controlled to account for changes in radar velocity and antenna look angle. The I and Q outputs of the synchronous detector that performs this shift are likewise sampled at intervals on the order of the transmitted pulse width[1] and digitized.

However, to reduce the dynamic range required of the A/D converter, automatic gain control is provided ahead of the converter. For this, the converter's output is monitored and a continuously updated profile of the output over the course of the interpulse period is stored. On the basis of this profile, a gain control signal is produced and applied to the amplifiers ahead of the A/D converter.

By reducing the gain when the mainlobe clutter and strong close-in sidelobe clutter are being received, the control signal keeps the converter from being saturated, yet maintains the input to the converter well above the local noise level when weaker return is coming through. Since

1. Compressed pulse width, when pulse compression is used.

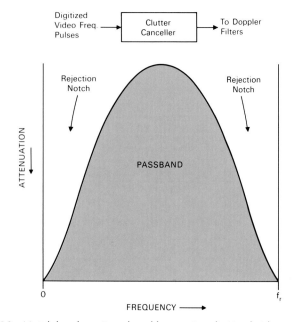

10. Mainlobe clutter is reduced by passing digitized video-frequency output of receiver through simple clutter canceller having characteristic such as this.

the control signal is derived after the return is digitized, this is called *digital automatic gain control* (DAGC).

To reduce the dynamic range required in the subsequent processing, once the output of the A/D converter has been sorted into range bins, an optional next step is to get rid of the bulk of the mainlobe clutter in each bin. This may be accomplished with a clutter canceller (Fig. 10).

The return in each bin is next applied to a bank of doppler filters. At the end of every integration period, the magnitude of each filter's output is detected. If the scan of the radar antenna and the bandwidth of the doppler filters are such that the filter integration time is less than the time-on-target, magnitude detection may be followed by postdetection integration.

In either event, the integrated return passed by each doppler filter during every time-on-target is applied to a separate threshold detector. The threshold of this detector is adaptively set to keep the probability of clutter producing false alarms acceptably low. The setting may be based on the average level of the clutter for (a) several range increments on either side of that in question, (b) several doppler frequencies on either side of that in question, (c) several integration periods before and after that in question, or (d) some combination of these. In general, the optimum averaging scheme for a clutter background is different from that for a noise background.

When a target is detected, we can tell its apparent range by observing which bin (or adjacent bins) it was detected in. Similarly, we can tell its apparent doppler frequency, hence range rate, by observing which doppler filter (or adjacent filters) the detection occurred in.

The observed range and doppler frequency will, of course, be ambiguous. Range ambiguities are resolved by PRF switching, as outlined in Chap. 12. Doppler ambiguities may be resolved by the methods described in Chap. 21.

Rejecting Ground Moving Targets (GMTs)

GMTs are not nearly the problem they are at low PRFs. For, at medium PRFs, the mainlobe clutter lines are spread sufficiently far apart that GMTs appear only near the ends of the region between lines. Targets with positive closing rates appear at the lower end; targets with negative closing rates, at the upper end ($f_r - f_d$). GMTs, therefore, can be eliminated without losing an unreasonable fraction of the target return simply by discarding any return in the frequency bands where GMTs may appear.

The width of these bands depends upon the wavelength and the velocities of the targets. Most surface vehicles travel at less than 65 miles per hour. At X-band (30 hertz per mile per hour of closing rate), the maximum doppler shift of the

GMTs relative to the center frequencies of the mainlobe clutter lines would be about 2 kilohertz (65 x 30 = 1950 Hz).

So, for an X-band radar, those GMTs having a component of velocity toward the radar (positive doppler shift) can be eliminated by discarding all return whose frequency is less than 2 kilohertz above the center of each mainlobe clutter line (Fig. 11) And those GMTs having a component of velocity away from the radar (negative doppler shift) can be eliminated by discarding all return whose frequency is less than 2 kilohertz below the center of each line. At the same time, the mainlobe clutter residue will also be eliminated.

When this approach is taken to the problem of GMTs, the anticipated maximum doppler frequency of the GMTs relative to the doppler frequency of the mainlobe clutter usually puts the lower limit on the selection of PRF. Suppose that to provide a reasonable amount of room in which to look for airborne targets, we establish a design criterion that at least 50 percent of the doppler spectrum be clear, i.e., not covered by blind zones. If to eliminate GMTs we discard all return whose frequency is within 2 kilohertz of the center of each clutter line (Fig. 12), we must make the filter bank's passband at least 4 kilohertz wide. To accomplish this, the PRF must be at least 2 + 4 + 2 = 8 kHz.

Since the doppler shift is inversely proportional to wavelength ($f_d = 2\dot{R}/\lambda$), the shorter the wavelength, the higher the minimum PRF will be, and vice versa. Take a wavelength of 1 centimeter, for example. At this wavelength, the maximum relative doppler shift for a 65 mile per hour vehicle is 6 kilohertz as opposed to 2 kilohertz. Consequently, if we apply the above design criterion to a 1-centimeter radar, the minimum PRF is 6 + 12 + 6 = 24 kHz.

Eliminating Blind Zones

Blind zones, too, are still a problem at medium PRFs. In fact, because of range ambiguities, the radar must contend not only with blind zones in the doppler spectrum but with blind zones in the range interval being searched, as well.

Doppler Blind Zones. Because mainlobe clutter covers a much smaller portion of the doppler frequency spectrum, doppler blind zones are far less severe at medium PRFs than at low PRFs and so can be eliminated by switching among fewer PRFs. However, additional PRFs are required to resolve range ambiguities and eliminate ghosts.

Typically, the radar is cycled through a fixed number of fairly widely spaced PRFs (Fig. 13). If a target is in the clear on any three of these and its echoes exceed the detection threshold on all three, the target will be deemed to have been detected. The range ambiguities will then be resolved and "de-ghosted." The optimum number of PRFs varies

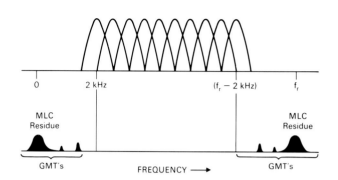

11. Most ground moving targets (GMTs), as well as residue of mainlobe clutter (MLC) passed by clutter canceller, can be rejected by discarding the return between 0 and 2 kilohertz and between (f_r – 2 kHz) and f_r.

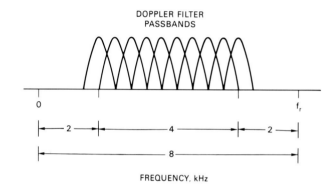

12. If 4 kilohertz of doppler spectrum is discarded to eliminate GMTs, the PRF must be at least 8 kilohertz to meet criterion that 50 percent of doppler spectrum be clear.

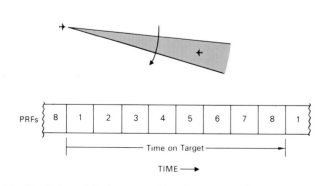

13. To eliminate blind zones and resolve range ambiguities, a radar may cycle through a number of widely spaced PRFs.

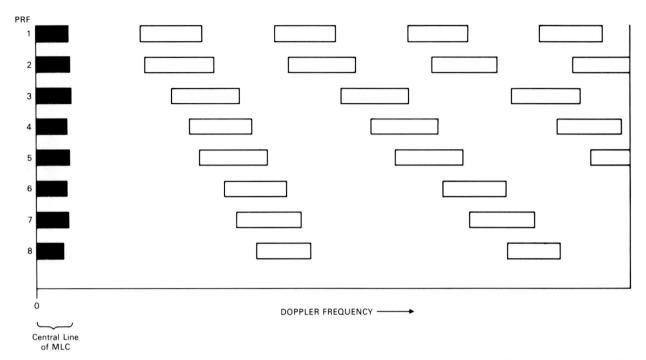

PRF
1
2
3
4
5
6
7
8

0

DOPPLER FREQUENCY ⟶

Central Line
of MLC

14. Doppler blind zones for eight widely spaced PRFs. Any target within the frequency range shown here will be "in the clear" for at least three PRFs.

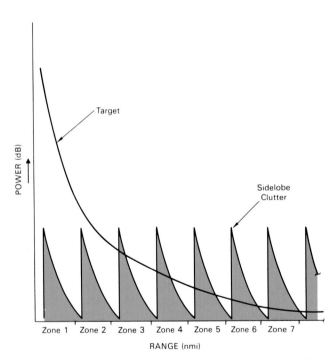

Target

Sidelobe
Clutter

POWER (dB)

Zone 1 Zone 2 Zone 3 Zone 4 Zone 5 Zone 6 Zone 7

RANGE (nmi)

15. Variation in the strength of target return with range. Strength of the sidelobe clutter with which the target must compete is superimposed. Blind zones occur where the plots overlap.

with the operational situation. A typical waveform, called 3:8, cycles through 8 PRFs, any 3 of which must be clear for detection (Fig. 14, above).

Range Blind Zones. These zones bracket the ranges at which targets will generally not be detected because their echoes are drowned out by sidelobe clutter simultaneously received from shorter ranges or are eclipsed by the transmitted pulses.

Just how the blind zones due to sidelobe clutter come about is best illustrated by a graph such as that shown in Fig. 15. It contains a plot of the strength of a target's echoes as the target's range increases from a relatively few miles out to the maximum range of interest. Superimposed over this plot is a periodically repeated plot of the strength of the sidelobe clutter received over the course of the interpulse period. Each repetition of the sidelobe clutter plot represents the clutter background against which the target echoes must be detected when the target is in a different one of the ambiguous range zones into which the true range profile is divided.

For the particular clutter spectrum illustrated, if a target is in the first or second range zone, it will be substantially stronger than any of the sidelobe clutter. If it is in the third range zone, it will still be stronger than most of the clutter, but not as strong as the peak produced by the altitude return and the sidelobe return immediately following it. If the target is in the fourth range zone, it will be stronger than the clutter over a smaller portion of the zone, and so on.

At those ranges where the clutter is as strong or stronger than the target echoes, the target will go undetected, just as it would if masked by receiver noise. The radar is thus "blind" to the target.

Obviously, the extent of the range blind zones increases with the strength of the clutter. The stronger the clutter, the wider the blind zone will be, and vice versa.

The strength of the clutter, in turn, depends on several things: the gain of the sidelobes, the nature of the terrain, the altitude of the radar, etc.

Added to the range blind zones due to strong sidelobe clutter are the blind zones due to eclipsing (Fig. 16). While the radar is transmitting (and for a very short recovery time thereafter), the receiver is blanked. Consequently, if a target's echoes are received at such times that they overlap these periods—as they invariably will be if the target's range is a multiple of R_u—not all of the target return will get through the receiver, and the target may not be seen. The resulting blind zones may be narrow enough to be inconsequential. But they can become significant if the pulses are long, as they are in some medium PRF radars.

The combined range blind zones due to sidelobe clutter and eclipsing for a representative radar are illustrated in Fig. 17.

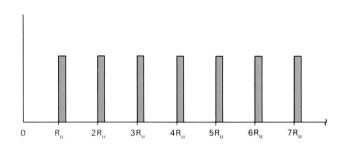

16. Range blind zones due to eclipsing by the transmitted pulses. As the width of transmitted pulses is increased, these zones become appreciable.

RANGE BLIND ZONES

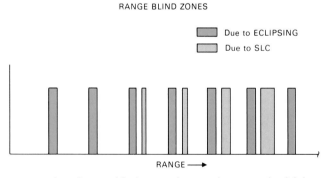

■ Due to ECLIPSING
▨ Due to SLC

RANGE ⟶

17. Combined range blind zones due to eclipsing and sidelobe clutter for a representative medium PRF radar.

As with doppler blind zones, the positions of the range blind zones shift with changes in the PRF. Fortunately, the shift is such that the same PRF switching as is used to reduce doppler blind zones will also largely reduce range blind zones (Fig. 18).

Bear in mind, though, that it is not enough for a target to be in a doppler clear region for one set of PRFs and in a range clear region for another. For a target to be detected, it must be in both a doppler clear region and a range clear region for the same set of PRFs. If the target's doppler frequency falls in a doppler blind zone, its echoes will not get through a doppler filter and be detected even though it is in a range-clear region. And if the target is in a range blind

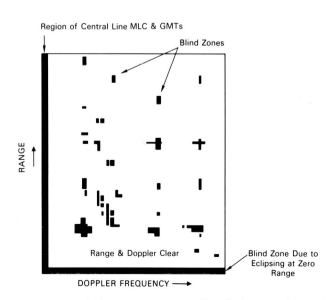

Region of Central Line MLC & GMTs

Blind Zones

RANGE

Range & Doppler Clear

Blind Zone Due to Eclipsing at Zero Range

DOPPLER FREQUENCY ⟶

18. Region in which a representative radar is both range clear and doppler clear for at least three of eight widely spaced PRFs.

zone, even though its echoes may get through a filter, they will be buried in the accompanying sidelobe clutter, which will drive the detection threshold up to a point where the target will not be detected.

It should be noted that the foregoing discussion of blind zones all pertains to search. In single-target tracking, there are many PRFs to choose from to avoid blind zones.

Minimizing Sidelobe Clutter

Clearly, at medium PRFs, sidelobe clutter must be kept to a minimum. For it not only determines the extent of the range blind zones, but limits the maximum detection range.

Since most of the sidelobe clutter comes from relatively short ranges, the background of clutter against which targets must be detected is generally stronger than the background noise falling in the passband of a doppler filter. Therefore, no matter how powerful the radar or how great a target's radar cross section, if the target's range is continuously increased (Fig. 19), a point will ultimately be reached where its echoes become lost in the clutter. The stronger the clutter, the shorter this range will be.

What, then, can be done to minimize the sidelobe clutter? Several things (Fig. 20). Without question, the most important measure is to design the radar antenna so that the gain of its sidelobes is low. In fact, this is essential. As described in Chap. 8, sidelobes can be reduced by tapering the distribution of radiated power across the antenna.

For a given level of sidelobe clutter in the receiver output, the amount of clutter with which a target's echoes must compete can be further reduced by narrowing the radar's pulses and correspondingly narrowing the range gates. If, for example, the pulse width is reduced by a factor of 10, the sidelobe clutter will be reduced by roughly the same ratio. Narrowing the pulses, of course, requires adding more range gates and forming more doppler filters—a separate bank of filters being required for every range gate.

By employing pulse compression, the narrowing can be accomplished without reducing the average transmitted power (Fig. 21). A common practice is to maximize the average power by making the transmitted pulses as wide as possible without incurring an unacceptable loss due to eclipsing. Enough pulse compression is then provided to achieve the desired range resolution.

An alternative approach, which minimizes eclipsing, is to transmit very narrow pulses of higher peak power.

The sidelobe clutter with which a target must compete can be still further reduced by narrowing the passbands of the doppler filters. For that, the return from more pulses must be integrated by the filters—the time-on-target must

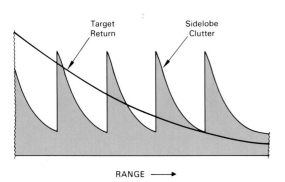

19. As a targets' range is increased its echoes may eventually be engulfed in sidelobe clutter, unless special measures are taken to minimize it.

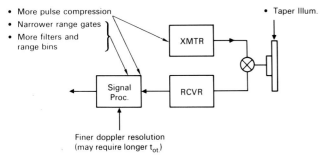

- More pulse compression
- Narrower range gates
- More filters and range bins
- Taper Illum.
- Finer doppler resolution (may require longer t_{ot})

20. Measures that can be taken to reduce sidelobe clutter.

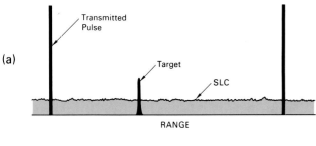

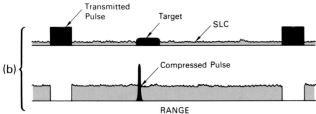

21. Two ways of reducing sidelobe clutter through range resolution: (a) transmit very narrow pulses of high enough peak power to provide adequate detection range; (b) transmit wider pulses of the same average power and use pulse compression to provide the desired range resolution.

be sufficient to permit this—and more filters, magnitude detectors, and threshold detectors must be provided.

To retain all of the power in the target return, of course, the filters must still be wide enough to pass the spectrum of frequencies over which the power is spread. The width of this spectrum varies from target to target, but is on the order of 10 to 20 hertz. Usually, the filters must be made wider than this to minimize filter straddle loss, as discussed in Chap. 18. Also, it may be necessary to allow for possible changes in doppler frequency due to acceleration of one or both aircraft during the filter formation time (Fig. 22).

Through techniques such as those outlined in the foregoing paragraphs, sidelobe clutter may be reduced to a point where it falls below the noise level at the ends of the interpulse period (Fig. 23). The detection ranges of targets appearing there will then be limited only by noise.

By switching among a large enough selection of PRFs, these clutter-free regions can be shifted about so that the detection range of virtually all targets will be limited only by noise. As more PRFs are added, of course, the available integration time for each PRF decreases, and this decrease limits the detection range.

Detection ranges achievable with medium PRFs are thus invariably somewhat less than those achievable under similar conditions with high PRFs against nose-aspect targets or with low PRFs in those situations where mainlobe clutter is not a problem.

Sidelobe Return from Targets of Large RCS

One important form of sidelobe return we have not yet considered is that from structures on the ground—buildings, trucks, etc. having exceptionally large radar cross sections. As explained in Chap. 22, even when in the sidelobes, such structures can return echoes every bit as strong as those received from an aircraft in the mainlobe (Fig. 24). If the ground target's doppler frequency falls in the filter bank's passband—as it most often will when the ground target is in a sidelobe—the ground target will be detected no differently than if it were an aircraft in the mainlobe.

Since these unwanted targets are nearly point reflectors, no amount of range or doppler resolution will make them less likely to be detected. On the contrary, the greater the resolution provided, the greater the extent to which the surrounding sidelobe clutter will be attenuated and the more prominently the point targets will protrude above it.

While a radar is vulnerable to such targets when operating at low PRFs, it is much more vulnerable when operating at medium PRFs because of the more severe range ambigui-

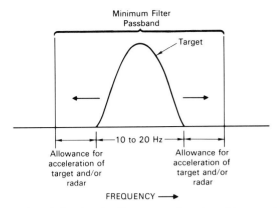

22. Passband of a doppler filter must at least be wide enough to accommodate the target return and allow for changes in doppler frequency during the integration period.

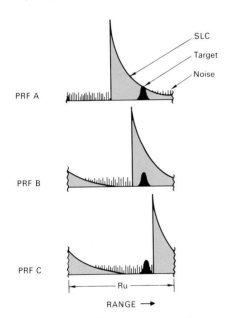

23. By switching among enough PRFs, the sawtooth pattern of sidelobe clutter can be shifted about so that virtually every target may be detected against a background only of noise.

24. Because of range and doppler ambiguities at medium PRFs, a radar is vulnerable to unwanted ground targets.

25. Antenna of a medium PRF radar. Note horn antenna for guard receiver.

2. Mainlobe return from the unwanted ground target will, of course, fall in the rejection notch of the clutter canceller. If the return is extremely strong, however, a detectable fraction of it may get through.

ties. Some special means, therefore, must be provided to keep these targets from reaching the radar display.

One way of dealing with these unwanted targets is to provide the radar with a guard channel. In essence it consists of a separate receiver whose input is supplied by a small horn antenna mounted on the radar antenna (Fig. 25).

The width of the horn's mainlobe is sufficient to encompass the entire region illuminated by the radar antenna's principal sidelobes, and the gain of the horn's mainlobe is greater than that of any of the sidelobes (Fig. 26).

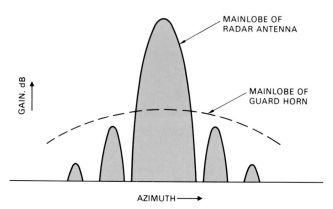

26. Gain of horn's mainlobe is greater than that of the radar antenna's sidelobes but less than that of radar antenna's mainlobe.

Any detectable target in the radar antenna's sidelobes, therefore, will produce a stronger output from the guard receiver than from the main receiver.

On the other hand, because the gain of the radar antenna's mainlobe is much greater than that of the horn, any target in the radar antenna's mainlobe will produce a much stronger output from the main receiver than from the guard receiver.

Consequently, by comparing the outputs of the two receivers and inhibiting the output of the main receiver when the output of the guard receiver is stronger (Fig. 27), we can prevent any targets that are in the sidelobes from appearing on the radar display.[2]

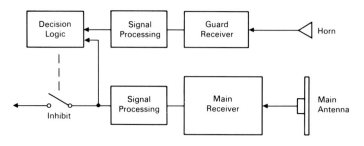

27. Output of main receiver is inhibited when a target is detected simultaneously through guard channel and main receiver channel.

Summary

In medium PRF operation, the PRF is usually set just high enough to spread the mainlobe clutter lines so that mainlobe clutter and any ground moving targets (GMTs) can be rejected without rejecting the return from an unreasonably high percentage of targets. Range ambiguities are then still sufficiently mild that, through a combination of range and doppler discrimination, the background of sidelobe clutter against which the target echoes must be detected can be reduced to an acceptable level.

Because of the increased separation of the mainlobe clutter lines, doppler blind zones can be largely eliminated by switching among a few fairly widely spaced PRFs. Because distant targets must compete with close-in sidelobe clutter, the peaks of this clutter produce range blind zones. If the clutter is not too strong, these as well as blind zones due to eclipsing can largely be eliminated by the same PRF switching as is used to eliminate doppler blind zones. But even in the doppler clear regions, sidelobe clutter usually limits detection range.

It is essential that a low sidelobe antenna be used. Sidelobe return can be further reduced by increasing the range and doppler resolution. To eliminate sidelobe return from ground targets of exceptionally large radar cross section, a guard channel may be provided.

MEDIUM PRFs

ADVANTAGES	LIMITATIONS
1. Good all-aspect capability—copes satisfactorily with both mainlobe and sidelobe clutter.	1. Detection range against both low and high closing-rate targets can be limited by sidelobe clutter.
2. Ground-moving targets readily eliminated.	2. Must resolve both range and doppler ambiguities.
3. Pulse delay ranging possible.	3. Special measures needed to reject sidelobe return from strong ground targets.

GENERAL DYNAMICS FB-111A (1969)

This bomber is a derivative of the F-111A fighter-bomber which was the first aircraft to enter service with a variable sweep wing. It was equipped with a General Electric attack radar for weapon delivery and a Texas Instruments terrain-following radar for low altitude penetration of hostile territory.

High PRF Operation

28

A high PRF is one for which the observed doppler frequencies of all significant targets are unambiguous. The observed ranges, however, are generally highly ambiguous.

High PRF operation has three principal advantages. First, since doppler frequencies are unambiguous, mainlobe clutter can be rejected without rejecting any target echoes whose doppler frequencies are different from that of the clutter. Second, by employing a high enough PRF, the "lines" (more realistically, bands) of the clutter spectrum can be spread far enough apart to open up an entirely clutter-free region between them, where high closing rate[1] (nose-aspect) targets will appear (Fig. 1). Third, transmitter duty factors can be increased by increasing the PRF rather than the pulsewidth, thereby enabling high average powers to be obtained without the need for large amounts of pulse compression or very high peak powers.

Detection range, of course, increases with the ratio of the signal energy to the energy of the background noise and clutter. By employing a high duty factor, high PRF waveform, therefore, long detection ranges can be obtained against nose-aspect targets even in a clutter environment. However, where strong sidelobe clutter is encountered, detection ranges against low-closing-rate (tail-aspect) targets may be impaired because of range ambiguities.

In this chapter, we will consider a high duty factor, high PRF waveform, see what must be done to separate targets from ground return, and learn how the signal processing is done. We'll then take up the problem of range measurement, eclipsing loss, and the steps which may be taken to improve performance against low-closing-rate targets.

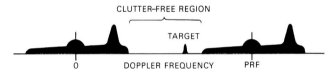

1. High PRF operation spreads the clutter bands far enough apart to open up a clutter-free region in which high closing rate targets will appear.

1. Targets whose closing rates are greater than the radar's ground speed.

369

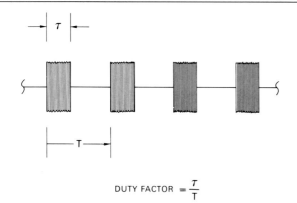

$$\text{DUTY FACTOR} = \frac{\tau}{T}$$

2. High duty factor, high PRF waveform typical of those used in radars for fighter aircraft. Duty factor is generally somewhat less than 50 percent.

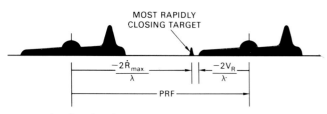

3. For the doppler clear region to encompass all high-closing-rate targets, the PRF must exceed the doppler frequency of the most rapidly closing target, plus the maximum sidelobe clutter frequency.

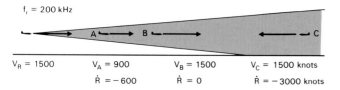

4. Representative flight situation. Target A has a low-closing-rate; target B, zero-closing-rate; target C, high-closing-rates.

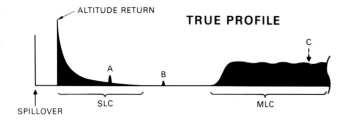

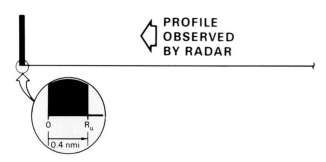

5. Range profile for the representative flight situation. Returns from virtually all ranges are collapsed into a band of observed ranges less than half a mile wide.

High PRF Waveform

A representative high duty factor, high PRF waveform is shown in Fig. 2. Because the radar receiver must be blanked during transmission and the duplexer has a finite recovery time, the maximum useful transmitter duty factor is generally somewhat less than 50 percent.

As for the PRF, if the clutter-free ("doppler clear") region is to encompass all significant high-closing-rate targets, the PRF must be greater than the sum of the

- Doppler frequency of the most rapidly closing target

- Maximum sidelobe clutter frequency (determined by the radar's velocity)

The maximum sidelobe clutter frequency is twice the radar velocity divided by the wavelength. The target doppler frequency is twice the target closing rate divided by the wavelength (Fig. 3).

The shorter the wavelength, of course, the higher the doppler frequencies of the clutter and the target will be; hence, the higher the required PRF. Typically, in fighter applications at X-band frequencies the PRF is on the order of 100 to 300 kilohertz.

Isolating the Target Returns

To get a clear picture of the problem of differentiating between target echoes and ground clutter and of isolating the echoes from the clutter and as much of the background noise as possible, we'll look at the range and doppler profiles for the representative flight situation shown in Fig. 4.

We will assume this time that the radar is operating at a PRF of 200 kilohertz with a duty factor of 45 percent, that the aircraft carrying the radar has a ground speed of 1500 knots, and that echoes are being received from three targets. Targets A and B are flying in the same direction as the radar. Target A has a closing rate of 600 knots; Target B has zero closing rate. The third target, C, is approaching the radar from long range and has a closing rate of 3000 knots.

Range Profile. This profile is illustrated in Fig. 5. Its width as observed by the radar (R_u) is less than half a nautical mile ($80 \div 200 = 0.4$ nmi). Into this narrow interval is collapsed (telescoped) the return from every 0.4 mile increment of range out to the maximum range from which return is received: all of the mainlobe clutter, the altitude return, all of the remaining sidelobe clutter, plus the transmitter spillover and the background noise. Buried in the midst of this pileup are the echoes from the three targets. Virtually the only way they can be separated from it, or from each other for that matter, is to sort out the return by doppler frequency.

Doppler Profile. This profile is shown in Fig. 6. As with low and medium PRF operation, the profile is a composite of the entire true doppler spectrum of the radar return, repeated at intervals equal to the PRF. But there is one important difference. Since the width of the spectrum is less than the PRF, the repetitions (bands) do not overlap.

Also, in bands on either side of the central one, the amplitude of the radar return is noticeably reduced. With a duty factor of 0.45, the nulls in the envelope within which the spectral lines fit are only $2.2f_r$ above and below the central line. So, for each component of the return, there are only two spectral lines between the central one and the nulls on either side. The amplitudes of these lines are considerably less than that of the central line.

Examining the central band closely, Fig. 7, we can clearly identify the following features: transmitter spillover, altitude return, sidelobe clutter, and mainlobe clutter. The width of the sidelobe clutter region varies with the radar velocity. The width and frequency of the mainlobe clutter line vary continuously with the antenna look angle as well as with the radar velocity.

Barely poking up above the sidelobe clutter are the echoes from the low-closing-rate target, Target A. In the clear between the high frequency end of the sidelobe clutter region and the low frequency end of the next higher band are the echoes from the high closing rate target, Target C. Provided the other clutter is removed, this target need only compete with thermal noise to be detected.

The echoes from the zero-closing-rate target (Target B) are nowhere to be seen. Actually, they are there, But they have merged with the combined altitude return and transmitter spillover, which has zero doppler frequency.

Rejecting the Strong Clutter. One needn't contemplate Fig. 7 very long to conclude that a logical first step in isolating the target return is rejecting the spillover and strong ground return—mainlobe clutter (MLC) and altitude return. In fact, this is essential. Why?

Where little or no range discrimination is provided, this return may be as much as 60 dB stronger than a target's echoes (Fig. 8). Sixty dB, remember, is a power ratio of 1,000,000 to 1; a doppler filter bank alone simply cannot cope with such strong clutter. Even though the clutter may be widely separated from a target's frequency, the attenuation that a doppler filter provides outside its passband is insufficient to keep the clutter from drowning out the target, albeit centered in the passband.

If we wish to search for targets in both sidelobe clutter and doppler clear regions, the spillover and altitude return, which have essentially zero doppler frequency, must be

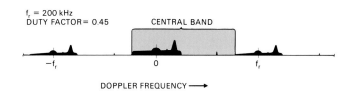

6. Doppler profile for the representative flight situation. Repetitions of the true profile do not overlap.

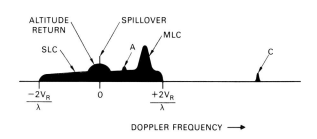

7. Central band of the doppler profile.

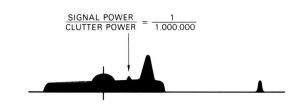

8. Power of mainlobe clutter and altitude return may be 60 dB stronger than that of a target's echoes.

HIGH—CLOSING—RATE TARGET

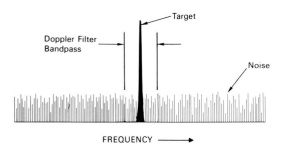

9. Doppler filter isolates the high-closing-rate target from all other returns and all but the immediately surrounding noise.

LOW—CLOSING—RATE TARGET

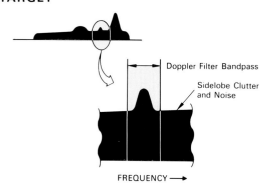

10. Low-closing-rate target must compete with immediately surrounding sidelobe clutter, much of which may come from a far closer angle than the target's.

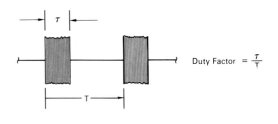

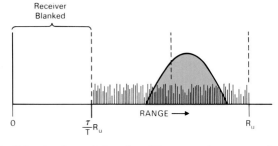

11. If the duty factor is less than 50 percent, the amount of noise or clutter with which a target must compete may be reduced by providing more than one range gate.

rejected separately from the mainlobe clutter, which has a widely varying frequency.

Doppler Resolution. Once the strong clutter has been removed, the target echoes can be isolated through doppler filtering—just as in medium PRF operation. The role of the doppler filters, though, is slightly different for high-closing-rate targets than for low.

In the case of high-closing-rate targets—those having closing rates greater than the radar's ground speed—the doppler filters serve three basic functions. First, they separate the target echoes from all of the remaining sidelobe clutter, as well as from the residual mainlobe clutter. Second, by reducing the spectral width of the background noise accompanying the echoes of any one target, the filters reduce the amount of noise with which the echoes must compete (Fig. 9). Third, they isolate the echoes received from different targets—provided they have sufficiently different doppler frequencies. It is this noise that ultimately limits the maximum range at which high-closing-rate targets may be detected. The more the noise is reduced, the greater the detection range will be.

In the case of low-closing-rate targets, the doppler filters perform the same target isolation function. But they cannot completely separate a target's echoes from the sidelobe clutter (Fig. 10). For some of this clutter has the same doppler frequency as the target. As with both high- and low-closing-rate targets in medium PRF operation, it is generally this clutter which ultimately limits the range at which low-closing-rate targets may be detected.

Because of the more severe range ambiguities, however, the competing clutter is much stronger than that encountered under the same conditions in medium PRF operation. For this reason, when high PRFs are used in situations where appreciable sidelobe clutter is received, detection ranges against low-closing-rate targets are degraded.

Range Gating. With duty factors approaching 50 percent, there is little or no possibility of isolating the return from different ranges with range gates. Receiver blanking, in fact, serves the function of a single range gate; none other need be provided.

However, if the duty factor is much less than 50 percent, i.e., if the interpulse period is much more than twice the pulse width, the opportunity for employing additional range gating arises (Fig. 11). By providing more than one range gate, the amount of noise—or sidelobe clutter—with which a target must compete may be reduced, and the loss in signal-to-noise (or clutter) ratio due to targets not being centered in the gate may be cut. The lower the duty factor, the greater the improvement that may be realized by adding range gates. In fact, if the average transmitted power is held constant (by increasing the peak power), the maximum detection range can be significantly increased by backing off from a 50 percent duty factor and then range gating. The reduction in duty factor reduces the eclipsing loss, and the range gates reduce the competing noise or clutter.

Adding range gates, however, is costly. Since range gating must precede doppler filtering, the entire doppler filter bank and all subsequent signal processing must be duplicated for every range gate that is provided (Fig. 12). Increasing the number of range gates from one to just two, for instance, entails forming twice as many doppler filters, detecting the magnitudes of twice as many filter outputs, setting twice as many detection thresholds, and so on. Moreover, where the value of the PRF is very high (as in most fighter applications), a great many more doppler filters are required to provide the same doppler resolution for high PRF operation as for medium. The cost of range gating, therefore, can be much higher in a high PRF radar than in a medium PRF radar.

Mechanization

Just how the signal processing functions outlined in the preceding paragraphs are actually performed varies widely from one radar to another. The mechanization, for instance, may be either analog or digital. Or, to reduce the dynamic range required of the analog-to-digital converter, the initial filtering may be analog and only the final doppler filtering, digital. Also, in some cases the processor may be designed to look for targets in both sidelobe clutter and doppler clear regions; in others, only in the doppler clear region.

Analog Mechanization. In analog processors, the IF output of the receiver is applied at the outset to a bandpass filter which passes only the central band of doppler frequencies.[2] The pulsed return is thereby converted to a continuous wave signal (Fig. 13), and all subsequent processing is handled as in a CW radar.

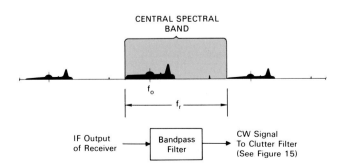

13. For analog processing, the IF output of the receiver is fed to a filter which passes only the central spectral band, thereby converting the return to a CW signal.

SIGNAL PROCESSING FOR THREE RANGE GATES

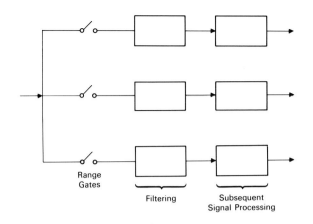

12. For every range gate that is provided, all filtering and subsequent signal processing must be duplicated.

SOME OPTIONS FOR SIGNAL PROCESSING

Filtering		Regions Processed	
Initial	Remainder	Sidelobe Clutter	Doppler Clear
Analog	Analog	X	X
Analog	Digital	X	X
Analog	Digital		X
Digital	Digital		X

2. The central band is selected because it contains the most power. Loss of signal power in the other bands is no problem since the noise and clutter they contain is rejected too.

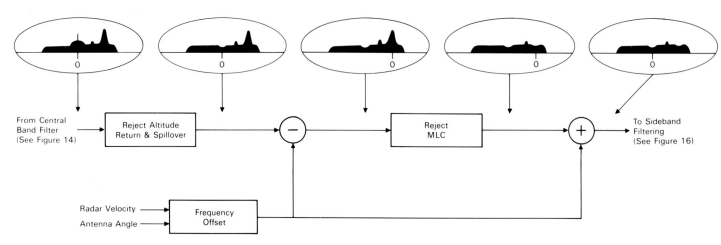

14. First steps in signal processing: (1) reject altitude return and transmitter spillover; (2) offset spectrum to track mainlobe clutter; (3) reject mainlobe clutter; (4) remove frequency offset.

The output of the central band filter (Fig. 14, above) is passed through a filter which removes the clutter having zero doppler frequency —transmitter spillover and altitude return. Along with them, the echoes of any zero-closing-rate targets will unavoidably be rejected.

The entire doppler spectrum is then shifted in frequency so as to center the mainlobe clutter in the rejection notch of a second rejection filter. As with low and medium PRF operation, this shift must be dynamically controlled to match the changes in clutter frequency due to changes in radar velocity and antenna look angle. Once the mainlobe clutter has been removed, the same frequency offset may be applied in reverse to center the doppler spectrum again at a fixed frequency.

Up to this point, processing is normally done at very low signal levels. To keep the stronger return in the target regions from saturating subsequent stages, its relative amplitude is now reduced through automatic gain control (AGC). For this, the doppler spectrum is commonly broken into contiguous subbands (Fig. 15). One or more subbands may span the sidelobe clutter region, and one or more may span the doppler clear region. By applying the AGC separately in each subband, strong return (or jamming) in one band is prevented from desensitizing the other bands.

Signal levels equalized, the subbands are, in effect, recombined and applied to a single, long bank of doppler filters (Fig. 16).

At the very end of every filter integration time, t_{int}, the amplitude of the signal that has built up in each filter is detected. If t_{int} is less than the time-on-target for the radar antenna, the detector outputs for the complete time-on-target are added up in postdetection integrators. Finally, the integrated output of each doppler filter is supplied to a separate threshold detector.

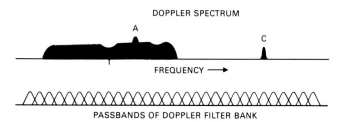

15. To avoid desensitization by strong signals, return is segregated into subbands for the application of automatic gain control, AGC.

DOPPLER SPECTRUM

FREQUENCY ⟶

PASSBANDS OF DOPPLER FILTER BANK

16. After the altitude return, transmitter spillover, and mainlobe clutter are filtered out, the radar return is applied to a bank of doppler filters.

Digital Mechanization. In digital signal processors, as with medium PRF operation, the IF output of the receiver is applied to an I/Q detector (Fig. 17). Its outputs (video) are then sampled at intervals equal to the (compressed) pulse width by an analog-to-digital converter. To prevent saturation, digital automatic gain control (DAGC) is applied to amplifiers ahead of the converter.

In contrast to the output of an analog central band filter, the samples taken by the A/D converter represent the power in *all* of the doppler bands passed by the receiver's IF amplifier. However, since the power includes both signal and noise plus clutter, the signal-to-noise ratio is essentially the same as when central band processing is employed.

Following analog-to-digital conversion, all of the same steps may be performed as in analog processing. The only difference is that they are performed by digital rather than analog filters.

Ranging

Because of the difficulty of pulse-delay ranging at high PRFs, FM ranging is generally employed. The accuracy of FM ranging is proportional to the ratio of the frequency resolution of the doppler filter bank to the rate of change of the transmitter frequency, $\dot{f}$ (Fig. 18). The finer the frequency resolution and the greater $\dot{f}$, the more accurately the ranging time can be measured. Frequency resolution roughly equals the 3-dB bandwidth of the doppler filters, so

$$\text{Range accuracy} \approx \frac{BW_{3\,dB}}{\dot{f}}$$

To illustrate, let's say that the 3-dB bandwidth of the doppler filters is 100 hertz and the rate of change of the transmitter frequency $\dot{f}$ = 3MHz per second. The accuracy with which the ranging time t_r can be measured then is $100 \div (3 \times 10^6)$ = 33 μs. At 12.4 microseconds per nautical mile of range, this corresponds to an accuracy of about 2.7 miles.

One might suppose that virtually any degree of range accuracy could be obtained simply by narrowing the filters and/or increasing $\dot{f}$. But there are practical limits on both.

Filter bandwidth is limited by the integration time ($BW_{3\,dB} \cong 1/t_{int}$). Furthermore, with three slope modulation, the maximum integration time is less than 1/3 of the time-on-target (Fig. 19). The minimum possible filter bandwidth, therefore, is roughly $3/t_{ot}$ hertz.

Not so apparent, $\dot{f}$ is limited by the spreading of the clutter spectrum. The spreading is due to the clutter being received from a wide span of ranges. To keep spreading within acceptable bounds, the maximum shift in the frequency of the radar return due to range must be no more than a small fraction of the maximum doppler shift of the clutter.

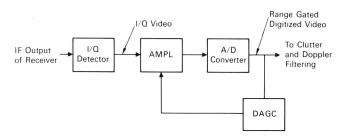

17. For digital processing, output of the I/Q detector is sampled at intervals equal to the (compressed) pulse width and digitized. Digital AGC prevents saturation of the A/D converter.

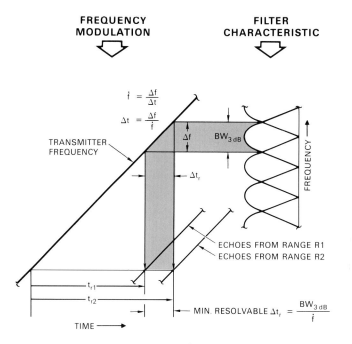

18. Slope of the modulation curve ($\dot{f}$) and bandwidth of the filters ($BW_{3\,dB}$) determine the minimum resolvable difference in ranging time (Δt_r).

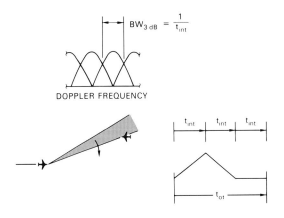

19. Filter bandwidth is inversely proportional to the integration time. With three-slope ranging, the integration time is less than 1/3 of the time on target.

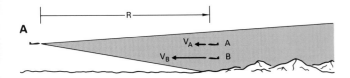

a) Two targets approaching from long range. One has much higher closing rate than the other.

b) Doppler profile, without FM ranging—no clutter spreading.

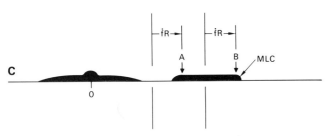

c) Doppler profile, with FM ranging and *large* value of ḟ. Mainlobe clutter shifts out in frequency and spreads over doppler clear region, obscuring targets.

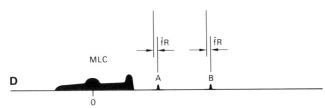

d) Doppler profile, with FM ranging and *small* value of ḟ. Clutter spreads only moderately, leaving targets in the clear.

20. How clutter spreading limits the rate (ḟ) at which the transmitter frequency may be changed for FM ranging.

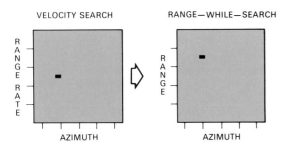

21. For maximum detection range, a velocity-search mode may be provided. Once the target is detected, the operator may switch to the range-while-search mode.

The reason is illustrated (for the descending slope of the modulation cycle) in Fig. 20. It shows a radar detecting two long range targets. One has a high doppler frequency. The other, a doppler frequency only slightly higher than the highest clutter frequency. The antenna's mainlobe illuminates the ground for a very great distance.

Beneath the situation diagram are three frequency profiles. The first, is a plot of the doppler frequencies of the targets and the ground return.

The second profile shows what happens if the frequency shift corresponding to the targets' range is comparable to the doppler shift of the higher closing rate target. As you can see, the mainlobe clutter not only shifts into the normally clutter-free region, but spreads to the point where it blankets both targets.

In the third profile, the frequency shift corresponding to range is a small fraction of the doppler shift. Although the clutter still spreads, it does not spread enough to interfere with target detection.

It turns out that for a typical fighter application the constraints on minimum filter bandwidth and maximum rate of change of transmitter frequency are such that the range accuracy is on the order of a few miles. This is substantially poorer than can be obtained with pulse-delay ranging.

While range information is always highly desirable, it is not essential when searching for targets at extremely long ranges. What is most important then is detecting targets and knowing their direction. Determining whether a target in a given direction is 100 or 150 miles away can come later.

Because a target must be detected on all three slopes of the modulation cycle to be detected at all and the integration time per slope is only one-third of what it would be without FM ranging, the price one pays for range measurement is a reduction in detection range.

Accordingly, for situations where extremely long detection ranges are desired, a special mode may be provided in which range is simply not measured. In this mode, called *velocity search* or *pulse-doppler search*, targets are displayed in range rate versus azimuth (Fig. 21). Once a target is detected, the operator can switch to the more standard range-while-search mode, in which range is measured by FM ranging and the targets are presented on a range versus azimuth display.

Problem of Eclipsing

When operating at very high duty factors, a considerable amount of target return is lost as a result of eclipsing—i.e., echoes being received in part or in whole when the radar is transmitting and the receiver is blanked.

22. YF-12A of the early 1960s. First airplane capable of sustained Mach-3 flight; first interceptor to be equipped with a high PRF pulse-doppler radar.

Eclipsing, however, is not always as severe a problem as it might at first seem. A target is totally eclipsed only when its range is such that the period during which its echoes are received exactly coincides with the period during which the receiver is blanked. Otherwise, at least a portion of the return gets through (Fig. 23). As the degree of coincidence is reduced, so is the eclipsing loss.

Even so, eclipsing reduces the signal-to-noise ratio sufficiently to leave periodic holes of appreciable size in the radar's range coverage. Fortunately, when searching for targets approaching from very long ranges, one is concerned mainly with the cumulative probability of detection—the probability that a target will be seen at least once before it has approached to within a given range. Moreover, once a target has been detected, it generally need not be detected continuously. A rapidly approaching target will not remain at an eclipsed range very long. And as the range decreases and the signal strength increases, the gaps in range coverage tend to fill in (Fig. 24).

In applications where closing rates may be comparatively low and/or more nearly continuous detection is required, the length of time any one range remains eclipsed may be reduced by switching the PRF among different values, much as it is switched to eliminate blind zones in medium

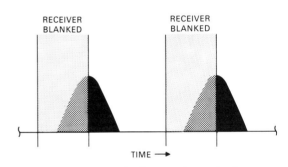

23. As long as the received echoes and periods of receiver "blanking" do not coincide exactly, a portion of the return will get through.

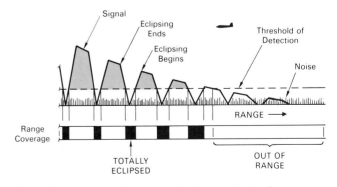

24. Reduction in signal-to-noise ratio due to eclipsing for an approaching target. As range decreases, gaps in range coverage grow narrower.

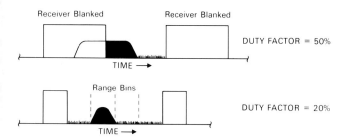

25. Eclipsing loss may be reduced by reducing the duty factor and employing multiple range gates.

PRF operation. In single-target tracking, by periodically changing the PRF at appropriate times, a target can be kept largely in the clear. At high duty factors, though, the holes in range coverage are not easily eliminated, particularly at short ranges. Furthermore, PRF switching introduces losses which reduce the maximum detection range in the doppler clear region.

Eclipsing may also be reduced by lowering the duty factor. This, of course, will reduce the average transmitted power. But that reduction can be compensated for by using multiple range gates. Suppose, for example, that the duty factor is reduced from 50 percent to 20 percent and four range gates are provided (Fig. 25). If the peak transmitted power remained the same, the average power hence the total received energy would be *decreased* by a factor of 0.2 ÷ 0.5 = 0.4. But as can be seen from the figure, with four range gates the noise energy with which the signal would have to compete at any one time would be reduced by the same factor, so these two effects would cancel. For a continuously closing target, then, the signal-to-noise ratio would increase in direct proportion to the increase in the fraction of the time the receiver is not blanked. In this case, the increase would be on the order of 0.5 ÷ 0.2 = 2.5.

Thus, by reducing the duty factor somewhat and providing multiple range gates, not only may the detection range be increased, but the holes in range coverage due to eclipsing may be correspondingly narrowed. As noted earlier, though, providing multiple range gates substantially increases the cost of implementation.

Improving Tail Aspect Performance

Several approaches may be taken to improving performance against low-closing-rate targets in severe clutter. Since the root of the problem is sidelobe clutter, a logical first step is to minimize the antenna sidelobes.

For a given sidelobe level, the amount of sidelobe return with which a low-closing-rate target must compete may be further reduced by narrowing the passbands of the doppler filters (Fig. 26). This, of course, entails adding more filters and, as noted earlier, there are practical limits on how narrow the passbands can be made.

At the expense of still greater complexity and a lower duty factor, the competing return may be still further reduced by narrowing the pulses and employing more range gates.[3] Even then, because of the transmitter spillover and altitude return, the radar will be blind to zero closing rate targets—those being pursued at constant range.

A particularly attractive solution to the problem is to employ high PRFs when long detection range against nose hemisphere targets is essential and to interleave high and

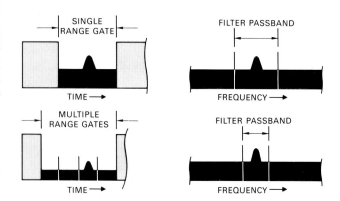

26. Signal-to-clutter ratio, hence performance against low closing rate targets, may be improved by reducing the duty factor and providing multiple range gates or by narrowing the passband of the doppler filters.

3. For a given duty factor, hence degree of eclipsing, the number of range gates may be increased still further, without loss of signal, through pulse compression.

medium PRFs when long detection ranges are required against both nose and tail hemisphere targets.

An effective way of accomplishing this is illustrated in Fig. 27. High and medium PRF modes are employed on alternate bars of the antenna scan. The bars assigned to the high PRF mode in one frame are assigned to the medium PRF mode in the next frame, and vice versa. Since adjacent bars overlap, virtually complete solid-angle coverage is achieved in both modes. Rapidly approaching targets which are beyond reach of the medium PRF mode are detected in the high PRF mode. Low-closing-rate targets, as well as any shorter range targets which may be eclipsed in the high PRF mode, are detected in the medium PRF mode.

When PRF interleaving is used, the complexity of the signal processor for the high PRF mode can be substantially reduced by processing only the return falling in the doppler clear region (Fig. 28). This return, of course, must first be

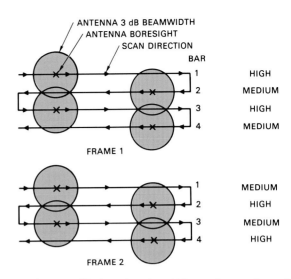

27. Interleaving of high and medium PRFs on alternate bars of the search scan to achieve maximum detection range in both nose hemisphere aspects and tail chases.

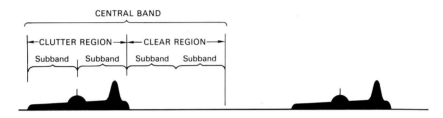

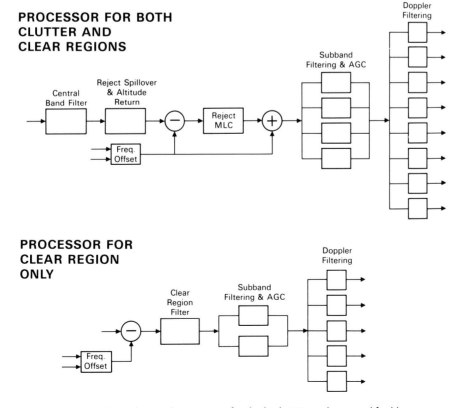

28. When medium and high PRFs are interleaved, signal processing for the high PRF mode is simplified by processing only the return in the clutter-free region.

ILLUMINATING TARGETS FOR SEMIACTIVE MISSILE GUIDANCE

If the pulsed transmission of a high PRF radar is used to illuminate targets for semiactive missiles to home on, both the PRF and the duty factor may be made somewhat higher than they would be for normal searching and tracking.

Increased PRF. As explained in detail in Chapter 15, because of the high velocity of the missile relative to the launch aircraft, a target's doppler frequency is generally much higher as seen by the missile than as seen by the radar in the launch aircraft. To ensure that the velocity data obtained by the missile is unambiguous, in computing the minimum acceptable PRF one must add to the maximum target closing rate half the velocity of the missile relative to the radar.

Increased Duty Factor. Since for a given peak transmitted power, detection range increases with duty factor, in the case of a semiactive missile which must be launched at long ranges it is desirable to make the duty factor as high as possible. Although the maximum useful duty factor which a radar may employ for detecting and tracking targets is limited by the eclipsing loss to somewhat less than 50 percent, when the radar's pulsed transmission is used to illuminate a target for a missile, this limitation does not necessarily hold. Because the missile is remote from the radar, blanking is not needed to keep transmitter noise from leaking directly into the missile receiver when the radar is transmitting. Consequently, if the missile's seeker must be capable of long detection range, the duty factor may be made considerably higher than 50 percent. The maximum acceptable duty factor is, of course, limited by eclipsing losses in the radar.

What about the directly received signal from the radar? Because of the doppler shift, the frequency of the radar's transmitted signal is sufficiently different from the frequency of the target echoes that the seeker in the missile can separate the two. However, care must be taken in the design of the radar to minimize the radiation of transmitter noise, some of which invariably has the same frequency as a target's echoes.

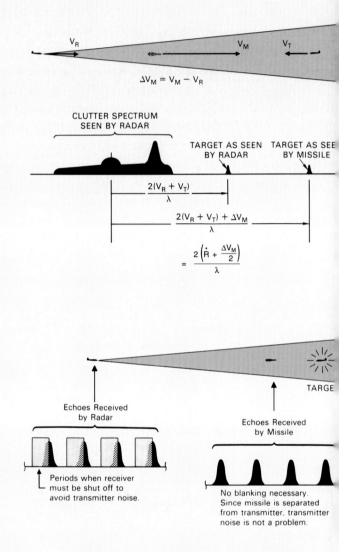

isolated from the clutter—mainlobe, sidelobe, and altitude return. But that can readily be accomplished by passing the receiver output through one or more broad bandpass filters. After performing automatic gain control, their outputs are supplied to a suitably long bank of doppler filters.

Summary

To maximize detection range, high PRF waveforms for fighter applications usually have duty factors approaching 50 percent. Even higher duty factors may be used when illuminating targets for long range semiactive missiles.

To provide an adequate doppler clear region, the PRF must at least equal the maximum sidelobe clutter frequency plus the doppler frequency of the highest closing rate target. For semiactive missile guidance an allowance must be added for the velocity of the missile relative to the radar.

In analog signal processing, the receiver output is usually converted at the outset to a CW signal by a central band filter. Next, successive filters reject the combined spillover and altitude return and the mainlobe clutter. The remaining return may then be divided into subbands for the application of AGC. The return is then applied to a bank of doppler filters. If a target's closing rate is greater than the radar's ground speed, the target echoes compete only with the noise passed by the same doppler filter that passes the echoes. But if the closing rate is less than this, they must compete with sidelobe clutter passed by the filter, much of which may come from comparatively close range.

In many high duty factor fighter radars, the only range gating is that provided by receiver blanking. At the cost of increased complexity, noise and sidelobe clutter may be reduced by using multiple range gates.

Range must generally be measured with FM ranging. Because the rate at which the transmitter frequency may be changed is limited by the spreading of the clutter spectrum, range accuracy is poor. Since range measurement reduces detection range, where maximum detection range is desired, a special velocity search mode may be provided in which range is not measured.

When operating at high duty factors, eclipsing losses are significant. They may be minimized by switching PRFs, and/or providing multiple range gates. PRF switching, though, reduces detection range in the doppler clear region, and employing multiple range gates increases the cost of implementation.

Performance against low closing rate targets may be improved by providing a low sidelobe antenna, greater doppler resolution, and multiple range gates. One attractive approach is to interleave high and medium PRF operation on alternate bars of the antenna scan.

HIGH PRFs

ADVANTAGES	LIMITATIONS
1. Good nose-aspect capability—high-closing-rate targets appear in clutter-free region of spectrum.	1. Detection range against low-closing-rate targets may be degraded by sidelobe clutter.
2. High average power can be provided by increasing PRF. (Only moderate amounts of pulse compression, if any, are needed to maximize average power.)	2. Precludes use of simple, accurate pulse delay ranging.
3. Mainlobe clutter can be rejected without also rejecting taget echoes.	3. Zero-closing rate targets may be rejected with altitude return and transmitter spillover.

GRUMMAN A-6E INTRUDER (1972)

The original A-6A became operational with the USN in 1965 and was used successfully in all-weather attack missions during the Vietnam conflict. This later version was equipped with the Norden APQ-156 multimode attack radar.

Automatic Tracking 29

In the preceding chapters, we became acquainted with various approaches to target detection. In this chapter, we'll take a closer look at the techniques for tracking the targets that are detected: the single-target track (STT) and track-while-scan (TWS) modes introduced in Chap. 2.

Single-Target Tracking

The goal of single-target tracking is to continuously and accurately provide current data on a given target's position, velocity, and acceleration—all of which may be continuously changing. Toward that end, separate semi-independent tracking loops are typically established for range, range rate (or doppler frequency), and angle.

Functions Included in a Tracking Loop. Each tracking loop includes four basic functions: measurement, filtering, control, and response (Fig. 1).

Measurement is the determination of the difference between the actual value of the parameter (e.g., the target's range) and the radar's current knowledge of the parameter: in short, the tracking error.

Filtering is the processing of successive measurements to minimize the random variations (noise) due to target scintillation, thermal agitation, and other corrupting interference. Needless to say, tracking accuracy depends critically on how effectively filtering is done. A tracking filter may be thought of as a low-pass filter (Fig. 2) whose key parameters—cut-off frequency, gain, etc.—are constantly adjusted in light of the signal-to-noise ratio, the target's potential maneuvers, and the radar-bearing aircraft's actual maneu-

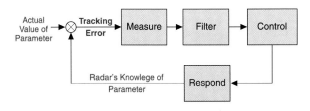

1. Basic functions performed by a single-target tracking loop.

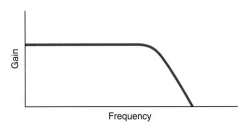

2. A tracking filter may be thought of as a low-pass filter whose gain and cut-off frequency are adjusted to eliminate as much noise as possible without introducing excessive lag.

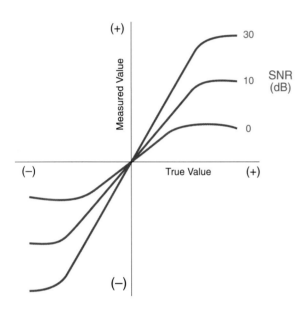

A tracking discriminant may be represented by a normalized plot of the measured value of the tracking error versus the true value. The steeper the linear portion of the discriminant, the more sensitive the measurement.

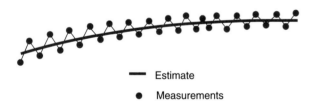

— Estimate

• Measurements

4.   The value of any parameter that can only be determined on the basis of successive measurements which are corrupted by noise or other interference is termed an estimate.

vers, to eliminate as much noise as possible without introducing excessive lag.

Control is the generation of a command calculated on the basis of the filter's outputs to reduce the tracking error as nearly as possible to zero.

Response is the response of the hardware and/or software to which the command is given. The difference between the response and the current actual value of the parameter feeds back to the input, closing the loop, and the entire process repeats. Through successive iterations, the parameter may be tracked with extreme precision.

Special Terminology. Before proceeding further, it will be well to introduce two important technical terms used by tracking-loop designers: *discriminant* and *estimate*.

Discriminant is the term for the calibration of the measurement function. It is commonly represented by a plot of the output of the hardware and/or software that performs the measurement versus the true value of the tracking error (Fig. 3). The slope of the linear portion of the plot determines the sensitivity of the measurement. Typically, the slope increases as signal-to-noise ratio increases.

An important feature of discriminants is that they are dimensionless (normalized). Consequently, precise measurement of voltage or power levels isn't required. Moreover, except for the influence of signal-to-noise ratio, the measured values of the tracking error don't vary with signal strength. They are independent of the target's size, its range, its maneuvers, and fluctuations of its RCS. If desired, though, a discriminant can be given a dimension simply by multiplying it by a precomputed constant.

Estimate is the term applied to the value of any parameter that is

(a)   measurable only in combination with corrupting interference—e.g., thermal noise (Fig. 4), or

(b)   not directly measurable, e.g., range rate based on a sequence of range measurements.

According to this definition, virtually every parameter measured or computed by a radar, no matter how precisely, is an estimate.

With these definitions in mind, let us take a quick look at the angle-tracking loops commonly incorporated in single-target tracking modes.

Range-Tracking Loop. This loop has two primary goals: to continuously and accurately determine the target's current range, and to keep a range-gate—actually two adjacent sampling times—centered on the target's echoes to isolate them for doppler and angle tracking.

To facilitate forming the range discriminant, the video output of the receiver is passed through a low-pass filter, stretching the target's echoes to roughly twice the radar's pulse width and giving them a more "rounded" shape.[1] Assuming that the video is sampled at intervals equal to the pulse width, this results in two samples being taken of each target echo and in the amplitudes of the samples differing in proportion to the displacement of the range gate from the center of the echo (Fig. 5). Because successive samples are stored in separate range bins, the first sample is called the *early range bin*; the second, the *late range bin*.

The goal being to keep the range gate centered on the target echoes, the range discriminant is formed by measuring the difference between the amplitudes of the two samples: $R_L - R_E$. The measurement is normalized by dividing it by the sum of the amplitudes (Fig. 6).

RANGE DISCRIMINANT, ΔR

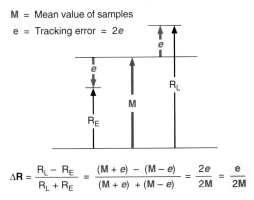

M = Mean value of samples

e = Tracking error = 2e

$$\Delta R = \frac{R_L - R_E}{R_L + R_E} = \frac{(M + e) - (M - e)}{(M + e) + (M - e)} = \frac{2e}{2M} = \frac{e}{2M}$$

6. Range-tracking error is proportional to the difference between the magnitudes of the samples stored in the early and late range bins. Dividing by their sum yields a nondimensional ratio of the error to twice the mean of the samples.

On the basis of the range discriminant and the previous range-gate command, the range filter produces best estimates of the target's range and range rate, a measure of the range acceleration, and a new range-gate command (Fig. 7).

The range-gate command is essentially a prediction of what the target's range will be when the next target echo is sampled. Typically, the command is formed by taking the filter's latest estimates of the target's range and range rate and linearly extrapolating the range.

To carry out the range-gate command, the predicted target range is first corrected for radar peculiarities (such as sampling-time granularity) and distortion of the pulse-shape in going through the receiver and pulse-stretching low-pass filter. The prediction is then converted into units of time measured from the trailing edge of the immediately preceding transmitted pulse, hence to the estimated arrival time of the next echo (Fig. 8).

R_E = Magnitude of sample in early range bin

R_L = Magnitude of sample in late range bin

e = Tracking error

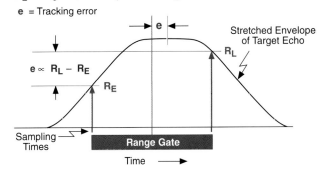

5. The range gate is centered between two adjacent sampling times. To track a target in range, the sampling times must be shifted to center the range gate on the target's echoes. The tracking error, e, is proportional to the difference between the early and late samples, R_E and R_L.

1. The filter removes the pulse's higher frequency components, which contribute to the sharpness of its leading and trailing edges.

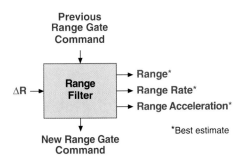

7. Inputs and outputs of the range filter. ΔR is the range discriminant.

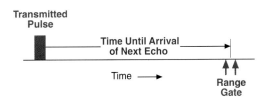

8. Positioning of the range gate in response to range-gate command. For this, the predicted range is converted to time.

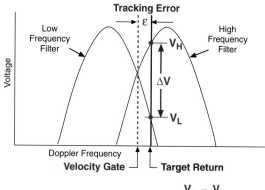

Tracking Error

Low Frequency Filter

High Frequency Filter

Voltage

V_H

ΔV

V_L

Doppler Frequency

Velocity Gate

Target Return

$$\text{Velocity Discriminant} = \frac{V_H - V_L}{V_H + V_L}$$

9. The simplest velocity gate is the intersection of two adjacent doppler filters. The velocity discriminant is the difference between the output voltage the target return produces from the two filters divided by the sum of the two voltages.

2. Two separate banks of filters are formed by integrating the samples collected in the early and late range bins. The velocity gate may be formed in either or both of them.

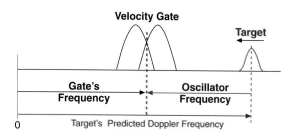

Velocity Gate

Target

Gate's Frequency

Oscillator Frequency

0   Target's Predicted Doppler Frequency

10. When the oscillator has moved the target into the gate, the sum of the oscillator's frequency and the velocity gate's fixed frequency is the target's predicted doppler frequency.

3. If the PRF is less than the target's doppler frequency, some multiple, n, of the PRF must be added to this sum. See Chap. 21, page 286.

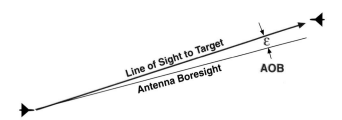

ε

Line of Sight to Target

Antenna Boresight

AOB

11. What the angle tracking loop measures is the angle, AOB, between the line of sight to the target and the antenna boresight line.

Doppler (Range-Rate) Tracking Loop. The purpose of this loop is two-fold: (a) to provide a directly measured, more accurate value of the target's range rate than is available from the range-tracking loop, and (b) to isolate the target's returns for angle tracking by keeping a so-called "velocity gate" centered on the target's doppler frequency.

The simplest velocity gate is the crossover point of two adjacent doppler filters,[2] called the low- and high-frequency filters. Any error in the alignment of the velocity gate, of course, shows up as a difference between the outputs of these filters. The discriminant is formed by taking the difference between the magnitudes of the outputs, $V_H - V_L$, and normalizing it by dividing by their sum (Fig. 9). The result is supplied to the velocity filter.

The functions of this filter almost exactly parallel those of the range filter. The velocity filter's outputs are simply more accurate estimates of the target's range rate and range acceleration.

Based on the velocity filter's most current range-rate and range acceleration estimates, a velocity-gate command is produced. It is essentially a prediction of what the target's doppler frequency will be when the next set of doppler filters is formed.

The command is applied to a variable-frequency RF oscillator. Its output is mixed with the received signal, thereby shifting its frequency so that the target's predicted doppler frequency will be centered in the velocity gate. The sum of the oscillator's frequency and the velocity gate's fixed frequency then is the target's predicted doppler frequency (Fig. 10).[3]

Angle-Tracking Loop. The role of this loop is to (a) accurately determine the target's direction (angle) relative to a chosen coordinate system, (b) determine the target's angle rate, and (c) keep the antenna boresight precisely trained on the target. Commonly used coordinate systems are defined in the panel on the facing page.

What the angle tracking loop measures is the angle between the antenna boresight and the line of sight to the target. This angle, ε, is called the *angle off boresight*, AOB (Fig. 11), and is generally resolved into azimuth and elevation coordinates.

Previous chapters introduced three techniques for sensing the AOB: sequential lobing, amplitude-comparison monopulse, and phase-comparison monopulse. Since they're basically quite similar, we'll consider only one here: amplitude-comparison monopulse. For it, you'll recall, during reception, the antenna's radiation pattern is split into two lobes which cross at their half power points.

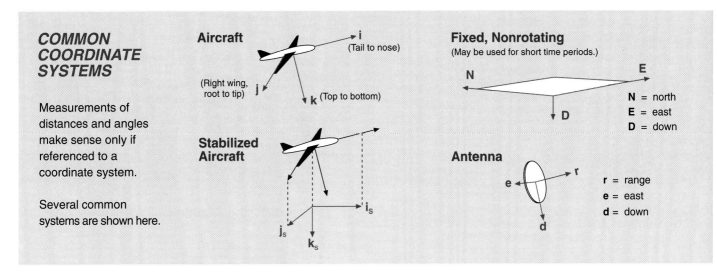

COMMON COORDINATE SYSTEMS

Measurements of distances and angles make sense only if referenced to a coordinate system.

Several common systems are shown here.

As can be seen from Fig. 12, the difference between the amplitude of the target's echoes as received through the left and right lobes, $V_L - V_R$, is roughly proportional to the AOB. Dividing this difference by the sum of the two amplitudes yields a dimensionless discriminant for the azimuth component of the AOB. A discriminant for the elevation component is similarly formed.

The measured components of the AOB are supplied to the angle-tracking filter along with the following environmental information:

- Signal-to-noise ratio

- Radar-bearing aircraft's velocity

- Target range and range rate

- Antenna's current angle rate

From these inputs, the filter produces best estimates of the azimuth and elevation components of the AOB, the angle rate of the line of sight to the the target, and the target's acceleration (Fig. 13).

To reduce the AOB and keep the antenna boresight trained on the target, azimuth and elevation rate commands are generated. Each of these is the algebraic sum of (a) the filter's best estimate of the respective line-of-sight rate and (b) a rate proportional to the filter's best estimate of the respective component of the AOB.

The rate commands are fed to the antenna stabilization system (Fig. 14). There they control the rate of precession of gyros that inertially establish azimuth and elevation axes in space to which the antenna is tightly slaved.

In the case of an electronically steered antenna, steering commands for both angle tracking and space stabilization must be provided. To continuously correct for changes in aircraft attitude, no matter how small, new commands must be computed and fed to the antenna at a very high rate.

ANGLE DISCRIMINANT

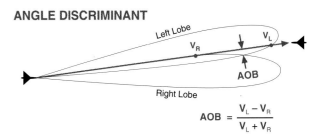

$$AOB = \frac{V_L - V_R}{V_L + V_R}$$

12. Angle-tracking discriminant for amplitude comparison monopulse. The antenna lobes cross on the boresight line; so the angle AOB is roughly proportional to the difference between the voltage of returns received through the two lobes.

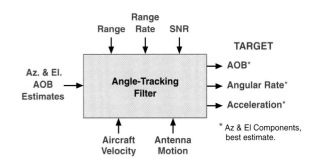

13. Inputs and outputs of the angle-tracking filter.

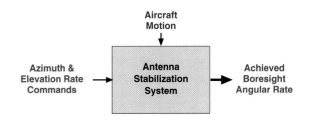

14. The antenna is stabilized against changes in aircraft attitude by slaving it to azimuth and elevation axes established by rate-integrating gyros mounted on it. The rate commands precess the gyros.

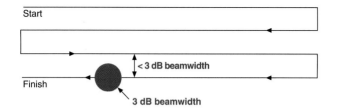

15. A representative four-bar raster scan. So that targets won't be missed, spacing of bars is less than the 3-dB beamwidth. Consequently, the same target may often be detected on more than one bar—one of several conflicts TWS resolves.

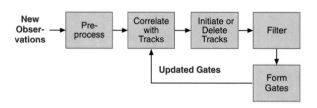

16. The five basic steps in track-while-scan processing.

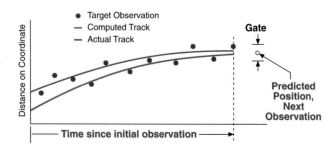

17. Representative track of one component (N, E, or D) of one of a target's parameters, illustrating its predicted value at the time of the next observation and the gate for correlating the observation with the track.

Track-While-Scan

Track-while-scan (TWS) is an elegant combination of searching and tracking. To search for targets, the radar repeatedly scans a raster of one or more bars (Fig. 15). Each scan is independent of all the others. Whenever a target is detected, the radar typically provides both the operator and the TWS function with estimates of the target's range, range rate (doppler), azimuth angle, and elevation angle. For any one detection the estimates are referred to collectively as an observation.

In pure search, the operator must decide whether targets detected on the current scan are the same as those detected on a previous scan or scans. With TWS, however, this decision must be made automatically. The algorithm used to make it is one of the most complex algorithms in the radar.

In the course of successive scans, TWS maintains an accurate track of the relative flight path of each valid target. This process is iteratively carried out in five basic steps: preprocessing, correlation, track initiation and deletion, filtering, and gate formation (Fig. 16).

Preprocessing. In this step, two important operations may be performed on each new observation. First, if a target having the same range, range rate, and angular position has been detected on a preceding, overlapping bar of the scan, the observations are combined. Second, if not already so referenced, each observation is translated to a fixed coordinate system, such as the NED. The angle estimates are conveniently formulated as direction cosines—cosines of the angles between the direction of the target and the N, E, and D axes. Range and range rate may be projected onto the N, E, and D axes simply by multiplying them by the respective direction cosines.

Correlation. The purpose of this step is to determine whether a new observation should be assigned to an existing track. On the basis of the observations assigned to the track thus far, tracking filters accurately extend the values of the N, E, and D components of each parameter of the track to the time of the current observation. The filters then predict what the values of these components will be at the time of the next observation.

On the basis of accuracy statistics derived by the filters, a gate scaled to the maximum error in measurement and prediction is placed around each component of the prediction for the track, as illustrated in Fig. 17. If the next observation falls within all of the gates for the track, the observation is assigned to the track.

Naturally, when closely spaced observations are received, conflicts in assignments are likely to occur. To facilitate their resolution, a statistical distance of each observation from the track or tracks involved is computed by normalizing and combining the differences between measurement and prediction for all components of the observation. Each track is centered in a gate, the radius of which corresponds to the maximum possible statistical distance between measurement and prediction.

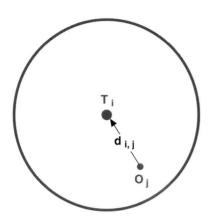

18. Gate for correlating an observation, O_j, with a track, T_i. Size of the gate corresponds to the maximum possible statistical distance, d, a valid observation may be from the track.

A representative conflict is illustrated in Fig. 19. Observation O_1 falls within the gates of two different tracks: T_1 and T_2. Observations O_2 and O_3 both fall within the gate of track T_2. Conflicts such as this are typically resolved as follows.

- Observation O_1 is assigned to track T_1 because it is the only observation within the gate of T_1, while T_2 has other observations, O_2 and O_3, within its gate.

- Observation O_2 is assigned to track T_2 because its distance, $d_{2,2}$, from the center of the gate is less than that of O_3.[4]

Track Creation or Deletion. When a new observation, such as O_4 in Fig. 19 does not fit in the gate of an existing track, a tentative new track is established. If, on the next scan (or possibly the next scan after that) a second observation correlates with this track, the track is confirmed. If not, the observation is assumed to have been a false alarm and is dropped. Similarly, if for a given number of scans no new observation correlates with an existing track, the track is deleted.

Filtering. This is similar to the filtering performed in single-target tracking. On the basis of the differences between the predictions and new measurements for each track, the track is updated, new predictions are made, and accuracy statistics for both observations and predictions are derived.

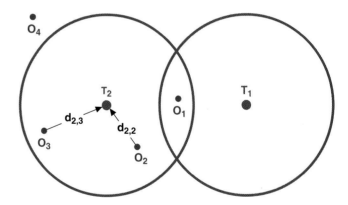

19. Typical conflicts arising when targets are closely spaced. Here, gates for tracks T_2 and T_1 overlap. Observation O_1 falls in both gates, and observations O_2 and O_3 both fall in the gate for track T_2.

4. A restriction applied in this case is that a tentative track cannot be initiated for an observation that falls within the gate of an existing track. Accordingly, because a competing observation is assigned to the track O_3 falls in, O_3 is discarded.

Gate Formation. From the prediction and accuracy statistics derived by the filter, new gates are formed and supplied to the correlation function.

As a result of the filtering, the longer a target is observed, the more accurately the new gates are positioned, and the closer the computed track comes to the actual track.

Summary

For single-target tracking, semi-independent tracking loops are generally provided for range, doppler frequency, azimuth, and elevation. Each loop includes four basic functions: measurement, filtering, control, and system response.

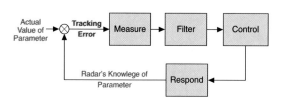

The range-tracking error is measured by taking the difference between early and late samples of the target echoes; the doppler-tracking error, by taking the difference between the outputs of two adjacent doppler filters; the angle-tracking errors, by taking the difference between the returns received through two antenna lobes.

The "scale factor" of each measurement, commonly represented by a plot of the measured value of the tracking error versus the true value is called a discriminant. So that the measurement will be largely independent of signal strength and precise measurement of voltages or powers won't be required, the discriminant is normalized.

R_E = Magnitude of sample in early range bin
R_L = Magnitude of sample in late range bin
e = Tracking error

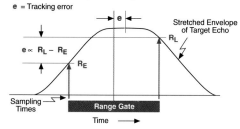

Successive measurements are, in effect, passed through a low-pass filter whose gain and cut-off frequency are constantly adjusted in light of the SNR, potential target maneuvers, and the aircraft's own maneuvers to eliminate as much noise as possible without introducing excessive lag.

From the filter outputs, a command calculated to reduce the tracking error to zero is produced. For range tracking, the command adjusts the radar's sampling times; for doppler tracking, it shifts the frequency of the received echoes; for angle tracking it precesses the rate gyros of the antenna stabilization system.

In track-while-scan, targets detected in successive search scans are accurately tracked by filtering their parameters, much as in single-target tracking. For each track, gates based on the filtered parameters are used to determine whether new detections should be assigned to existing tracks or tentative tracks should be established for them, and whether any existing tracks should be dropped.

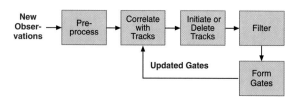

PART VII

High-Resolution Ground Mapping & Imaging

LOCKHEED TR-1A (1981)

This is an advanced version of the original U-2, subsequently redesignated U-2. Its mission: to make high-resolution SAR maps for tactical reconnaissance.

Meeting High Resolution Ground Mapping Requirements

30

An increasingly important airborne radar application is making radar maps of sufficiently fine resolution that topographic features and objects on the ground can be recognized.

In this chapter, we will learn how ground map resolution is defined and see what the optimum resolution is for various uses; then, review the approaches taken to providing it.

How Resolution Is Defined

The quality of the ground maps produced by a radar is gauged primarily by the ability of the radar to resolve closely spaced features of the terrain. This ability is generally defined in terms of *resolution distance* and *cell size*.

Resolution distance is the minimum distance by which two points on the ground may be separated and still be discerned individually by the radar. The separation is usually expressed in terms of a range component, d_r, and an azimuth or cross range component, d_a—the component at right angles to the line of sight from the radar.

A resolution cell, or *"pixel"* (for picture element),[1] is a rectangle whose sides are d_r and d_a (Fig. 1). Because features of the terrain may be oriented in any direction, ideally d_r and d_a are equal, making the cell a square.

As a rule, however, one does not deliberately restrict the resolution in one direction to make the cells square. In real-beam mapping for instance, where fine azimuth resolution is difficult to obtain, d_r is typically a small fraction of d_a (see Fig. 7). Nor is the resolution cell a sharply delineated rectangle, as shown in Fig. 1. Rather, it is usually a rounded rectangular "blob" whose brightness falls off at the edges.

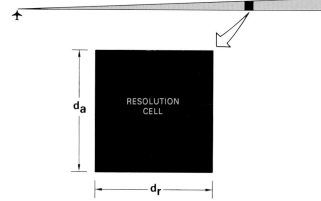

1. Resolution distance is the minimum distance two points on the ground can be separated and still be discerned separately. A resolution cell is a rectangle whose sides are the range and azimuth resolution distances.

1. Pixel and resolution cell are not exactly synonymous. Their dimensions may differ considerably depending upon how the radar's signal processor and display are mechanized.

RESOLUTION REQUIRED FOR VARIOUS MAPPING APPLICATIONS

Features to be Resolved	Cell size
Coast lines, large cities, and the outlines of mountains	500 ft
Major highways, variations in fields	60–100 ft
"Road map" details: city streets, large buildings, small airfields	30–50 ft
Vehicles, houses, small buildings	5–10 ft

Factors Influencing Choice of Cell Size

Among the more important considerations influencing the choice of cell size are the sizes of the objects that must be resolved, the amount of signal processing required to produce the maps, cost, and finally the task of interpreting the maps once they have ben made.

Size of Objects to Be Resolved. How large the resolution cells can be and still provide useful ground maps depends upon what the maps are used for. For discerning gross features of the terrain such as coastlines and the outlines of cities and mountains, a resolution of 500 feet or so will do. For recognizing major highways, variations in the texture of fields, and the like, a resolution of around 100 feet is needed. To recognize city streets, large buildings, and small airfields—the sort of details commonly included in a road map—resolution on the order of 30 to 50 feet is required.

To recognize the shapes of objects on the ground—such as vehicles, houses, and small buildings—the resolution must be considerably finer. Exactly how fine varies with both the sizes and the shapes of the objects. As a rule, the required resolution distance is somewhere between 1/5th and 1/20th of the major dimension of the smallest object to be recognized.

This is illustrated in Fig. 2 (facing page). It shows two silhouettes of the same airplane. Over one is superimposed a grid of resolution cells whose sides are 1/5th of the wingspan. Over the other is superimposed a grid of cells whose sides are 1/20th of the wingspan.

Alongside each silhouette is a simplified representation of the ground map corresponding to the indicated cell size. In these maps, cells that are filled completely by the silhouette are shown as yellow; cells that are partly filled are shown in shades of green corresponding to the percentage filled; cells that do not include the airplane at all are shown as dark green. For this particular shape, a resolution of 1/5th of the major dimension enables some shape recognition, while a resolution of 1/20th of this dimension enables good recognition.

It should be pointed out, though, that in preparing Fig. 2 all elements of the airplane were assumed to reflect radio waves in the radar's direction equally. Actually, for any one combination of look angle, radio frequency, and polarization, only a few bright scattering centers might be mappable. So even though the cell size was only 1/20th of the major dimension, the airplane's shape might still be difficult to recognize. However, as we shall see, by repeatedly mapping the same area from different directions and with different radio frequencies and polarizations, we can substantial-

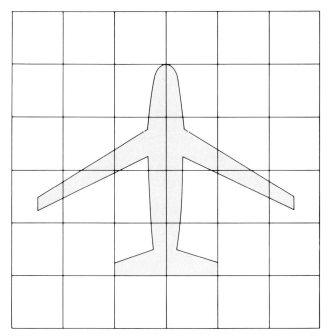

CELL SIZE: 1/5 MAJOR DIMENSION

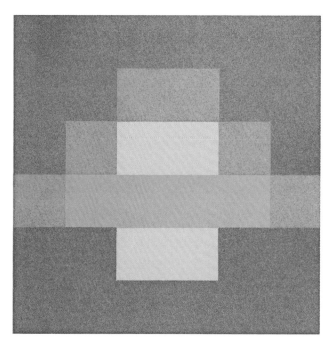

CORRESPONDING MAP

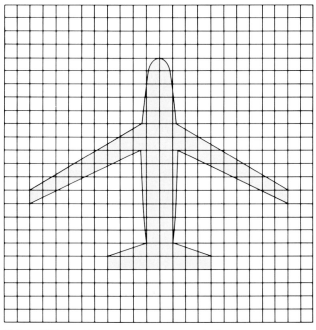

CELL SIZE: 1/20 MAJOR DIMENSION

CORRESPONDING MAP

2. Cell size required for shape resolution. Silhouettes (left) are identical. Radar maps, right, are simplified representations. Assuming that all elements of the plane reflect equally in the radar's direction, a cell size of 1/20 of the silhouette's major dimension enables good shape recognition.

ly increase the fraction of an object's surface from which mappable reflections are received. Through such techniques, we can come quite close to realizing the kind of shape recognition illustrated in Fig. 2 (above).

Amount of Signal Processing Required. A major constraint on the fineness of resolution that one would like to provide is the amount of signal processing it requires.

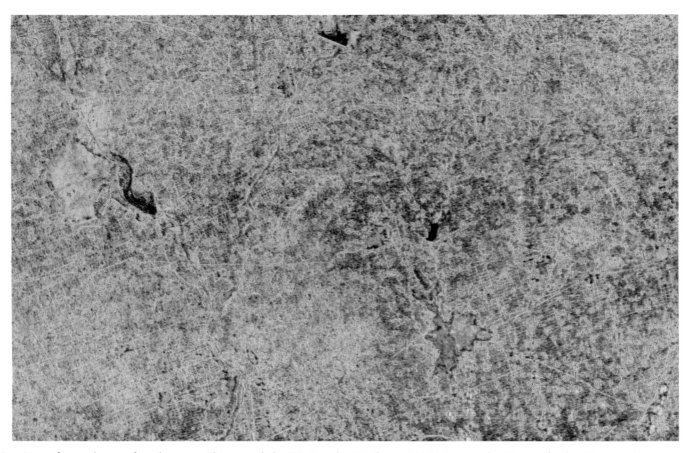

3.   Map of a rural area of northeastern China, made by SIR–A radar (similar to SEASAT) carried in Space Shuttle. (Courtesy Jet Propulsion Laboratory)

2. By comparison, the total area of North America is only slightly more than nine million square miles.

In general, to map an area of a given size (Fig. 3), the amount of processing goes up in proportion to the number of resolution cells the area contains. If the cells are square, the number of cells is inversely proportional to the *square* of the resolution distance. Cutting it in half, for example, quadruples the number of cells.

Consider, for example, the SEASAT radar. Orbiting the earth at a height of 500 miles, in just 100 days of operation it mapped a total area of 48,000,000 square miles.[2] The cell size was on the order of 80 by 80 feet, bringing the total number of cells to around 200 billion. Had the cell size been reduced to say 10 feet by 10 feet—as would have been necessary to resolve objects such as houses—the amount of processing would have increased by roughly 64 times.

Cost. It is not easy to generalize regarding this important parameter. About all one can say without getting into considerable detail is that, as the resolution is made finer and the complexity of the signal processing increases, cost goes up to various degrees. Depending on the situation, at some point a further increase in resolution becomes prohibitively expensive. Yet with technological advances, the cost of providing a given resolution tends to decrease.

Task of Interpreting the Maps. Superficially, this would hardly seem an important consideration, but it is. A great many features of the terrain as well as the objects on it appear quite differently in a radar map than they do visually. [Objects on the ground, for instance, are often recognized as much by the size and shape of the shadows they cast (Fig. 4) as by the brightness and shape of their images.] Consequently, the amount of time required to interpret the details in a map of a given region also increases as the resolution is made finer. How much time is available for this depends on the application.

At one extreme is an application such as SEASAT. Since this was a research project, the prodigious amount of information the radar gathered could reasonably be analyzed and interpreted over a period of months and years.

At the other extreme are applications such as target location in a single-seat attack aircraft. Streaking across the countryside at a speed of say 800 knots (1350 feet per second), its radar is called upon to map selected regions forward of the aircraft in real time with resolutions that may be as fine as a few feet. In addition to other duties, the pilot must analyze a map in a matter of seconds. To make his job manageable, only relatively small patches of ground are mapped, and the maps are temporarily frozen on his display. When the resolution is increased to enable positive identification of specific points on the ground, the area covered by the individual maps is correspondingly reduced.

Thus, resolution requirements, as well as the sizes of the areas mapped, vary widely. So, too, do the approaches to implementation.

Achieving Fine Resolution

In general, fine resolution is more readily obtained in range than in azimuth. So we'll consider range resolution first.

Range Resolution. As we saw in Chap. 9, the resolution that may be obtained in range amounts to about 500 feet per microsecond of pulse width. Fine range resolution, therefore, can be obtained simply by narrowing the pulses. Whereas a 1 microsecond pulse yields a resolution of only about 500 feet, a 0.1 microsecond pulse yields a resolution of about 50 feet, and a 0.01 microsecond pulse, a resolution of about 5 feet.

The principal limitation on how narrow the pulses may be made is the width of the band of frequencies that can be passed by the transmitter and receiver (Fig. 5). To pass the bulk of the power contained in the pulses, the 3-dB bandwidth must be on the order of $1/\tau$ hertz, which means that for a 0.01 microsecond pulse width, the bandwidth must be on the order of 100 megahertz.

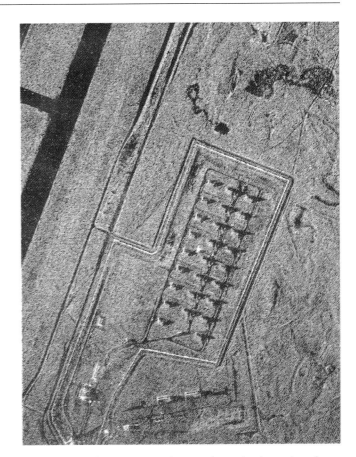

4. High resolution map made in real time by the radar of a small civil aircraft.

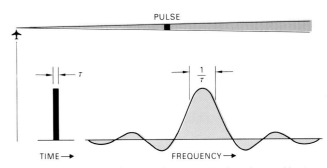

5. Range resolution distance decreases with pulse width. As pulse is narrowed, required bandwidth increases.

How readily wide bandwidths may be obtained depends primarily upon the radar's operating frequency. For any one frequency, as the required bandwidth is increased, a point is ultimately reached beyond which the hardware becomes increasingly difficult, hence costly, to design and build. As a very crude rule of thumb, depending upon the particular situation this point lies somewhere between 3 and 10 percent of the operating frequency. At 10,000 megahertz (X-band), a bandwidth of 100 megahertz would be only about 1 percent. At 1000 megahertz (L-band), it would be 10 percent. Among hardware items for which bandwidth is more critical are the antenna (if a planar array is used) and various radio frequency components, such as coupled-cavity TWTs (see page 27).

Naturally, if the peak power and PRF are kept the same, transmitting extremely narrow pulses greatly reduces the average transmitted power. But this problem can be avoided by employing pulse compression (as discussed in Chap. 13). With a pulse compression ratio of 1000:1, a radar can transmit 10 microsecond pulses and after compression (10 ÷ 1000 = 0.01) still achieve a range resolution of 5 feet. The required bandwidth is of course determined by the compressed pulse width, so it remains the same—in this example, 100 megahertz.

Azimuth Resolution. Depending upon the application, the approaches taken to obtaining fine azimuth resolution vary considerably.

Azimuth resolution distance is roughly equal to the 3-dB beamwidth of the antenna times the range. The 3-dB beamwidth (in radians) roughly equals the wavelength divided by the length of the antenna in units consistent with the wavelength. So, for a given range, fine resolution can be obtained by operating at a very short wavelength or by employing a long antenna, or both. In the atmosphere, because of severe attenuation at the shorter wavelengths, the minimum practical wavelength for long-range mapping is around 3 centimeters (see page 89). In airborne applications, the length of the radar antenna is usually severely limited by the dimensions of the aircraft.

Even so, if the maximum range of interest is reasonably short and the resolution requirements are not too demanding, an antenna of practical size can provide a narrow enough "real" beam to yield quite adequate results. At ranges of up to 10 or 12 nautical miles, for instance, a sidelooking array radar (SLAR) having a 16-foot-long antenna and operating at X-band (3 centimeters) can provide resolution adequate for identifying such features as oil slicks and resolving small craft (Fig. 6, top of next page).

EXAMPLE: AZIMUTH RESOLUTION

For a Real Antenna:

$$\theta_{3dB} \cong \frac{\lambda}{L} \text{ radians}$$

$$d_a \cong \theta_{3dB} R \cong \frac{\lambda R}{L}$$

Conditions:

Wavelength (λ).................... 0.1 ft.
Length of Antenna (L)........ 10 ft.
Range (R)........................ 50 nmi
(300,000 ft.)

Calculation:

$$d_a \cong \frac{0.1 \times 300,000}{10} \cong 3,000 \text{ ft.}$$

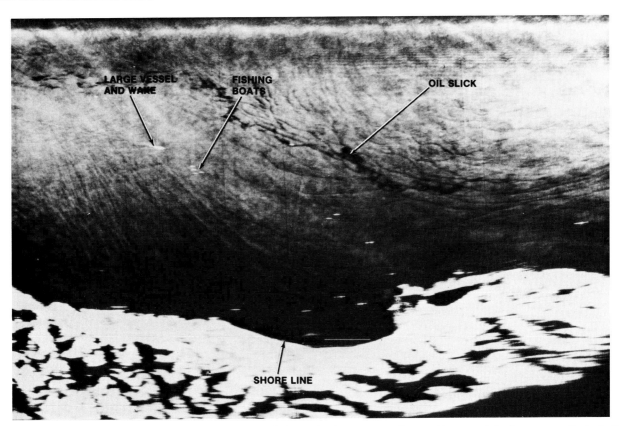

6.   Map of oil seepage off Santa Barbara, California, made by radar having a 16-foot real-beam sidelooking array. Radar's flight path is along top of map. Range to coast is about 15 nmi. At 5 nmi, azimuth resolution is roughly 200 feet. (Courtesy Motorola Inc.)

However, to obtain resolutions fine enough for recognizing the shapes of even fairly large objects at long ranges, we must resort either to an impractically long antenna or use wavelengths so short that the radar must contend with severe attenuation in the atmosphere. The answer to this dilemma is to create an antenna of the desired length synthetically—the process called *synthetic array (aperture) radar*, SAR.

Synthetic Array (Aperture) Radar

SAR takes advantage of the forward motion of the airborne radar to produce the equivalent of an array antenna which may be thousands of feet long. Moreover, as will be explained in the next chapter, the beamwidth of this array is *roughly half that of a real array* of the same length. The outputs of the array are synthesized in a signal processor from the returns received by the real radar antenna over periods of up to several seconds or more. The processing may be done either optically or digitally.[3]

Optical Processing. The first SAR systems (developed in the late 1950s and early 1960s) employed optical signal processors. These can produce very high quality maps and are intrinsically quite fast. But to date, the inputs and out-

3. In the 1960s and 1970s, some SAR systems were developed which processed the radar data with analog circuits. Such processors have since been supplanted by optical and digital processors.

THE EARLY OPTICAL SAR PROCESSOR

The first airborne SAR systems, developed more than two decades ago, employed optical processing. For this, an intensity-modulated scanner photographically records the coherent video output of the radar receiver in a two-dimensional raster format on film. This recording is essentially a hologram of the radar map.

After the film has been developed, coherent light from a laser is projected through it. An elegant system of lenses focuses the light onto a second film in such a way as to combine the range and doppler information contained in the recorded video into an image.

Although digital processors have compelling advantages over the conventional optical processors—versatility, small size, high speed—optical processing is by no means dead. From the Apollo orbiter, it was used to map the lunar subsurface, and from the SEASAT satellite and the space shuttle it produced spectacular maps of vast areas of the Earth's surface. (In the case of SEASAT, the radar echoes were radioed directly to ground stations especially equipped to record them.)

Currently, work is underway that promises to eliminate the need for photographic recording and thereby make real-time onboard operation in spacecraft practical.

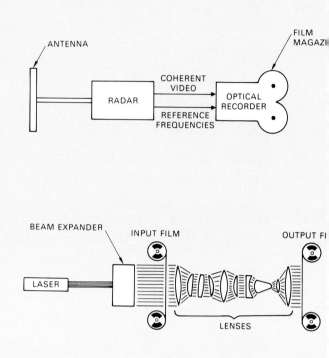

7.   Real-time SAR mapping in small aircraft was made possible by the advent of integrated solid-state circuits, such as used in this programmable signal processor.

puts have had to be recorded photographically (see panel above). This requirement has introduced a time lag in the processing and made the equipment heavy and bulky. Also, since the optics must be aligned with precision, they are sensitive to vibration, which can be a problem in an aircraft. Accordingly in many airborne and all spaceborne applications, the radar data has been returned to the ground for processing.

Digital Processing. With the advent of low-cost, high-speed integrated solid-state circuits in the 1970s, it became possible to process the video signals digitally in real time, with lightweight equipment compact enough to be incorporated in small airborne radars (Fig. 7). This advance greatly expanded the list of possible SAR applications (see Chap. 3).

Besides solving the problems of speed and equipment size, digital processing has the advantages of being extremely accurate and flexible. Once the video signal from the radar receiver has been accurately digitized and stored, it can readily be processed to meet a host of operational requirements.

Literally with the flick of a switch, the range of the area being mapped can be changed by an order of magnitude, detailed large-scale maps can be made of areas of special interest, resolution can be increased (Fig. 8) or decreased as desired, and maps can be displayed in a variety of formats. Indeed, the potential capabilities of digital SAR systems seem limited only by the ever increasing speeds of digital devices and the ability of the operator to interrupt the immense amount of data that lies at his fingertips.

Summary

The quality of ground maps is gauged by the size of the resolution cell—a rectangle (actually a rounded blob) whose sides are the minimum resolvable difference in range, d_r, and azimuth, d_a. How large the cells can be and still provide adequate resolution is determined primarily by the size of the smallest objects that must be recognized.

Resolution requirements are tempered by such considerations as the amount of signal processing that must be done, the task of interpreting the details of the maps, and cost. In general, these vary inversely with the square of the resolution distance.

Fine range resolution may be obtained with reasonable levels of peak power by using large amounts of pulse compression. Fine azimuth resolution may be obtained by using short wavelengths and long antennas.

At short ranges, azimuth resolution adequate for many high resolution applications can be obtained with real-beam antennas. But to recognize the shapes of even fairly large objects at long ranges, sufficient resolution can only be obtained by synthesizing the output of a long array antenna from the returns received over a period of time by the real antenna—SAR. The equivalent of an antenna thousands of feet long may thus be realized.

SAR processing may be performed either optically or digitally. Optical processors are capable of producing very high quality maps. But to date these processors have required intermediate photographic film recording. Digital processing has the advantages of being extremely accurate, versatile, and fast, and can be implemented with hardware that is small, rugged, and lightweight.

8. SAR map made in real time at long range as evidenced by long radar shadows of a plant's stacks (center).

Some Relationships To Keep In Mind

- **Minimum resolution requirements:**
  Road map details: 30 to 50 feet
  Shapes: 1/5 to 1/20 of major dimension

- **Achievable resolution**

  $d_r = 500\ \tau$ feet

  τ = compressed pulse width

  Required bandwidth = $1/\tau$

  $d_a \approx \dfrac{\lambda}{L}\ R$ (for real array)

  $d_a \approx \dfrac{\lambda}{2L}\ R$ (for synthetic array)

  (L = array length, same units as λ)

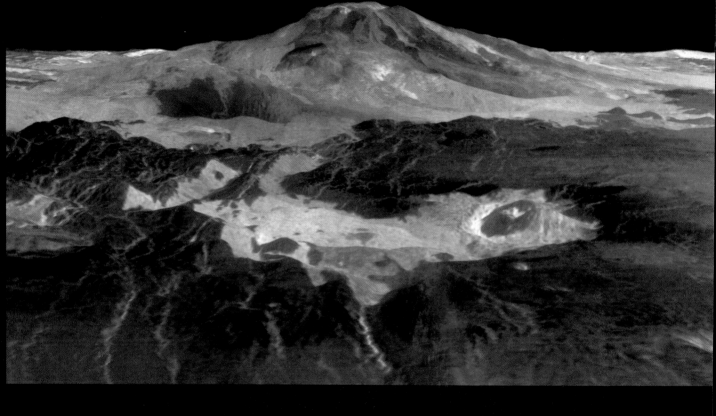

RADAR MAP OF VENUS

Computer-generated perspective view of 26,000-foot Maat Mons volcano on perpetually cloud-shrouded Venus, whose surface temperatures are hotter than the melting point of lead. Picture is composed of SAR imagery collected by Magellan radar in more than a dozen orbits.

Principles of Synthetic Array (Aperture) Radar 31

In the last chapter, we saw how synthetic array radar (SAR) solves the problem of providing fine azimuth resolution, even at very long ranges. The signal processing, we learned, may be performed either optically or digitally—digital processing having the advantages of being extremely flexible and not requiring film processing.

Regardless of which method is used, however, the SAR principles are the same. They are founded primarily on a combination of antenna theory and signal processing concepts. In addition, certain aspects of SAR design—such as resolution, focusing, and the correction of distortion—are rooted in the theory of optics.

In this chapter, we will examine the SAR principles more closely and become acquainted with the basic digital processing techniques. We will see (1) how the equivalent of a long array antenna may be synthesized from returns gathered over a period of several seconds by a comparatively small real antenna, (2) how the array may be focused, (3) what determines the angular resolution of such an array, and (4) how the computing load can be reduced by processing the returns with doppler filtering techniques.

Basic SAR Concept

As explained in Chap. 30, SAR takes advantage of the forward motion of the radar to produce the equivalent of a long antenna. Each time a pulse is transmitted, the radar occupies a position a little farther along on the flight path. By pointing a reasonably small antenna out to one side and summing the returns from successive pulses, it is possible to synthesize a very long sidelooking linear array.

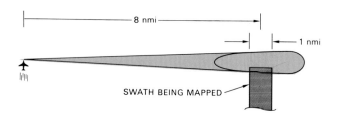

1. Hypothetical operational situation for a SAR radar. With real antenna trained at fixed azimuth angle of 90° to flight path, radar maps a 1-mile wide swath at range of 8 miles.

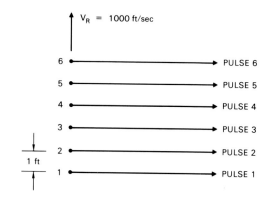

2. Points representing positions of center of antenna when successive pulses are transmitted. Each point constitutes one "element" of synthetic array.

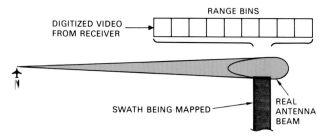

3. Returns received by successive elements of the synthetic array are summed in bank of range bins spanning the range interval being mapped.

Rudimentary Example (Unfocused Array). Just how the array is synthesized is perhaps most easily visualized by considering an extremely simple SAR system in a hypothetical operational situation.

An aircraft carrying an X-band radar is flying in a straight line at constant speed and altitude. The radar antenna is pointed downward slightly and aligned at a fixed angle of 90° relative to the flight path (Fig. 1).

As the aircraft progresses, the beam sweeps across a broad swath of ground parallel to the flight path. Only a relatively narrow portion of this swath, however, is of immediate interest. That portion, we'll say, is a strip 1 nautical mile wide, offset from the flight path by about 8 nautical miles.

The aircraft's mission requires that the ground within this strip be mapped with a resolution of about 50 feet. As will be explained shortly (page 414), to provide 50-foot resolution at a range of 8 miles, our hypothetical SAR radar must synthesize an array roughly 50 feet long.

The aircraft's ground speed, let's say, is 1000 feet per second (600 knots) and the PRF is 1000 pulses per second. Consequently, every time the radar transmits a pulse, the center of the radar antenna is one foot farther along the flight path. The synthetic array can thus be thought of as consisting of a line of elemental radiators one foot apart (Fig. 2). To synthesize the required 50-foot long array, 50 such elements are required. In other words, the returns from 50 consecutive transmitted pulses must be summed.

Typically summing is done after the receiver's output has been digitized. A bank of range bins is provided which just spans the 1 mile range interval being mapped (Fig. 3). Following every transmission, the return from each resolvable range increment within this interval is added to the contents of the appropriate bin.

This operation corresponds to the summing performed by the feed structure that interconnects the radiating elements of a real array antenna. The fundamental difference is that, with the real array, the return from each range increment is received simultaneously by all array elements every time a pulse is transmitted; whereas, with the synthetic array, the return is collected by the individual elements serially over the period of time the radar takes to traverse the array.

The return from the first pulse is received entirely by element number one; the return from the second pulse is received entirely by element number two; and so on.

The result, however, is substantially the same. Provided that the range is long compared to the array length, the distance from a patch of ground on the boresight line (perpendicular to the flight path) to each array element is essentially the same. So the echoes received from the patch by all elements have nearly the same radio frequency phase. When added up in the range bin corresponding to the range of the patch, they produce a sum (Fig. 4).

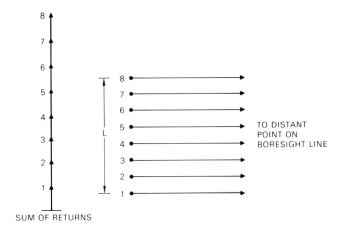

4. Distances from successive array elements to a distant point on the boresight line are equal, so returns from the point add up in phase.

On the other hand, for a ground patch that is *not* quite on the boresight line, the distance from the patch to successive array elements is progressively different. So the echoes received from the patch by successive elements have progressively different phases and tend to cancel. The equivalent of a very narrow antenna beam is thus produced (Fig. 5).

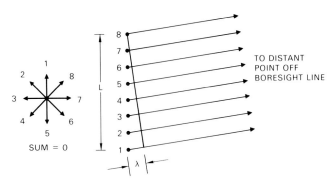

5. Distances from successive array elements to a point off the boresight line are progressively different, so returns from the point tend to cancel. Null condition is shown here.

When the returns from the 50 pulses required to form the array have been integrated, the sum that has built up in each range bin comes quite close to representing the total return from a single range/azimuth resolution cell (Fig. 6). The contents of the bank of bins, therefore, represent the returns from a single row of resolution cells spanning the 1 mile wide range swath being mapped.

At this point, the contents of the individual range bins are transferred to corresponding locations in the memory

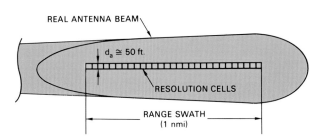

6. When returns from 50 pulses have been integrated, contents of range bins represent the returns from a single row of range/azimuth resolution cells.

405

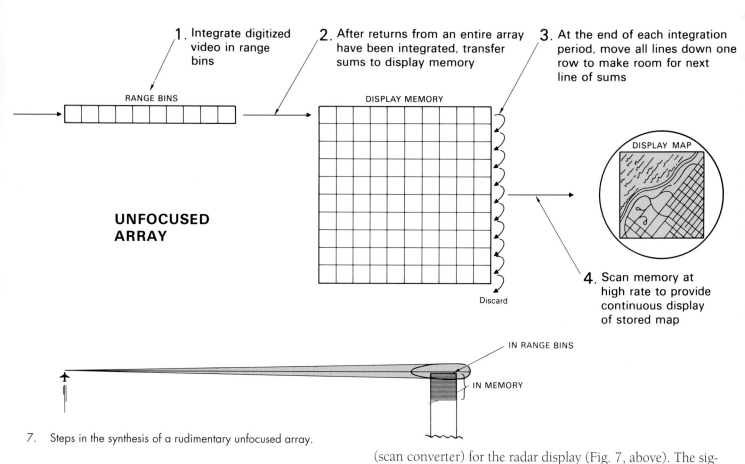

1. Integrate digitized video in range bins

RANGE BINS

2. After returns from an entire array have been integrated, transfer sums to display memory

DISPLAY MEMORY

3. At the end of each integration period, move all lines down one row to make room for next line of sums

UNFOCUSED ARRAY

Discard

DISPLAY MAP

4. Scan memory at high rate to provide continuous display of stored map

IN RANGE BINS

IN MEMORY

7.   Steps in the synthesis of a rudimentary unfocused array.

(scan converter) for the radar display (Fig. 7, above). The signal processor thereupon begins the formation of a new array, the beam of which will cross the 1-mile-wide swath immediately ahead of the row of cells that has just been mapped. Since the map is formed a line at a time, this method of SAR signal processing is called line-by-line processing.

The display memory stores the integrated returns from as many rows of cells as can be presented at one time on the radar display. As the returns from each new row of cells are received, the stored returns are shifted down one row to make room for the new data, and the data in the bottom row is discarded. Throughout the comparatively slow array-forming process, the data stored in the display memory is repeatedly scanned at a high rate and presented as a continuous picture on a TV-type display. The operator is thus provided with a strip map that moves through his display in real time as the aircraft advances.

To keep the explanation simple, the range to the swath mapped in this example was deliberately chosen so as to make the array length equal the desired resolution distance, d_a. If the array were longer than d_a, as it probably would be, additional storage capacity would have to be provided in each range bin so that an entire array could be synthesized every time the radar advanced a distance equal to d_a.[1]   In essence, though, the operations would be the same as described here.

1.   If the array were $2d_a$ long, the first 50 returns would be summed in one memory position, the second in another, and the sum of the two sums would be transferred to the display memory. The first sum would then be dumped and the next 50 returns would be summed. The sum of that sum plus the second sum would be transferred to the display memory, and so on.

SIGNAL PROCESSING FOR UNFOCUSED ARRAY

The signal processing required to synthesize an unfocused array antenna can be summarized mathematically as follows:

Inputs: For each resolvable range (R_r), N successive pairs of numbers are supplied.

$$x_n, y_n \qquad n = 1, 2, 3, \ldots \ldots N$$

Each pair represents the I and Q components of the return received from range R_r by a single array element.

Integration: To form the beam (azimuth processing), the I and Q components are summed.

$$I = \sum_{n=1}^{N} x_n \qquad Q = \sum_{n=1}^{N} y_n$$

Magnitude Detection: The magnitude or the vector sum of I and Q is computed and output.

$$S = \sqrt{I^2 + Q^2}$$

S is the amplitude of the total return from a single resolution cell on the boresight line at range R_r.

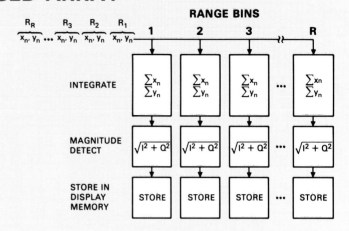

(An intermediate step not shown in the above diagram is the scaling of the detected magnitudes to the values of intensity – gray levels– that are to be displayed.)

Signal Processing Required. In the foregoing discussion, the inputs to the SAR processor were referred to only as the digitized radar returns. In an actual radar, the inputs would be the digitized I and Q components (x_n, y_n) of the returns (translated to video frequencies). The sums that build up in the range bins, then, would be the vector sums of the accumulated values of x_n and y_n. And the quantities transferred from the range bins to the display memory would be the magnitudes of the vector sums. The signal processing required to synthesize a simple array of this sort is summarized in the panel at the top of the page. Note that the equations shown there are identical to those which must be solved to form a doppler filter tuned to zero frequency (or to the PRF).

Limitation of Unfocused Array. The rudimentary array just described is called an "unfocused" array. For it must be short enough in relation to the range to the swath being mapped that the lines of sight from any one point at the swath's range to the individual array elements are essentially parallel. In this respect, the array is similar to a pinhole camera: both are focused at infinity.

Now, if the array length is an appreciable fraction of the swath's range, the lines of sight from a point at that range to the individual elements will diverge slightly. Then, even if the point is on the boresight line, the distances to the ele-

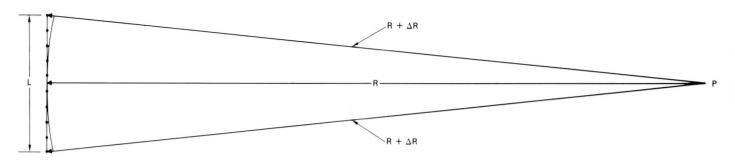

8.   Distance from a point (P) on the boresight line of an array antenna to the individual array elements. If array length (L) is an appreciable fraction of the range (R), the distances to the end elements will be appreciably greater than the distances in the central element.

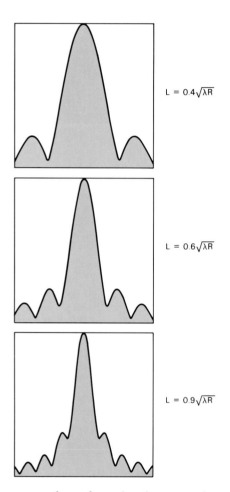

$L = 0.4\sqrt{\lambda R}$

$L = 0.6\sqrt{\lambda R}$

$L = 0.9\sqrt{\lambda R}$

9.   Radiation pattern of an unfocused synthetic array showing increase in relative gain of sidelobes and merging of sidelobes with mainlobe as array length (L) is increased.

ments will not all be the same (Fig. 8, above). Since the wavelength is generally fairly short, very small differences in these distances can result in considerable differences in the phases of the returns which the individual array elements receive from the point.

Because these phase errors are not compensated in the unfocused array, its capabilities are quite limited. While the azimuth resolution distance the array provides at any given range can initially be reduced by increasing the length of the array, a point is soon reached beyond which any further increase in length only degrades performance.

Degradation begins with a gradual increase in gain of the sidelobes relative to the mainlobe and a merging of the lower order sidelobes with the mainlobe (Fig. 9).

This effect continues increasingly and is accompanied by a progressive fall-off in the rate at which the mainlobe gain increases with array length.

The reason for the fall-off in gain can be seen if we examine Fig. 8 again. It shows the distance from point P on the boresight line to each element of an array of length L. For elements near the array center, there is very little difference in this distance. But for elements farther and farther removed from the center, the difference grows increasingly. As the array is lengthened, therefore, the phase of the returns received by the end-most elements falls increasingly far behind the phase of the sum of the returns received by the other elements.

This progressive phase rotation and its effect on the mainlobe gain of the synthetic array is illustrated in Fig. 10 (top of facing page). The phasors shown there represent the returns received by the individual elements of a 27-element array from a distant point (P) on the boresight line. The phase of the return received by the middle element (14) is taken as the reference. The gain in the boresight direction corresponds to the sum of the phasors.

The returns received by the central elements (9 through 19) are so close to being in phase that their sum is virtually undegraded by the lack of focus. But the phases of the returns received by elements farther and farther out are rotated increasingly. The returns received by elements 4 and 24 are nearly 90° out of phase with the sum of the returns received by the elements closer in, hence, contribute only negligibly to that sum. The returns received by elements 1, 2, and 3 and 25, 26, and 27 are actually subtractive. Obviously, under the conditions for which Fig. 10 was drawn, the gain would have its maximum value if the array were only 21 elements long (elements 4 through 24).

The degradation of *beamwidth* closely parallels that of mainlobe gain. Initially, the gain at angles slightly off bore-sight is degraded by the lack of focus to nearly the same extent as the gain in the boresight direction. So at first, there is little reduction in the rate at which the beam narrows as the array is lengthened. But when the length reaches the point where the gain in the boresight direction stops *increasing*, the beamwidth stops *decreasing*. If we lengthen the array beyond this point, the beam starts spreading. From the standpoint of both gain and beamwidth, the maximum effective length has been reached.

In Fig. 11, the gain and beamwidth for an unfocused array are plotted versus array length.

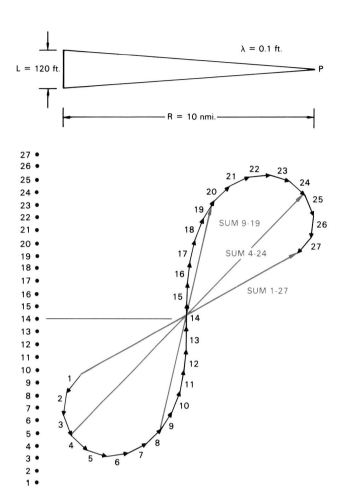

10. Degradation in the gain of an unfocused array. Phasors represent returns received from a distant point P by individual array elements. Gain is the sum of the phasors. In this case, the gain could be increased by decreasing the length of the array.

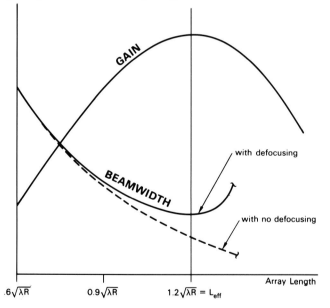

11. Effect of increased array length on gain and beamwidth of an unfocused synthetic array. Gain is maximum and beamwidth is minimum when length, $L = 1.2 \sqrt{\lambda R}$.

It can be shown from the geometry of the situation that the array length for which given values of gain and beamwidth are obtained varies in proportion to $\sqrt{\lambda R}$, where λ is the wavelength and R is the range. To make the graph of Fig. 11 applicable to any combination of λ and R, the array length is plotted there in terms of $\sqrt{\lambda R}$. As you can see, the maximum effective array length is

$$L_{eff} = 1.2 \quad \lambda R.$$

Another way of looking at the effects of defocusing is this. Imagine that you are approaching an unfocused array of a given length from a long enough distance that the lines of sight to each "elements" of the array are essentially parallel. The beamwidth of the array at your range, therefore, does not change as you advance.

However, as you approach the range for which the array length is optimum and defocusing comes into play, the beamwidth starts increasing.

The azimuth resolution distance, hence the finest achievable resolution at that range, turns out to be roughly 40 percent of the array length.

$$d_{a_{min}} \cong 0.4 \, L_{eff}$$

Moreover, beyond that range, we cannot make the azimuth resolution of the radar independent of range, as we would like. For when we lengthen the array further, the resolution distance increases as the square root of the range (Fig. 12).

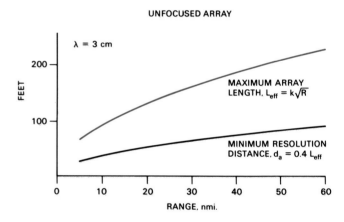

12. Maximum effective length of unfocused array increases only as square root of range. Resolution distance at range for which length is optimized is roughly 40 percent of array length.

Focused Array

The limitation on array length may largely be removed by focusing the array. Then, by suitably increasing the length of the array in proportion to the range, virtually the same resolution may be obtained at any desired range.

How Focusing Is Done. In principle, to focus an array all you need to do is apply an appropriate phase correction (rotation) to the returns received by each array element. As illustrated in Fig. 13 (top of facing page), the phase error for any one element, hence the phase rotation needed to cancel the error, is proportional to the square of the dis-

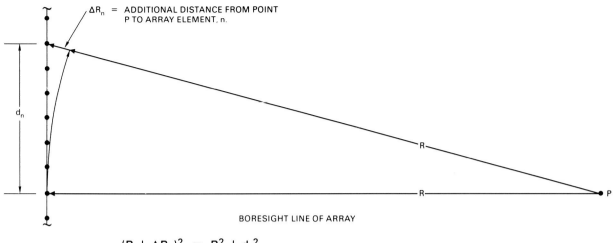

$$(R + \Delta R_n)^2 = R^2 + d_n^2$$

$$R^2 + 2R\Delta R_n + (\Delta R_n)^2 = R^2 + d_n^2$$

$$2R\Delta R_n \left(1 + \frac{\Delta R_n}{2R}\right) = d_n^2$$

$$\Delta R_n \cong \frac{d_n^2}{2R} \qquad \text{Assuming } \Delta R_n \ll 2R$$

$$\therefore \quad \text{PHASE ERROR, } \phi_n = \frac{2\pi}{\lambda} (2\Delta R_n) \approx \frac{2\pi}{\lambda} \left(\frac{d_n^2}{R}\right) \text{ Radians}$$

Accounts for Round-Trip Travel

13. Phase error for return received by any one array element (n) is proportional to the square of the distance (d_n) from the element to the array center. Factor of two by which ΔR_n is multiplied accounts for the phase error being proportional to the difference in *round-trip* distance from the element to point P (see page 416).

tance of the element from the center of the array.

$$\text{Phase correction} = -\frac{2\pi}{\lambda R} d_n^2$$

where

d_n = distance of element n from array center
λ = wavelength (same units as d_n)
R = range to area being mapped (same units as λ)

In some cases, it may be possible to presum the returns received by blocks of adjacent elements without impairing performance. By rotating only the phases of the sums, computing and storage requirements may be eased.

To simplify the description, however, we will omit presumming here. Also, we will assume that the combination of array length, PRF, and range is such that the resolution distance, d_a, is roughly equal to the spacing between array elements.

To focus an array when no presumming is done, we must provide as many rows of storage positions in the bank of

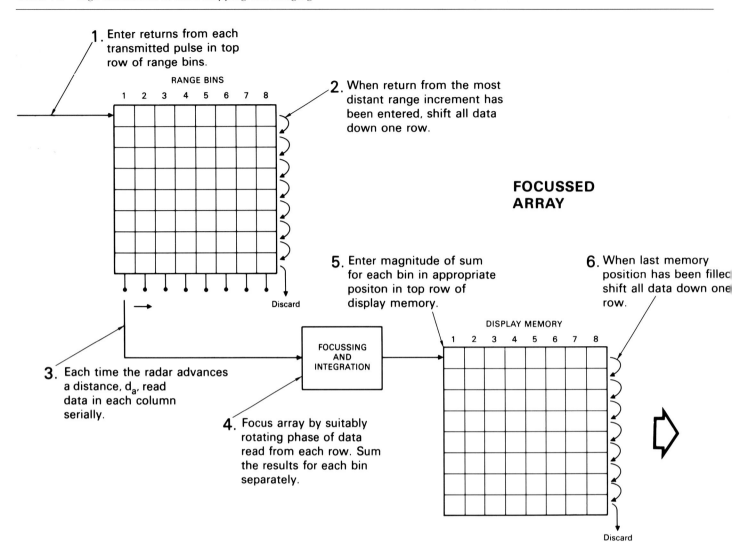

1. Enter returns from each transmitted pulse in top row of range bins.

2. When return from the most distant range increment has been entered, shift all data down one row.

FOCUSSED ARRAY

5. Enter magnitude of sum for each bin in appropriate positon in top row of display memory.

6. When last memory position has been filled shift all data down one row.

3. Each time the radar advances a distance, d_a, read data in each column serially.

4. Focus array by suitably rotating phase of data read from each row. Sum the results for each bin separately.

14. How array is focused in a line-by-line processor. To simplify the description, conditions are assumed to be such that the resolution distance, d_a, equals the spacing between array elements. Hence, every time a pulse is transmitted a new array must by synthesized.

range bins as there are array elements (Fig. 14, above). As the returns from any one transmitted pulse (array element) come in, they are stored in the top row. When the return from the most distant range increment has been received, the contents of every row are shifted down to the row below it to make room for the incoming returns from the next transmitted pulse. The contents of the bottom row are discarded.

Between these shifts, the column of numbers in each bin is read serially and the numbers are appropriately phase shifted and summed—a process called *azimuth compression*. The magnitude of the sum for each range bin is entered in the appropriate range positions in the top row of the display memory. Thus (for the conditions assumed in this example), every time the returns from another transmitted pulse have been received—i.e., every time the radar has advanced a distance equal to the spacing between array elements—another array is synthesized.[3]

3. If the resolution distance, d_a were greater than the spacing between array elements, an array would be synthesized only after the radar had advanced a distance d_a.

412

SIGNAL PROCESSING FOR A FOCUSED ARRAY

To focus an array, for every range bin the signal processor must mathematically perform the equivalent of rotating the phasor representation (A) of the return received by each successive array element (n) through the phase angle, ϕ_n (the value of which was derived in Figure 13).

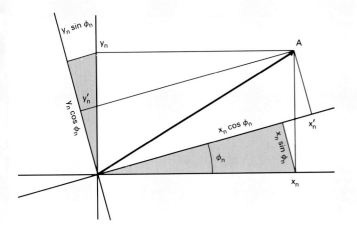

The I and Q components of the phasor before rotation are the inputs, x_n and y_n. After rotation, we'll represent the components by x'_n and y'_n. To perform the rotation, the following algorithms must be computed.

$$x'_n = x_n \cos \phi_n + y_n \sin \phi_n$$

$$y'_n = y_n \cos \phi_n - x_n \sin \phi_n$$

The values of x'_n and y'_n for the total number of array elements (N) must then be summed separately

$$X = \sum_{n=1}^{N} x'_n \qquad Y = \sum_{n=1}^{N} y'_n$$

and the magnitude of the vector sum of x and y must be calculated.

$$S = \sqrt{X^2 + Y^2}$$

These, you'll recognize, are the same algorithms that are computed when forming a digital filter with the DFT.

Signal Processing Required. The computations (per range bin) required to perform the phase rotation and summing for a focused array are shown in the panel, above. They are exactly the same, you will notice, as the computations required to form a doppler filter with the DFT.

Azimuth Resolution. With focusing, the length of the array can be greatly increased. But as with all things, a point is ultimately reached where a further increase in length does not improve resolution. In the case of an array whose azimuth angle is fixed—that is, one which looks out at a constant angle relative to the flight path—this limit is established by the physical size of the elemental radiator, the real antenna. Surprisingly, the smaller this antenna is, the longer the array can be made.

The reason is simple enough. For return to be received from any one ground patch by all elements of the array, the beam of the real antenna must be wide enough for the patch to fall within the beam for every position of the antenna in the entire length of the array (Fig. 15). For that condition to be satisfied, the width of the beam at the range of the patch must at least equal the length of the array. The smaller the real antenna is, the wider its beam; hence, the longer the array can be made and the finer the resolution that can be achieved.

For a given real-antenna size, how fine can that be? Before answering this question, we must consider an impor-

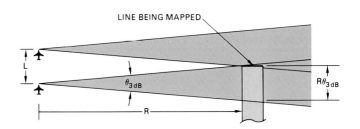

15. Each line that is mapped must be within the mainlobe of the real antenna while the radar traverses the entire length of the array, L.

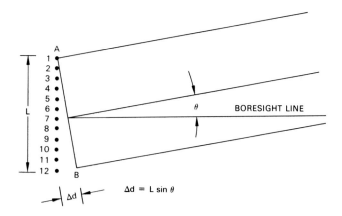

16. One-way radiation pattern of a real array is formed during transmission as a result of progressive difference in distance from successive array elements to observation point.

REAL VS. SYNTHETIC ARRAY ANTENNA GAIN PATTERN

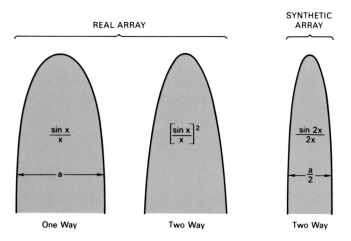

17. Comparison of mainlobes of real and synthetic arrays of same length. Synthetic array has no one-way radiation pattern since the array is synthesized from the radar return.

tant difference between the beam of a synthetic array and the beam of a real array.

A real array has both a one-way and a two-way radiation pattern. The one-way pattern is formed upon transmission as a result of the progressive difference in the distances from successive array elements to any point off the boresight line (Fig. 16). (The phases of the radiation from the individual elements arriving at that point differ in proportion to the differences in distance.) This pattern has a sin x/x shape. The two-way pattern is formed upon reception, through the same mechanism. Since the phase shifts are the same for both transmission and reception, the two-way pattern is essentially a compounding of the one-way pattern, and so has a $(sin\ x/x)^2$ shape.

A synthetic array, on the other hand, has only a two-way pattern. For the array is synthesized out of the returns received by the real antenna, which sequentially assumes the role of successive array elements. Because each element receives only the returns from its own transmissions, however, the element-to-element phase shifts in the returns received from a given point off the boresight line correspond to the differences in the *round-trip* distances from the individual elements to the point and back (see panel, page 416). This is equivalent to saying that the two-way pattern of the synthetic array has the same shape as the one-way pattern of a real array of *twice* the length, sin 2x/2x (Fig. 17).

For a uniformly illuminated *real array*, the one-way 3-dB beamwidth is 0.88 times the ratio of the wavelength to the array length. Consequently, for a uniformly illuminated *synthetic* array, the two-way 3-dB beamwidth is

$$\theta_{3\ dB} = 0.44\ \frac{\lambda}{L}\ \text{radians}$$

The point on the radiation pattern where the beamwidth is measured, of course, is fairly arbitrary. It turns out that by measuring the beamwidth at a point 1 dB lower down, the factor 0.44 can be increased to one half. To simplify the beamwidth equation, therefore, the minus 4-dB point is commonly used.

$$\theta_{4\ dB} = \frac{\lambda}{2L}\ \text{radians}$$

The azimuth resolution distance, then is

$$d_a = \frac{\lambda}{2L}\ R$$

Armed with this expression, we can now go back and answer the question raised earlier: if the length of the synthetic array is limited to the width of the beam of the real antenna at the range being mapped, for a given sized antenna how fine can the resolution of the synthetic array be?

If the real antenna is a linear array and its length is ℓ, then its one-way 4-dB beamwidth is λ / ℓ, where ℓ is the array length. Multiplying this expression by the range, R, of the swath being mapped gives the maximum length of the real array.

$$L_{max} = \frac{\lambda}{\ell} R$$

Substituting L_{max} for L in the expression for azimuth resolution distance (Fig. 18), we find that the minimum resolution distance $d_{a_{min}}$ is half the length of the real antenna.

$$d_{a_{min}} = \frac{\text{Length of real antenna}}{2}$$

This, then, is the ultimate resolution of a synthetic array whose beam is positioned at a fixed angle relative to the flight path, as in strip map radars. As we will see in the next chapter, this limitation is removed in the "spotlight" mode, by keeping the beam of the real antenna continuously trained on the area being mapped.

Reducing the Computing Load: Doppler Processing

As is clear from Fig. 14 (page 412), if the array is very long, an immense amount of computing is required for line-by-line processing of a focused array. In the simple example illustrated there, every time the radar transmits a pulse, it must perform the phase correction all over again for every pulse that has been received over an entire array length and sum the results.

Put another way, if there are N elements in the array, every time the radar advances a distance equal to the array length, it must phase correct and sum (N x N) returns for every range bin. This load may be reduced somewhat by presumming (if conditions permit presumming). But it is still formidable.

The computing load can, however, be dramatically reduced by processing the data in parallel for many lines of the map at one time, rather than serially, a line at a time.

For parallel processing, the returns from different azimuth angles are isolated with doppler filters. But before getting into the details of that, we must see how doppler frequency is related to azimuth angle.

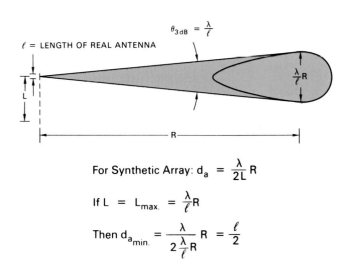

For Synthetic Array: $d_a = \frac{\lambda}{2L} R$

If $L = L_{max.} = \frac{\lambda}{\ell} R$

Then $d_{a_{min.}} = \frac{\lambda}{2\frac{\lambda}{\ell}R} R = \frac{\ell}{2}$

18. If array length, L, is made equal to beamwidth of real antenna at range, R, azimuth resolution of synthetic array will equal 1/2 length of real antenna.

WHY THE ELEMENT–TO–ELEMENT PHASE SHIFT IS DOUBLED IN A SYNTHETIC ARRAY

A linear-array radar antenna achieves its directivity by virtue of the progressive shift in the phases of the returns received by successive array elements from points off the antenna boresight line. For any one angle off boresight, this shift is twice as great for a synthetic array as for a real array having the same inter-element spacing. The reason can be seen by considering the returns received from a distant point, P, displaced from the boresight line by a small angle, θ.

REAL ARRAY

1. TRANSMISSION: Each successive pulse (A, B, C, D, E) is radiated simultaneously by all array elements (1, 2, 3, 4, 5).

2. REFLECTION: Phasors representing reflections of pulses A, B, C, D, and E from a distant point P.

3. RECEPTION: Phasors representing returns received by each array element (1, 2, 3, 4, 5) from pulses A, B, C, D, and E.

Real Array. In the case of a real array, transmission from all array elements is simultaneous. Every time a pulse is transmitted, the radiation from all elements arrives at P simultaneously—albeit staggered in phase as a result of the progressive differences in the distances from successive elements to P. The phase differences naturally reduce the amplitude of the sum of the radiation received at P from the individual elements. (This reduction gives the one-way radiation pattern its sin x/x shape.) But for each pulse, the phase shift of the sum is determined by the distance from the central element (3) to P. Therefore, if the position of the antenna is not changed, all of the pulses reflected by P have the same phase.

The portions of each reflected pulse that are received by the individual array elements similarly differ in phase as a result of the progressive difference in the distances from P to the elements. As with transmission, the phase differences reduce amplitude of the sum of the outputs the received pulse produce from the individual elements. This reduction, compounded with the reduction in the amplitude of the pulses reflected from P, gives the two-way radiation pattern its $(\sin x/x)^2$ shape. But the phase differences are again due only to the differences in the *one-way* distances from P to the elements.

SYNTHETIC ARRAY

1. TRANSMISSION: Pulses A, B, C, D, and E are radiated sequentially by array elements 1, 2, 3, 4, and 5.

2. REFLECTION: Phasors representing reflections of pulses A, B, C, D, and E from point P.

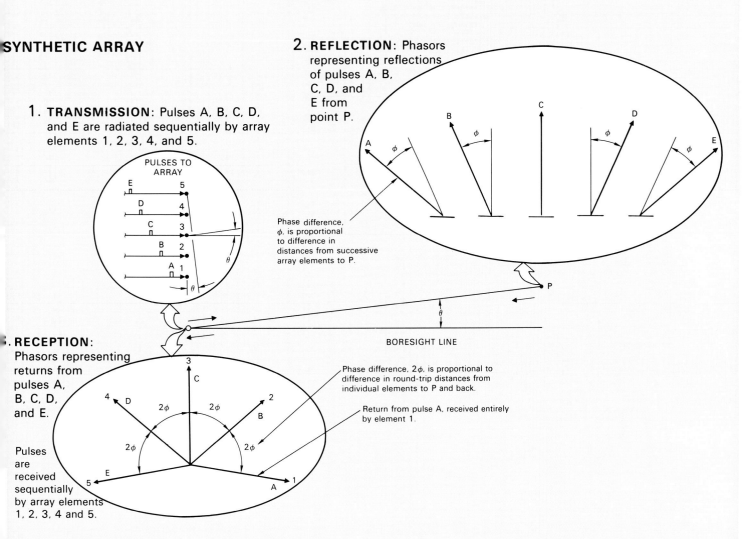

Phase difference, ϕ, is proportional to difference in distances from successive array elements to P.

3. RECEPTION: Phasors representing returns from pulses A, B, C, D, and E.

Pulses are received sequentially by array elements 1, 2, 3, 4 and 5.

BORESIGHT LINE

Phase difference, 2ϕ, is proportional to difference in round-trip distances from individual elements to P and back.

Return from pulse A, received entirely by element 1.

Synthetic Array. In the case of a synthetic array, transmission from the individual array elements is sequential. The first pulse is transmitted and received entirely by the first element; the second pulse, entirely by the second element; and so on. Consequently, the returns received by successive elements differ in phase by amounts proportional to the differences in the *round-trip* distance from each element to P and back to the element again.

Thus, for any one angle off the boresight line, the progressive shift in the phases of the returns received by successive array elements is twice as great for a synthetic array as for a real array. (The doubling of phase shift gives the beam of the synthetic array its sin 2x/2x shape.)

Significance. Because of the doubling of phase shifts, the null-to-null beamwidth of a synthetic array is only half the null-to-null beamwidth of a real array of the same length. And the 3-dB beamwidth is roughly 70 percent of the two-way 3-dB beamwidth of the real array. (At the -3 dB points, sin (2x/2x) $\simeq$ 0.7 (sin x/x)^2.)

The doubling of the phase shifts must, of course, also be kept in mind when calculating such factors as the phase corrections necessary to focus an array and the angles at which grating lobes (see Chapter 32) will occur.

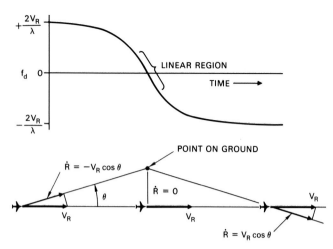

19. As a radar passes a point on the ground, its doppler frequency decreases at a nearly linear rate, passing through zero when the point is at an angle of 90° to the radar's velocity.

Doppler Frequency Versus Azimuth Angle. Figure 19 shows the doppler history of the return from a point of ground offset from the radar's flight path. When the point is a great distance ahead, its doppler frequency corresponds very nearly to the full speed of the radar and is positive. When the point is a great distance behind, its doppler frequency similarly corresponds to the full speed of the radar, but is negative.

As the radar goes by the point, its doppler frequency decreases at virtually a constant rate, passing through zero when the point is at an angle of 90° to the radar's velocity. If the radar antenna has a reasonably narrow beam and is looking out to the side at a reasonably large azimuth angle, the point will be in the antenna beam only during this linearly decreasing portion of the point's doppler history.

A plot of this portion of the doppler histories of several evenly spaced points at the same offset range is shown in Fig. 20.

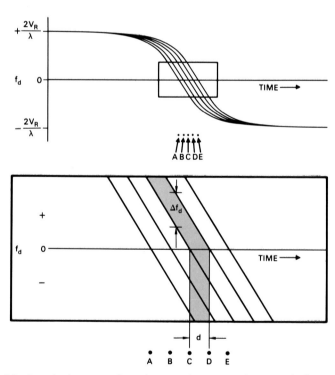

20. Doppler histories of evenly spaced points on the ground. The instantaneous frequency difference, Δf_d, is proportional to the azimuthal distance between points, d.

As you can see, the histories are identical—the frequency decreases at the same constant rate—except for being staggered slightly in time. Because of this stagger, at any one instant, the return from every point has a slightly different frequency. The difference between the frequencies for adjacent points corresponds to the azimuth separation of the points. We can isolate the return received from each point, therefore, by virtue of this difference in doppler frequency.

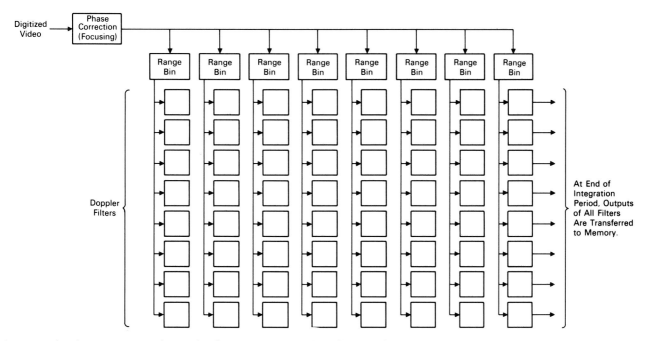

21. How doppler processing is done. After focusing corrections have been made, returns are sorted by range. When returns from a complete array have been received, a separate bank of filters is formed for each range bin.

Implementation. The block diagram of Fig. 21 (above) indicates in general how doppler processing is done.

At the outset, a phase correction is made to the returns received from each pulse to remove the linear slope of the doppler histories (i.e., to focus the array). This process, called *focusing,* converts the return from each point on the ground to a constant doppler frequency (Fig. 22). That frequency corresponds to the azimuth angle of the point, as seen from the center of the segment of the flight path over which the return was received.

Every time the aircraft traverses a distance equal to the length of the array that is to be synthesized, the phase-corrected returns which accumulate in each range bin are applied to a separate bank of doppler filters. Thus, for every array length, as many banks of filters are formed as there are range bins. The integration time for the filters is the length of time the aircraft takes to fly the array length. The number of filters included in each bank correspondingly depends upon the length of the array. The greater it is (hence the longer the filter integration time), the narrower the filter passbands and the greater the number of filters required to span a given band of doppler frequencies. The narrower the filters, of course, the finer the azimuth resolution.

Since the frequencies to be filtered are relatively constant over the integration time and (for uniformly spaced points on the ground) are evenly spaced, the fast Fourier transform (FFT) can be used to form the filters, greatly reducing the amount of computation. Herein lies the advantage of doppler processing.

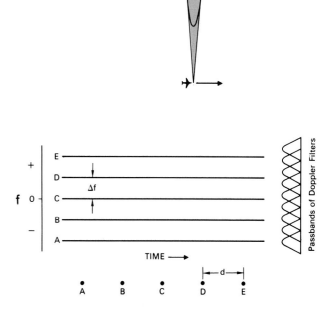

22. A phase correction converts the return from each point on the ground to a constant frequency, enabling the doppler filters to be formed with the FFT.

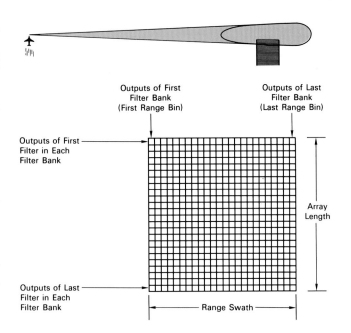

Outputs of First Filter Bank (First Range Bin)

Outputs of Last Filter Bank (Last Range Bin)

Outputs of First Filter in Each Filter Bank

Outputs of Last Filter in Each Filter Bank

Array Length

Range Swath

23. The outputs of each filter bank represent the return from a single column of range/azimuth resolution cells.

As required for the FFT, the filters are formed at the end of the integration period, i.e., after the radar has traversed an entire array length. The outputs of each bank of filters represent the returns from a single column of resolution cells at the same range—the range of the range bin for which the bank was formed (Fig. 23). The outputs of all of the filter banks, therefore, can be transferred as a block, in parallel, directly to the appropriate positions in the display memory. The radar, meanwhile, has traversed another array length thereby accumulating the data needed to form the next set of filter banks, and the process is repeated.

Incidentally, as illustrated in the panel on the facing page, the focusing and azimuth compression process just described is strikingly similar to the stretch-radar deramping and range compression process for decoding chirp pulses.

Reduction in Arithmetic Operations Achieved. Having gained (hopefully) a clear picture of the doppler-filtering method of azimuth compression, let's see what kind of saving in arithmetic operations it actually provides. To simplify the comparison, we'll assume that no presumming is done by either processor.

In the doppler processor, phase rotation takes place at two points: (1) when the return is focused, and (2) when the doppler filtering is done. For focusing, only one phase rotation per pulse is required for each range bin. As was explained in Chap. 20, in a large filter bank the number of phase rotations required to form a filter bank with the FFT is $0.5N \log_2 N$, where N is the number of pulses integrated. The total number of phase rotations per range bin for parallel processing, then, is $N + 0.5N \log_2 N$. For line-by-line processing, as we just saw, the number of phase rotations per pulse per range gate is N^2.

Processing	Phase Rotations
Line-by-line	N^2
Parallel (doppler)	$N(1 + 0.5 \log_2 N)$

To get a feel for the relative sizes of the numbers involved, let's take as an example a synthetic array having 1024 elements. With line-by-line processing a total of $1024 \times 1024 = 1,048,576$ phase rotations would be required. With parallel processing, only $1024 + 512 \log_2 1024 = 6,144$ would be required. The number of additions and subtractions would similarly be reduced. Thus, by employing parallel processing the computing load would be reduced by a factor of roughly 170!

Correspondence to Conventional Array Concepts. Superficially, doppler processing may seem like a funda-

SIMILARITY OF AZIMUTH COMPRESSION TO RANGE COMPRESSION WITH STRETCH RADAR

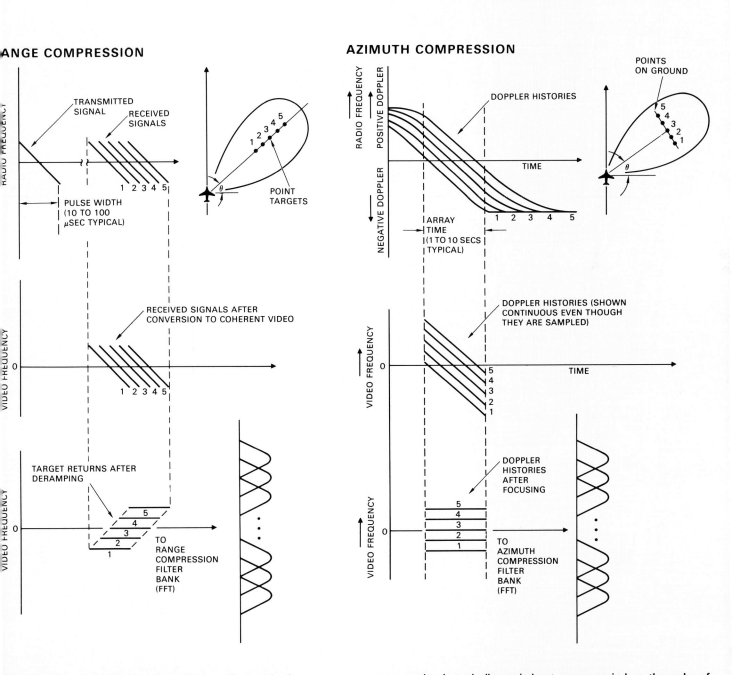

RANGE COMPRESSION

AZIMUTH COMPRESSION

The focusing and azimuth compression performed in the Doppler processing of SAR signals are strikingly similar to the deramping and range compression performed in the stretch-radar decoding of chirp pulses (when that method of pulse compression is used). The chief difference lies in the rate at which the compression is performed. Whereas azimuth

compression is typically carried out over a period on the order of 1 to 10 *seconds*, range compression is typically carried out over a period on the order of 10 to 100 *microseconds*. In both cases, deramping (focusing) may be performed either digitally, as described here, or by analog means as described in Chapter 13.

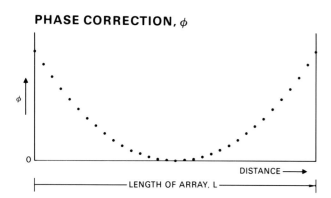

PHASE CORRECTION, ϕ

ϕ

0

DISTANCE ⟶

⟵ LENGTH OF ARRAY, L ⟶

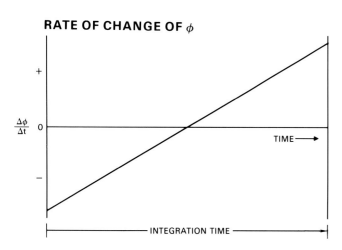

RATE OF CHANGE OF ϕ

+

$\frac{\Delta\phi}{\Delta t}$ 0

TIME ⟶

−

⟵ INTEGRATION TIME ⟶

24. Focusing corrections, ø, made to return from successive blocks of pulses. Rate of change of ø has same slope as doppler history of point on ground, but is rising rather than falling.

mental departure from conventional array concepts. But it is not. As we learned in Chap. 15, a doppler frequency is nothing more nor less than a progressive phase shift. To say that a signal has a doppler frequency of one hertz is but to say that its phase is changing at a rate of 360° per second. If the PRF is 1000 hertz, the pulse-to-pulse phase shift is 360° ÷ 1000 = 0.36°. Viewed in this light, the doppler histories we have been considering are really phase histories. Virtually every aspect of the doppler processor's operation, therefore, directly parallels that of the line-by-line processor described earlier.

The phase corrections used to remove the slope of the doppler history curves are exactly the same as the corrections used to focus the array in the line-by-line processor. This is illustrated by the graphs of Fig. 24. The "U" shaped curve is a plot of the focusing corrections applied to the returns received by successive array elements in line-by-line processing. The straight diagonal line is a plot of the rate of change of these corrections. Its slope, you will notice, is identical to the slope of the doppler history of a point on the ground but is rising, rather than falling. The same focusing correction that is used by the line-by-line processor, therefore, converts the linearly decreasing frequency of the return from each point on the ground to a constant frequency.

While not identical, the beams synthesized by the two processors are virtually the same. The only difference is in their points of origin (Fig. 25).

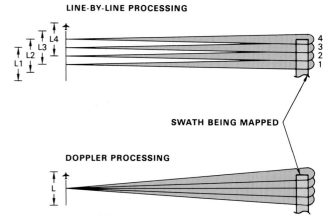

LINE-BY-LINE PROCESSING

L4
L3
L2
L1

4
3
2
1

SWATH BEING MAPPED

DOPPLER PROCESSING

L

25. Synthetic array beams formed with line-by-line processing and doppler processing differ only in their points of origin.

With the line-by-line processor, every time the radar advances one azimuth resolution distance, d_a, a new beam is synthesized. Whereas, with the doppler processor, every time the radar advances one array length, each doppler filter bank synthesizes a new beam.

The beams formed by the line-by-line processor all have the same azimuth angle (90° in the example we have been considering). But because of the radar's advance, at the range being mapped they overlap only at their half power points.

The beams formed by the doppler filters, on the other hand, all originate at the same point (center of the array). But they fan out at azimuth angles such that they overlap at their half power points.

And how do the azimuth resolutions provided by the two processors compare?

The 3-dB bandwidth of the doppler filters is roughly equal to one divided by the integration time (BW$_{3dB}$ $\cong$ 1/t$_{int}$). As shown in Fig. 26, the difference between the doppler frequencies of two closely spaced points on the ground at azimuth angles near 90° is twice the radar velocity times the azimuth separation of the points, divided by the wavelength.

$$\Delta f_d = \frac{2V_R \Delta\theta}{\lambda}$$

Equating Δf_d to BW$_{3\ dB}$ and substituting 1/t$_{int}$ for it, we obtain the following expression for the width of the beam synthesized by the doppler processor.

$$\Delta\theta = \frac{\lambda}{2V_R\ t_{int}}$$

where

$\Delta\theta$ = beamwidth
V_R = radar velocity
t_{int} = integration time

The product of the radar's velocity and integration time, $V_R t_{int}$, is the distance flown during the integration time. As illustrated in Fig. 27, that is the length of the array, L.

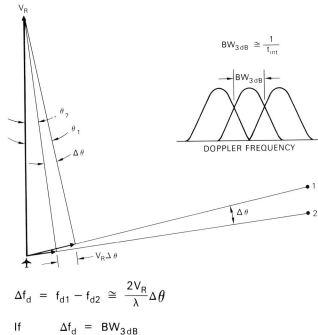

$$BW_{3dB} \cong \frac{1}{t_{int}}$$

DOPPLER FREQUENCY

$$\Delta f_d = f_{d1} - f_{d2} \cong \frac{2V_R}{\lambda}\Delta\theta$$

If $\qquad \Delta f_d = BW_{3dB}$

Then $\qquad \dfrac{2V_R}{\lambda}\Delta\theta = \dfrac{1}{t_{int}}$

And $\qquad \Delta\theta \cong \dfrac{\lambda}{2V_R t_{int}}$

26. The 3-dB bandwidth of a doppler filter is one divided by the integration time. The difference in doppler frequency, Δf_d, of the returns from two points on the ground is proportional to their angular separation, $\Delta\theta$. Equating BW$_{3\ dB}$ to Δf_d yields an expression for the angular resolution.

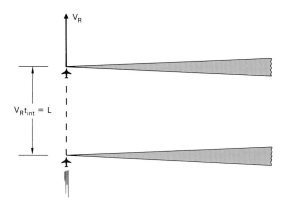

27. The length of the array synthesized with doppler processing is the distance flown during the integration time of the doppler filters.

423

Substituting L for $V_R t_{int}$ and multiplying by the range, R, we find the azimuth resolution distance to be:

$$d_a = \frac{\lambda}{2L} R$$

This is exactly the same as the azimuth resolution distance for the line-by-line processor: one half the resolution distance for a real array of the same length.

So, whether you think of a synthetic array radar in terms of doppler processing or of conventional array concepts is largely a question of which view makes the particular aspect of the array you are concerned with easier to visualize.

Summary

Fine azimuth resolution may be obtained by pointing a small radar antenna out to one side, storing the returns received over a period of time, and integrating them so as to synthesize the equivalent of a long array antenna—SAR. The points at which successive pulses are transmitted can be thought of as the elements of this array.

Phase errors due to the greater range of a point on the ground from the ends of the array than from the center limit its useful length. The limitation may be removed through phase correction, a process called "focusing."

With focusing, azimuth resolution can be made virtually independent of range by increasing the array length in proportion to the range of the region being mapped.

Since that region must lie within the beam of the real antenna throughout the entire time the array is being formed, the length of an array having a fixed look angle is limited to the width of the beam of the real antenna at the range being mapped. (This limitation is removed in the spotlight mode.) The smaller the real antenna, the wider its beam will be, hence the longer the synthetic array can be made.

Computation may in some cases be reduced by presumming (when possible) the returns received by blocks of array elements and applying the phase corrections for focusing only to the sums.

In any event, computation may be substantially reduced by integrating the phase-corrected returns in a bank of doppler filters, with the FFT.

Some Relationships To Keep In Mind

- Azimuth (cross-range) resolution distance for a synthetic array

$$d_a = \frac{\lambda}{2L} R$$

λ = wavelength
L = array length (same units as λ)
R = range (same units as λ)

- Maximum effective length of unfocused array

$$L_{eff} = 1.2 \sqrt{\lambda R}$$

- Minimum resolution distance for unfocused array

$$d_{a_{min}} \approx 0.4 \, L_{eff}$$

- Minimum resolution distance for unfocused array having fixed look angle

$$d_{a_{min}} = \frac{l}{2} \text{ where } l = \text{length of real antenna}$$

SAR Design Considerations

32

In the last chapter, we saw how SAR takes advantage of a radar's forward motion to synthesize a very long linear array from the returns received over a period of up to several seconds by a small real antenna. We learned how the array may be focused at virtually any desired range and how the immense amount of computing required for digital signal processing may be dramatically reduced through doppler filtering techniques.

In this chapter, we will consider certain critical aspects of SAR design which, if not properly attended to, may seriously degrade the quality of the maps or perhaps even render them useless: selection of the optimum PRF, sidelobe reduction, compensation for phase errors resulting from deviation of the radar bearing aircraft from a perfectly straight constant-speed course—called *motion compensation*—and the minimization of other phase errors.

Choice of PRF

The PRF must be set low enough to avoid range ambiguities, yet high enough to avoid doppler ambiguities—or, in terms of antenna theory, high enough to avoid problems with grating lobes.

Avoiding Range Ambiguities. The maximum value of the PRF is limited by the requirement that returns from the ranges being mapped not be received simultaneously with mainlobe returns from any other ranges. This requirement may be readily met by setting the PRF so that the echo of each pulse from the far edge of the real antenna's footprint is received before the echo of the following pulse from the

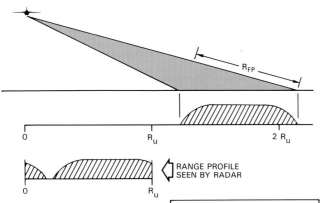

near edge—in other words, so that the unambiguous range, R_u, is at least as long as the slant range, R_{FP}, spanning the footprint (Fig. 1). That criterion will be satisfied if the PRF is less than

$$PRF_{max} = \frac{c}{2R_{FP}}$$

where

c = speed of light (162,000 nmi/second)

R_{FP} = range spanning footprint of real antenna

1. Range ambiguities may be avoided by making R_u greater than the slant range from the near edge to the far edge of the real antenna's footprint, R_{FP}.

$$R_u = \frac{cT}{2} = \frac{c}{2PRF}$$

$$\text{If} \quad R_u \geq R_{FP}$$

$$\text{Then} \quad \frac{c}{2PRF} \geq R_{FP}$$

$$\therefore \quad PRF_{max} = \frac{c}{2\,R_{FP}}$$

Sample Computation of PRF$_{max}$

Find:
The maximum PRF a SAR radar can have and still avoid range ambiguities under these conditions:

- Range segment being mapped may lie anywhere within footprint of antenna beam.
- Slant range, R_{FP}, spanning footprint = 20 nmi
- Speed of light = 162,000 nmi/sec

Calculation:

$$PRF_{max} = \frac{c}{2R_{FP}}$$

$$= \frac{162,000}{2 \times 20}$$

$$= 4050 \text{ Hz}$$

As you may be thinking, if only a small segment of R_{FP} is being mapped, cannot higher PRFs in some cases be used? Certainly. Within a narrow segment in the center of R_{FP}, for instance, ambiguities may be avoided even with PRFs approaching twice the maximum given by the above expression.

Avoiding Doppler Ambiguities. The minimum PRF is generally limited by the requirement that the "lines" of mainlobe ground return must not overlap. To meet this requirement, the PRF must exceed the maximum spread between the doppler frequencies of points on the ground at the leading and trailing edges of the mainlobe of the real antenna.

Therefore,

$$PRF_{min} = f_{d_L} - f_{d_T}$$

where f_{d_L} and f_{d_T} are the doppler frequencies at the mainlobe's leading and trailing edges (Fig. 2).

In the case of a narrow azimuth beamwidth, the doppler spread is approximately equal to $2\,V_R\theta_{NN_a}/\lambda$ times the sine

2. To avoid doppler ambiguities, the PRF must exceed the difference between the doppler shifts at the leading and trailing edges of the real antenna's mainlobe.

of the look angle, ε (Fig. 3). So the minimum acceptable PRF is

$$\mathrm{PRF}_{min} \cong \frac{2\,V_R\,\theta_{NN_a}}{\lambda}\;\sin\varepsilon$$

where

$\quad V_R \quad$ = velocity of the radar

$\quad \theta_{NN_a} \quad$ = null-to-null azimuth beamwidth of real antenna in radians

$\quad \varepsilon \quad$ = azimuth look angle

$\quad \lambda \quad$ = wavelength, units consistent with V_R

In typical airborne applications, V_R is between 800 and 1500 feet per second.[1]

Grating Lobes. Some people find the limitation on minimum PRF easier to visualize in terms of antenna theory. In those terms, what determines the minimum PRF is the distance, d_e, between successive array elements. Now, d_e equals the radar's speed times the interpulse period, $1/f_r$. If d_e is greater than half a wavelength, as it generally will be in most SARs, so-called *grating lobes* will be produced. These are replicas of the mainlobe, occurring at increasingly large intervals on either side of the mainlobe[2] (Fig. 4).

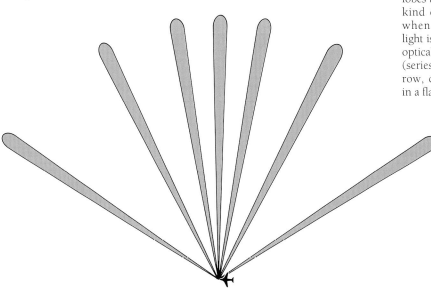

4.   Grating lobes are replications of an array's mainlobe occurring at increasingly large intervals on either side of the mainlobe.

Grating lobes are not unique to synthetic arrays. But they are more of a problem in these arrays then in real arrays. There are two reasons for this. First, because of the restrictions on maximum PRF, the array elements generally cannot be placed as close together in a synthetic array as in a real array. Second, as we saw earlier, for a reflector at any one angle off boresight, the difference in the phase shift of the returns received by successive array elements is twice as great in a synthetic array as in a real array.

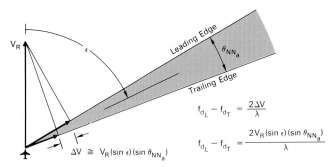

$$f_{d_L} - f_{d_T} = \frac{2\Delta V}{\lambda}$$

$$f_{d_L} - f_{d_T} = \frac{2V_R(\sin\epsilon)(\sin\theta_{NN_a})}{\lambda}$$

$$\Delta V \cong V_R(\sin\epsilon)(\sin\theta_{NN_a})$$

3.   Geometric relationships determining the spread between doppler frequencies of return from leading and trailing edges of the real antenna's mainlobe.

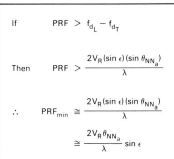

If $\qquad \mathrm{PRF} > f_{d_L} - f_{d_T}$

Then $\qquad \mathrm{PRF} > \dfrac{2V_R(\sin\epsilon)(\sin\theta_{NN_a})}{\lambda}$

$\therefore \qquad \mathrm{PRF}_{min} \cong \dfrac{2V_R(\sin\epsilon)(\sin\theta_{NN_a})}{\lambda}$

$\qquad\qquad \cong \dfrac{2V_R\theta_{NN_a}}{\lambda}\sin\epsilon$

1.   If we assume that $\theta_{NN_a} = 2\lambda/l$, where l is the length of the real antenna, then at an azimuth look angle of 90°, $\mathrm{PRF}_{min} = 4V_R/l$.

2.   They are called grating lobes because they are the kind of lobes produced when monochromatic light is passed through an optical diffraction grating (series of extremely narrow, closely spaced slits in a flat plate).

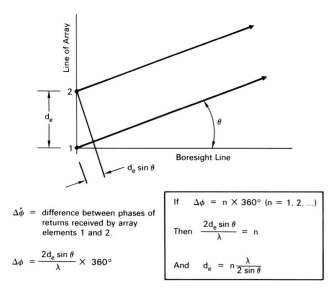

$\Delta\dot\phi$ = difference between phases of returns received by array elements 1 and 2.

$\Delta\phi = \dfrac{2d_e \sin\theta}{\lambda} \times 360°$

If $\Delta\phi = n \times 360°$ (n = 1, 2, ...)

Then $\dfrac{2d_e \sin\theta}{\lambda} = n$

And $d_e = n\dfrac{\lambda}{2\sin\theta}$

5. Conditions under which grating lobes are produced. If spacing between array elements (d_e) times sine of angle θ off boresight is a multiple of half a wavelength, returns received from angle θ by successive elements will be in phase.

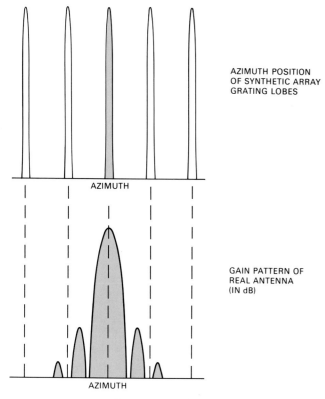

AZIMUTH POSITION OF SYNTHETIC ARRAY GRATING LOBES

GAIN PATTERN OF REAL ANTENNA (IN dB)

6. Minimum acceptable PRF places first grating lobe between first and second sidelobes of real antenna.

3. Since the angle θ_N between this null and the boresight line equals θ_{NN}, the PRF that places the first grating lobe here is the same as PRF_{min} derived on page 427.

How the lobes are produced can be explained as follows. If a reflector is gradually moved away from the boresight line of the array, a phase difference develops between the returns received by successive array elements (Fig. 5). This difference is proportional to twice the spacing of the elements times the sine of the azimuth angle (θ) of the reflector relative to the boresight line. The familiar pattern of nulls and sidelobes is thus observed.

However, if the spacing of the array elements is much greater than half a wavelength, as the azimuth of the reflector is increased, a point is soon reached where the phase shift is 180°. Beyond this point, the amplitudes of successive lobes start *increasing* (sin [180 + θ] = – sin θ). The increase continues until the element-to-element phase shift reaches 360°. At this point, the returns received by all of the elements add up, once again, exactly as they did when the reflector was in the center of the mainlobe. This "replica" of the mainlobe is the first grating lobe.

If the azimuth of the reflector is increased further, the process repeats and successive grating lobes appear.

In a real antenna, the gains of grating lobes fall off gradually from that of the mainlobe with increasing azimuth angles; but in a synthetic array, the fall off is much greater. The reason for this is that the synthetic array is formed from returns received through the real antenna. The strength of the returns received from any one direction, of course, is proportional to this antenna's two-way gain in that direction. In general, that gain decreases rapidly as the azimuth angle increases. If the azimuth angle of the first grating lobe can be made sufficiently large, the amount of energy received through the grating lobes can be reduced to negligible proportions.

Generally, the restriction imposed on the PRF by the requirement that range not be ambiguous is reasonably loose. So the PRF can usually be set high enough to place the first grating lobe well outside the mainlobe of the real antenna. If it can't be, it is placed in a null between sidelobes (Fig. 6). But it must not be placed closer to the mainlobe than the second null.[3]

To maintain the desired spacing between array elements, as well as to keep the number of pulses that must be processed in a given array length constant, a common practice is to adjust the PRF to the speed of the radar.

Minimizing Sidelobes

Performance of a synthetic array radar may be degraded by both range sidelobes due to pulse compression and the sidelobes of the synthetic array. The sidelobes affect the radar maps in two different ways. First, the peaks of the stronger sidelobes may cause a string of progressively weaker false targets to appear on either side of a strong target (Fig. 7).

Second, the combined power of all sidelobes—called the *integrated sidelobe return*—together with receiver noise, tends to fog or wash out the detail of the maps.

The effect of the integrated sidelobe return can be visualized by imagining an area of ground the size of a resolution cell which produces no return—a smooth surfaced pond, for example—in the middle of a region of uniform backscattering—say a grassy field. The signal output when the pond is being mapped is the sum of the simultaneously received power of the range sidelobes and the azimuth sidelobes (Fig. 8)—plus the receiver noise. To the extent that this power is comparable to that received from the surrounding terrain, the "hole" in the map corresponding to the pond will be filled in. If nothing is done to reduce the sidelobes, the integrated sidelobe return alone may contain up to 10 percent as much power as the mainlobe return. Consequently, the loss of contrast can be considerable.

Like the sidelobes of a real array, the sidelobes of a synthetic array are produced by the elements at the ends of the array. Consequently, just as the sidelobes of a real antenna may be reduced through illumination tapering (see Chap. 8), the sidelobes of the synthetic array can be reduced by weighting the returns received by the individual array elements (i.e., the returns from successive transmitted pulses) so as to de-emphasize the returns received by the end elements relative to the returns received by the central elements. The cost of this reduction, of course, is a slight loss of resolution. The weighting can be conveniently accomplished when the focusing corrections are applied to the stored returns. The loss in resolution can be avoided by increasing the array length (integration time), at the expense of increased computational load.

Having reduced the range and azimuth sidelobes to acceptable levels, the remaining loss in contrast due to noise can be reduced by increasing the gain of the real antenna or the average transmitted power.

Motion Compensation

In Chap. 31, it was assumed that the aircraft was flying at constant speed in a perfectly straight line. But this is virtually never the case. Since the whole SAR concept revolves around the effect of very slight differences in the phases of signals received over a comparatively long period of time, typically 1 to 10 seconds, it is essential that any acceleration of the aircraft during that period be compensated. The acceleration may be measured either by accelerometers mounted on the antenna or by a separate inertial navigation

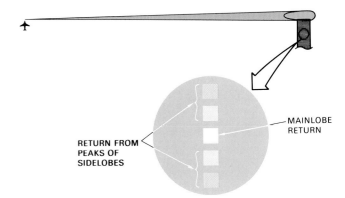

RETURN FROM PEAKS OF SIDELOBES

MAINLOBE RETURN

7.   Close-up of a portion of a radar map showing effect of unsuppressed lower-order sidelobes when a strong point target is mapped.

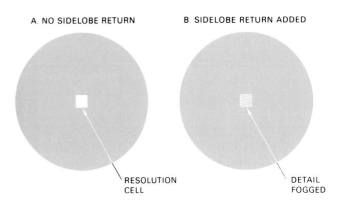

A. NO SIDELOBE RETURN

B. SIDELOBE RETURN ADDED

RESOLUTION CELL

DETAIL FOGGED

8.   How integrated sidelobe return washes out detail. Cell in center represents area from which no return is received.

system, the outputs of which are referenced to the phase center of the antenna.

On the basis of the measured acceleration, phase corrections are computed. These may be applied to the received signals at virtually any point in the radar system, from local oscillator to final integration. Where presumming is employed, the corrections may, for example, be applied on a sample-by-sample basis after the presumming.

PHASE ERRORS

COMMON SOURCES

- Unmeasured velocity error
- Unmeasured acceleration along line of sight during array time
- Non-linear motion of aircraft
- Equipment imperfection
- Processing approximations
- Atmospheric disturbances

EFFECTS

- Increased sidelobe levels
- Degraded resolution
- Reduced antenna peak gain
- Beam wander

Limit on Uncompensated Phase Errors

The common sources of phase errors and their effects on the radar's performance are listed in the table.

The significance of these errors cannot be overstressed. It can be shown mathematically that an uncorrected *low-frequency* phase error of only 114° from the center to the ends of an array will result in a 10 percent spreading of the synthetic beam. At X-band wavelengths, 114° amounts to only about 3/8 of an inch. As you might expect from looking at Fig. 9 of the last chapter, though, the predominant effect of uncompensated phase errors usually is increased sidelobe levels. As a rule, to keep these within acceptable bounds, the total random (high-frequency) phase error must be held to within 2 to 6°. Yet at X-band, 6° of phase is equivalent to an antenna motion of only 0.01 inch!

Summary

If not duly considered, certain aspects of SAR design may seriously degrade the radar maps. Among the more important are choice of PRF, sidelobe reduction, motion compensation, and uncompensated phase errors.

Two primary factors influence the choice of PRF. The *maximum* value is limited by the requirement that no range ambiguities occur within the span of ranges from which mainlobe return is received. The *minimum* value is limited by the requirement that there be no doppler ambiguities within the band of frequencies spanning the central spectral line of the mainlobe return. In terms of antenna theory, this same minimum PRF places the first grating lobe between the first and second sidelobes of the real antenna.

A synthetic array's stronger sidelobes may cause weak, false targets to appear on either side of strong targets. And the combined sidelobe return from all targets—integrated sidelobe return—may wash out detail. The sidelobes may be reduced through amplitude weighting.

Particularly important is motion compensation—measuring the radar's acceleration and introducing phase corrections to compensate for deviations from a straight, constant-speed course.

SAR Operating Modes

Operationally, SAR has several striking advantages. First, with a small physical antenna operating at wavelengths suitable for long-range mapping, SAR can provide azimuth resolutions as fine as a foot or so. Second, by increasing the length of the array in proportion to the range of the area to be mapped, the resolution can be made independent of range. Third, since the array is formed in the signal processor, the basic SAR technique can conveniently be adapted to a wide variety of operational requirements.

Added to these advantages are those of all radar mapping. Maps can be made equally well day or night, through smoke, haze, fog, or clouds. The maps are plan views and can be made even at shallow grazing angles (Fig. 1, below).

While simple strip-mapping described in Chap. 31 is quite useful, it has been improved upon and adapted to special requirements in a variety of modes. Some of the more important of these are squinted array, multilook mapping, spotlighting, doppler beam sharpening, moving-target display, and inverse SAR (ISAR) imaging. Each is briefly described in the following paragraphs.

1. Comparison of real-time SAR map with an aerial photo of the same region. Map was made from range many times as great as that from which photo was taken. Radar not only sees through haze, but provides a plan view.

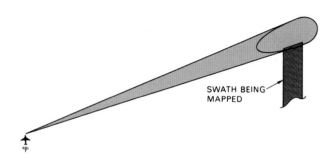

2.   If beam of real antenna is squinted forward and appropriate focusing correction and coordinate rotation are made, region ahead can be mapped.

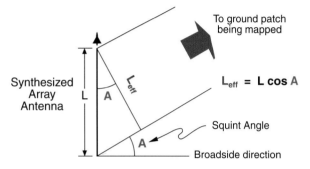

3.   Foreshortening of synthesized arrary that occurs when radar beam is squinted forward (or backward) by an angle A for SAR mapping.

Squinted Array

In real-time SAR mapping, if the radar's azimuth look angle is 90°, the leading edge of the map may lag behind the radar by as much as two full array lengths. This limitation can readily be eliminated. By training the beam of the real antenna forward and making an appropriate focusing correction and coordinate rotation, the synthetic array can be squinted ahead by any desired amount within a fairly wide zone to the side of the aircraft (Fig. 2). It can similarly be squinted behind. The radar can thus be made to map not only territory which it has long since passed, but territory which lies far ahead.

A price, of course, must be paid for this versatility. For, as viewed from the patch being mapped, the effective length, L_{eff}, of the synthesized array is foreshortened in proportion to the cosine of the squint angle, A (Fig. 3). The azimuth (cross-range) resolution distance, d_a, is correspondingly reduced.

$$d_a = \frac{\lambda R}{2L \cos A}$$

But this reduction is generally a small price to pay for the increased utility of the maps obtained.

Multilook Mapping

Sometimes the beam of the real antenna may be wide enough to enable the same area to be mapped several times without changing the antenna's look angle. This is called *multilook mapping*. When the maps are superimposed (i.e., when successive returns from each resolution cell are averaged), the effects of scintillation are reduced.

Most of the maps used as illustrations in these chapters were made with more than one look (Fig. 4).

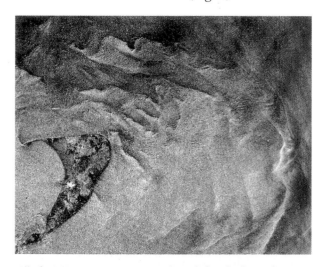

4.   All of SEASAT's maps were made with four looks. In this one, surface waves around Nantucket Island reflect the contours of the ocean floor. (Courtesy Jet Propulsion Laboratory)

Spotlight Mode

By gradually changing the look angle of the real antenna as the radar advances and making appropriate phase corrections, the radar can repeatedly map a given region of interest. This mode, called *spotlight*, not only enables the operator to maintain surveillance over an area for an appreciable period of time but can produce maps of superior quality.

Quality may be improved in three basic ways. First, since the beam is continuously trained on the area being mapped, the length of the synthetic array is not limited by the beamwidth of the real antenna.

Second, the size of the real antenna can be increased without reducing the array length. By using a larger antenna, the mainlobe gain can be increased and the signal-to-noise ratio correspondingly improved (Fig. 5).

The third way in which spotlighting improves the quality of a map is by filling in gaps in the backscatter from points on the ground. In Chap. 30, you may recall, it was pointed out that when a radar illuminates an object on the ground —such as a parked airplane—from a given angle, mappable returns may be received from only a few main scattering centers. The reason (explained on page 394) is that in terms of fractions of wavelengths, hence radio frequency phase, the distances from the radar to the various scatterers comprising the airplane may differ in such a way that much of the scatter does not combine constructively in the radar's direction. The net result is that the airplane's shape is not necessarily as easily recognized as one would expect from the ratio of the aircraft's principal dimensions to the size of th radar's resolution cells.

In the case of distributed targets, such as fields, grasslands, paved areas, etc., this same effect makes the backscatter spotty. The result in this case is a general graininess of the images.

Since the wavelength is usually comparatively short, the relative distances to the individual scatters (in fractions of wavelengths) can change markedly when the same area is viewed from slightly different angles. Consequently, the graininess may be considerably reduced by repeatedly mapping the same region from points progressively farther along on the flight path and averaging successive returns from each resolution cell.

The quality of the maps may be further enhanced by periodically switching from one to another of several different radio frequencies. These should be separated by at least the bandwidth of the transmitted pulses, and the switching should be done at points that are one or more array lengths apart. By similarly changing the polarization of the antenna, the quality may be improved even further.

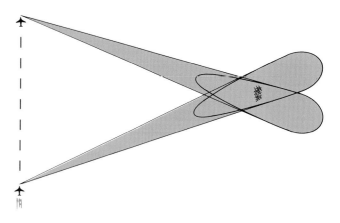

5. In spotlight mode, the beam of the real antenna is held on a given region of interest so that it can be mapped repeatedly from different angles and the array length can be increased, without decreasing the size, hence the gain, of the real antenna.

Doppler Beam Sharpening (DBS)

This mode differs from other SAR modes in that the length of the array is not increased in proportion to the range of the area to be mapped but is the same for all ranges. The maps, therefore, are the same as those produced by a real array having an extremely narrow beam—hence the name *doppler beam sharpening*.

Typically, the antenna continuously scans the region of interest on one side of the flight path or the other, or both. Because the integration time is limited to the length of time a ground patch is in the antenna beam—or, if you prefer, the length of the array that can be synthesized is so limited—the resolution is coarser than can be achieved with a nonscanning antenna. Moreover, since the integration time is the same for the returns from all ranges, the azimuth resolution distance increases with range, rather than being independent of it.

Nevertheless, except for the region directly ahead (Fig. 6), where there is little or no spread in the doppler frequencies within the mainlobe, the resolution is much finer than could be achieved by the real antenna. Also, in contrast to the higher resolution modes, DBS can provide a continuously updated map of a large expanse of ground.

Implementation is similar to that of the other SAR modes. In one advanced design, the gap in the region ahead is filled in by scanning it in a phase-comparison-monopulse detection mode providing substantially finer resolution than conventional real-beam mapping (see APN-241, Part X).

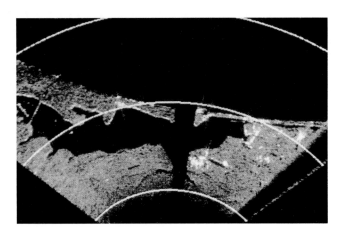

6. With doppler beam sharpening, the antenna scans a wide region—in this case on both sides of the flight path. Except for the region directly ahead, resolution is comparable to that achieved by a real antenna having an exceptionally narrow beam.

Moving Target Display

Frequently, it is desired to show ground-moving targets on the SAR map. Most of these are essentially point-source reflectors: cars, trucks, etc. Because of their motion while the radar is collecting the returns from which to synthesize the array, they tend to wash out in the map. To detect them, a ground-moving-target-indication (GMTI) mode is generally interleaved with SAR mapping. Markers indicating the targets' positions (Fig. 7)—and in some cases their range rates as well—are then superimposed on the SAR map. In one intriguing design, GMTI and SAR mapping are performed simultaneously with the same antenna (see APG-76, Part X).

Bigger, larger-RCS targets, such as trains, are clearly visible on most SAR maps. If a train is moving and has a component of velocity toward or away from the radar, however, the resulting doppler shift will normally be interpreted by the radar as indicative of a displacement in the cross-range direction. As a result, the train will be displayed as though it is traveling off its tracks. In more advanced SAR systems, the error is sensed and the train is put back on its tracks.

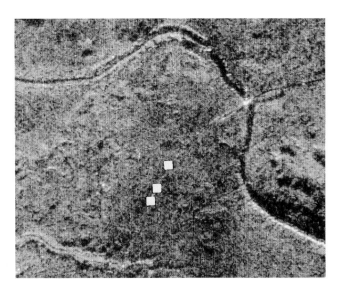

7. Markers indicate locations of moving targets.

Inverse SAR (ISAR) Imaging

The classic SAR is ill-suited for imaging targets such as ships and aircraft which have rotational motion. For unless the differential doppler shifts which such motions produce are accurately predicted and compensated, they tend to defocus the array and blur the image.

With slightly different algorithms, however, these shifts, rather than those due to the radar's forward motion can be used to provide the angular resolution needed for imaging a target (Fig 8). The technique then is called *inverse SAR*, or ISAR.

ISAR is most easily explained by starting with the SAR spotlight mode. As we have seen, in it the radar spotlights the ground patch that is to be mapped and records the returns received over a period of time. Returns simultaneously received from points at different azimuth angles are then separated on the basis of differences in their doppler (phase) histories due to the corresponding differences in the points' range rates.

With ISAR, the principle is the same. But the differences in range rate are those due to rotation of the spotlighted target about its yaw, pitch, and roll axes, as seen by the radar. This difference is illustrated in Fig. 9.

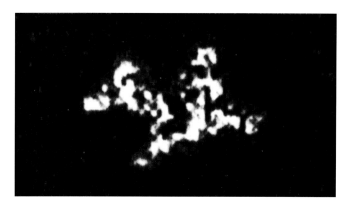

8.   An ISAR image of a fighter aircraft.

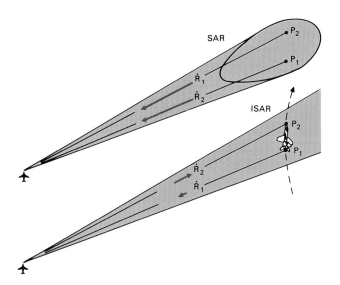

9.   With conventional SAR, the differences in doppler frequency which enable fine angular resolution to be obtained are due to the forward motion of the radar-bearing aircraft. With ISAR, the differences are due to the angular rotation of the target as seen by the radar

Since the target shown in Fig. 9 is turning away from the radar (rotating clockwise) the range rate of point P_1 on the tail is slightly lower than the range rate of point P_2 on the nose. Consequently, the doppler frequency of the returns from P_2 and all points on the target between it and P_1 differ in proportion to their distances from P_1.

CROSS-RANGE RESOLUTION OF ISAR IMAGES

Assuming that

$\dot{\theta}$ = Target's rate of rotation.

d_n = Desired cross-range resolution

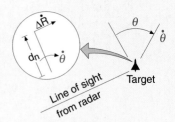

As is clear from the diagram,

$\Delta\dot{R}$ = incremental increase in range rate of points on the target separated by the distance, d_n, normal to the line of sight from the radar

$$\Delta\dot{R} = d_n\dot{\theta}$$

The resulting difference in the doppler frequencies of the radar returns from successive points on the target is

$$\Delta f_d = \frac{2\Delta\dot{R}}{\lambda} = \frac{2 d_n\dot{\theta}}{\lambda}$$

The minimum difference in doppler frequency the radar can resolve equals the 3-dB bandwidth of its doppler filters. Accordingly,

$$\frac{2 d_n\dot{\theta}}{\lambda} = BW_{3dB}$$

making the cross-range resolution distance,

$$d_n = \frac{BW_{3dB}}{2\dot{\theta}}\lambda$$

To make an image of the target, phase corrections must first be made to compensate for the displacement of the target relative to the radar while it collects the returns from which the image will be produced. This is called motion compensation. The returns from each resolvable range increment are then applied to a separate bank of doppler filters, and an image is produced from their outputs just as in conventional SAR mapping.

As shown in the panel (left), the image's cross-range resolution, d_n, is proportional to the ratio of the doppler filters' 3-dB bandwidth, BW_{3dB}, to the target's rate of rotation, $\dot{\theta}$.

$$d_n = \frac{BW_{3db}}{2\dot{\theta}}\lambda$$

The cross-range dimension, though, is not necessarily horizontal, as with conventional SAR, but perpendicular to the axis about which the target happens to be rotating. No image is formed, of course, if that axis is colinear with the radar's line of sight, or if the target has no rotational motion as viewed from the radar.

Besdies imaging targets having rotational motion, ISAR has another important advantage over conventional SAR. This advantage is illustrated in Fig. 10 (below).

With SAR, the cross-range resolution distance, d_a, is inversely proportional to the angle, $\boldsymbol{\theta}$, through which the radar flies during the doppler filters' integration time, t_{int}.

With ISAR, the cross-range resolution distance d_n is inversely proportional to the angle, $\boldsymbol{\theta}$, through which the target rotates during t_{int}. The radar need not fly through any angle to obtain an image of a target. In fact, the ISAR image of a fighter aircraft shown in Fig. 8 on the preceeding page was made by a stationary radar on the ground.

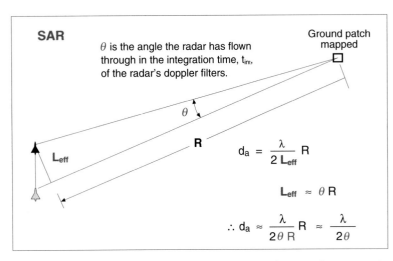

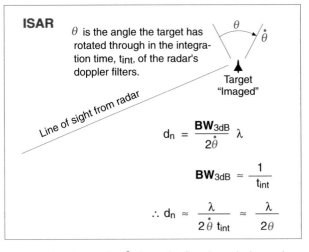

10. With conventional SAR, the cross-range resolution distance, d_a, is inversely proportional to the angle, $\boldsymbol{\theta}$, the radar flies through during the doppler filters' integration time, t_{int}. With ISAR, the cross-range resolution distance, d_n, is proportional to the angle, $\boldsymbol{\theta}$, the target rotates through in t_{int}.

Summary

Operationally, SAR has many compelling advantages. It:

- Affords excellent resolution with a small antenna even at very long ranges

- Where desired, provides resolution as fine as a foot or so (Fig. 11)

- Enables resolution to be made independent of range

- Produces recognizable images of ships and aircraft on the basis of their rotational motions, and

- Is exceptionally versatile.

11. Real-time, 1-foot resolution SAR map made from long range in the spotlight mode. (Crown copyright DERA Malvern)

Among the more important operational modes are strip mapping, forward and backward squinted array, multilook mapping, spotlighting, doppler beam sharpening, moving target display, and ISAR imaging.

Multilook mapping improves map quality by averaging out the scintillation of the radar returns.

Spotlighting removes the limitation on the size of the real antenna, enabling both antenna gain and array length to be increased and more uniform radar reflections to be obtained.

Doppler beam sharpening provides high quality continuously updated maps of large expanses of ground, at the expense of (a) greater cross-range resolution distance and (b) cross-range resolution distance increasing with range.

Whereas, with conventional SAR, angular-resolution distance is inversely proportional to the angle the *radar flies* through, with ISAR, it is inversely proportional to the angle the *target rotates* through.

An Instrumentation Application of ISAR. To detect "hot spots" in a target's RCS, a nose-mounted radar in an A-3 testbed makes ISAR images of the target from various positions within the target's rear hemisphere. For forward hemisphere imaging, a tail-mounted radar is used and positions of the aircrafts are juxtaposed.

McDONNELL DOUGLAS F-15A EAGLE (1973)

This premier air superiority fighter and later versions of it have been in service with the USAF worldwide. It was equipped with the Hughes APG-63 pulse doppler radar—the first to enable look-down shoot-down attacks. Armament includes Sparrow and Sidewinder missiles plus a 20mm cannon. AMRAAM missiles have subsequently been added.

Electronic Countermeasure (ECM) Techniques

34

In this chapter, we will be introduced to the six basic types of countermeasures—chaff, noise jamming, false targets, gate stealers, angle deception, and decoys. We will see how each is used, and learn how it is implemented and what its limitations are.

Chaff

The simplest of all countermeasures, and historically the earliest to be used, is chaff. Originally strips of metal foil, chaff today consists of metal-coated dielectric fibers, billions of which can be stored in a small space (Fig. 1). Injected into the air stream, they hang in the air for long periods. When dispensed in large numbers, they can produce strong radar echoes.

Upon being dispensed, the chaff rapidly decelerates and, except for turbulence, soon has little motion. Consequently, its echoes are rejected by radars employing moving target indication (MTI), just as weather clutter is. Against radars without MTI, however, chaff can be highly effective. Dispensed by a few escorting aircraft, chaff can screen an entire raid. If fired forward, it may screen the dispensing aircraft, as well.

Even against radars with MTI, chaff has important uses. It may screen surface targets, which have little or no motion. Against an approaching radar guided missile, chaff at the very least introduces tracking noise in the missile's seeker. If dispensed in conjunction with an evasive maneuver,[1] it can break the seeker's lock on the aircraft's echoes and so cause the missile to miss.

The strength of the radar returns produced by chaff

1. Chaff consists of thin metal-coated dielectric fibers, billions of which can be stored in a small space.

1. One such maneuver is to turn so the aircraft's velocity is normal to the missile's, making the doppler frequency of the aircraft the same as that of the chaff and the seeker will transfer look to it.

439

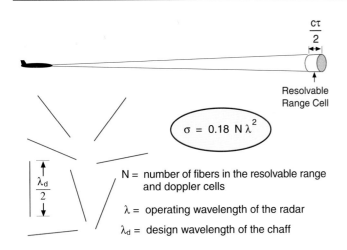

$$\sigma = 0.18 \, N \, \lambda^2$$

N = number of fibers in the resolvable range and doppler cells

λ = operating wavelength of the radar

λ_d = design wavelength of the chaff

2.   Radar cross section, σ, of randomly distributed and randomly oriented chaff fibers. Since the fibers are light and small, very large radar cross sections can readily be achieved.

varies with the length and orientation of the individual fibers and the number of fibers in the resolvable range and doppler cells (Fig. 2).

The bandwidth of the returns is inversely proportional to the fibers' aspect ratio (ratio of length to diameter). This ratio is generally quite high. Nevertheless, since the fibers are resonant at wavelengths which are multiples of half their length, a wide frequency band can be covered by dispensing chaff of several different, well chosen lengths (Fig. 3).

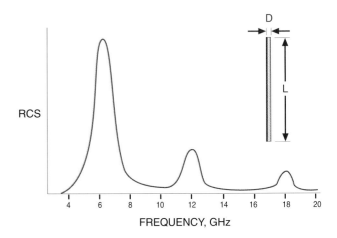

3.   Bandwidth of chaff fibers one-half wavelength long at 6 GHz. The greater the fibers' ratio of length, L, to diameter, D, the narrower the peaks will be. A wide band can be covered, however, by dispensing chaff of several different lengths.

Noise Jamming

Though simple to produce, noise jamming can be highly effective. Similar to thermal noise, it raises the level of the background against which target returns must be detected, swamping out all but very strong returns.

Undesirably from the user's standpoint, the jammer also serves as a beacon, revealing both the presence and the direction of the jamming aircraft. Moreover, since the jamming travels only one way — from the jammer to the radar — the range at which a radar can detect the jammer is often limited only by the horizon.

Nevertheless, by preventing detection of the radar returns from the jamming aircraft, the jammer denies the enemy knowledge of the aircraft's range and range rate, thus preventing accurate calculation of missile launch zones, lead angles, and fuse times, and forcing the enemy either to waste missiles by launching them at too great a range or to give up valuable intercept range.

When used to screen the aircraft of a raid, noise jamming not only enables the raid to penetrate farther in safety, but prevents the defending forces from accurately assessing the raid's size. Consequently, noise jamming is especially useful for protecting multiple attacking groups.

Mechanization. A simple responsive noise jammer is shown in Fig. 4. It consists of four basic elements: an RF noise source, a bandpass filter, an RF amplifier, and a broad-beamed antenna—plus control circuitry. Cued by a separate ECM receiver, the control circuitry tunes the filter to the victim radar's frequency and turns on the noise source and amplifier.

Periodically, the jamming is interrupted so that the ECM receiver can tell if the radar's frequency has changed. If it has, the filter is quickly tuned to the new frequency, a process called "*set-on,*" and jamming is resumed.

Effectiveness of the Jamming. Those factors which determine the effectiveness of noise jamming in masking targets from a radar may be seen most clearly by deriving a simple equation for the power of the jamming in the output of the victim radar's receiver. This has been done in the panel below.

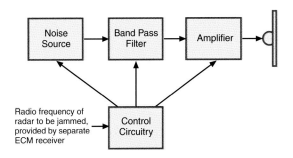

4. Basic elements of a simple responsive noise jammer. Control circuitry tunes filter to victim radar's frequency. Antenna typically has a broad beam to simplify handling multiple threats at different angles. Against surface-based radars, which at least temporarily are stationary, a directional antenna may be used. The victim radar's direction, then, would also be provided by the ECM receiver.

Power of Noise Jamming on the Output of a Victim Radar's Receiver

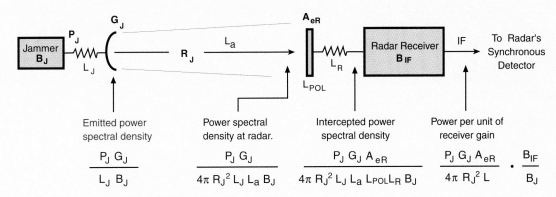

Emitted power spectral density	Power spectral density at radar.	Intercepted power spectral density	Power per unit of receiver gain
$\dfrac{P_J \, G_J}{L_J \, B_J}$	$\dfrac{P_J \, G_J}{4\pi \, R_J^2 \, L_J \, L_a \, B_J}$	$\dfrac{P_J \, G_J \, A_{eR}}{4\pi \, R_J^2 \, L_J \, L_a \, L_{POL} L_R \, B_J}$	$\dfrac{P_J \, G_J \, A_{eR}}{4\pi \, R_J^2 \, L} \cdot \dfrac{B_{IF}}{B_J}$

Assuming:
- Spectrum of jamming closely approximates thermal noise
- $B_J > B_{IF}$
- Jammer is in center of radar antenna's main lobe

Mean Power of the Jamming in the Receiver's output, per unit of receiver gain is:

$$P_{JR} = \frac{P_J \, G_J \, A_{eR}}{4\pi \, R_J^2 \, L} \cdot \frac{B_{IF}}{B_J} \text{ watts}$$

P_J = Power output of the jammer

G_J = Gain of jammer's antenna in radar's direction

A_{eR} = Equivalent area of radar antenna

B_{IF} = Bandwidth of receiver IF amplifier

R_J = Range from jammer to radar

L = Total losses: $L_J \, L_a \, L_{POL} \, L_R$

B_J = Bandwidth of jammer's output

L_J = RF losses in jammer feed and antenna

L_a = Atmospheric loss (function of radar's operating frequency and R_J)

L_{POL} = Loss due to antenna polarization misalignment

L_R = RF losses in radar antenna & receiver front end

Note: If the radar antenna is not trained on the jammer, A_{eR} will be reduced by the ratio of the antenna gain in the jammer's direction to the gain at the center of the antenna's mainlobe.

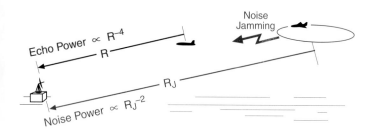

As derived on the preceding page, the power of the noise jamming in the receiver's output, per unit of receiver gain, is

$$P_{JR} = \frac{P_J \, G_J \, A_{eR} \, B_{IF}}{4\pi \, R_J^2 \, L \, B_J}$$

where $P_J \, G_J / B_J$ is the power spectral density of the jammer's radiation, R_J is the jammer's range, A_{eR} is the equivalent area of the victim radar's antenna, and B_{IF} is the bandwidth of the receiver.

For an aircraft which is screening itself, the range, R_J, from the jammer to the victim radar is the same as the range, R, of the screened aircraft from the radar. However, for an aircraft which is screening another aircraft from a standoff position (Fig. 5), R_J may be much longer than R.

In either case, whereas the received signal power varies as $1/R^4$, the power of the jamming (which travels only one way) varies only as $1/R_J^2$. As the range, R, of the screened aircraft decreases, therefore, the signal-to-jamming ratio rapidly increases. Eventually, a point may be reached where the signal "burns through" the jamming (Fig. 6).

5. For an aircraft which is screening another aircraft from a standoff position, R_J may be much longer than R. This difference is generally more than made up for by the jamming traveling only one way, whereas the radar signal travels both out and back.

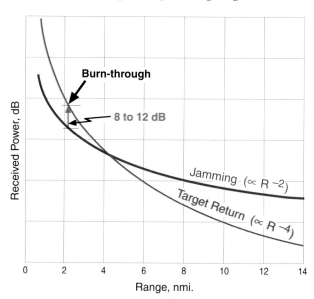

6. As the range of a target decreases, a point eventually is reached where the power of the target return exceeds the power of the received jamming by enough—8 to 12 dB—to "burn through" the jamming and be detected.

Assuming that the jammer's noise quality is high (i.e., *equivalent to thermal noise*), we can determine the burn-through range by modifying the radar range equation as follows. To the expression for *thermal* noise power ($F_n k T_0 B_{IF}$ or equivalent) in the denominator of the equation, add the expression for the mean *jamming* power in the receiver output (P_{JR}). Provided the jamming is sufficiently noiselike, beyond the receiver the jamming and thermal noise will be processed identically, regardless of the radar's design.

A radar's detection range, you will recall, varies inversely as the one-fourth power of the mean thermal-noise power in the receiver's output. Consequently, we can determine the *fraction* to which noise jamming will reduce the radar's detection range, simply by taking the one-fourth power of the ratio of $F_n k T_0 B_{IF}$ to $(F_n k T_0 B_{IF} + E_{JR})$.

If the jamming is strong and the jammer is in the radar's mainlobe or close-in sidelobes, burn-through ranges may be negligible (Fig. 7). However, if the jammer is in the radar antenna's far sidelobes, targets in the vastly higher-gain mainlobe may burn through the jamming at appreciably long ranges.

Cooperatively Blinked Noise Jamming. If several closely grouped aircraft equipped with noise jammers are operating together, as in a coordinated raid, the effectiveness of their jamming may be greatly enhanced by turning each aircraft's jammer on and off sequentially in accordance with a pre-arranged plan (Fig. 8).

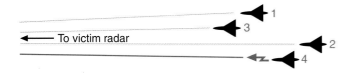

8. Cooperatively blinking the noise jamming from several closely grouped aircraft causes the centroid of the jamming as seen by the victim radar to oscillate erratically in angle.

The blinking handicaps a victim radar in several significant ways:

- If the radar is searching, it seriously degrades resolution of the aircraft in angle

- If the radar is operating in a track-while-scan or search-while-track mode, it may also create false target tracks, possibly saturating the radar's track file

- If multiple jammers are in the radar's mainlobe, it makes the radar's angle tracking oscillate erratically

- If the radar is employing passive ranging, it seriously degrades that.

Jamming More Than One Radar. So far, we have considered only the jamming of a single radar. For that, the spectral density of the jamming power is maximized by making the passband of the jammer's narrowband filter only wide enough to effectively jam that radar's operating frequency—a technique called *spot jamming* (Fig.9).

If more than one radar is to be jammed and the radars are operating at different radio frequencies, any of three alternative techniques may be used.

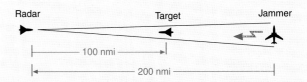

Problem: Suppose a given radar has a 50% probability of detecting a given target at 100 nmi. Find the range at which the target returns will burn through the jamming when the radar is jammed by an aircraft in the radar's main lobe at a range of 200 nmi.

Radar's Characteristics	Jammer's Characteristics
A_{eR} = 4 square ft.	P_J = 1000 Watts
B_{IF} = 1 MHz	G_J = 20
$F_n k T_0 B_{IF}$ = 8 x 10^{-21} watt	R_J = 200 nmi
Total Losses	B_J = 10 MHz
L = 3 dB	Note: 1 nmi = 6000 ft

Solution

$$W_{JR} = \frac{P_J \, G_J \, A_{eR} \, B_{IF}}{4\pi \, R_J^2 \, L \, B_J}$$

$$W_{JR} = \frac{1000 \times 20 \times 4 \times 10^6}{4\pi \times (200 \times 6000)^2 \times 2 \times 10^7} = 2.2 \times 10^{-9} \text{ watt}$$

$$\text{Reduction Factor} = \left[\frac{F_n k T_0 B_{IF}}{F_n k T_0 B_{IF} + W_{JR}}\right]^{1/4} = 0.0014$$

Burn-Through Range = 0.0014 x 100 x 6000 = ⬭840 ft⬭

7. Reduction in detection range produced by a noise jammer in a radar's mainlobe. Although in this case burn-through range is negligible, it would be significant if either the jammer were in the radar's sidelobes or the radar were in the jammer's sidelobes.

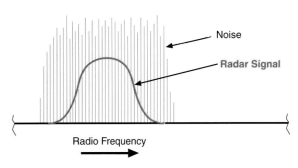

9. Spot noise jamming. Maximum efficiency may be achieved by making the bandwidth of the jamming only slightly wider than the spectrum of the radar signal to be jammed. Because of mechanization limitations, however, the bandwidth is generally made much wider—between 3 and 20 MHz.

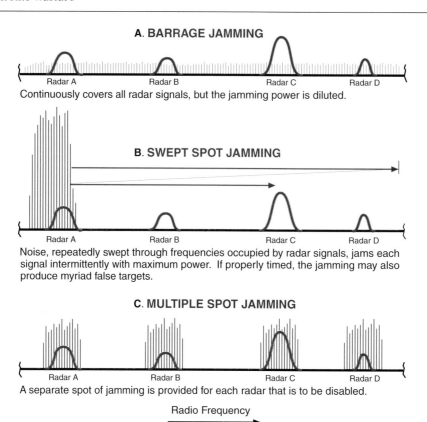

A. BARRAGE JAMMING

Radar A   Radar B   Radar C   Radar D

Continuously covers all radar signals, but the jamming power is diluted.

B. SWEPT SPOT JAMMING

Radar A   Radar B   Radar C   Radar D

Noise, repeatedly swept through frequencies occupied by radar signals, jams each signal intermittently with maximum power. If properly timed, the jamming may also produce myriad false targets.

C. MULTIPLE SPOT JAMMING

Radar A   Radar B   Radar C   Radar D

A separate spot of jamming is provided for each radar that is to be disabled.

Radio Frequency

10. Alternatives for jamming more than one radar operating on different frequencies. Although requiring complex RF switching, multiple spot jamming is the most effective.

The simplest, called *barrage jamming*, is to spread the jammer's power over a broad enough frequency band to simultaneously blanket the frequencies of all the radars (Fig.10A). A large number of radars can thus be jammed. The spreading, however, greatly reduces the spectral density of the jamming power with which each radar must contend. This may, in fact, result in burn-through ranges becoming unacceptably long.

To get around that problem, spot jamming may be repeatedly swept through the band of frequencies occupied by the victim radars (Fig.10B), a technique, called *swept-spot jamming*. Although not delivering any more average power than barrage jamming, swept-spot jamming *periodically* brings the maximum possible power to bear on each radar. If a radar has a long-time-constant AGC loop, the jamming may drive the receiver gain down to such an extent that the radar will not have recovered its full sensitivity by the time the jamming sweeps over the radar's frequency again.

In practice, though, swept-spot jamming has proven to be more useful in producing false targets. For this, the best results are obtained by making the jamming "spiky," rather than uniform, and by adjusting the sweep rate to keep the jamming in each radar's passband for a period roughly equal to the width of the radar's transmitted pulses. Against scanning radars, swept-spot jamming may produce enough creditable false targets to prevent detection of real aircraft.

Be that as it may, if the threat radars are widely spaced in frequency, swept-spot jamming will leave them uncovered much of the time.

Consequently, a more efficient technique is *multiple spot jamming* (Fig. 10C). That is, jamming enough spots to continuously desensitize each threat radar. The cost of implementation may be mitigated by using the same noise source for all of the spots. Cost may be further reduced by jamming only a few spots at a time and optimally jumping each of them from one radar's frequency to another's at a very high rate.[2]

Bin Masking. Even spot jamming uses power inefficiently. For the jamming power is blindly spread over the entire interpulse period and over all possible doppler frequencies. The waste can be reduced with a technique called *bin masking*. This form of jamming is of two basic types: range bin masking (RBM) formerly called "cover pulse" and *velocity (doppler) bin masking* (VBM) also known as "doppler noise."

Against low and medium PRF radars—which resolve target returns primarily in range—range bin masking is the more effective. For it, the jamming is transmitted in short bursts timed to fall within the range interval in which the aircraft to be screened may lie (Fig.11). If started early enough in a radar's coherent integration period, the jamming can completely mask any targets in the selected interval.[3]

Against high PRF radars—which resolve returns in doppler frequency—velocity bin masking is the more effective. It is useful, too, against medium PRF radars. One of the most efficient implementations is a *"straight-through" repeater*. Designed to receive the transmissions from the victim radar, shift their radio frequency, and retransmit them to the radar, the repeater consists of a receiving antenna, a modulator, a traveling-wave-tube (TWT) amplifier, and a transmitting antenna (Fig. 12).

2. The jammer might, for example, cycle through up to four spots at a rate of 100 to 250 kHz.

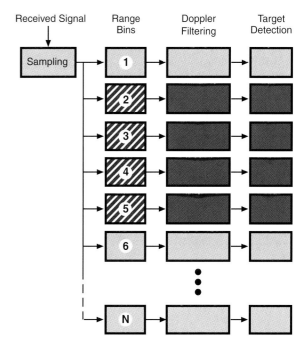

11. With range bin masking, the jamming is timed to fall within a block of range bins covering the range interval in which the aircraft to be screened may lie.

3. Range-bin masking is especially useful against radars employing PRF jittering. For, despite the jitter, the jamming will always cover the target's echoes.

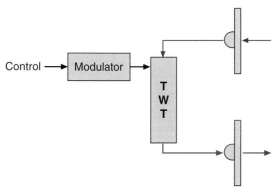

12. A straight-through repeater such as used to provide coherent jamming signals for velocity bin masking. Modulator sweeps frequency of retransmitted signals through band of doppler frequencies to be masked.

Serrodyne Modulation

The time a signal takes to pass through a TWT depends to some extent on the velocity of the tube's electron beam, hence on the voltage applied to the anode of its electron gun. The phase, ϕ, of the tube's output, therefore, can be varied by modulating the anode voltage.

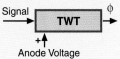

$$f = \frac{d\phi}{dt}$$

In essence, frequency, f, is a continuous phase shift, e.g., a phase shift of 360° per second is a frequency of 1 cycle per second.

By linearly advancing the phase of the TWT's output, therefore, the signal's frequency can be increased.

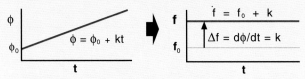

By advancing the phase at a geometrically increasing rate, the signal's frequency can be linearly swept through a band of frequencies.

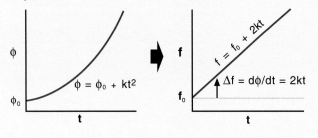

The modulator shifts the frequency of the signal passing through the TWT by appropriately varying the voltage applied to the tube's anode, a process called *serrodyne modulation*—see panel (left). For bin masking, the frequency generally is swept in a sawtooth pattern through that portion of the doppler spectrum in which returns from the aircraft to be protected may lie (Fig. 13).

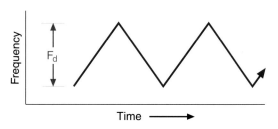

13. One approach to doppler bin masking. Continuously sweep a straight-through repeater's frequency through the desired band of doppler frequencies, F_d, in a triangular pattern.

The repeated signals thus saturate the block of doppler bins spanning those frequencies.

Another useful approach is to transmit multiple false targets whose doppler frequencies are staggered to cover the desired frequency band (Fig. 14).

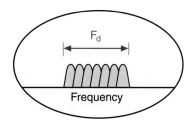

14. Another approach to doppler bin masking. Transmit multiple false targets whose doppler frequencies are staggered to cover the desired band.

Transponder For Producing False Targets

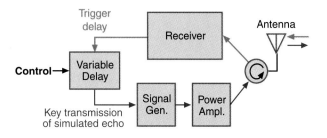

15. Upon receiving a pulse from a threat radar, the transponder delays for a period corresponding to the desired difference in range of the false target; then, transmits an RF pulse simulating a target echo back to the radar.

False Targets

With the exception of those false targets produced by swept-spot noise jamming, most false targets are produced with transponders and repeaters.

A *transponder* for false-target generation (Fig. 15) consists of a receiver, a variable delay circuit, a signal generator, a power amplifier, and an antenna. Upon receiving a pulse from a threat radar, the transponder waits for a period corresponding to the desired additional range of the false target; then, transmits back to the radar an internally generated signal simulating a target echo.

A *repeater* for generating false targets generally includes a memory, enabling it to produce much more realistic targets. The memory stores the actual pulse received from the radar.

After the desired delay, the pulse is read out, amplified, and transmitted back to the radar (Fig. 16).

Repeater For Producing False Targets

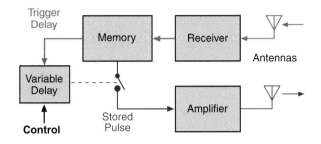

16. A repeater produces more realistic false targets. When a pulse from a threat radar is received, it is stored in the repeater's memory. After the desired time delay, the pulse is read out of memory, amplified, and transmitted back to the radar. A time-shared antenna may be used, but isolation of transmission and reception is simplified by using separate antennas.

With either a repeater or a transponder, by providing multiple time delays, it is possible to create any number of false targets and make them appear on the victim radar's display at widely different ranges. By making the time delays enough longer than the radar's interpulse period, false targets may be made to appear at shorter, as well as longer, ranges than the originating aircraft.

With a repeater, the false targets may also be given widely different apparent doppler frequencies.

The repeater's memory may be simply a recirculating delay line. However, much higher fidelity can be obtained with a *digital RF memory (DRFM)*. It temporarily stores a digitized sample[4] of each received pulse. From these samples, the repeater may synthesize highly realistic, deceptively timed, and doppler-shifted false echoes.

Going a step further, by sensing the victim radar's search scan and delaying the repeater's response until sometime after the radar has scanned past the repeater-bearing aircraft, false targets can be "injected" into the radar's side lobes and so be made to appear on the radar's display at angles offset from the repeater's direction.[5]

Against CW and high-PRF radars, which don't employ pulse-delay ranging (and even against some medium-PRF radars, which do), a straight-through repeater, such as described earlier (Fig. 13), may be used.

By creating large numbers of false targets having different doppler frequencies and appearing at different ranges and different azimuths, a repeater can greatly increase an opposing force's response time and may also prevent the detection of true targets.

4. These samples may be the usual I and Q components of the pulse, or samples taken at twice the normal sampling rate to likewise fully define the pulse's phase.

5. This requires lots of memory, especially when multiple radars are encountered and many different signals must be stored simultaneously.

Gate Stealing Deception

If, despite noise jamming, bin masking, and false targets, a threat radar manages to lock onto a screened target, a gate stealer may keep the radar from usefully tracking the target. In essence, the stealer disrupts tracking by transmitting false target returns contrived to capture the gate which the radar places around the aircraft's skin return for clutter reduction and tracking.

Having captured the gate, the stealer may do one of the following:

- "Walk" it off the skin return, causing the radar to provide false range and range-rate data

- Break lock, by pulling the gate off the skin return, and dropping or transferring it to chaff return or clutter

- Facilitate angle deception countermeasures by increasing the jamming-to-signal ratio

By repeatedly breaking lock every time the victim radar relocks on the skin return, the stealer can drastically reduce the radar's tracking accuracy.

Gate stealers are of two basic types: *range-gate stealers* (RGS) and *velocity-gate stealers* (VGS).[6]

Range Gate Stealers. These are typically used against radars operating at low or medium PRFs.

Against noncoherent radars (low PRF only), the stealer may be mechanized with a transponder. It detects the leading edge of each radar pulse and, after a delay, transmits an RF pulse[7] back to the radar.

Initially the delay is made short enough that successive pulses cover the skin return. Being very much stronger than it, they capture the range gate. The time delay is then gradually increased, pulling the gate out in range and off the skin return (Fig. 17).

6. Another type of gate stealer is the so-called *chirp-gate stealer.* It shifts the chirp frequency used for pulse compression up or down, thereby moving the range gate out or in, in range.

7. Or possibly gated spot noise.

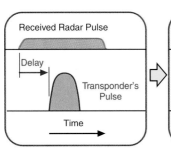

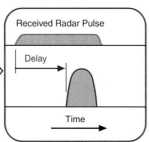

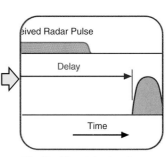

Initially. Delay is set so that successive transponder pulses cover the skin return and thus capture the radar's range gate.

Then: The delay is gradually increased, so the transponder pulse will pull the radar's range gate out in range.

Finally: The delay has been increased enough for the range gate to be pulled completely off the skin return.

17. Against a noncoherent radar, the range-gate stealer may be mechanized with a transponder. Upon receipt of each radar pulse, the transponder transmits a delayed RF pulse to the radar.

If the radar's PRF is known or has been measured by the stealer's logic, by initially making the delay equal to the interpulse period and then gradually reducing it, the gate can instead be pulled in in range.

Against coherent radars—for which the doppler frequency of the skin return must be matched—the range-gate stealer is implemented with a repeater.

Older designs, using circulating-delay-line memories,

- Sample the leading edge of each received pulse

- Delay the sample for the desired length of time

- Amplify and beam the sample back to the radar

Since only the leading edge of the pulse is stored, any pulse compression coding is not repeated.

Newer designs repeat the coding with a DRFM[8] (Fig. 18).

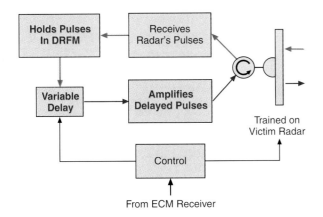

Trained on Victim Radar

From ECM Receiver

18. A more capable range-gate stealer for use against a coherent radar. DRFM stores each received pulse enabling stealer to match doppler frequency and pulse-compression coding of skin return. Antenna is trained on radar by ECM receiver.

Velocity-Gate Stealers. These are typically used against high-PRF and CW radars and missile guidance seekers. Consisting of a straight-through repeater, such as was illustrated in Fig. 12, the velocity-gate stealer performs essentially the same function in the frequency domain as a range-gate stealer does in the time domain.

Initially the received radar signal is amplified and transmitted back without modification. Thus synchronized with the skin return in doppler frequency, the much stronger repeater signal captures the gate the radar uses to isolate and track the skin return in doppler frequency (velocity). The radio frequency of the repeated signal is then gradually shifted up, or down, pulling the gate off the skin return.

In some mechanizations, the retransmitted pulses are automatically beamed toward the victim radar by a retrodirective antenna (see panel, right). It, unfortunately, requires much more space than a simple antenna, such as a spiral, and so has limited applicability.

8. Or a pulse-compression code memory.

RETRODIRECTIVE REPEATER

Operation of the Retrodirective Repeater is best explained by first considering a simple passive retrodirective antenna. It consists of a linear array of radiating elements, interconnected in pairs by coaxial cables.

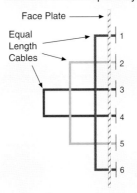

Radiation received by each element is reradiated by the other element of the pair. In the array shown here, for instance, radiation received by element #1 is reradiated by element #6, and radiation received by element #6 is reradiated by element #1.

The delay incurred in passing through the cables is equalized by making all of the cables the same length. Thus, the progressive phase lag in the radiation received by successive elements from a direction not normal to the array is reversed in

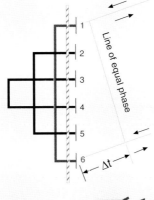

the reradiated signal. To illustrate, the radiation *emitted* from element #6 leads the radiation emitted from element #1 by the same length of time (Δt)—hence phase—that the radiation *received* by element #6 lags the radiation received by #1. Accordingly, the composite radiation from all elements propagates in a direction exactly opposite that of the received radiation.

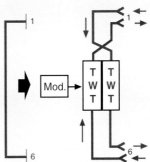

By replacing each pair of radiators and its interconnecting cable in the above-described antenna with a pair of repeaters, a retrodirective repeater may be implemented.

Coordinated Range/Velocity Gate Stealing. While described here singly, range and velocity gate stealing may be performed in concert. By employing a repeater having a DRFM, the combined techniques may be made much more difficult to counter than either technique alone.

Angle Deception

The object of this countermeasure is to introduce angle-tracking errors in an enemy's fire-control radar or radar-guided missiles, causing his weapons to miss.

Errors were introduced in early angle-tracking systems that employed lobing simply by sending back suitably timed false returns.

Several techniques have been devised for introducing errors in the more advanced, monopulse tracking systems. All of these, however, require fine-grain information on the victim radar's parameters, which may not be available.

More robust techniques capable of defeating both monopulse and lobing are terrain bounce, crosseye, cross polarization, and double cross.

Terrain Bounce Jamming (TBJ). Intended for low-altitude short-range engagements, terrain bounce is an effective defense against an approaching radar guided missile. A repeater in the threatened aircraft is equipped with a directional antenna whose beam is deflected downward to bounce false returns off the terrain in front of the missile (Fig. 19).

Overpowering the directly received skin returns,[9] the bounced signal causes the missile to head for a virtual target image beneath the surface and miss the aircraft.

Crosseye. For this deception, the aircraft to be protected is equipped with a repeater having exceptionally high gain and receiving and transmitting antennas installed as close as practical to each wing tip (Fig. 20). The repeater is mechanized in such a way that

- Signals received from the threat radar by the receiving antenna on the *left* wing tip are shifted in phase, amplified, and returned to the radar by the transmitting antenna on the *right* wing tip

- Signals simultaneously received from the radar by the receiving antenna on the *right* wing tip are similarly shifted in phase, amplified, and returned to the radar by the transmitting antenna on the *left* wing tip

- Phase shifts incurred in passing through the repeater are such that the signal retransmitted from one wing tip is very nearly 180° out of phase with the signal retransmitted from the other wing tip.

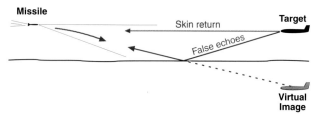

19. Terrain bounce. Downward deflected antenna in target bounces false echoes off terrain in front of missile, causing it to steer for a virtual image.

9. Plus directly received sidelobe radiation from the repeater.

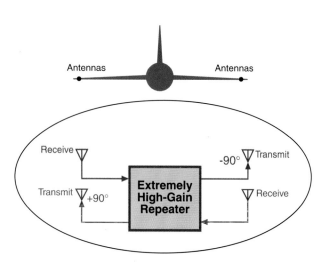

20. Crosseye is implemented with a repeater having transmit and receive antennas on both wing tips. Signals received at right wing tip are retransmitted from the left wing tip and vice versa. To ensure extreme stability of gain and phase, both signals time share the same TWT amplifier chain.

② To left of the phase center, path A is longer; path B is shorter. So signals are partially in phase and produce a sum.

① At antenna's phase center, distances traveled via paths A and B are equal. So, signals are out of phase. Sum ≈ 0

③ To right of the phase center, path B is longer; path A is shorter. So signals are partly in phase, but sum is reversed.

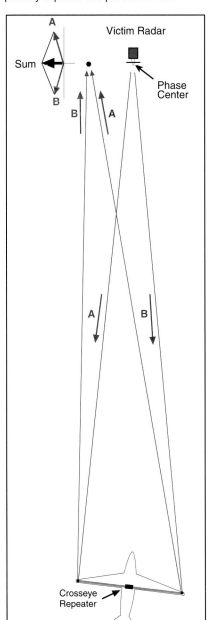
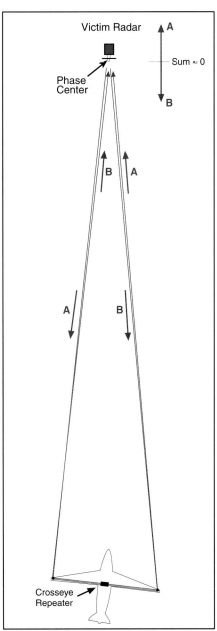
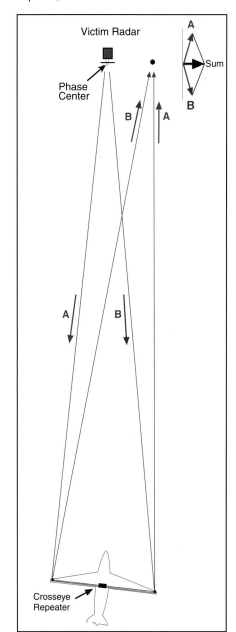

21. How the reradiated crosseye signals combine upon returning to the victim radar.

As illustrated in Fig. 21, in going from the radar antenna's phase center, through the repeater, and back to the phase center, the two signals traverse exactly the same round-trip distance. Because of the nearly 180° phase difference imparted by the repeater, they essentially cancel.[10]

But at points to the right and left of the phase center, the round-trip distances traversed are increasingly different. As a result, the phase difference imparted between the signals by the repeater is correspondingly reduced, and they combine to produce an appreciable sum. The magnitude of the

10. The reason for making the signals nearly but not exactly 180° out of phase is to ensure that there will be some output from the "sum" channel of a *monopulse* radar's antenna. Otherwise, crosseye would not be able to drive the antenna off the target.

sum increases with the distance of the points from the phase center. What's more, the sum on the right is 180° out of phase with the sum on the left.

Consequently, when the crosseye signals merge with the skin return, they warp the return's phase front so that it is not quite normal to the line of sight to the Crosseye-bearing aircraft. Consequently, in aligning the face of the antenna with the warped phase front, the radar's angle tracking system trains the antenna in a direction offset to one side.

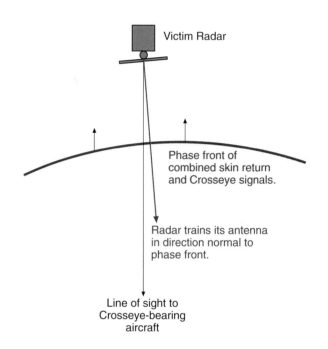

22. When crosseye's signals combine with the skin return, they warp its phase front, causing the victim radar to train its antenna off to one side of the crosseye-bearing aircraft.

Up to a limit that depends primarily on the separation of the crosseye antennas, the stronger the crosseye signals are relative to the skin return, the greater the warp; hence, the greater this offset will be. By slowly varying the amplitude or phase of the crosseye signals, it is possible to walk the radar antenna off the target.

Cross Polarization. *"Crosspol,"* or *polarization-exchange cross modulation (PECM)* as this countermeasure is also called, takes advantage of a distortion in the polarization of a radar's received signals due to several possible causes: the curvature of the radome; the diffraction occurring at the edges of the antenna; and, in parabolic reflector antennas, the curvature of the reflector.

Because of this distortion, when an antenna is illuminated with a very strong signal whose polarization is rotated 90° relative to that of the antenna, the antenna's receive pattern becomes distorted (Fig. 23). As a result, large and erratic tracking errors build up.

NORMAL RESPONSE

0 dB = 36.4 dBi

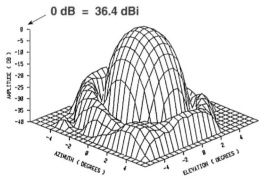

CROSSPOL RESPONSE

0 dB = 11.0 dBi

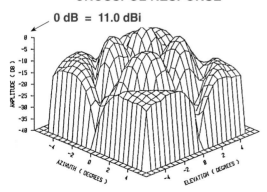

23. Effect of a strong cross-polarization—such as crosspol's—on the receive pattern of a planar array antenna in its radome. Mainlobe is replaced by four lobes, on the diagonal axes, whose peak gain is reduced by more than 25 dB. Such distortion results in large and erratic tracking errors.

Crosspol is implemented with a repeater employing a high-gain TWT-amplifier chain and oppositely polarized receiving and transmitting antennas.

If the victim radar is linearly polarized, in order for the repeater's operation to be independent of the direction of the polarization, circularly polarized antennas may be used—right-hand for reception; left-hand for transmission; or vice versa (Fig. 24a).

If the victim radar is circularly polarized, the repeater may employ two channels of roughly equal gain and orthogonal linearly polarized antennas (Fig. 24b). So that the deception can be unobtrusively introduced, the repeaters' gain is adjustable.

Double Cross. As the name implies, double cross is a combination of crosseye and crosspol. Though more complex, it can be more difficult to counter.

Radar Decoys

Radar decoys may be deployed to confuse an enemy and draw his radar, or the seeker of an approaching radar guided missile, away from the deploying aircraft. Decoys are of two basic types: towed and expendable.

A towed decoy is attached to a thin cable, which can gradually be reeled out as much as 300 feet behind the aircraft (Fig. 25).

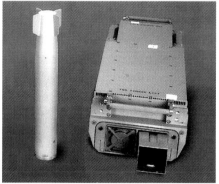

25. A towed decoy and the launcher/launch controller used in the F-16. Decoy is packaged in a sealed canister which also contains the payout reel (Courtesy of Raytheon Company)

Towing has the advantage that the decoy is reusable, but restricts the aircraft's maneuverability. The restriction is minimized by designing the decoy to have very little drag, and possibly by incorporating control surfaces in the decoy to control its position relative to the towing aircraft.

Expendable decoys (Fig. 26) are more versatile. They can, for example, pull ahead of the deploying aircraft, fall behind it, or gradually assume a radically different course. This capability is gained at the expense of providing self-contained propulsion and navigation systems and of the decoys not being recoverable.

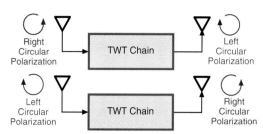

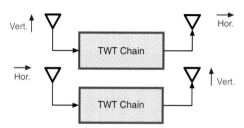

(a) Against a linearly polarized radar, Crosspol would use circularly polarized antennas of opposite hand.

(b) Against a circularly polarized radar, Crosspol would use linearly polarized antennas of opposite sense.

24. Possible implementations of crosspol. To make implementation independent of direction of victim radar's polarization, two channels are used.

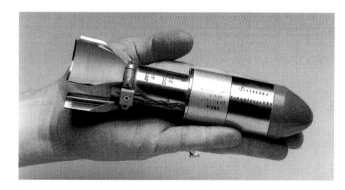

26. An active expendable decoy used by the U.S. Navy and RAF. (Courtesy of Raytheon Company)

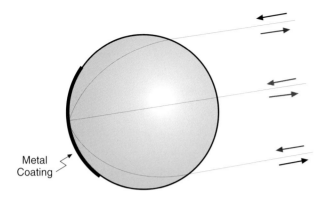

26. In simplest form, a Luneberg lens reflector consists of a dielectric sphere. Its index of refraction increases from 1 at the surface to a maximum at the center, bending the rays of an incoming plane wave so that they converge at a point on the opposite surface. There, a metal coating reflects the wave back in the direction from which it came.

Decoys of both types may be designed to produce the desired radar returns either passively or actively. Since a decoy is necessarily quite small, for passive operation its radar cross-section must generally be augmented. This may be accomplished with a corner reflector or a Luneberg lens (Fig. 26), both of which are comparatively simple and inexpensive.

Active decoys generally carry a repeater and need a control system and power supply. Also, for small decoys, isolating transmit and receive antennas is a challenging problem—all of which makes the decoys more expensive.

Regardless of the mechanization, to achieve its purpose a decoy must:

- Have an RCS greater than twice that of the aircraft

- Initially, match the deploying aircraft's speed

- For tracking-gate pull-off, initially appear in conjunction with the deploying aircraft as a single target

- Not exceed reasonably expected accelerations

Provided these conditions are met, an appropriately controlled decoy deployed in synchronism with a critically timed evasive maneuver may save an aircraft from almost certain destruction by a radar guided missile.

Future Trends

As radar capabilities grow, they will, as always, be matched by more severe and increasingly sophisticated ECM.

The RF coverage and responsiveness of noise jammers will increase. Their effectiveness in standoff and escort missions will undoubtedly grow.

Deception ECM will similarly advance. False targets will become increasingly deceptive, electronically flying realistic profiles and exhibiting the electronic signatures of friendly, neutral, or hostile aircraft. The present gate-stealing, terrain-bounce, crosseye, and crosspol techniques will be refined. New angle deception techniques, not presently envisioned, may also evolve.

ECM systems will become more intelligent, more responsive. They will adjust agilely to changes in the encounter scenario, to changing radar characteristics, even to new waveforms and ECM thwarting radar responses.

Summary

Chaff, the simplest of all ECM, can screen an entire raid from radars operating over a wide range of frequencies. But moving-target indication rejects chaff return.

Noise jamming screens targets by swamping out all but

the strongest target returns. For maximum efficiency it must be concentrated at the radar's frequency (spot jamming) and in those range or doppler bins where the returns to be masked may appear (bin masking).

Against multiple radars operating on different frequencies, the jamming may be spread over the entire operating band (barrage jamming), swept through that band (swept spot jamming), or concentrated at each radar's frequency (multiple spot jamming).

Jamming prevents a threat radar from measuring target range and range rate and assessing raid size. By cooperatively blinking their jamming, closely grouped aircraft may confound the enemy's attempts both to track the jamming accurately in angle and to passively measure range.

To delay and possibly prevent acquisition by an enemy, multiple false targets may be produced with swept-spot jamming or be realistically simulated with repeaters having digital RF memories.

If a threat radar achieves lockon, its range or velocity tracking gates may be captured by a gate-stealing repeater and pulled off the target's skin return.

Should these measures fail, the radar's tracking may be compromised through these robust deceptions:

- Terrain bounce—a repeater in a low-flying aircraft deceives an approaching missile by bouncing false returns off the ground

- Crosseye—a time-shared repeater with receiving and transmitting antennas on opposite wing tips, warps the phase front of the aircraft's skin return

- Crosspol—a repeater returns a strong cross-polarized signal which distorts a hostile radar's receive pattern

- Double cross—combines crosseye and crosspol.

As a last resort, a well timed burst of chaff coupled with a maneuver may disrupt a missile's tracking, or a towed or expendable decoy may draw the missile off.

ACRONYMS OF ECM

Bin Masking Techniques
- RBM — Range-Bin Masking, or "Cover Pulse"
- VBM — Velocity-Bin Masking, or "Doppler Noise"

Gate Stealing
- RGS — Range-Gate Stealer
- VGS — Velocity Gate Stealer
- VGPO — Velocity-Gate Pull-Off
- VGWO — Velocity-Gate Walk-Off

False Targets
- DRFM — Digital Radio Frequency Memory

Angle Deception
- TBJ — Terrain Bounce Jamming
- PECM — Polarization-Exchange Cross-Modulation (Crosspol)

HAWKER SIDDELEY SEA HARRIER

The Harrier is unique among fighters in its ability to take off and land verti-cally. The USMC version, built by McDonnell Douglas is designated AV-8B and equipped with the APG-73 radar.

Electronic Counter Countermeasures (ECCM)

35

I n the previous chapter, we examined the principal types of electronic countermeasures (ECM). We learned how each type is implemented and what its limitations are. In this chapter, we will examine some of the important electronic counter-countermeasures (ECCM) which have been devised to exploit the limitations of ECM and so defeat them. We will begin by examining the conventional techniques for combating noise jamming, gate stealing, and angle deception. We will then look at some significant advanced ECCM developments which promise quantum jumps in a radar's ability to contend with severe noise jamming, as well as with various other ECM.

Conventional Measures for Countering Noise Jamming

Over the years three basic techniques have been used in airborne radars to counter noise jamming:

- Frequency agility

- Detection and angle tracking on the jamming

- Passive ranging

These techniques and certain conventional clutter reduction features which also reduce vulnerability to noise jamming, are discussed briefly in the following paragraphs.

Frequency Agility. Prior to the advent of coherent pulse-doppler radars, a common means of countering noise jamming was frequency agility. At the low PRFs used by noncoherent radars, the interpulse period is sufficiently long that even a simple magnetron transmitter can be tuned to widely different operating frequencies from one pulse to the next. While an enemy's ECM receiver can quickly determine the frequency of each pulse it receives,

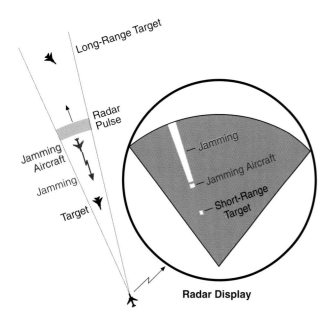

1. By changing its operating frequency from pulse to pulse, a noncoherent radar can keep a jamming aircraft from masking both itself and targets at shorter ranges. But it cannot keep the jammer from masking targets at longer ranges.

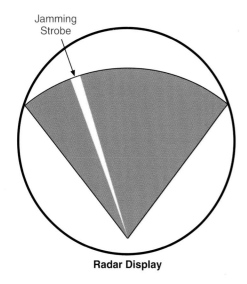

2. In angle-on jamming, as the radar beam scans across a jamming aircraft in search, the jamming produces a bright line (strobe) on the radar display in the jammer's direction.

it cannot predict the frequency of the next pulse the radar will transmit. To jam the radar, therefore, the enemy has but two options, neither of which is entirely effective.

The first is to quickly tune the jammer to the frequency of the last received pulse. The jamming then will mask the returns from targets at greater ranges than the jammer from the radar (Fig. 1). But it cannot mask the jamming aircraft itself or targets at shorter ranges.

The enemy's second option is to use barrage jamming—i.e., spread the jammer's power throughout the entire band of frequencies over which the radar happened to be operating or, in the case of a simple preset jammer, is known to be capable of operating. The jamming then will similarly mask the weak returns from long-range targets. But, if the jammer is in a stand-off position, unless the jamming is extremely powerful, it generally will be spread so thin that the returns from shorter-range targets would burn through.

In a coherent radar, however, frequency agility is of limited value in countering jamming. For a coherent radar's ability to perform predetection integration depends upon the operating frequency remaining constant throughout the integration period, which frequently is comparatively long. A fast-set-on jammer can concentrate its power at the radar's frequency during virtually all of this period.

Detection and Angle-Tracking on the Jamming. Although coherent radars cannot easily avoid noise jamming, they can exploit it. Early on, a mode—variously called *angle-on jamming (AOJ)*, *jam angle track (JAT)*, and *angle track-on jamming (ATOJ)*—was provided which is still implemented in radars today.

In this mode, the radar's automatic detection function is adjusted so that the jamming produces a bright line, or strobe, on the radar display as the antenna scans across the jammer in search (Fig. 2). By observing the strobe, the operator can determine the jamming aircraft's direction and, by locking the radar onto the jamming, track the aircraft in angle.

By then launching IR-guided missiles or radar-guided missiles capable of homing in on the jamming, a feature called *home-on jamming (HOJ)*, the pilot has a good chance of shooting the aircraft down.

But, to avoid blindly wasting missiles, by launching them at too long a range, or unnecessarily extending the attack and increasing the risk of getting shot down, the crew of the launch aircraft must at least have a rough idea of the target's range. One way of obtaining that is through passive ranging.

Passive Ranging. Of various passive techniques for estimating range, four are listed in Table 1. While all have limitations, the limitations are all different.

The first technique, *angle-rate ranging*, is attractive for being quick and autonomous—though applicable only at short ranges.

It takes advantage of the relationship between the target's range, R, and the angular rate of rotation, ω, of the line of sight to the target. As illustrated in Fig. 3, R is equal to the component of the target's relative velocity normal to the line of sight to the target, V_n , divided by ω.

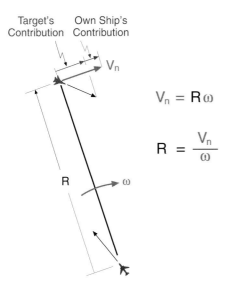

$$V_n = R\omega$$

$$R = \frac{V_n}{\omega}$$

3. The angle-rate ranging technique takes advantage of the relationship between a target's range, R, and the angular rate of rotation, ω, of the line of sight to the target.

TABLE 1. PASSIVE RANGING TECHNIQUES

Type	Basis for Range Estimate	Limitations
Angle-Rate	Change in jamming strobe's angular rate of rotation in response to change in direction of own-ship's velocity.	Practical only at short ranges. Also, jammer's velocity may change unpredictably.
Triangulation (Own ship only)	Change in jammer's bearing due to own-ship course deviation. Deviation is measred by INS*. Change in bearing is adjusted for measured angular rate.	Jammer's velocity may change unpredictably during own-ship's maneuver.
Triangulation (With other aircraft)	Bearing of jammer measured in own ship and in another aircraft (received via secure data link). Positions of both aircraft measured with INS.*	A suitably equipped aircraft may not be present, or in a location enabling accurate triangulation.
Signal-Strength	Rate of increase of target's RF or IR signal strength, both of which vary as $1/R^2$.	Factors besides range (e.g., multipath or change in look angle) also affect signal strengths.
Off-board Data	Target coordinates obtained via secure data link from ground-based tracking radar or other source. Own-ship's position obtained by INS.*	Suitably equipped and located ground-based radars may not be available.

* Preferably GPS supervised

While V_n is not known, a change in the radar-bearing aircraft's contribution to V_n can readily be determined. Knowing that and measuring the resulting change in ω, the range, R, can be computed. In essence, the procedure is this:

1. The radar-bearing aircraft maneuvers to change the direction of its velocity

2. The resulting change in the component of the aircraft's velocity normal to the line of sight to the target, ΔV_n , is computed

3. The concomitant change in angular rate, Δ_ω, is sensed

4. From ΔV_n and Δ_ω, the range, R, is then computed

$$R = \frac{\Delta V_n}{\Delta_\omega}$$

While for clarity the technique is described here as a series of incremental steps, it is actually performed continuously.

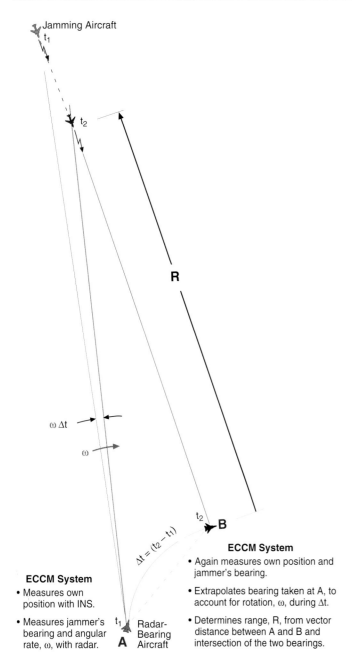

Jamming Aircraft
t₁

t₂

R

ω Δt

ω

t₂ ▶ B

Δt = (t₂ – t₁)

t₁ ▶ Radar-
A Bearing
Aircraft

ECCM System

- Measures own position with INS.

- Measures jammer's bearing and angular rate, ω, with radar.

ECCM System

- Again measures own position and jammer's bearing.

- Extrapolates bearing taken at A, to account for rotation, ω, during Δt.

- Determines range, R, from vector distance between A and B and intersection of the two bearings.

4.   How range is determined by triangulation from own ship only.

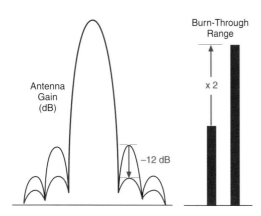

Burn-Through
Range

Antenna
Gain
(dB)

x 2

–12 dB

5.   Reducing the sidelobe gain by 12 dB doubles target burn-through range.

At longer ranges, $\Delta\omega$ may be immeasurably small. If it is, then, the second technique listed earlier in Table 1 might logically be used: *triangulation, own-ship only.*

With it, the radar-bearing aircraft deviates from its course for a considerably longer period, Δt, than for angular-rate range measurement. As illustrated in Fig. 4, the aircraft's own position is measured by the aircraft's inertial navigation system (INS) both before and after the deviation. The range to the jammer is then estimated by triangulation on the basis of:

1. The true bearing of the jammer at the start of the maneuver (extrapolated for Δt seconds in accordance with the initially measured angular rate, ω)

2. The true bearing of the jammer Δt seconds later

3. The vector distance between the two measured positions

The range estimate obtained with either this or the angle-rate ranging technique is of questionable accuracy. For there is nothing to prevent the target itself from simultaneously changing its velocity. Still, to a pilot faced with determining when a target is within an acceptable launch range and what settings of missile-gain and g-bias to use, a crude estimate of a target's range is far better than none at all.

Depending upon the tactical situation, of course, a more accurate estimate may be obtained with one of the other methods listed in Table 1.

Clutter Reduction Features That Reduce Vulnerability to Noise Jamming. In modern radars, vulnerability to noise jamming is materially reduced by certain conventional design features provided to enhance the radars' ability to contend with strong ground clutter:

- Low antenna sidelobes

- Wide dynamic range, with fast-acting AGC

- Constant false alarm rate (CFAR) detection

- Sidelobe blanking

Just as reducing antenna sidelobes reduces vulnerability to strong sidelobe clutter, so too it reduces vulnerability to sidelobe jamming. A reduction in sidelobe gain of 12 dB, for example, doubles target burn-through ranges (Fig. 5).

Insuring wide dynamic range throughout the receive chain reduces the possibility of the receiver being saturated, hence desensitized, by strong jamming. In addition, making the automatic gain control (AGC) fast-acting prevents desensitization following the receipt of periodic strong pulses or bursts of jamming.

Constant false alarm rate (CFAR) detection—described in detail in Chap. 10—keeps all but short spikes of jamming from being detected, hence in a jamming environment makes targets easier to see. Bear in mind, though, that since CFAR keeps jamming strobes from being detected, when it is employed, a separate jamming detector must be provided for the ECCM system.

Sidelobe blanking (described in detail in Chap. 27) is a mixed blessing in so far as countering jamming is concerned. This feature inhibits the output of the radar receiver when the amplitude of the signal received through a broad-beamed low-gain "guard" antenna exceeds the amplitude of the signal simultaneously received through the main antenna. Blanking thus eliminates false targets injected into the radar antenna's sidelobes. It also clears from the display the jamming strobes produced during search, as the radar antenna's sidelobes sweep across a jammer.

But since the guard antenna has little directivity and has a higher gain than the strongest sidelobes (Fig. 6), the radar's blanking logic must be sufficiently intelligent to keep jamming in the far sidelobes that otherwise might not be a problem from blanking the display and preventing the weak echoes of long-range targets from being detected.

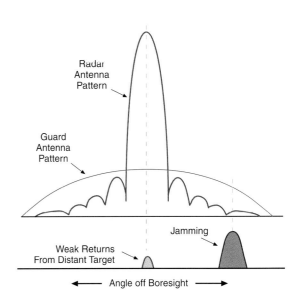

6. Sidelobe blanking eliminates false targets injected into radar antenna's sidelobes. But it must be intelligent enough to keep weak echoes from distant targets from being blanked as a result of jamming in the far sidelobes that otherwise would not be a problem.

Conventional Counters to Deception ECM

Measures have been devised for countering virtually every deception ECM developed to date. Within the limits of military security, the following paragraphs describe those ECCM for countering range- and velocity-gate stealers and certain angle-deception ECM.

Countering Range-Gate Stealers. The primary technique for countering range-gate stealers has long been leading-edge tracking. It takes advantage of two characteristics of a simple stealer. First, because of the stealer's finite response time, at the very earliest the stealer's pulse will arrive at the radar slightly after the leading edge of the skin return. Second, the simple stealers will always pull the tracking gate off the skin return to greater ranges.

Therefore, the stealer's pulse can be kept from capturing the gate by (a) passing the receiver's video output through a differentiation circuit to provide a sharp spike at the skin return's leading edge, (b) narrowing the tracking gate, and (c) locking the gate onto the spike (Fig. 7).

In noncoherent radars, the possibility of a more capable stealer sensing the PRF and pulling the gate off the skin return to shorter ranges may be forestalled by jittering the PRF. Unable, then, to accurately predict when successive pulses will be transmitted, the stealer cannot transmit puls-

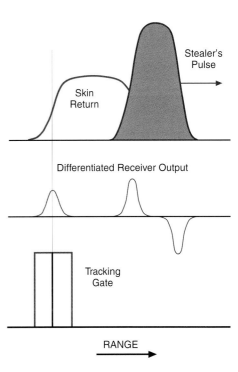

7. By differentiating the receiver output to produce a sharp spike at the skin return's leading edge, narrowing the tracking gate, and locking it onto the spike, a simple gate stealer can be kept from capturing the gate and pulling it off to longer range.

es that will deceptively precede the skin return.

In coherent radars, however, PRF jittering is not practical. For, the PRF can't be changed during the coherent integration period. Consequently, in these radars other measures have been taken to reduce vulnerability to the more capable range-gate stealers. They include:

- Limiting the maximum speed at which the position of the gate can be change once locked onto a target

- Providing an automatic means of quickly detecting pull-off

- When pull-off is detected, extrapolating the target's range on the basis of the last doppler measurement of range rate

- Designing the tracking system to rapidly relock on the skin return

Pull-off may be detected by sensing abnormally large range rates, range accelerations, or changes in signal strength. Against transponders and those repeaters that do not duplicate the radar's pulse compression coding, pull-off may be detected by sensing the spreading of otherwise compressed pulse widths. (Spreading, though, may be due to other causes.)

Rapid relock—a feature commonly called *snapback*—takes advantage of the sluggish response of the tracking loop to the gate-stealer's pulses plus the inherent time lag in the stealer's performance. The longer these lags and the faster the relock, the greater the fraction of the time the radar will be accurately measuring the target's range and the less it must depend upon extrapolation (Fig. 8).

In situations where none of the above features prove effective, the range-gate stealer may possibly be avoided by switching to a high PRF mode which does not depend upon range gating. An intelligent ECM system, however, can sense the changes and switch to velocity-gate stealing.

Countering Velocity-Gate Stealers. Much as in countering range-gate stealers, *velocity-gate pull-off (VGPO)* may be detected by sensing abnormally high accelerations and tracking rates or the abnormal spreading of the received signal in the velocity gate. If pull-off is detected, the radar may either be rapidly relocked on the skin return, or—against a not-so-intelligent ECM system—be switched to a low-PRF mode where tracking in velocity is not essential.

Countering Deception of Lobing Systems. The deception of lobing systems for angle tracking may be countered by *lobing on receive only (LORO)*, a technique also called *passive*

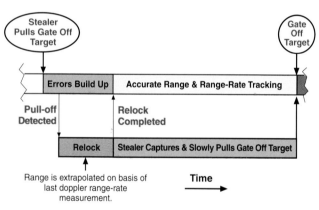

8. How sluggish response of range-tracking gate plus rapid-relock counter the more capable range-gate stealers. The longer the stealer takes to capture and pull the gate off the target and the more rapidly the radar detects pull-off and relocks the gate on the target, the greater the percentage of time the radar will be accurately tracking the target.

lobing or *silent lobing.* Deception of LORO may be made more difficult by varying the lobing frequency and may be circumvented by employing simultaneous lobing (monopulse tracking).

Countering Terrain Bounce. Against terrain bounce, the simplest ECCM is leading-edge tracking, such as used against simple range-gate stealers. In this case, advantage is taken of the deception signal traversing a slightly longer path than the skin return, hence arriving at the radar a fraction of a pulse width behind the leading edge of the skin return (Fig. 9). By tracking it, therefore, the deception signal is kept out of the tracking gate.

Countering Crosseye and Crosspol. Because of military security restrictions, advanced techniques for countering these deceptions cannot be described here. The techniques may be helped, however, by providing a good ECCM against gate-stealing.

The reason: both crosseye and crosspol require high jam-to-signal (J/S) ratios. To get a sufficiently high J/S ratio, gate stealing may be necessary. Consequently, a good counter to gate stealing may help defeat these two formidable ECM threats.

ECCM Used by Surface-Based Radars. Before moving on to advanced ECCM developments, it may prove instructive to consider the ECCM listed in the panel (right) that are used by surface-based radars to contend with jamming.

Advanced ECCM Developments

With continuing technological advances and dramatic increases in available processor throughputs, during the 1980s and early 1990s ECCM development broadened into several new areas:

• Sidelobe jamming cancellation, already widely used in surface-based radars

• Mainlobe jamming cancellation

• Vastly increased radio frequency bandwidths

• Sensor fusion

• Offensive ECCM

• Application of artificial intelligence to ECCM development and utilization

Within the constraints of military security, these developments are touched on briefly in the following paragraphs.

Sidelobe Jamming Cancellation. Besides sidelobe reduction, one of the most effective ways to counter sidelobe

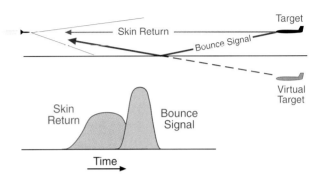

9. Countering terrain bounce. Because of the greater distance the bounce signal travels, it arrives at the radar a fraction of a pulse width behind the leading edge of the skin return; hence, deception can be avoided by leading-edge tracking.

HOW GROUND-BASED RADARS COUNTER JAMMING

• **Increased ERP** Use higher antenna gain and/or higher transmitted power.

• **Vertical Triangulation** Angle track on jamming; compute range on basis of elevation angle, estimated target altitude, and earth curvature charts.

• **Multiple Radar Triangulation** Simultaneously track jamming in angle with one or more widely separated radars; compute range on basis of measured angles and radars' known locations.

• **Second Radar Assist** Track jamming in angle with main radar; briefly transmit on another frequency with a co-located second radar to determine range of target in noise strobe.

jamming is to introduce notches in the radar antenna's receive pattern in those directions from which the jamming arrives. The essence of this technique is illustrated for the simple case of a single jammer in Fig. 10.

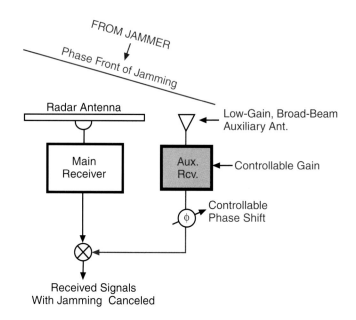

10. Essence of approach to canceling sidelobe jamming. Gain and phase shift of auxiliary receiver are adjusted so that jamming cancels when receiver outputs combine.

The radar antenna is supplemented with a low-gain broad-beamed auxiliary receiving antenna—such as a small horn—having the same angular coverage but displaced laterally to provide directional sensitivity. Signals received by the auxiliary antenna are fed to a separate receiver, having controllable gain and controllable phase shift. Its output is added to the main receiver's output.

As illustrated in the panel (left), by adjusting the gain of the auxiliary receiver, the difference in the gains of the two antennas in the jammer's direction is compensated. By adjusting the phase shift of the auxiliary receiver, the jammer's signal in its output is made 180° out of phase with the jammer's signal in the output of the main receiver. Consequently, when the outputs of the two receivers are combined, the jamming cancels—in effect producing a notch in the radar antenna's receive pattern in the direction of the jammer.

This process—broadened to include interactive insertion of notches in the directions of several jammers—is the basis for an ECCM technique called *coherent sidelobe cancellation (CSLC)*. For each desired notch, a separate auxiliary antenna and receiver must be provided. To ensure best results, a radar is typically provided with between $1^1/_2$ and 2 times as many auxiliary antennas and receivers as the expected number of jammers to be canceled. The auxiliary antennas

How Sidelobe Jamming is Canceled

Signals Received From Jammer

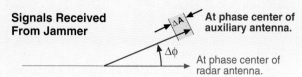

At phase center of auxiliary antenna.

At phase center of radar antenna.

Amplitude difference, ΔA, is due to difference in gains of auxiliary antenna and main antenna in jammer's direction.

Phase difference, Δφ, is due to difference in distance from jammer to the two phase centers.

Amplitude Adjustment

By adjusting the gain of the auxiliary receiver, the amplitude difference is removed.

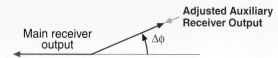

Main receiver output

Adjusted Auxiliary Receiver Output

Note: In this example, the jammer is assumed to be in the radar antenna's first sidelobe. So, the phase of jamming is reversed in the output of antenna.

Phase Adjustment

By adjusting the phase shift in the output of the auxiliary receiver, Δφ is removed.

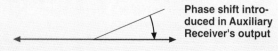

Phase shift introduced in Auxiliary Receiver's output

Result

Because the jammer's signal in the output of the auxiliary receiver is now equal to and 180° out of phase with the jammer's signal in the output of the main receiver, they cancel when the outputs combine.

Another way of looking at this: a notch has been produced in the radar antenna's sidelobe pattern in the jammer's direction.

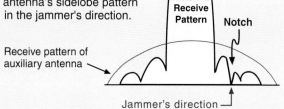

Radar Antenna Receive Pattern

Notch

Receive pattern of auxiliary antenna

Jammer's direction

must all cover the field of regard of the radar antenna. And they must be positioned so that their phase centers are displaced from one another, as well as from the phase center of the radar antenna.

A quickly converging algorithm adaptively adjusts the amplitude and phase of each auxiliary receiver to place notches in those directions from which jamming is being received. Phase rotation and signal combination may take place in the radar's RF, IF, or digital processing sections. Although requiring lots of throughput, digital processing works best and is the most flexible.

Mainlobe Jamming Cancellation. Jamming received through the radar antenna's mainlobe may be canceled with an adaptation of the GMTI notching technique described in Chap. 24. With this technique, sometimes called adaptive beam forming (ABF), a single notch is produced in the mainlobe receive pattern in the jammer's direction by adaptively shifting the relative phases of the outputs of the monopulse antenna's right and left halves so that when they combine, radiation arriving from the jammer's direction cancels (Fig. 11).

As with sidelobe cancellation, phase rotation and signal combination are generally performed in the radar's digital processing section. However, with the advent of the active ESA and its highly adaptive beam-forming capability, both mainlobe and sidelobe jamming cancellation may be performed entirely within the main antenna.

Exceptionally Broad RF Bandwidths. Another approach to countering severe noise jamming is to simultaneously employ widely spaced multiple operating frequencies, each of which is itself spread over a very broad band (Fig. 12, below). Against a spot jammer capable of jamming only a limited number of spots, a multifrequency radar can defeat the jammer by simultaneously transmitting on more channels. Against a barrage jammer, the radar's broad, widely spaced channels may overcome the jammer by forcing it to spread its power ever more thinly.

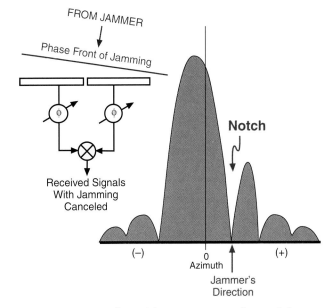

11. Basic concept of mainlobe jamming cancellation. Relative phases of radiation received through right and left halves of monopulse radar antenna are shifted so they are 180° out of phase for radiation coming from the jammer's direction.

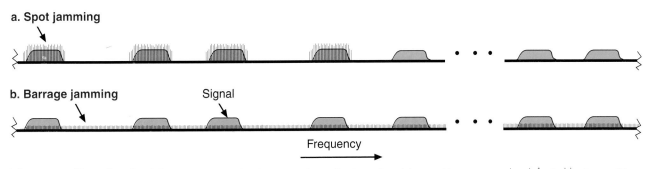

12. Advantages of broadband multifrequency operation in countering noise jamming: (a) a spot jammer may be defeated by transmitting on more channels than it can jam; (b) a barrage jammer may be defeated by forcing further dilution of its jamming power.

How, you may ask, does the radar come out ahead if, to force the jammer to spread its power, the radar must spread its own power over the same broad band. Apart from the corollary improvement in single-look probability of detection due to frequency diversity, the answer is integration. Being coherent and being spread out in frequency largely through pulse-compression coding, the radar returns can be decoded by the radar and integrated into very strong narrowband signals, containing virtually all of the energy received over the coherent integration period. Being neither coherent nor properly coded, the jamming doesn't build up in this way. Consequently, the integrated returns from a target need only compete with the mean level of the jamming.

Sensor Fusion. This is essentially the melding of data obtained by the radar with data obtained by the other onboard sensors, as well as data received via secure communication links from offboard resources. Onboard sensors (Fig. 13) have complementary capabilities and are not all vulnerable to the same kinds of countermeasures. Offboard resources have the additional advantage of viewing the battle scene from different locations and different perspectives. Consequently, even the most severe ECM may be circumvented by analyzing all available sensor data and extracting less contaminated information from it.

The chief technical challenge in fusing data from multiple sensors is associating the incoming data with the target tracks being maintained. The most applicable correlation techniques are *nearest neighbor (NN) correlation* and *multiple hypothesis tracking (MHT)*.

Nearest neighbor has long been used in track-while-scan modes (see Chap. 29). It works well if the targets are fairly widely spaced. But if they are not, because of the randomness of measurement errors from one observation to another, observations may be correlated incorrectly. Some tracks may be erroneously terminated and some false tracks may be initiated.

These problems are largely obviated in multiple hypothesis tracking. With it, incoming observations are similarly correlated with existing tracks. But instead of irrevocably assigning the observation to a single track, every reasonable combination of tracks with which the observation may be correlated is hypothesized. The individual tracks are then graded, and each hypothesized combination of tracks (called a *hypothesis*) is given a grade equal to the sum of the grades of the individual tracks it includes.

A process of combining and pruning is then carried out. Similar tracks or tracks with identical updates over the recent past are combined and so are similar hypotheses. Tracks and hypotheses whose scores fall below a certain threshold are

---

RADAR
• Long range search and track.
• All Weather.
• Accurately measures range, range rate, and angle.
• Can break out closely spaced targets in range (except in conventional High PRF modes).
• **Active; may indicate its presence and direction to enemy.**
• **Subject to RF countermeasures.**
• Even when jammed, it can track the jamming aircraft in angle and passively estimate its range.

IR SEARCH TRACK SET
• Long range search and track.
• Detects subsonic and supersonic targets plus missile launches.
• Measures angle precisely. Measures range crudely with angle-rate method.
• Can break out closely spaced targets in angle.
• Passive; hence does not alert enemy to its presence or location.
• Not affected by RF countermeasures.
• **Can only operate in clear weather.**
• **Has poor look-down performance.**

RADAR WARNING RECEIVER
• Long range detection (in some cases).
• 360° azimuth coverage; very broad frequency coverage.
• All weather.
• Measures angle (usually crudely).
• May give very crude estimate of range and indicate whether range is closing or opening.
• Identifies type of emitter.
• Passive.
• **Target must radiate.**

FORWARD LOOKING IR
• Detects targets in same way as IRST.
• Provides image of target, enabling ID.
• Passive; hence, doesn't alert enemy.
• Not subject to RF countermeasures.
• **Can only operate in clear weather.**

LASER RANGE FINDER
• Trained on target by IRST or radar.
• Precisely measures range.
• Not subject to RF countermeasures.
• **Active, may indicate its presence and direction to enemy.**

13. The complementary capabilities of an aircraft's onboard sensors. Characteristics limiting a sensor's utility or making it vulnerable to ECM are set in bold type. Since these are not the same for all of the sensors, a weapon system's vulnerability to ECM can be materially reduced by selectively combining the sensor's outputs.

deleted. All tracks are then smoothed, and the process is repeated when the next set of observations comes in.

With each iteration, the accuracy of the established tracks is updated. At any one time, the hypothesis having the highest score is output as the current most likely partitioning of all observations into target tracks.

Offensive ECCM. Unlike the counter-countermeasures discussed so far, offensive ECCM are designed not just to defeat an enemy's countermeasures, but to do so in such a way as to confuse the opponent and confound his attempts to optimally employ his ECM.

A simplistic example is simultaneous multifrequency operation, in which the radar transmits on a large number of frequencies, spread over a very broad spectrum, but receives on only a few, adaptively selected ones where ECM are minimal.

Artificial Intelligence Applied to ECCM. Electronic warfare is by no means a static art. To maintain an edge, the radar designer must: (1) quickly develop robust new ECCM to counter emerging ECM, and (2) provide the radar with the ability to optimally employ its existing ECCM repertoire when confronted with new countermeasures during combat.

Toward these ends, designers are hard at work on the application of knowledge-based systems, multiple hypothesis testing, and neural networks to ECCM development.

The Most Effective ECCM of All

Without question, the most effective ECCM of all is simply not to be detected by the enemy. If the enemy cannot detect the radiation from your radar, he also cannot

- Concentrate his jamming power at the radar's operating frequency

- Increase his jamming power in the radar's direction with high-gain antennas

- Mask the range or doppler bins in which his radar returns will be collected

- Respond to the radar's pulses with false target returns

- Steal the radar's tracking gates

- Deceive the radar's range or angle tracking systems

To hope to completely avoid detection of one's radar signals by the enemy is patently absurd. But by employing the low probability of intercept (LPI) techniques described in Chap. 42, the possibility of avoiding useful detection by the enemy and still being able to use your radar to advantage is very real and practical.

ACRONYMS OF ECCM

Tracking In Angle On A Target's Jamming
- **TOJ** – Track On Jamming
- **JAT** – Jam Angle Track
- **ATOJ** – Angle Track On Jamming
- **HOJ** – Home On Jamming
  (for radar-guided missiles)

Jamming Cancellation
- **CSLC** – Coherent Side Lobe Cancellation
- **ABF** – Antenna Beam Forming
  (main-lobe cancellation)

Countering ECM Used Against Lobing Systems
- **LORO** – Lobe On Receive Only
  (passive lobing.)
- **COSRO** – Conical Scan On Receive Only
  (silent lobing)

Countering Range-Gate Stealers and Terrain-Bounce
- **LET** – Leading Edge Tracking

Summary

Over the years, many ECCM techniques have been devised which are still viable today.

Among those for countering noise jamming are detection and angle tracking on the jamming, and several passive ranging techniques, of which angle-rate ranging for short ranges and various triangulation techniques for longer ranges are attractive. In addition, many radar system improvements for reducing vulnerability to strong ground clutter also reduce vulnerability to ECM: sidelobe reduction, wide dynamic range; fast-acting AGC, constant false-alarm rate (CFAR) detection, and, to some extent, sidelobe blanking.

To counter deceptive ECM, leading-edge tracking has been provided for simple range-gate stealers and terrain bounce; rapid relock, for more capable range-gate and velocity-gate stealers; and still others, which cannot be described here.

Meanwhile, dramatic increases in processor throughputs, have led to several newer ECCM developments:

- Coherent sidelobe cancellation—adaptive introduction of nulls in the antenna receive pattern in directions from which jamming is received

- Adaptive beam forming—introduction of a similar null in the mainlobe receive pattern

- Broadband multifrequency operation—to counter noise jamming

- Sensor fusion—melding the radar's capabilities with those of other sensors, both onboard and offboard

- Offensive ECCM—countering ECM in such a way as to confound the enemy's attempts to optimally employ his countermeasures

Finally, artificial intelligence is being applied both to the optimal employment of existing ECCM and to the rapid development of counters for emerging ECM.

Electronic Warfare Intelligence Functions

36

With the continual advances in radar technology and the increasing complexity of aerial combat, the effectiveness of ECM and ECCM has become increasingly dependent on three levels of intelligence:

- Knowledge of the capabilities and operating parameters of hostile systems which may be encountered—*what's potentially out there*

- Knowledge of the electronic order of battle (EOB) of the hostile force about to be engaged—*what's out there today and where*

- Real-time threat warning—*what's after me now*

Answers to these questions are provided by ELINT, ESM, and the RWR, respectively. This chapter briefly introduces them and explains what functions they perform.

Electronic Intelligence (ELINT)

ELINT is the gathering of information on the radars and associated electronics of potential hostile threats. It is typically performed by government intelligence agencies. The continually gathered data from various sources—including both human agents and sensitive radio receivers—is thoroughly analyzed and used as a basis for the design of ESM systems.

Electronic Support Measures (ESM)

Carried in certain tactical aircraft ESM, systems are designed to collect, in advance, information on the elec-

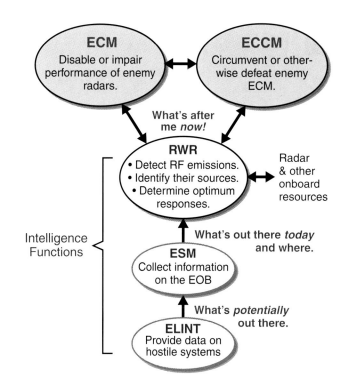

BASIC ESM FUNCTIONS

- Detect enemy's RF emissions
- Measure their parameters
- Identify their sources

1. For economy, though, a smaller number of wider channels may be used.

2. Which entails providing wideband antenna and other RF hardware.

tronic order of battle (EOB) for the radar warning receivers (RWRs) and flight crews of the aircraft about to be deployed on a mission.

In essence, the ESM system performs three main functions: (1) detects the enemy's RF emissions; (2) measures their key parameters; (3) from them, identifies the sources of the emissions.

Detecting RF Emissions. Combat aircraft may encounter threats over a broad spectrum of radio frequencies. The ESM system must cover all of it, yet have the RF selectivity to separate simultaneously received signals that are closely spaced in frequency. In the past, this difficult combination of requirements was satisfied with scanning superheterodyne receivers, which are comparatively slow.

Today, the requirements are satisfied much more rapidly through *channelization*, that is, by dividing the spectrum to be monitored into a great many partially overlapping channels[1] (Fig. 2). Each channel is made wide enough to accommodate the spectra of extremely short pulses, with enough margin to enable accurate measurement of their times and angles of arrival, yet narrow enough to separate individual signals.[2]

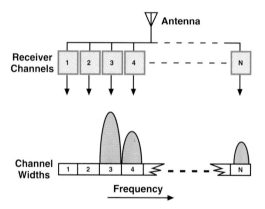

2. With channelization, the spectrum to be monitored is divided into partially overlapping channels, each just wide enough to pass the spectra of very short pulses with sufficient margin to enable measurement of time of arrival and angle of arrival.

Most of the radars whose radiation the ESM system must detect will have their antennas trained on the aircraft carrying the system only fleetingly. Consequently, the ESM receivers must be sensitive enough to detect even very weak sidelobe emissions. Hundreds of radars, therefore, may be within the system's detection range at any one time. Considering that some of these radars may be operating at high PRFs—a vast number of pulses and other signals may be received from all directions. So that their sources may be identified, every received signal—be it a short pulse or a continuous wave—must be individually detected.

Extracting Key Signal Parameters. The principal steps in extracting the parameters of the detected signals are outlined in Fig. 3. The first step is to record their times of arrival (TOA) and measure their angles of arrival (AOA) and radio frequencies (RF).

The angles of arrival may be measured virtually instantaneously by either of two methods. One is to provide a separate antenna and receiving system for each quadrant in azimuth and to sense the difference in amplitude of each signal as received by the four antennas (Fig. 4).

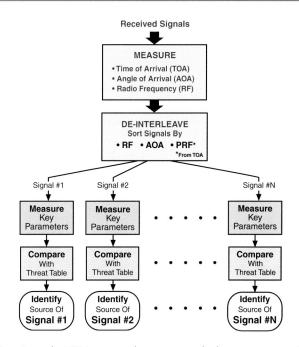

3. Steps the ESM system takes to extract the key parameters of the signals it detects and characterize their sources.

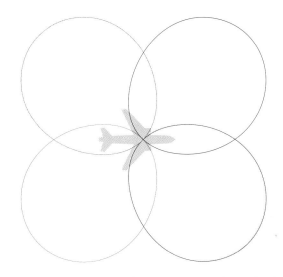

4. One way to instantaneously measure a signal's angle of arrival (AOA): sense the difference in amplitude of the outputs it produces from four antenna beams.

The other method is to place three or four antennas in each quadrant and to sense the difference in phase of each signal as received by the individual antennas.

Frequency also may be measured instantaneously. Coarse frequency is determined from the channel the signal is received through. Fine frequency may then be determined by a frequency discriminator or a special *instantaneous frequency-measurement* circuit (IFM), such as is illustrated in Fig. 5 in the output of each channel. Less sophisticated systems may instead make the fine measurements with a scanning narrowband superheterodyne receiver in each channel.

By sorting the signals according to angle of arrival, frequency, and PRF (obtained from the recorded times of arrival), the ESM system quickly separates—"de-interleaves"—the signals received from different sources. It then precisely measures key parameters—such as interpulse modulation, intrapulse modulation (pulse compression coding), beam width, scan rate, polarization, and pulse width[3]—of the signals from each source.

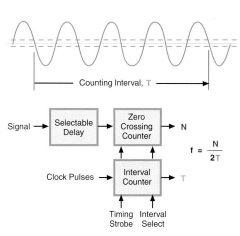

5. Innovative approach to instantaneously measuring the radio frequency of a received signal. Number, N, of signal's zero crossings in interval, T, is counted and divided by 2T. Selectable delay compensates for short time it takes to detect signal and generate a timing strobe.

3. Pulse width is difficult to measure accurately; for reflections may be received from the ground which are staggered relative to the directly received pulses.

Identifying the Sources. Finally, by comparing the measured signal parameters with the parameters of all known threats, stored in "threat tables," the ESM system identifies each source. For mobile surface-based threats, the system also determines current location. These data, together with ELINT data, enable the mission to be planned to avoid unnecessary exposure to lethal threats.

If the ESM system detects previously unknown waveforms or variations of known waveforms it stores the measured parameters for post-flight analysis and subsequent permanent entry into the threat tables of the radar warning receivers.

Radar Warning Receiver (RWR)

As a rule, RWRs are less comprehensive and far more numerous than the ESM systems. Intended primarily to warn the air crew of imminent attack, they generally are sensitive only to the mainlobe emissions of systems tracking the aircraft.

Much as in an ESM system, the RWR detects these emissions and identifies the threats they represent by comparing their characteristics with those stored in a threat table. It then evaluates and prioritizes the threats. Through expert systems techniques, the modern RWR (Fig. 6) may even determine the optimum responses to be made by the pilot and/or the appropriate electronic combat (EC) systems—radar, ECM, ECCM, IR search track set, FLIR, etc. The RWR may also control the timing and execution of the EC responses under close oversight of the air crew who are alerted to the RWR's actions and can override any of them.

Summary

Effective employment of both ECM and ECCM depends on the ability of (a) ELINT to determine the capabilities of the radars of potential hostile forces, (b) the ESM system to determine the electronic order of battle, and (c) the ability of the RWRs in the individual aircraft to detect the RF emissions of any enemy system that threatens the aircraft, identify the sources of the emissions, and determine optimum responses.

6. While most RWRs are comparatively simple, an advanced RWR, such as the ALR-67 V3/4, may perform virtually all of the functions of a highly capable ESM system.

Electronically Steered Array Antennas (ESAs)

37

E lectronically steered array antennas, ESAs, have been employed in surface based radars since the 1950s.[1] But, because of their greater complexity and cost, they have been slow to replace mechanically steered antennas in airborne applications.

However, with the advent of aircraft of extraordinarily low radar cross section and the pressing need for extreme beam agility, in recent years avionics designers have given the ESA more attention than virtually any other "advanced" radar concept.

In this chapter, we will briefly review the ESA concept, become acquainted with the two basic types of ESAs, and take stock of the ESA's many compelling advantages, as well as a couple of significant limitations.

Basic Concepts

An ESA differs from the conventional mechanically steered array antenna in two fundamental respects:

- It is mounted in a fixed position on the aircraft structure

- Its beam is steered by individually controlling the phase of the radio waves transmitted and received by each radiating element (Fig. 1)

A general purpose digital processor, referred to as the *beam steering controller (BSC)* translates the desired deflection of the beam from the broadside direction (normal to the plane of the antenna) into phase commands for the individual radiating elements.

The incremental phase difference, $\Delta\phi$, which must be applied from one radiating element to the next to deflect

1. In surface-based radars, they were called "phased arrays"— a name which has carried over to airborne applications. They are frequently called electronically "scanned," as opposed to "steered" arrays. In light of the versatility of the technique, the more general "steered" is used here.

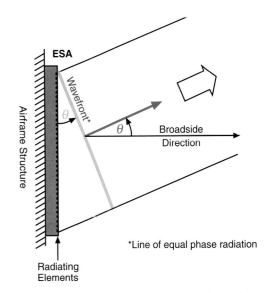

1. The ESA is mounted in a fixed position on the airframe. Its beam is steered by individually controlling the phase of the waves transmitted and received by each radiating element.

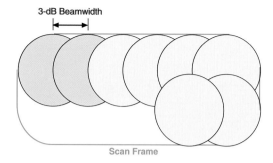

Scan Frame

2. For search, the beam steps ahead in increments nominally equal to the 3-dB beamwidth, dwelling in each position for a period equal to the desired time-on-target.

PHASE SHIFT NEEDED TO STEER THE BEAM

To steer the beam θ degrees off broadside, the phase of the excitation for element **B** must lead that for element **A** by the phase lag, $\Delta\phi$, that is incurred in traveling the distance, ΔR, from radiator **B**.

In traveling one wavelength (λ) a wave incurs a phase lag of 2π radians. So, in traveling the distance ΔR, it incurs a phase lag of

$$2\pi \frac{\Delta R}{\lambda} \text{ radians}$$

As can be seen from the diagram,

$$\Delta R = d \sin \theta$$

Hence, the element-to-element phase difference needed to steer the beam θ radians off broadside is

$$\Delta\phi = 2\pi \frac{d \sin \theta}{\lambda}$$

(Diagram labels: Radiating Elements, A, Line of Equal Phase Radiation, d, θ, Broadside, B, ΔR)

---

the beam by a desired angle, θ, is proportional to the sine of θ (see panel, left center).

$$\Delta\phi = \frac{2\pi d \sin \theta}{\lambda}$$

where d is the element spacing and λ is the wavelength.

For search, the beam is scanned by stepping it in small increments from one position to the next (Fig. 2), dwelling in each position for the desired time-on-target, t_{ot}. The size of the steps—typically on the order of the 3-dB beam width—is optimized by trading off such factors as beam shape loss and scan frame time.

Types of ESAs

ESAs are of three basic types: passive, active, and a variant of the active ESA, called the true-time-delay (TTD) ESA.

Passive ESA. Though considerably more complex than a mechanically steered array (MSA), the passive ESA is far simpler than the active ESA. It operates in conjunction with the same sort of central transmitter and receiver as the MSA. To steer the beam formed by the array, an electronically controlled phase shifter is placed immediately behind each radiating element (Fig. 3, below left), or each column of radiating elements in a one-dimensional array. The phase shifter is controlled either by a local processor called the *beam steering controller (BSC)* or by the central processor.

Active ESA. The active ESA is an order of magnitude more complex than the passive ESA. For, distributed within it, are both the transmitter power-amplifier function and the receiver front-end functions. Instead of a phase shifter, a tiny dedicated transmit/receive (T/R) module is placed directly behind each radiating element (Fig. 4).

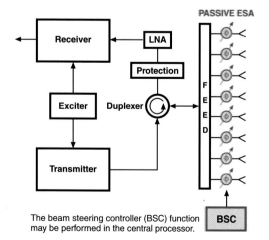

PASSIVE ESA

The beam steering controller (BSC) function may be performed in the central processor.

3. The passive ESA uses the same central transmitter and receiver as the MSA. Its beam is steered by placing an electronically controlled phase shifter immediately behind each radiating element.

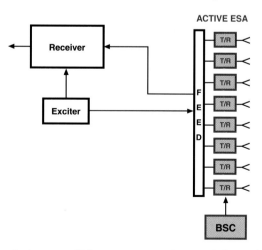

ACTIVE ESA

4. In the active ESA, a tiny transmit/receive (T/R) module is placed immediately behind each radiating element. The centralized transmitter, duplexer, and front-end receiving elements are thereby eliminated.

474

This module (Fig. 5) contains a multistage high power amplifier (HPA), a duplexer (circulator), a protection circuit to block any leakage of the transmitted pulses through the duplexer into the receiving channel, and a low-noise pre-amplifier (LNA) for the received signals. The RF input and output are passed through a variable gain amplifier and a variable phase shifter, which typically are time shared between transmission and reception. They, and the associated switches, are controlled by a logic circuit in accordance with commands received from the beam steering controller.

To minimize the cost of the T/R modules and to make them small enough to fit behind the closely spaced radiators, the modules are implemented with integrated circuits and miniaturized (Fig. 6).

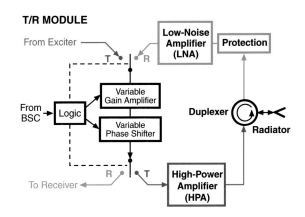

T/R MODULE

5. Basic functional elements of a T/R module. Variable gain amplifier, variable phase shifter, and switches are controlled by the logic element. They may be duplicated for transmit and receive, or time shared as shown here.

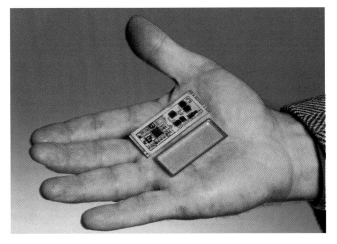

6. A representative T/R module. Even a fairly small ESA would include two to three thousand such modules.

TTD ESA. This is an active ESA in which the phase shifts for beam steering are obtained by varying the physical lengths of the feeds for the individual T/R modules. Drawing on the photonic techniques that have proved so valuable in communications systems, a fiber-optic feed is provided for each module. The time delay experienced by the signals in passing through the feed—hence their phase—is controlled by switching precisely cut lengths of fiber into or out of the feed. By avoiding the limitations on instantaneous bandwidth inherent in electronic phase shifting, the photonic technique makes possible extremely wide instantaneous bandwidths.

Since TTD is still in its infancy, it will be described in Chap. 40, Advanced Radar Techniques, rather than here.

Advantages Common to Passive and Active ESAs

Both passive and active ESAs have three key advantages which have proved to be increasingly important in military aircraft. They facilitate minimizing the aircraft's RCS. They enable extreme beam agility. And they are highly reliable.

Facilitating RCS Reduction. In any aircraft which must have a low RCS, the installation of a radar antenna is of critical concern. For even a comparatively small planar array can have an RCS of several thousand square meters when illuminated from a direction normal to its face (i.e., broadside). With an MSA, which is in continual motion about its gimbal axes, the contribution of antenna broadside reflections to the aircraft's RCS in the threat window of interest cannot be readily reduced. With an ESA, which is fixed relative to the aircraft structure, it can be. How that is done is explained in Chap. 39.

Extreme Beam Agility. Since no inertia must be overcome in steering the ESA's beam, it is far more agile than the beam of an MSA. To appreciate the difference, consider some typical magnitudes. The maximum rate at which an MSA can be deflected, hence the agility of its beam, is limited by the power of the gimbal drive motors to between 100 and 150 degrees per second. Moreover, to change the direction of the beam's motion takes roughly a tenth of a second.

By contrast, the ESA's beam can be positioned anywhere within a ±60 degree cone (Fig. 7) in less than a millisecond! This extreme agility has many advantages. It enables:

- Tracking to be established the instant a target is detected

- Single-target tracking accuracies to be obtained against multiple targets

- Targets for missiles controlled by the radar to be illuminated or tracked by the radar even when they are outside its search volume

- Dwell times to be individually optimized to meet detection and tracking needs

- Sequential detection techniques[2] to be used, significantly increasing detection range

- Terrain-following capabilities to be greatly improved

- Spoofing to be employed anywhere within the antenna's field of regard

These capabilities have given rise to a whole new, highly versatile and efficient approach to allocating the radar front-end and processing resources and to controlling and interleaving the radar's various modes of operation (see Chap. 41).

High Reliability. ESAs are both reliable and capable of a large measure of graceful degradation. They completely eliminate the need for a gimbal system, drive motors, and rotary joints—all of which are possible sources of failure.

In a passive ESA, the only active elements are the phase shifters. High quality phase shifters are remarkably reliable.

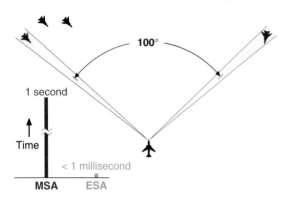

7. To jump the antenna beam from one to another of two targets separated by 100°, an MSA would take roughly a second. An ESA could do it in less than a millisecond.

2. Such as alert-confirm detection. See Chap. 40.

Moreover, if they fail randomly, as many as 5% can fail before the antenna's performance degrades enough to warrant replacing them.

The active ESA yields an important additional reliability advantage by replacing the central transmitter with the T/R modules' HPAs. Historically, the central TWT transmitter and its high-voltage power supply have accounted for a large percentage of the failures experienced in airborne radars. The active ESA's T/R modules, on the other hand, are inherently highly reliable. Not only are they implemented with integrated solid-state circuitry, but they require only low-voltage dc power.

In addition, like the phase shifters of the passive ESA, as many as 5% of the modules can fail without seriously impairing performance. Even then, the effect of individual failures can be minimized by suitably modifying the radiation from the failed element's nearest neighbors. As a result, the *mean time between critical failures* (MTBCF) of a well designed active ESA may be comparable to the lifetime of the aircraft!

Additional Advantages of the Active ESA

The active ESA has a number of other advantages over the passive ESA. Several of these accrue from the fact that the T/R module's LNA and HPA are placed almost immediately behind the radiators, thereby essentially eliminating the effect of losses not only in the antenna feed system but also in the phase shifters.

- Neglecting the comparatively small loss of signal power in the radiator, the duplexer, and the receiver protection circuit, the net receiver noise figure is established by the LNA (Fig. 8). It can be designed to have a very low noise figure.

- Loss of transmit power is similarly reduced. This improvement, though, may be offset by the difference between the modules' efficiency and the potentially very high efficiency of a TWT.

- Amplitude, as well as phase, can be individually controlled for each radiating element on both transmit and receive, thereby providing superior beam-shape agility for such functions as terrain following and short-range SAR and ISAR imaging.

- Multiple independently steerable beams may be radiated by dividing the aperture into sub apertures and providing appropriate feeds.

- Through suitable T/R module design, independently steerable beams of widely different frequencies may simultaneously share the entire aperture.

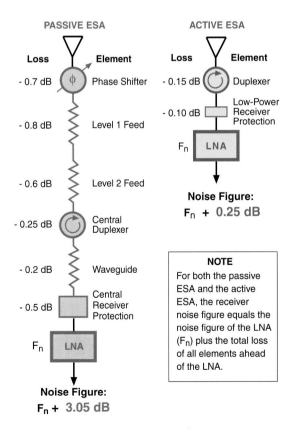

8. By eliminating sources of loss ahead of the LNA, the active ESA achieves a dramatic reduction in receiver noise figure over that obtainable with a comparable passive ESA.

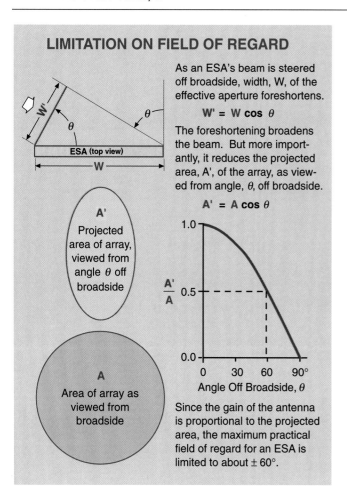

LIMITATION ON FIELD OF REGARD

As an ESA's beam is steered off broadside, width, W, of the effective aperture foreshortens.

$$W' = W \cos \theta$$

The foreshortening broadens the beam. But more importantly, it reduces the projected area, A', of the array, as viewed from angle, θ, off broadside.

$$A' = A \cos \theta$$

ESA (top view)

A'
Projected area of array, viewed from angle θ off broadside

A
Area of array as viewed from broadside

Since the gain of the antenna is proportional to the projected area, the maximum practical field of regard for an ESA is limited to about ± 60°.

Key Limitations and Their Circumvention

Along with its many advantages, the ESA—whether active or passive—complicates a radar's design in two areas which are handled relatively simply with an MSA: (a) achieving a broad field of regard, and (b) stabilizing the antenna beam in the face of changes in aircraft attitude. These complications and the means for circumventing them are outlined briefly in the following paragraphs.

Achieving a Broad Field of Regard. With an MSA, to whatever extent the radome provides unobstructed visibility, the antenna's field of regard may be increased without in any way impairing the radar's performance. With an ESA, however, as the antenna beam is steered away from the broadside direction, the width of the aperture is foreshortened in proportion to the cosine of the angle off broadside, increasing the azimuth beam width (see panel, left).

More importantly, the projected *area* of the aperture also decreases in proportion to the cosine of the angle, causing the gain to fall off correspondingly. At large angles off broadside, the gain falls off still further as a result of the lower gain of the individual radiators at these angles.

Depending upon the application, the fall-off in gain may be compensated to some extent by increasing the dwell time—at the expense of reduced scan efficiency. Even so, the maximum usable field of regard is generally limited to around ±60°.

While ±60° coverage is adequate for many applications,[2] wider fields of regard may be desired. More than one ESA may then be provided—at considerable additional expense. In one possible configuration, a forward-looking main array is supplemented with two smaller "cheek" arrays, extending the field of regard on either side (Fig. 10).

2. In many applications, because of radome restrictions, ±60° is about all that can be obtained, even with an MSA.

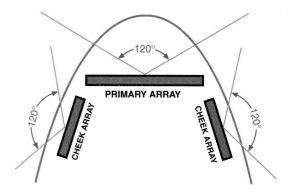

9. Where a broad field of regard is desired, more than one ESA may be used. Here, a central primary array is supplemented with two smaller, "cheek" arrays providing short-range coverage on both sides, for situation awareness.

Beam Stabilization. With an MSA, beam stabilization is not a problem. For the antenna is mounted in gimbals and slaved to the desired beam-pointing direction in spatial coordinates by a fast-acting closed-loop servo system incorporating rate-integrating gyros on the antenna. If the antenna and gimbals are dynamically balanced, this system effectively isolates the antenna from changes in aircraft attitude. The only beam steering required is that for tracing a search scan pattern or tracking a target—neither of which necessitate particularly high angular rates.

With an ESA, stabilization is not so simple. Since the array is fixed to the airframe, every change in aircraft attitude—be it in roll, pitch, or yaw—must be inertially sensed. Phase commands for steering out the change must be computed for each radiator, and these commands must be transmitted to the antenna's phase shifters or T/R modules and executed. The entire process must be repeated at a high enough rate to keep up with the changes in aircraft attitude.

If the aircraft's maneuvers are at all severe, this rate may be exceptionally high. For a nominal "resteer" rate of 2,000 beam positions per second, the phase commands for two to three thousand radiating elements must be calculated, distributed, and executed in less than 500 microseconds!

Fortunately, with advanced airborne digital processing systems, throughputs of this order can be provided.

Summary

Mounted in a fixed position on the aircraft structure, the ESA produces a beam which is steered by individually controlling the phase of the signals transmitted and received by each radiating element.

A passive ESA operates with a conventional central transmitter and receiver; while an active ESA has the transmitter and the receiver front end functions distributed within it at the radiator level. The passive ESA is considerably more complex than a mechanically steered array (MSA); the active ESA is an order of magnitude more complex than the passive ESA.

Both types have three prime advantages: (1) the contribution of their reflectivity to the aircraft's RCS in the threat window of interest can readily be reduced; (2) their beams are extremely agile; (3) they are highly reliable and capable of graceful degradation. The active ESA also has the advantages of providing an extremely low receiver noise figure, affording beam-shaping versatility, and enabling radiation of independent multiple beams of different frequencies.

The principal limitations of the ESAs are (a) restriction of the maximum field of regard to roughly ±60° by the foreshortening of the aperture and consequent reduction in gain at large angles off broadside and (b) the requirement for a substantial amount of processor throughput to stabilize the pointing of the antenna beam in the face of severe aircraft maneuvers.

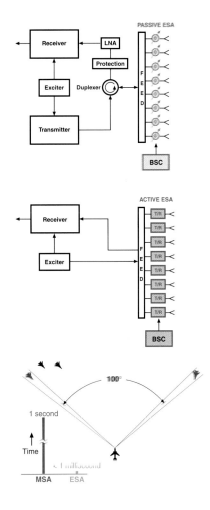

SAR INDOOR TESTING

Passive ESA of the ultrahigh-resolution SAR radar for the U-2 reconnaissance aircraft undergoes tests in an indoor range.

ESA Design

38

To fully realize the compelling advantages of the ESA, its design and implementation must meet a number of stringent requirements, not the least of which is affordable cost.

This chapter begins by discussing those design considerations common to both passive and active ESAs. It then takes up the considerations pertaining primarily to passive ESAs and, finally, those pertaining solely to active ESAs.

Considerations Common to Passive and Active ESAs

The cost of both passive and active ESAs increases rapidly with the number of phase shifters or T/R modules required, hence with the number of radiators in the array.

Consequently, a key design requirement common to both types of ESAs is to space the radiators as widely as possible without creating grating lobes and—if stealth is required—without creating Bragg lobes either. The number of radiators may in some cases be further reduced through judicious selection of radiator lattice.

Avoiding Grating Lobes. Grating lobes (Fig. 1) are repetitions of an antenna's mainlobe[1] which are produced if the spacing of the radiating elements is too large relative to the operating wavelength. They are undesirable because they rob power from the mainlobe, radiate this power in spurious directions, and from these directions receive returns which are ambiguous with the returns received through the mainlobe. Also, ground return or jamming received through the grating lobes may mask targets of interest or desensitize the radar by driving down the automatically controlled gain (AGC).

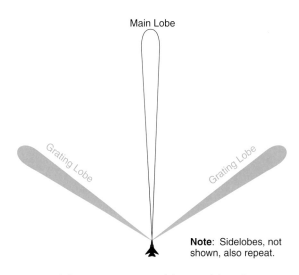

1. Grating lobes are repetitions of the mainlobe. They are produced if the spacing of the radiated elements is too large in comparison to the wavelength.

---

1. And sidelobes, as well.

Grating lobes are not unique to ESAs. They may be produced by any array antenna if the radiators are too widely spaced. Like the mainlobe, they occur in those directions for which the waves received by a distant observer from all of the radiators are in phase. As illustrated by the panel on the facing page, in the case of a mechanically steered array, where the phases of the waves radiated by all radiators are the same, grating lobes can be avoided even if the radiators are separated by as much as a wavelength.

In an ESA, however, the element spacing cannot be this large. For the angles at which the waves from all radiating elements are in phase depend not only upon the element spacing but also upon the incremental element-to-element phase shift, $\Delta\phi$, which is applied for beam steering. As the mainlobe is steered away from broadside (i.e., as $\Delta\phi$ is increased from 0), a grating lobe whose existence was precluded by the radiators being no more than a wavelength apart, may materialize on the opposite side of the broadside direction and move into the field of regard (Fig. 2).

For an ESA, therefore, the greater the desired maximum look angle, the closer together the radiating elements must be. The maximum acceptable spacing is

$$d_{max} = \frac{\lambda}{1 + \sin \theta_0}$$

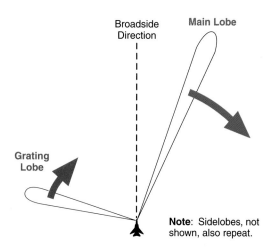

2. With an ESA, if the radiator spacing is not less than 1 wavelength, as the mainlobe is steered away from broadside, a grating lobe will appear and move into the field of regard.

where λ is the wavelength and θ_0 is the maximum desired look angle. As illustrated in the example (left), for a maximum look angle to 60°, the maximum radiator spacing is little more than half the operating wavelength.

Radiator Spacing Example

If the maximum look angle, θ_0, is 30°, what radiator spacing can be used and still avoid grating lobes?

$$d_{max} = \frac{\lambda}{1 + \sin 30°} = \frac{\lambda}{1.5} = 0.67\,\lambda$$

If θ_0 is increased to 60°, what must d_{max} be reduced to?

$$d_{max} = \frac{\lambda}{1 + \sin 60°} = \frac{\lambda}{1.87} = 0.54\,\lambda$$

Incidentally, while the possible locations and movement of grating lobes may be readily visualized for a one-dimensional array, many people find visualizing them for a two-dimensional array annoyingly difficult. The difficulty may be avoided, by plotting the lobe positions in so-called *Sine Theta Space*, as explained in the panel on page 484.

Avoiding Bragg Lobes. *Bragg lobes* are retrodirective reflections[2] which may occur if an array is illuminated by another radar from certain angles off broadside. If stealth is required, they must be avoided. As explained in Chap. 39, avoiding Bragg lobes may require a much tighter radiator lattice than is necessary to avoid grating lobes.

Choice of Lattice Pattern. For an ESA, the choice of radiator-lattice pattern may also influence the number of radiators required.

The most common lattice patterns are rectangular and triangular or diamond shaped (Fig. 3). With a diamond lattice, the number of radiators may be reduced by up to 14% without compromising grating lobe performance. The

2. Energy reflected back in the direction from which it came.

Rectangular Lattice

Diamond Lattice

3. Common radiator lattice patterns. With the diamond pattern, the number of radiators may be reduced by up to 14% without compromising grating lobe performance.

AVOIDING GRATING LOBES

Where Grating Lobes Occur. Like the main lobe, grating lobes occur in those directions, θ_n,

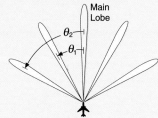

in which the waves received by a distant observer from all of the antenna's radiating elements are in phase.

For an MSA, where all radiating elements are excited in phase, θ_n is simply the direction in which the incremental difference in range, ΔR_θ, from successive radiating elements to a distant observer is a whole multiple, n, of the operating wavelength, λ.

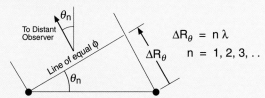

$$\Delta R_\theta = n \lambda$$
$$n = 1, 2, 3, ..$$

The direction, θ_n, is thus related to λ and the distance, d, between radiators by the sine function.

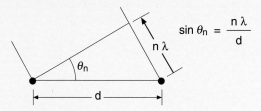

$$\sin \theta_n = \frac{n \lambda}{d}$$

Now, the gain of each radiator goes to zero as θ approaches 90°.

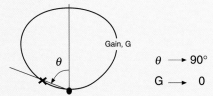

$$\theta \longrightarrow 90°$$
$$G \longrightarrow 0$$

And θ_1, the direction of the first grating lobe, approaches 90° as d is reduced to λ.

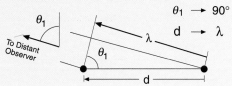

$$\theta_1 \longrightarrow 90°$$
$$d \longrightarrow \lambda$$

So, for an MSA, grating lobes can be avoided by reducing the spacing of the radiators to 1 wavelength or less.

$$\underline{\underline{d \leq \lambda}}$$

For an ESA, avoiding grating lobes is not quite so simple. For an incremental phase difference, $\Delta\phi$, is applied to the excitation of successive radiators to steer the main lobe to the desired look angle, θ_L.

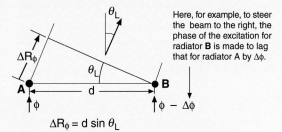

Here, for example, to steer the beam to the right, the phase of the excitation for radiator **B** is made to lag that for radiator A by $\Delta\phi$.

$$\Delta R_\phi = d \sin \theta_L$$

Consequently, for an ESA, grating lobes occur in those directions, θ_n, where the incremental distance, ΔR_θ, from successive radiating elements to a distant observer equals a whole multiple of a wavelength (nλ) **minus** the distance, ΔR_ϕ, corresponding to the phase lag, $\Delta\phi$.

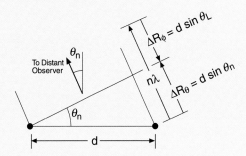

From this simple relationship,

$$d \sin \theta_n = n\lambda - d \sin \theta_L$$

we can obtain the positions of all possible grating lobes. Setting n equal to 1 and θ_L equal to the maximum desired look angle, θ_0, yields a "worst case" equation for the position of the first grating lobe.

$$d \sin \theta_1 = \lambda - d \sin \theta_0$$

As with an MSA, to avoid grating lobes the first grating lobe must be placed at least 90° off broadside. As illustrated in the diagram below, θ_1 approaches 90° as d is reduced to λ **minus** d $\sin\theta_0$.

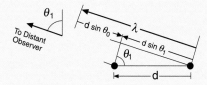

So, since sin 90° = 1, letting sin θ_1 equal 1 and solving the above equation for d yields the maximum spacing an ESA's radiators may have and avoid grating lobes.

$$\underline{\underline{d \leq \frac{\lambda}{(1 + \sin \theta_0)}}}$$

SINE THETA SPACE

For even a mechanically steered array, visualizing the possible positions of grating lobes is made difficult by the fact that their directions, θ_n, relative to the antenna broadside direction are related to the distance, d, between radiators and the wavelength, λ, by the sine function.

$$\sin \theta_n = n \frac{\lambda}{d} \qquad n = 1, 2, 3, \ldots$$

where n is the number of the lobe. (The main lobe is number 0.)

For an ESA, the difficulty is compounded by θ_n being determined not only by the radiator spacing, but also by the deflection, θ_0, of the main lobe from broadside.

$$\sin \theta_n = n \frac{\lambda}{d} \pm \sin \theta_0$$

In the case of a 2-D ESA, these difficulties are further compounded by the lobes existing in three-dimensional space.

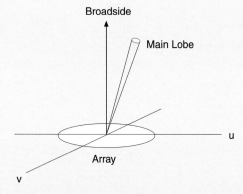

An engineer named Von Aulock elegantly solved all three problems in a single stroke by (a) representing the main lobe and each grating lobe with a unit vector (arrow one unit long) and (b) projecting the tip of this vector onto the plane of the array.

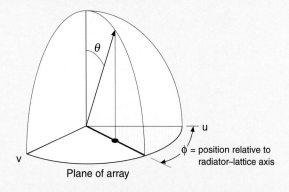

Since the distance from the center of the plane to each point projected onto it is $(1 \times \sin \theta_n)$, Von Aulock named the plane Sine Theta Space.

The beauty of Sine Theta Space is that the position of the main lobe can be plotted on it simply by scaling off (in the direction ϕ relative to the related lattice axis, u or v) a distance equal to the sine of the lobe's deflection, θ_0. The positions of any grating lobes can then be predicted by scaling off on either side of the main lobe distances equal to n λ divided by the radiator spacings d_u and d_v. Thus:

- Main lobe distance = $\sin \theta_0$ (at angle ϕ)

- Grating lobe distances = $\pm \dfrac{\lambda}{d_u}$ and $\pm \dfrac{\lambda}{d_v}$

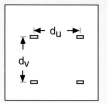

Radiator Lattice

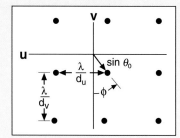

Grating Lobe Diagram
Plotted in Sine Theta Space

Since lobes cannot exist at angles greater than 90° off broadside, a circle of radius 1 (the sine of 90°) is drawn around the origin. The area within this circle is termed "real space"; the area outside it, "imaginary space."

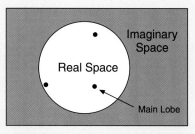

When evaluating radiator lattice patterns and radiator spacing, potential grating lobe positions are often plotted in imaginary space.

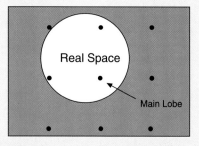

One can then readily see whether any of these lobes will materialize—i.e., move into real space—when the main lobe is steered to the limits of the desired field of regard.

choice of lattice pattern, though, is also influenced by other considerations, such as RCS-reduction requirements.

The number of radiators may be reduced still further by selectively thinning the density of elements near the edges of the array. In assessing thinning schemes, however, their effects on sidelobes and their interaction with edge treatment for RCS reduction must be carefully considered.

In short, no matter what the scheme, some price is always paid for reducing the number of radiators beyond what is achieved by simply limiting their spacing to d_{max}.

Design of Passive ESAs

Among basic considerations in the design of passive ESAs are the selection of phase shifters, the choice of feed type, and the choice of transmission lines.

Selection of Phase Shifters. In a two-dimensional array employing 2000 or more radiators, phase shifters (Fig. 4) typically account for more than half the weight and cost of the array. Consequently, it is critically important that the individual devices be light weight and low cost. Also, so as not to reduce the radiated power and not to increase the receiver noise figure appreciably, the phase shifters' insertion loss must be very low. Other critical electrical characteristics of the phase shifters are accuracy of phase control, switching speed, and voltage standing-wave ratio.

Choice of Feed Type. The feeds used in passive ESAs are of two basic types: constrained and space. Constrained feeds may be either traveling-wave or corporate.

In a traveling-wave feed, the individual radiating elements, or columns of radiating elements, branch off of a common transmission line (Fig. 5). This type of feed is comparatively simple. But it has a limited instantaneous bandwidth. The reason is that the electrical length of the feed path in wavelengths, hence also the phase shift from the common source to each radiator is different.

The difference may be compensated by adding a suitable correction to the setting of the phase shifter for each radiator. But since the required correction is a function of the wavelength of the signals passing through the feed, any one phase setting generally provides compensation over only a narrow band of frequencies.[3]

A corporate feed has a pyramidally shaped branching structure (Fig. 6). It can readily be designed to make the physical length, hence also the electrical length, of the feed paths to all radiating elements the same, thereby eliminating the need for phase compensation. The instantaneous bandwidth then is limited only by the bandwidths of the radiators and of the phase shifters, transmission lines, and connectors making up the feed system.

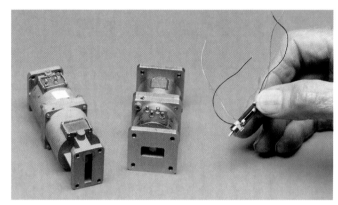

4. Ferrite phase shifters of the sort used in passives ESAs: X-band (left); Ku-band (center); Ka-band, removed from its housing (right).

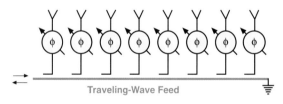

5. Traveling-wave feed is simple and inexpensive. But, since the electrical length of the path to each radiator is different, a phase correction must be made for each element, limiting the instantaneous bandwidth.

3. Some feeds get around this limitation but are impracticably bulky.

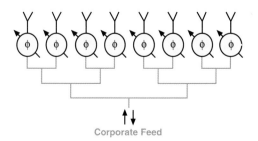

6. Corporate feed makes the electrical length of paths to all radiators the same, eliminating the need for phase corrections and widening the instantaneous bandwidth.

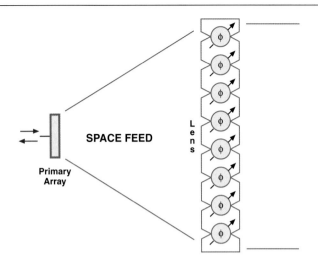

SPACE FEED

Primary
Array

Lens

7. The space feed is simple, inexpensive and has an instantaneous bandwidth comparable to a corporate feed's. But the focal length of the primary array adds to the depth of the antenna.

4. Strip line is more precisely defined as transverse electromagnetic mode (TEM) transmission line; hollow wave guide, as transverse electric/transverse magnetic (TE/TM) transmission line.

Space feeds vary widely in design. Figure 7 shows a representative feed. In it, a horn or a small primary array of radiating elements illuminates an electronic lens filling the desired aperture. The lens consists of closely spaced radiating elements, such as short open-ended wave guide sections, each containing an electronically controlled phase shifter.

The space feed is simple, lightweight, and inexpensive. It has low losses and an instantaneous bandwidth comparable to that of a corporate feed. But the focal length of the primary array adds considerably to the depth of the antenna. Also, sidelobe control is difficult to obtain without amplitude tapering at the radiator level.

Choice of Transmission Lines. The transmission lines commonly used in antenna feed systems are of two general types: strip line and hollow waveguide.[4]

Strip line consists of narrow metal lines (strips) sandwiched between metal surfaces. It is lightweight, compact, and low cost. Moreover, it can pass signals having instantaneous bandwidths of up to a full octave! It thus meets the requirements of applications ranging from ECCM and LPI to high resolution mapping.

Strip line is of two general types (see panel below). In one, the strips are insulated from the metal surfaces by a dielectric sheet, making this feed cheaper but lossy. In the other—called *"power" strip line*—losses are minimized by isolating the strips from the metal surfaces with an air gap.

REPRESENTATIVE STRIP LINE CONSTRUCTION

Dielectric Strip Line

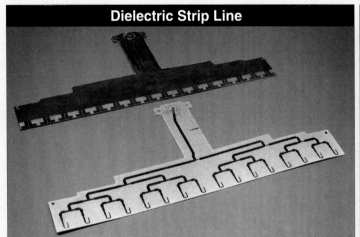

Air (Power) Strip Line

This type of strip line is made of two thin metalized dielectric sheets. The bottom sheet (foreground) is metalized on both sides. Metal on top is etched away leaving a strip-like conductor.

The upper sheet is metalized only on top. When the two sheets are put together, the conductor is sandwiched between the metal layers and insulated from them by the dielectric.

The result is lightweight, compact, low cost, and can pass wideband signals. It is lossy, but good for low-power and strong-signal applications.

Strip-like conductor, etched from the metalized surface of a dielectric sheet, is sandwiched between thin aluminized sheets into which matching grooves have been stamped. Supported by the dielectric, the conductor is separated from the metal by air in the groves. Also very wide band, it is more expensive than dielectric strip line but has much lower losses.

In another version of air strip line, conductor is supported at intervals by plastic standoffs in groves cut into light metal plates by an automated machine tool.

Hollow metal waveguide (Fig. 8) is heavier, more expensive, and has a limited instantaneous bandwidth. But it has very low losses. Consequently, it is required for high transmitted powers, weak signal detection, and long runs.

With advances in plastic molding and plating techniques, high-quality low-cost metal-coated hollow plastic wave guide has become an attractive option.

Design of Active ESAs

The key element of an active ESA is the T/R module. Among the many important considerations in its design, are the number of different types of integrated circuit chips required, the power output to be provided, the limits imposed on transmitted noise, and the required precision of phase and amplitude control. Not to be overlooked is the array's crucial physical design. Each of these considerations is discussed briefly below.

Chip Set. Ideally, all of a module's circuitry would be integrated on a single wafer. However, because of differences in the requirements of the various functional elements, technology for achieving this goal is not presently available. Consequently, the circuitry is partitioned by function and placed on more than one chip. The chips are then interconnected in a hybrid microcircuit (Fig. 9).

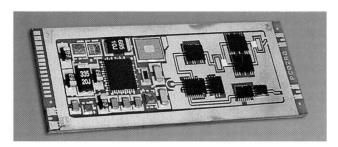

9.   Closeup of a representative T/R module (cover removed). Integrated circuit chips are interconnected in a hybrid microcircuit.

The basic chip set for a T/R module (Fig. 10) includes three *monolithic microwave integrated circuits*, called *MMICs*,[5] plus a digital VLSI (very large scale integrated circuit):

- High-power amplifier (MMIC)

- LNA plus protection circuit (MMIC)

- Variable-gain amplifier and variable phase shifter (MMIC)

- Digital control circuit (VLSI)

Depending upon the application, to these may be added other chips, such as a driver MMIC to amplify the input to the high-power amplifier when high peak powers are required, circuitry for built-in testing, and so on.

8.   A section of hollow metal waveguide. It is heavier and more expensive than stripline and has a limited instantaneous bandwidth. But, having very low losses, it is required in applications requiring high transmitter powers and/or weak signal detection.

5.   Circuits for millimeter wavelengths are called MIMICs.

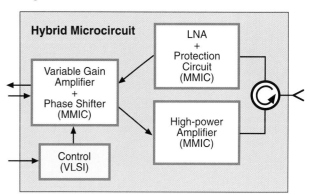

10. Basic chip set for a representative T/R module. Set consists of three monolithic microwave integrated circuits (MMICs) and one digital very large-scale integrated circuit (VLSI).

To date, virtually all MMICs for X-band and higher frequencies have been made of gallium arsenide (GaAs), since it is the only material yet proven capable of handling such high frequencies. One limitation of GaAs is its very low thermal conductivity. For the circuitry on a chip to be adequately cooled, the chip must either be ground very thin—making it fragile and difficult to handle—or mounted on the hybrid substrate face down (*flip-chip technique*).

Power Output. In general, for a given array size, the array's average power output is dictated by the desired maximum detection range. The realizable average power output, however, is usually constrained by (a) the amounts of primary electrical power and cooling the aircraft designer allocates to the ESA and (b) the module's efficiency. For a given primary power and cooling capacity, the higher the efficiency, the higher the average power can be.

Regarding module efficiencies, two terms which often come up are "power added efficiency" and "power overhead." These are explained in the panel on the facing page.

In designing the module's high-power amplifier, the required peak power is of greatest concern. It, of course, equals the desired average power per module divided by the minimum anticipated duty factor.

For a given peak power output from the array as a whole, the peak power per module is inversely proportional to the number of modules, hence to the area of the array. Consequently, to obtain the same peak power from an array having an area of 4 square feet, as from an array having an area of 8 square feet, the peak power of each module must be doubled (Fig. 11).

Transmitter Noise Limitations. As with a radar employing a central transmitter, noise modulation of the transmitted signal must be minimized. The principal sources of noise modulation in an active ESA are ripple in the dc input voltage and fluctuations in the input voltage due to the pulsed nature of the load. Because the voltages are low and the currents are high, adequately filtering the input power is a demanding task. It may require distributing the power conditioning function at an intermediate level within the array, or even including a voltage regulator in every T/R module.

Receiver Noise Figure. Since one of the main reasons for going to an active ESA is reduction of receive losses, to fully realize the ESA's potential it is essential that the T/R module have an extremely low receiver noise figure. Typically, the receiver noise figure is quoted for the module as a whole. It equals the noise figure of the LNA plus the losses ahead of the LNA—i.e., losses in the radiator, the duplexer, the protection circuit, and the interconnections (Fig. 12).

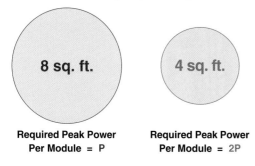

For the same peak power output:

8 sq. ft.   **4 sq. ft.**

**Required Peak Power   Required Peak Power
Per Module = P   Per Module = 2P**

11. Relationship between the peak power per module and the area of an array.

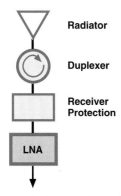

Radiator

Duplexer

Receiver Protection

LNA

12. Receiver noise figure equals the noise figure for the LNA plus the losses in the elements ahead of the LNA: radiator, duplexer, receiver protection circuit, and interconnections.

Phase and Amplitude Control. The precision with which the phase and amplitude of the transmitted and received signals must be controlled at the radiator level is dictated by the maximum acceptable peak sidelobe level of the full array. The lower it is, the

- Smaller the quantization step sizes of the phase and amplitude control circuits must be

- Wider the amplitude-control range needed to achieve the necessary radiation taper across the array for side lobe reduction

- Smaller the acceptable phase and amplitude errors

Array Physical Design. The performance and cost of an active ESA depend critically not only upon the design of the T/R modules, but also upon the physical design of the assembled array.

In general, the radiators must be precisely positioned and solidly mounted on a rigid back plane. This is essential if the antenna's RCS is to be minimized; for any irregularities in the face of the array will result in random scattering which cannot otherwise be reduced (see page 495).

The modules are typically mounted behind the back plane on cold plates, which carry away the heat they generate.

Behind the cold plates then are: (a) a low-loss feed manifold connecting each module to the exciter and the central receiver; (b) distribution networks providing control signals and dc power to each module; and (c) a distribution system for the coolant that flows through the cold plates.

Just how this general design is implemented may vary widely. One approach, called *stick architecture*, is illustrated in Figs. 13 and 14.

MEASURES OF MODULE EFFICIENCY

Power-Added Efficiency. Since a module's high-power amplifier (HPA) typically includes more than one stage, the efficiency of the final stage is generally expressed as power added efficiency, E_{PA}.

$$E_{PA} = \frac{P_o - P_i}{P_{dc}}$$

where

P_o = RF output power
P_i = RF input power
P_{dc} = DC input power.

If the gain of the final stage is reasonably high, the power added efficiency very nearly represents the efficiency of the entire amplifier chain.

Power Overhead. This is the power consumed by the other elements of the module—switching circuitry, LNA, and module control circuit. Because of this overhead, a module's efficiency may be considerably less than the HPA's efficiency, which typically is somewhere between 35 and 45%.

Since much of the overhead power is consumed continuously, while the RF output is pulsed, module efficiency may vary appreciably with PRF.

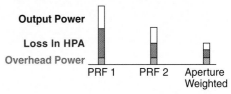

Output Power
Loss In HPA
Overhead Power
PRF 1   PRF 2   Aperture Weighted

Also, since overhead power is independent of output power, if all modules are identical, as they reasonably would be, aperture weighting can significantly reduce the efficiency of many modules. To minimize this reduction yet achieve extremely low sidelobes, special weighting algorithms have been developed for active ESAs.

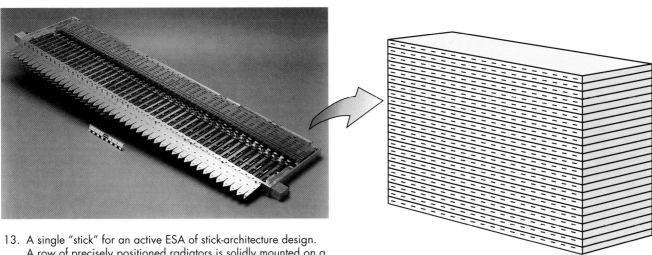

13. A single "stick" for an active ESA of stick-architecture design. A row of precisely positioned radiators is solidly mounted on a rigid structure serving as: (a) back plane for the radiators, (b) cold plate and housing for the T/R modules, and (c) housing for RF feed, power, and control-signal distribution network.

14. Sticks are rigidly mounted on top of each other to form the complete array.

Another approach to the physical design of an active ESA is a so called *"tile" architecture*. It employs dime-sized three-dimensional, four-channel modules (Fig. 15).

Enlarged

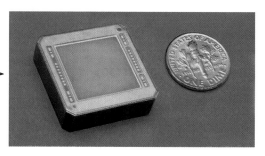

15. Dime-sized four-channel, three-dimensional T/R module.

16. Within the module, successive sections of four T/R circuits are placed on three boards, the heat from which is conducted out to the surrounding metal frame.

Within each module (Fig. 16), successive sections of four T/R circuits are placed on three circuit boards, mounted one on top of the other. Heat generated in the circuits on each board is conducted to the surrounding metal frame.

The modules are sandwiched between cold plates having feed-through slots for the RF signals, dc power, and control signals (Fig. 17).

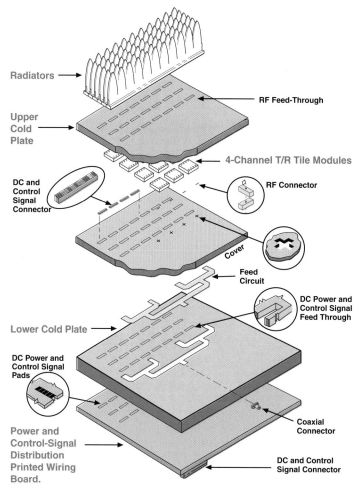

Radiators

Upper
Cold
Plate

RF Feed-Through

DC and
Control
Signal
Connector

4-Channel T/R Tile Modules

RF Connector

Cover

Feed
Circuit

DC Power and
Control Signal
Feed Through

Lower Cold Plate

DC Power and
Control Signal
Pads

Coaxial
Connector

Power and
Control-Signal
Distribution
Printed Wiring
Board.

DC and Control
Signal Connector

17. "Tile" array architecture. Four-channel three-dimensional T/R modules (such as shown in Fig. 10) are sandwiched between two cold plates. RF input and output signals, control signals, and dc-power feed through slots in the lower cold plate. RF signals to and from the radiators feed through slots in the upper cold plate.

For sidelobe reduction, precise control of phase and gain in each module is essential. Consequently, a comprehensive automatic self-test and calibration capability is provided. To account for manufacturing tolerances, the initial calibration correction for each module is set into a nonvolatile memory in the module's control circuit.

Finally, since more than the maximum acceptable number of modules may malfunction during the operational life of the aircraft, provisions must be included for removing and replacing individual modules—a difficult design task, to say the least.

Summary

To minimize the cost of an ESA—whether passive or active—the radiating elements must be spaced as far apart as possible without creating grating lobes. The maximum spacing is about half a wavelength. For stealth, still closer spacing may be required to avoid Bragg lobes.

The number of radiators may be reduced by up to 14% by using a diamond lattice. And it may be reduced still further by thinning the density of elements at the array's edges, but for such reductions, a price is paid in terms of sidelobe and RCS performance.

Key elements of a passive ESA are the phase shifters. They account for more than half the weight and cost of the array, hence must be lightweight and low cost. Also critical are the transmission lines and feed. For wideband operation, strip line and either a corporate or a space feed must be used. For high power and weak-signal detection, hollow waveguide is required.

The key element of an active ESA is the T/R module. It is implemented with a limited number of monolithic integrated circuits in a hybrid microcircuit. For X-band frequencies and higher, the monolithic circuits are made of gallium arsenide. Critical electrical characteristics are the module's peak power output, precision of phase and amplitude control, receiver noise figure, and noise modulation of the transmitted signal, which must be minimized through filtering of the dc input power.

To minimize the antenna's RCS, the radiators are mounted on an extremely rigid back plane. The T/R modules are mounted on cold plates, immediately behind the back plane. Self-test and self-calibration capabilities are essential.

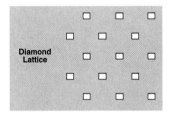

Diamond Lattice

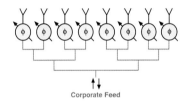

Corporate Feed

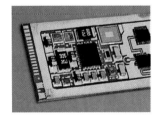

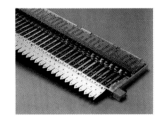

RCS TESTING

Preparations are made for evaluating the RCS of the antenna for a fighter's radar in the radome, installed on a low-RCS test body.

Antenna RCS Reduction

Viewed nose-on, a typical fighter aircraft has a radar cross section (RCS) on the order of one square meter. A similarly viewed low observable aircraft may have an RCS of only 0.01 square meter. Unless special RCS reduction measures are employed, even a comparatively small planar array antenna can have an RCS of up to several thousand square meters when viewed from a broadside direction! Since an aircraft's radome is transparent to radio waves, if stealth is required, steps must be taken to reduce the RCS of the installed antenna.

In this chapter, we will be introduced to the sources of reflections from a planar array antenna, learn what can be done to reduce or render them harmless, and see why these steps are facilitated in an ESA.

We will then take up the problem of avoiding so-called Bragg lobes, which are retrodirectively reflected at certain angles off broadside if the radiator spacing is too large compared to the radar's operating wavelength.

Finally, we will very briefly consider the critically important validation of an antenna's predicted RCS.

Sources of Reflections from a Planar Array

For our purposes here, a planar array antenna, regardless of whether it is an MSA or an ESA, can conveniently be thought of as consisting of a flat plate—referred to as the *"ground plane"*—containing a lattice of radiating elements (Fig. 1).

The backscatter from the antenna when illuminated by a radar in another aircraft—threat radar, we'll call it—is com-

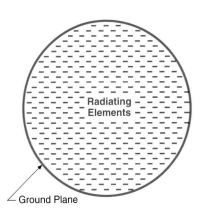

1. A planar array antenna, regardless of whether it is an MSA or an ESA, can conveniently be though of as a flat plate, termed the ground plane, containing a lattice of radiating elements.

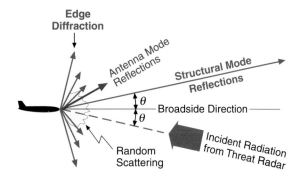

2. The four basic components of backscatter from a planar array antenna. Random scattering is the sum of the random components of the structural-mode and antenna mode reflections.

monly categorized as being comprised of four basic components (Fig. 2).

1. Specular (mirrorlike) reflections from the ground plane. These are called structural mode reflections.

2. Reflections of some of the received power by mismatched impedances within the antenna. Reradiated by the radiating elements, these are called antenna mode reflections.

3. Reflections due to the mismatch of impedances at the edges of the array, i.e., between the ground plane and the surrounding aircraft structure. These reflections are referred to as edge diffraction.

4. Random components of the structural mode and antenna mode reflections. These components are called random scattering.

In case you're wondering, there are two reasons for separately breaking out random scattering.

First, with the random scattering removed, the structural-mode and antenna-mode reflections can be characterized more simply.

Second, there is then a one-to-one relationship between the individual categories of reflections and the techniques for reducing or controlling them.

Reducing and Controlling Antenna RCS

By carefully designing and fabricating an antenna, each of the four components of backscatter may be acceptably minimized or rendered harmless.

Rendering Structural Mode Reflections Harmless. As may be seen from Fig. 3, these mirrorlike reflections may be controlled by physically tilting the antenna so that they are not directed back in the direction from which the illuminating radiation came. Although the tilt does not reduce the reflections, it prevents the threat radar from receiving them.

With an ESA, which is mounted in a fixed position in the aircraft, the antenna ground plane can be permanently tilted so that the incident radiation will be harmlessly reflected in the same direction as the irreducible "spike" in the pattern of reflections from the aircraft structure. The tilt reduces the antenna's effective aperture area somewhat, reducing the gain and broadening the beam about the axis of the tilt. But this is a small price to pay for the huge reduction in detectability that is achieved.

Minimizing Antenna Mode Reflections. At the radar's operating frequency, antenna mode reflections have a radia-

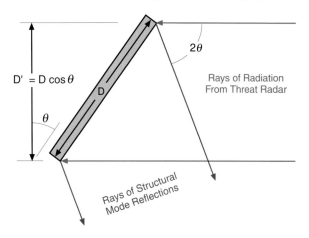

STRUCTURAL MODE REFLECTIONS

$D' = D\cos\theta$

3. Structural mode reflections may be rendered harmless by tilting the array. The tilt reduces the effective aperture somewhat but that is a small price to pay for the huge reduction in detectability achieved.

tion pattern similar to that of the transmitted signal: a main lobe, surrounded by sidelobes (Fig. 4). The direction of the main lobe is determined by the angle of incidence of the illuminating waves and the element-to-element phase shift occurring within the array. As is clear from the figure, these reflections are not necessarily rendered harmless by the tilt of the antenna.

They can be acceptably minimized, however, by employing well matched microwave circuitry in the antenna and by paying extremely close attention to design detail. In wideband MSAs and passive ESAs, even reflections from deep within the antenna must be eliminated. This may be accomplished by inserting isolators, such as circulators, at appropriate points in the feed.

Minimizing Edge Diffraction. Edge diffraction produces backscatter comparable to that which would be produced by a loop antenna having the same size and shape as the perimeter of the array. Since the dimensions of this loop are generally many times the operating wavelength of the radar, the radiation pattern of the loop typically consists of a great many lobes fanning out from the broadside direction (Fig. 5). Consequently, edge diffraction, too, is not rendered harmless by the antenna's tilt. Special measures must be taken to minimize it.

In some antenna installations, edge diffraction is rendered harmless by shaping the edge of the ground plane to disperse the diffracted energy so that it is beneath the threshold of detection of the threat radar.

In other installations, the diffraction is reduced by applying radar absorbing material (RAM) around the edges of the ground plane so that its resistivity smoothly tapers to that of the surrounding structure. To be effective, the treatment must be at least four wavelengths wide at the lowest threat frequency (Fig. 6). Consequently, it can seriously diminish the available aperture area, and so reduce the radar's performance. Accordingly, careful tradeoffs are necessary between radar performance and RCS performance.

In any event, the measures taken to reduce or render the diffraction harmless are greatly facilitated in an ESA, since it is permanently mounted in a fixed position on the aircraft structure.

Minimizing Random Scattering. The random components of structural mode and antenna mode reflections may be spread over a wide range of angles (Fig. 7). So, they are not rendered harmless by the antenna's tilt. To reduce them to acceptable levels, the antenna's microwave characteristics must be highly uniform across the entire array. This requires exceptionally tight manufacturing tolerances.

ANTENNA MODE REFLECTIONS

Incident Radiation from Threat Radar

4. Radiation pattern of these reflections is similar to that of transmitted signal. Since their direction is determined by internal phase shifts as well as by angle of incidence of illuminating waves, they are not necessarily rendered harmless by tilt of antenna.

EDGE DIFFRACTION

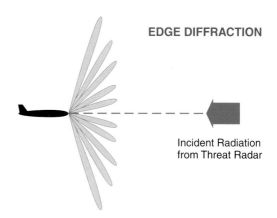

Incident Radiation from Threat Radar

5. Backscatter due to edge diffraction is comparable to that from a loop the size and shape of the array's perimeter. Since its diameter is many times the operating wavelength, the backscatter fans out in many directions.

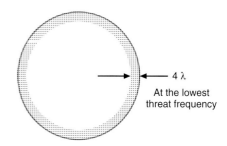

4λ

At the lowest threat frequency

6. Edge treatment must be at least four wavelengths wide. Depending on the antenna's size, this can seriously diminish the effective aperture.

RANDOM SCATTERING

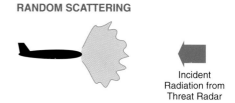

Incident Radiation from Threat Radar

7. The random components of structural mode and antenna mode reflections are spread over a wide span of angles.

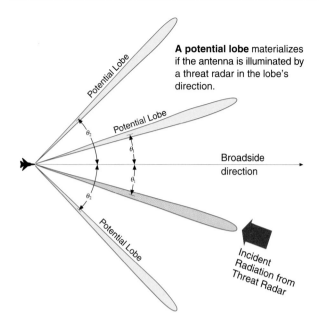

8. Bragg lobes are retrodirective reflections which may be received by an illuminating radar when it is a certain angle, θ_n, off broadside, if the spacing of the radiators is larger than half the wavelength of the illumination.

Avoiding Bragg Lobes

Bragg lobes are retrodirective reflections from the antenna's radiators, which may be received by an illuminating radar when it is in certain angular positions, θ_n, off broadside (Fig. 8). Depending on the antenna's design, besides direct reflections from the radiators, the lobes may also include energy reflected from within the antenna (i.e., antenna-mode reflections).

The lobes are due to the periodicity of the radiator lattice. They occur at those angles for which the phases of the waves reflected in the illuminator's direction by successive radiators differ by 360° or multiples thereof, hence are all in phase and add up to a strong signal.

While for simplicity Fig. 8 has been drawn for lobes in a single plane, bear in mind that for a two-dimensional array Bragg lobes occur about both lattice axes. As illustrated in the panel below, the directions of the lobes relative to the boresight direction are determined by the spacing of the radiators relative to the wavelength of the illumination. The greater the spacing and/or the shorter the wavelength, the closer the lobes will be to the broadside direction and the more lobes there will be.

CONDITIONS UNDER WHICH A BRAGG LOBE WILL BE PRODUCED

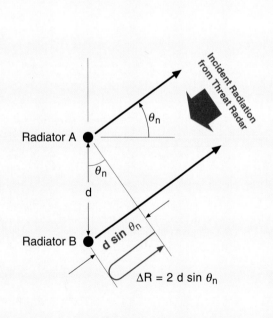

When adjacent radiators of an array antenna are illuminated by a threat radar, a Bragg lobe will be produced if the wave reflected in the radar's direction (θ_n) by the far radiator (B) is in phase with the wave reflected by the near radiator (A).

Assuming no regular radiator-to-radiator phase shift in reflections from within the antenna, that condition will occur if the additional round-trip distance, ΔR, traveled to and from radiator B is a whole multiple, n, of the incident radiation's wavelength, λ.

$$\Delta R = n\lambda$$

As is clear from the diagram ,

$$\Delta R = 2\,d \sin \theta_n$$

where d is the spacing between radiators. Thus, the relationship between radiator spacing and Bragg-lobe direction is

$$d = \frac{n\lambda}{2 \sin \theta_n}$$

To minimize the antenna's RCS, the first Bragg lobe (n = 1) must be placed 90° off broadside (sin θ_1 = 1). Substituting these values in the above equation yields:

$$d = \frac{\lambda}{2} \quad \text{for stealth.}$$

Like grating lobes, Bragg lobes can be avoided by spacing the radiators close enough together to place the first lobe 90° off broadside. As the panel shows, if the illuminator's wavelength is the same as the radar's, this may be accomplished with a spacing of half the operating wavelength.

But, if the illuminator's wavelength is shorter, the spacing must be proportionately reduced. Suppose, for instance, that the radar's wavelength is 3 centimeters and the illuminator is operating at 18 GHz (λ = 1.67 cm). To avoid Bragg lobes, the radiator spacing would have to be reduced to 1.67 ÷ 2 = 0.84 centimeters—little more than a quarter of the operating wavelength.

If such tight spacing is not economically feasible, the designer has three options. The first two are comparatively simple.

One is to use a diamond lattice such as that illustrated in Fig. 9. Despite the larger radiator spacing of this lattice, Bragg lobes may be rendered harmless.

The second option is simply to employ the tightest practical radiator spacing—at least along the axis of greatest concern.

The third and more costly option is to prevent any shorter wavelength radiation from reaching the array. One way of accomplishing this is to place a *frequency selective screen (FSS)* in front of the array (Fig. 10). The screen is designed to pass all wavelengths in the radar's operating band with little attenuation, yet reflect all out-of-band radiation. The screen may either be mounted externally as shown in Fig. 10 or be built into the antenna face. As with structural mode reflections, because of the tilt of the antenna—hence also of the screen—radiation reflected by the screen will be directed in a nonharmful direction.

In one possible implementation, the screen consists of a thin metal sheet containing a tight lattice of slots, mounted between two dielectric slabs. To be effective, the slots must be separated by no more than half the wavelength of the highest threat frequency.

Validating an Antenna's Predicted RCS

Because of the complexity of the factors contributing to an antenna's installed RCS, a key step in developing a low RCS antenna is validating the antenna's predicted RCS.

For this, one or more physical models of the radiating aperture are generally built. These are called *phenomenology models*, or *"phenoms."* Typically, they include not only the radiators and any covering that goes over them, but also the first few stages of internal circuitry. If the schedule allows, the phenoms may even be used to interactively refine the design.

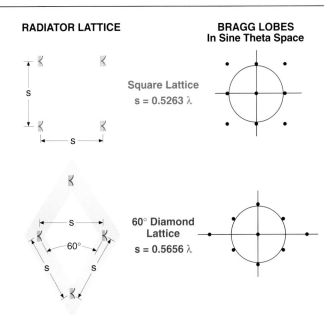

9. Bragg lobe patterns for square and 60° diamond radiator lattices. Despite the greater radiator spacing of the diamond lattice, all Bragg lobes except the central one are outside the boundary of visible (real) space. The central lobe is rendered harmless by the tilt of the antenna.

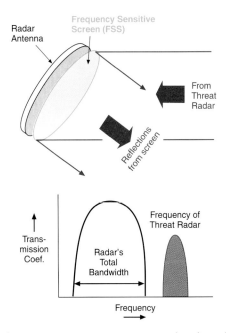

10. A frequency-sensitive screen acts as a bandpass filter, rejecting radiation of such high frequency that making the radiator lattice tight enough to avoid grating lobes is not practical.

11. The predicted RCS of the antenna for the radar of a fighter aircraft is verified in an anechoic chamber. Antenna in its radome is mounted on a low RCS test body.

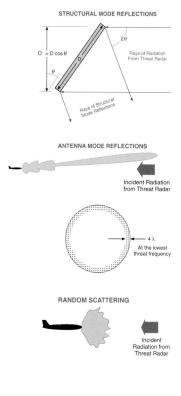

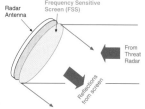

Measurements made on the phenoms, as well as on the complete antennas, include the following:

- Closed circuit measurements at the radiator level to isolate and quantify the complex reflections from each radiator and its internal circuitry—commonly referred to as "look-in" measurements.

- Angular "cuts" of the reflection pattern of the total array.

- Very high resolution ISAR (inverse synthetic aperture radar) images of the antenna, made to isolate individual reflection "hot spots" and to determine the effectiveness of the edge treatment.

To realistically evaluate the installed antenna's RCS, a full-scale model of the nose section of the aircraft including the phenom is generally tested in a large anechoic chamber (Fig. 11).

Summary

Unless special measures are taken to reduce the reflections from a planar array, its RCS may be several thousand square meters. The reflections are of four basic types, which may be reduced or rendered harmless as follows:

- Mirror-like reflections from the back plane (structural mode reflections)—may be rendered harmless by tilting the antenna.

- Reflections due to mismatched impedances within the antenna (antenna mode reflections)—may be reduced by minimizing the mismatches.

- Reflections due to mismatched impedances at the edges of the array (edge diffraction)—may be reduced by tapering the impedances with radar absorbing material or shaping the edges of the ground plane to disperse the diffracted energy.

- Random components of structural and antenna mode reflections (random scattering)—may be reduced by holding to extremely tight manufacturing tolerances.

To avoid retrodirective reflections from the radiator lattice—Bragg lobes—the radiator spacing must be less than half the wavelength of the illumination. If illuminators may be encountered whose wavelengths are shorter than the radar's, either the radiator spacing must be further reduced, or a frequency-sensitive screen must be placed over the array to keep out the shorter wavelength radiation.

Because of the complexity of factors contributing to antenna RCS, RCS predictions are validated with physical models (phenoms), and a full-scale model of the nose section is usually tested in an anechoic chamber.

Advanced Radar Techniques 40

The advent of active ESAs, the emergence of low RCS aircraft, and the growing threat of electronic countermeasures have given impetus to advanced work in several key areas of radar technology. This chapter, presents some significant developments spawned by that work:

- Innovative approaches to multiple-frequency operation—for reducing vulnerability to countermeasures and avoiding detection by the enemy

- Advanced signal integration and detection techniques—for small target detection

- Bistatic modes of radar operation—for increasing survivability and for circumventing the limitation on power-aperture product imposed by a tactical aircraft's small size

- Space-time adaptive processing—for efficiently rejecting external noise and jamming and compensating for the motion-induced clutter spread with which long range surveillance radars must contend

- True-time-delay beam steering—a technique still in its infancy which promises to broaden the instantaneous bandwidth of an active ESA sufficiently to enable simultaneous shared use of the same antenna for radar, electronic warfare, and communications

- Interferometric SAR—for making accurate high-resolution topographic maps

Most of these developments have only been made practical by the high throughput of advanced digital processors.

Approaches to Multiple Frequency Operation

Although the advantages of wideband multiple frequency operation in avoiding jamming were long realized, virtually all airborne radars developed in the first 50 years of radar history were comparatively narrowband. Many could be switched from one to another of several radio frequencies. But, with few exceptions, this agility was limited to a small fraction of the operating frequency. Moreover, no radars employed more than one operating frequency at a time.

Of many possible approaches to multifrequency operation, two are presented here. One, called *SIMFAR*, for *simultaneous frequency agile radar*, is a singularly convenient technique for generating a multifrequency drive signal for a radar transmitter in a way which simplifies both transmission and reception. The other approach, called *STAR*, for *simultaneous transmit and receive*, is a remarkably versatile multifrequency technique, which uniquely yields a duty factor of 100%.

Simultaneous Frequency Agile Radar (SIMFAR). This technique takes advantage of the unique characteristics of phase modulation to generate multiple frequencies from a single microwave source. Phase modulation, you'll recall, produces sidebands above and below the carrier at multiples of the modulating frequency. The number of sidebands is determined by the modulation index.

At a low value of the index, a pure sine-wave modulating signal produces two sidebands having the same amplitude as the carrier (Fig. 1); with the carrier, they contain 90% of the output power.

By increasing the modulation index and including harmonics of appropriate amplitude and phase in the modulating signal, the number of equal-amplitude sidebands may be increased and the power in the outer sidebands reduced to a negligible percentage (Fig. 2).

In this way, SIMFAR produces a constant-amplitude transmitter-drive signal composed of any desired odd number[1] of equally spaced, equal-power spectral lines from a single stable microwave source and a single stable offset-frequency source (Fig. 3).

1. Spectrum of a phase-modulated carrier. When the modulation index is low, the carrier and two equal-amplitude sidebands contain 90% of the output power.

2. By increasing the modulation index and including harmonics in the modulating signal, the number of equal-amplitude sidebands may be increased.

1. Carrier plus one or more pairs of sidebands.

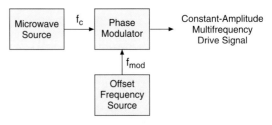

3. SIMFAR system. From a single stable microwave source and a single stable offset-frequency source, a constant-amplitude multifrequency transmitter drive signal is produced.

If desired, each line of the drive signal may be spread over a wide band by modulating the microwave source with phase or frequency coding, such as is used for pulse compression.

This drive signal may be applied either to a suitably broadband active ESA or to a TWT amplifier feeding a broadband passive ESA or MSA. Since the amplitude of the signal is constant, an important bonus is that a TWT driven by the signal may be operated at saturation without generating intermodulation products, which could limit the radar's detection sensitivity.

Upon reception, the composite signal can be handled by a single-channel receiver. To separate the spectral lines, the receiver's IF output is applied to a bank of bandpass filters, each of which is centered on a different line and has a bandwidth just wide enough to pass the line (Fig. 4).

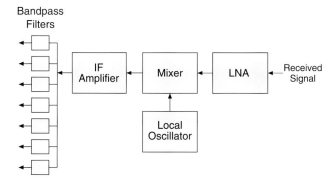

4.   Reception of the composite SIMFAR signal can be handled by a single-channel receiver. Following IF amplification, the individual spectral lines are separated by bandpass filters whose outputs are processed in separate channels.

Because all of the lines were produced by modulating the microwave reference signal with a single-offset frequency, coherent reference signals for I/Q detection of all the lines can be obtained simply by mixing a single reference frequency with the original offset-frequency.

Following coherent integration, the outputs of all channels are summed. For a point target, such as an aircraft, the net result is the same as if the total power in all of the lines had been transmitted at a single radio frequency with very much higher peak power and the received signal had been conventionally processed with a combination of pre- and post-detection integration.

STAR. In this technique, rather than transmitting several different radio frequencies simultaneously, the radar transmits continuously and switches from one frequency to the next at time intervals equal to the desired pulse width. In so doing, it in effect interleaves several pulse trains, each of which has a different radio frequency (Fig. 5).

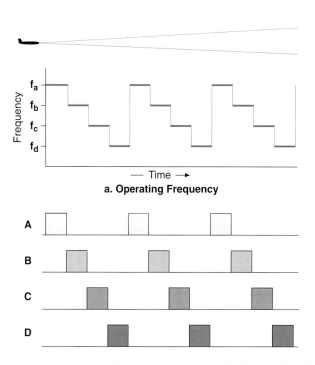

5.   STAR concept. Radar transmits continuously, but switches frequency at intervals equal to the desired pulse width, thus producing interleaved pulse trains having different radio frequency.

Every transmitted pulse inevitably has noise sidebands. They extend over a span of frequencies so broad that some of the noise has the same radio frequency as the returns from STAR's other pulse trains. Though this noise may be infinitesimally weak compared to the transmitted signal, it is many times stronger than the weak echoes from distant targets (Fig. 6).

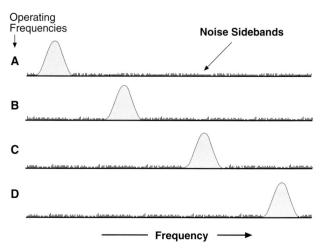

6. Spectra of pulse trains transmitted by STAR radar. As is inevitable in all radars, noise sidebands far stronger than the echoes of distant targets extend over a broad band on either side of each operating frequency.

To keep the noise from interfering with reception, as the transmitter switches from one frequency to the next, its output is switched from one to another of several bandpass filters, each of which passes a different one of the transmitter's frequencies, while stripping off its noise sidebands (Fig.7, below, left).

The frequencies are widely enough separated that the returns of each pulse train can be isolated by a bandpass filter. This filter also blocks any leakage of the transmitted pulses through the duplexer (Fig. 8).

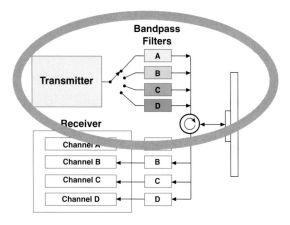

7. STAR implementation for transmission. As the transmitter switches from one operating frequency to the next, its output is switched to a filter which passes that frequency while stripping off its noise sidebands thus preventing them from interfering with reception of the echoes of the other pulse trains.

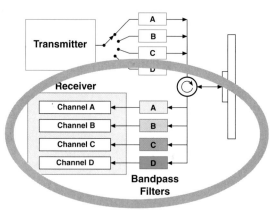

8. STAR implementation for reception. A bandpass filter is centered on the frequency of each pulse train. The frequencies are widely enough separated that each filter can pass only the returns of one pulse train and block any leakage of the other transmitted pulse trains through the duplexer.

502

Since each pulse train is both transmitted and received on a separate channel, even a radar having only a relatively narrow instantaneous bandwidth can operate simultaneously over an extremely broad total band (explained in the panel, right).

As with SIMFAR, the spectrum of each pulse train may itself be spread over a broadband by phase or frequency coding the transmitted pulses.

If the transmitter is not peak-power limited, in addition to having different radio frequencies, the pulse trains can have different PRFs. This capability further broadens the usefulness of STAR.

While the technique has been illustrated here for a radar employing a centralized transmitter, it is equally applicable to radars employing active ESAs. The configuration of a T/R module for a four-frequency STAR system is shown in Fig. 9.

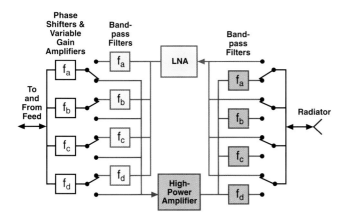

9.   T/R module for a four-frequency STAR system. Switch settings shown are for transmission on frequency f_a.

In an active ESA, besides spreading the transmitted signal over a broad spectrum, STAR has the advantage of facilitating simultaneous radiation of multiple beams.

To appreciate the tremendous potential of this capability, consider a four-frequency system which has been cued to detect a distant target in a given direction. To concentrate the radar's power in the target's direction, while both limiting peak power and gaining the advantage of frequency diversity, three beams search a narrow sector in the cued direction.

Meanwhile, the fourth beam maintains short-range situation awareness by rapidly searching a broad sector ahead (Fig. 10). Since the beams can be independently shaped, can employ common or diverse waveforms, can be transmitted at different power levels, and can have their functions instantly interchanged, the possibilities are virtually limitless.

The Difference Between
INSTANTANEOUS and TOTAL BANDWIDTH

A radar's **instantaneous** bandwidth is the widest band of radio frequencies the radar's antenna and RF circuits can pass without altering the relative amplitudes and phases of a signal's constituent frequency components or creating spurious modulation products. In other words, without distorting the radar's transmitted and received signals.

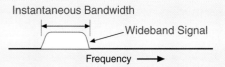

The radar's **total** bandwidth, is the span of frequencies within which its operating frequency can be set without the radar's signals being distorted or unacceptably attenuated.

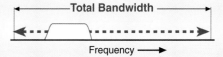

Generally, the total bandwidth is very much greater than the instantaneous bandwidth. While a certain amount of agility is thus allowed, the radar is constrained to shifting from one operating frequency to another within the total bandwidth at intervals of time no shorter than the coherent processing period for the received signals.

One way of circumventing this limitation is to simultaneously use several different operating frequencies, each of which is spread over the radar's full instantaneous bandwidth.

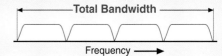

Wideband operation of this sort is possible with the STAR technique.

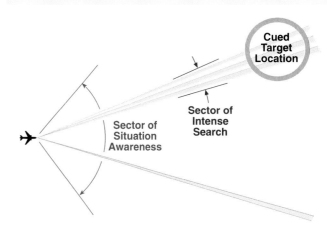

10.   Representative application of a four-frequency STAR system in a radar employing an active ESA. Four independent beams are produced. Three execute an intense low-peak-power cued search for an assigned long-range target. The fourth beam, meanwhile, maintains short range situation awareness.

Small Target Detection

Within the limits that the tactical aircraft imposes on a radar's power-aperture product, several approaches may be taken to increase the range at which targets of small RCS may be detected. Besides making refinements in conventional radar designs, as outlined in the panel on the facing page, detection sensitivity may be substantially increased through two advanced techniques: long coherent integration time, and sequential detection.

Long Coherent Integration Time. As is clear from the radar-range equation, for any given power-aperture product, detection range can be increased by limiting the scan volume and correspondingly increasing the antenna beam's time on target, t_{ot}. However, the extent to which the detection range may be increased thereby depends upon (a) the efficiency with which the increased energy received from the target is integrated and (b) the limit on scan frame time imposed by the application.

As explained in Chap. 10, signal integration is of two types:

- Coherent integration, which takes place in the doppler filters

- Post-detection integration (PDI), which takes place after the outputs of the doppler filters have been detected and phase information is no longer present

Both types can increase signal-to-noise ratios, hence detection ranges, substantially. However, coherent integration is considerably more efficient—*provided* the received signals retain their coherence throughout the entire integration period.

The key factor limiting the duration of a signal's coherence is target acceleration. In combat, a target is apt to change heading or speed continually. Unless this acceleration is compensated for, the target's doppler frequency may move out of the passband of the doppler filter that is integrating the target's returns. For long coherent integration times to be practical, therefore, acceleration compensation is essential.

Compensation can be provided by subtracting a continuously changing compensation frequency from the radio frequency of the target returns—much as a continuously changing reference frequency is subtracted in stretch-radar decoding of chirp (page 165). By making the compensation frequency equal to the change in doppler frequency due to the target's acceleration, the target's returns may be kept in the passband of the same doppler filter throughout the integration period.

INCREASING DETECTION SENSITIVITY
Through Conventional Design Refinements

Even though antenna size and average power may be limited, detection sensitivity can be enhanced considerably, by refining conventional radar features. Among key possibilities are employing frequency diversity, minimizing transmitted noise, widening the dynamic range of the receivers, minimizing quantization noise, and reducing the receiver noise figure.

Employing Frequency Diversity. As it closes on a target, a radar bearing aircraft may slip into one of the deep notches in the target's RCS pattern and remain there for some time. As a result, the target may not be detected by the radar until it has closed to a much shorter range than would be expected for the target's average RCS.

However, the locations of the notches vary with the radio frequency of the radar signal illuminating the target. The single-look probability of detection, therefore, may be substantially increased by changing the operating frequency at the end of each coherent integration period.

For best results, the frequencies should be separated by the bandwidth of a pulse whose length corresponds to the size of the target.

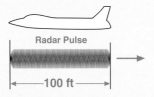

Radar Pulse

|◄——— 100 ft ———►|

The length of a pulse, you'll recall, is roughly 1000 feet per microsecond of pulse width. For a 100-foot target, for instance, the corresponding pulse width, τ, would be 0.1 μs. The bandwidth of a pulse being roughly $1/\tau$, the optimum separation of frequencies for this particular target would be $1 / 0.1$ μs = 10 MHz.

Minimizing Transmitter Noise. Inadvertent noise modulation of a radar's transmitted signal may produce ground clutter strong enough to limit the detection of weak signals. This clutter not only reduces detection sensitivity against tail aspect targets but, being inherently broadband, spreads over into the clutter-free spectral region in which nose-aspect targets are detected in high PRF operation.

Transmitter noise may be minimized by providing the following:
- An exciter that produces spectrally pure signals
- A ripple-free power supply
- A low-noise transmitter

Whatever noise is generated in the transmitter may largely be eliminated by adding a noise reduction loop around it.

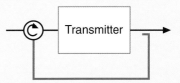

This loop detects any phase or amplitude variations in the transmitter output and adjusts the phase and amplitude of the input so as to reduce the variations toward zero.

Providing Wide Dynamic Range. Another common inadvertent source of clutter is saturation of the radar receiver or A/D converters by strong clutter, as a result of insufficient dynamic range. Saturation generates modulation products which—like transmitter noise—spread into otherwise clutter-free spectral regions.

Saturation of the receiver may be avoided by distributing the gain throughout the receiver chain with successive steps of automatic gain control.

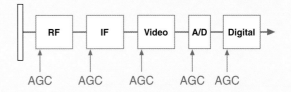

Minimizing Quantization Noise. Noise due to quantization of the received signals may be avoided by:
- Employing highly linear A/D converters
- Quantizing with a significant number of bits
- Employing high sampling rates
- Summing samples

Minimizing Receiver Noise Figure. Receiver noise may be minimized by employing very low noise preamplifiers (LNAs), minimizing all losses ahead of them, and placing the LNAs as close as possible to the radiating elements—as is done in active ESAs. With advanced solid state devices, remarkably low noise figures may be achieved.

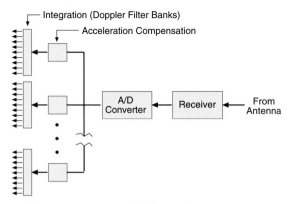

11. Acceleration compensation for long coherent integration times. For every possible acceleration, a separate filter bank is provided.

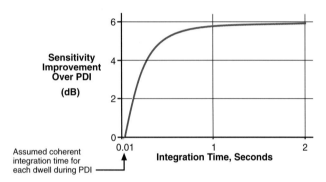

12. Increase in detection sensitivity obtained by employing coherent integration instead of PDI. Beyond a certain point, the advantage of coherent integration over PDI rapidly diminishes.

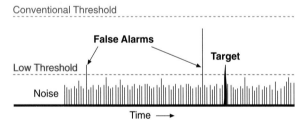

13. By lowering the detection threshold, weaker targets may be detected. But special measures must be taken to keep the increased number of false alarms from reaching the display.

2. Crossing of the detection threshold.

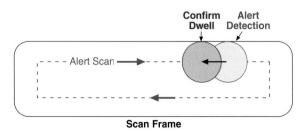

14. Alert/confirm technique. During the alert scan, the radar employs a low detection threshold. When a target "hit" occurs, the beam is immediately steered back to confirm the hit with a high threshold.

The radar, or course, has no way of knowing how much, if any, a target will accelerate during any one integration period. Moreover, during the same period returns may be received from targets accelerating at different rates. To get around these problems, the radar returns are integrated in a number of parallel channels. In each successive channel, compensation is provided for an incrementally greater acceleration within the range of possible values (Fig. 11). The increments are selected so that no matter what a target's acceleration is, its returns will be efficiently integrated in one of the channels.

As the coherent integration time is increased, two things happen. The required signal-processing throughput grows, and the passbands of the doppler filters narrow. Since target returns have a finite bandwidth and signal processing costs are not inconsequential, a point is ultimately reached where the advantage of coherent integration over PDI rapidly diminishes (Fig. 12). Consequently, if the dwell time, t_{ot}, is very long, it typically is broken into two or more coherent integration periods, and the outputs from each filter for successive periods are combined through PDI.

The increase in detection sensitivity obtained by efficiently integrating the received signals over long periods may be parlayed into a greater increase by increasing the radar's target-detection efficiency. One way of accomplishing this is through sequential detection.

Sequential Detection. As was explained in Chapter 10, a radar's detection threshold is conventionally set high enough to reduce the false alarm probability to an extremely low value. By lowering the threshold, the detection sensitivity can be considerably increased. But, then, the number of false alarms increases (Fig. 13). If the increased detection sensitivity is to be useful, the false alarms must be kept from reaching the operator's display. Two techniques for accomplishing that are *alert-confirm detection* and *track before detection*.

Alert-confirm takes advantage of the selective dwell capability of the ESA to break search operation into "alert" and "confirm" phases. During the alert phase, the radar scans the desired search volume, using a long coherent integration time and a low detection threshold.

Following every "hit",[2] the alert scan is temporarily interrupted for the confirm phase. In it, the antenna beam is instantly steered back to the direction of the hit. It dwells there long enough to verify, with a *high* detection threshold, whether the hit was a valid target and, if so, to accurately determine the target's location (Fig. 14). The alert scan is then resumed, and the target is passed on to the display.

Detection performance may be optimize by judiciously selecting the waveforms for the two phases. Since all of the desired target information need not be obtained in the alert phase, for it a waveform may be selected which maximizes detection sensitivity—such as *velocity search*. A waveform such as High PRF range-while-search (RWS) may then be selected for the confirm phase. To avoid wasting scan time on false alarms, however, only if the target is confirmed in the first FM ramp of this waveform would the dwell be extended to include the ramps for range measurement and ambiguity resolution.

Performance may be further optimized by adaptively selecting parameters of the confirm-phase waveform—PRF, pulse width, dwell time, FM-ramp slope, etc.—on the basis of data obtained in the alert phase. In the foregoing example, for instance, if the alert phase detection revealed that the target had a high doppler frequency, steep modulation ramps would be used in the confirm phase to achieve high range accuracy. On the other hand, if the target's doppler frequency were found to be low, shallow ramps would be used to keep the mainlobe clutter from smearing over the target returns.

If the density of targets is excessively high, special steps may be taken to keep the frame time from being stretched out. For example, by performing crude ranging in the alert phase, large long-range targets of no interest may be identified and their confirmation inhibited. For targets already in track, the confirm phase may be skipped and the tracks updated on the basis of data obtained from the alert detections.

The possibility of frame time being stretched out may be avoided completely with the track-before-detection technique. It uses only a low-detection threshold. Targets are confirmed if detected in several complete search frames (Fig. 15). Besides increasing detection sensitivity, this technique has the added advantage that advanced tracking information is already available when a detection is declared.

Bistatic Target Detection

In bistatic target detection, targets are illuminated by one radar and their returns are detected by one or more passively operating radars. Illumination may be provided either cooperatively, by a radar in a friendly aircraft, or inadvertently, by an enemy radar.

Cooperative Bistatic Detection. Cooperative operation has at least two particularly valuable applications.

One is enabling a fighter to get around the restrictions imposed on power-aperture product by the small diameter

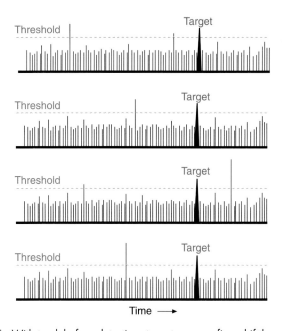

15. With track before detection, targets are confirmed if they cross a low detection threshold in several complete search scan frames. Noise spikes, too, will cross the threshold, but not necessarily at the same point in each frame.

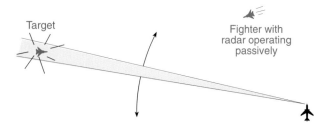

16. Limitations on the power-aperture product of a fighter's radar may be eased by illuminating targets with a high-power radar in a large aircraft safely flying in a standoff position.

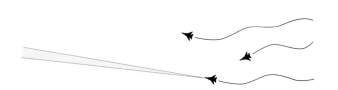

17. By cooperatively shifting radar transmission randomly among themselves, the fighters of a strike force may obtain almost complete protection from anti-radiation missiles.

of the aircraft's nose section, and the limited weight, cooling, and prime-power allocations that may reasonably be made to the radar. In this application, target illumination is safely provided from a standoff position by a large aircraft carrying a high-power radar having a large high-gain antenna. Target returns are received by the passively operating radar of the fighter (Fig. 16).

Besides bringing higher average power to bear on a target, the technique may also eliminate eclipsing loss. Except in trailing attacks, where the fighter may receive such strong pulses from the transmitting aircraft that they cannot be rejected, the receiver needn't be blanked during transmission. The net result: single-look probabilities of detection are significantly increased, and the fighter maintains radio silence.

A second cooperative bistatic application is protecting a strike force from anti-radiation missiles (ARMs). For this, the strike aircraft continuously maneuver ("S" turns or the like). Transmission, meanwhile, is shifted randomly from one aircraft to another, and the radars in all aircraft listen (Fig. 17). The radar seeker of an attacking ARM is thus presented with a shifting line of sight to the source of the radiation. If the aircraft spacing and the illumination shifting period are optimally selected for the parameters of the ARM, nearly complete protection may be provided.

For successful cooperative operation, precise synchronization is essential. To synchronize antenna scans and measure range accurately, the relative location and heading of the transmitting aircraft and the direction of its radar beam must be precisely known by the other aircraft. To extract the desired target information from the received signals, the transmitters and receivers of all radars must be tuned to within a few kilohertz of the same frequency. In addition, the transmit/receive timing and the start of the local oscillator sweeps for FM ranging (if used) must be synchronized to within a microsecond or less.

Difficult as these requirements appear, they can be readily met. Frequency synchronization can be obtained by sensing the "main bang" sidelobe radiation received through the receiving radars' antenna sidelobes. Timing may be adequately provided by a highly stable crystal oscillator in each aircraft, with but a single preflight synchronization. Locations and headings of adequate accuracy can be obtained from each aircraft's inertial navigation system, provided its alignment and positional output are periodically initialized. The only significant addition required to a typical avionics system is a secure data link to transmit the position, heading, and beam direction from the illuminating aircraft to the passively operating radars.

Noncooperative Bistatic Detection. Another possible source of target illumination may be an enemy radar of known location—such as an early warning radar (Fig. 18). This is an attractive possibility; for it completely obviates any friendly aircraft breaking radio silence.

However, several potential limitations must be kept in mind. Many such emitters are mobile or at least portable, so their locations may not be known with sufficient accuracy for good ranging. The bistatic geometry may be such that the difference in doppler shifts for the target echoes and the ground clutter are relatively small, making clutter rejection difficult. The illuminator may have a noisy transmitter, making good clutter rejection impossible. Finally, enemy emitters operate at their own convenience and so may be used only opportunistically.

Space Time Adaptive Processing (STAP)

STAP is a joint angle-doppler domain filtering technique applicable to long-range pulse-doppler surveillance radars employing phased array (ESA) antennas and clutter cancelers for mainlobe clutter rejection. The technique was conceived as an alternative to conventional means[3] of rejecting external noise and noise jamming and of compensating for aircraft-motion-induced spreading of the doppler spectrum of the ground clutter, which can severely degrade the detection of low closing-rate targets.

A simplified block diagram of a generic, fully adaptive implementation of STAP is shown in Fig. 19. A separate receive channel is provided for each element of the array antenna. The receivers' coherent video outputs are conventionally sampled and digitized. For every resolvable range interval, the samples taken during each *coherent processing interval (CPI)* are collected in a matrix. From it, weights for forming a filter "tuned" to pass potential target signals and reject the received noise and interference are adaptively computed. The samples are then weighted and summed.

Background. The concept of adaptive processing is by no means new. Radar engineers have long dreamed of adaptively minimizing virtually every type of external interference on the basis of its spatial and spectral characteristics. However, most of the early approaches to STAP proved to be impracticably slow in adapting to changes in the clutter and interference situation. But in the early 1970s, three pioneers in the field,[4] devised a remarkably fast-adapting algorithm, which has come to be called the *RMB*, an acronym coined from the initial letters of their last names. With computer simulations, they convincingly demonstrated the algorithm's effectiveness.

These results were published in 1974. For some 10 years, they received little attention. There were several reasons.

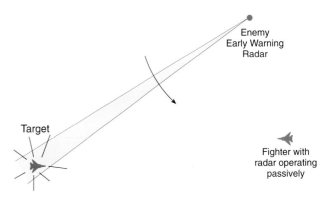

18. Target illumination for bistatic detection may be provided adventitiously by an enemy radar of know location, obviating the need for any friendly aircraft to break radio silence.

3. Ultra-low sidelobe beam forming, displaced phase center antenna, sidelobe jamming rejection, etc.

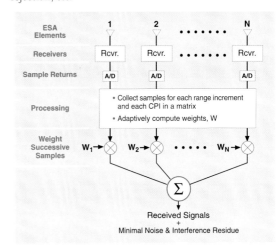

19. Generic, fully adaptive implementation of STAP. Filter formed by adaptively weighting and summing samples of receiver outputs is tuned to pass potential signals, reject noise jamming, and compensate for doppler spread of clutter spectrum.

4. I. S. Reed, J. D. Mallett, and L. E. Brennan.

For one, the required computer throughput was well beyond the capabilities of then current airborne processors. For another, the requirements for not only an ESA but also a separate receiver and A/D converter for each array element were plainly not affordable in the 1970s.

In the mid 1980s, however, in response to the anticipated need to detect emerging low-RCS aircraft, STAP became an active field of R&D and has remained so ever since.

RMB Weighting Algorithm. This algorithm takes advantage of the fact that coherent return from the ground generally has circular Gaussian statistics; hence, is completely characterized by the complex covariance matrix. Weights for implementing the algorithm are obtained in essentially two steps. At the outset, lacking a priori knowledge of the interference situation, an estimate of the covariance matrix of the received radiation is made using a well known statistical analysis device, called the *maximum-likelihood function.* The matrix is then inverted, thereby directly yielding weights for each receive channel. Thereafter, the matrix is continually updated (adapted) in light of the received noise and radar return, to accurately reflect the varying clutter and interference conditions.

Each update is based on separate and independent samples of received data obtained from range increments other than the one being processed (Fig. 20). The beauty of the algorithm is that only a relatively small number of samples is actually needed for an update, enabling most of the CPI to be devoted to efficiently filtering out the interference.

As a result, the filter's output contains a very high percentage of the received signal power. Provided the target density is not high and enough reasonably homogeneous independent samples of the interference are available for adaptive learning, the signal-to-noise ratio also is high.

Subsequent Development. Work on STAP since the mid-1980s has focused largely on the RMB algorithm. Among the primary goals have been the following:

- Make STAP more affordable

- Overcome its inherent dependence upon receipt of an homogeneous flow of independent, identically distributed interference data

- Enable STAP to handle a higher density of targets

- Cope with sophisticated forms of jamming, such as coherent-repeater jamming of randomized range, angle, and doppler frequency

- Get around the requirement of many STAP approaches for precisely matching receiver channels and calibrating them to match the antenna's characteristics.

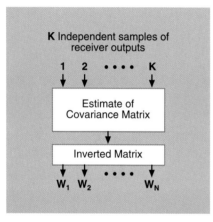

20. In RMB algorithm, weights are obtained directly by inverting the covariance matrix. Matrix is periodically updated (adapted) with a limited number of samples obtained from different range increments than the one currently being processed.

Measures proposed to date to satisfy these requirements have consisted primarily of reducing the number of spatial degrees of freedom (adaptability) by employing subarrays and various levels of analog beam forming.

While STAP was originally considered applicable only to radars employing expensive ESAs and multiple receiver channels, it has come to be viewed also as having potential applications as a relatively low-cost add-on to radars employing conventional antennas with sum-and-difference outputs, and possibly even having applications other than long-range surveillance.

Photonic True-Time-Delay (TTD) Beam Steering

TTD beam steering is a technique for greatly broadening the instantaneous bandwidth of an active ESA.

In conventional ESAs, which steer the antenna beam with phase shifters, instantaneous bandwidths are inherently limited. For phase shifts that are a linear function of carrier frequency cannot be provided simultaneously over a broad band of frequencies. Different phase shifts must be provided not only for each beam position, but also for each carrier frequency.

This limitation may be avoided by obtaining the phase shifts through the introduction of a controllable "true" time delay—TTD—in the feed for each T/R modules. As we shall see, by implementing the delays with fiber-optic and optoelectronic elements—photonics—they will, as desired, vary linearly with the frequencies of the RF signals passing through the feeds. Consequently, remarkably broad instantaneous bandwidths may be obtained.[5]

Photonic Implementation. In simplest form, a photonic feed for a T/R module consists of a single optical fiber (see panel, above right), having a laser diode attached to one end and a photo detector, to the other (Fig. 21).

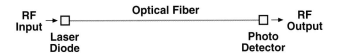

21. A simple fiber-optic feed for TTD beam steering. RF input signal is amplitude modulated on a beam of light emitted by the laser diode. The signal is delayed by the length of time it take to propagate through the fiber and is converted back to RF by the photo detector.

The radio-frequency signal to be fed through the fiber varies the bias voltage applied to the laser diode, thereby proportionally modulating the amplitude of the light emitted by the diode at the signal's radio frequency (Fig.22).

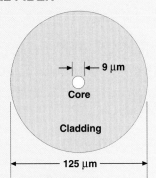

OPTICAL FIBER

Fiber of the type used for TTD beam steering is called single-mode fiber. The core, which carries the signal, is surrounded by cladding. Both are composed of pure silicon, with just enough doping added to give the cladding a slightly lower index of refraction than the core .

Advantages

- Low rf signal attenuation: 0.3 dB/km
- Flexible: bend radius several cm.
- Non-conducting
- Immune to EMI, cross-talk, and EMP; is secure
- Does not disrupt rf fields
- Large bandwidth: 10 GHz

- Stable transmission characteristics due to low ratio of bandwidth (e.g.,10 GHz) to carrier frequency (≥ 200 THz)
- Small size and light weight
- Can store wideband signals for 10s of milliseconds (duration limited only by length of fiber and acceptable loss)

5. TTD beam-steering can also be implemented with electronic devices and RF transmission lines, but because of the dispersion of waves of different frequency passing through them, instantaneous bandwidth is still limited.

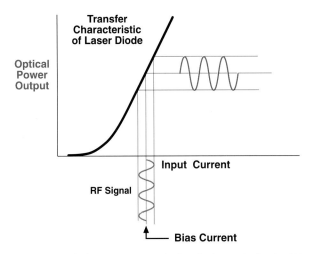

22. By varying the bias current applied to the laser diode, the RF signal amplitude modulates the light emitted by the laser diode.

6. This velocity is about 2/3 that in free space.

7. The optimum wavelength of the laser diodes for TTD beam steering is 200 to 250 THz (λ = 1.5 to 1.3 μm)

In passing through the fiber, the signal is delayed by a time, ΔT, equal to the fiber's length, L, divided by the signal's velocity of propagation, v, through the fiber.[6]

$$\Delta T = \frac{L}{v}$$

The photo detector converts the time-delayed signal back to radio frequencies. Since the ends of the fiber can be collocated, by adding switches at the input and output the same fiber can be used for both transmission and reception.

Because of the extremely high frequency of light[7] compared to that of radio waves, the feed can accommodate signals having exceptionally wide instantaneous bandwidths—up to 18 GHz or more.

Problem of Affordability. While the TTD concept is simple (Fig. 23), it is not at present affordable. There are three basic reasons why. First, since very little power can be conveyed by a fiber-optic feed, the antenna *must* be an active ESA which in itself is expensive. Second, except for the fibers, the photonic components required are currently quite expensive. Third, a great many components are required.

In a "brute-force" approach, each of the ESA's T/R modules would be provided with separate fiber-optic feeds cut to the correct lengths to provide the delays required for every potential look angle, θ (Fig. 24). The radar would then switch from one to another of these feeds as the desired look angle changes.

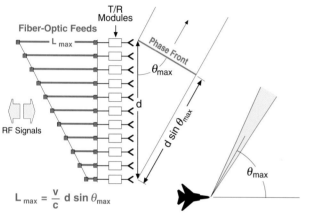

$$L_{max} = \frac{V}{C} \, d \sin \theta_{max}$$

V = velocity of propagation through the fiber
C = speed of light in free space

23. TTD concept. By progressively increasing length, L, of successive feeds, the antenna beam may be steered to any desired angle θ off broadside.

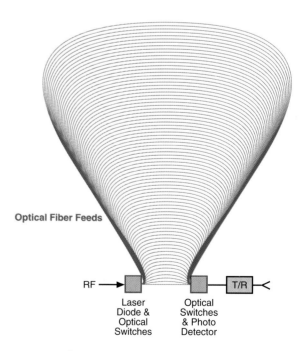

24. Brute force approach to photonic TTD. Each T/R module is provided with a separate feed of correct length for each resolvable look angle, and the radar switches among them as the desired look angle changes.

Needless to say, this approach is not particularly practical. Even a very small two-dimensional ESA may have as many as 400 radiators. Assuming a beamwidth of 3°, a desired angular resolution of half a beamwidth, and a field of regard of ± 60° in both azimuth and elevation, a total of 32,000 optical fibers, plus laser diodes and photo detectors would be required, not to mention switches and combiners. Considering that some ESAs may include up to two thousand or more radiators, the complexity and cost of TTD, if implemented as just described, could be staggering.

Fortunately, the required number of components may be reduced substantially. One way is to selectively switch precisely cut segments of fiber into or out of the feed for each T/R module. Another is to provide a portion of the required delay for each of a number of modules with the same delay line by means of wavelength-division multiplexing. Yet another approach, suitable only for certain applications, is to provide the smaller increments of delay with electronic circuitry. Each of these approaches is described briefly in the following paragraphs.

Switchable Fiber-Optic Delay Lines. A popular delay line of this sort consists of a number of successive fiber segments providing increments of delay equal to powers of two (2, 4, 8, . . .) times a basic increment, ΔT (Fig. 25). The desired total delay is obtained by switching appropriate segments of fiber into or out of the line with digitally controlled single-pole, double-throw switches.

The line may be made as long as necessary to provide the T/R module or modules that the feed serves with the number of different delays (R) needed to achieve the desired beam-steering resolution. The required number of fiber segments (N), hence circuit complexity, increases only as the logarithm to the base 2 of R,

$$N = \log_2 R$$

Accordingly, this general type of delay line is called a *binary fiber-optic delay line (BIFODEL)*. By using optical switches, the line may be made bidirectional, i.e., signals can be fed down it from either end.

Further reductions in complexity may be realized through wavelength-division multiplexing.

Wavelength-Division Multiplexing. Because of the extremely wide bandwidth available at optical frequencies, it is possible to simultaneously pass a large number of different optical carrier frequencies through the same delay line by optically filtering the outputs of the laser diodes and the inputs to the photo detectors.

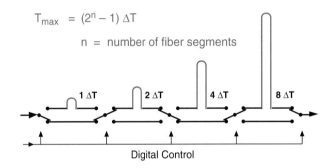

$$T_{max} = (2^n - 1) \Delta T$$

n = number of fiber segments

1 ΔT   2 ΔT   4 ΔT   8 ΔT

Digital Control

25. Binary fiber-optic delay line (BIFODEL).[8] Implemented with optical switches, this architecture not only significantly reduces the amount of hardware required but is bidirectional. Switches shown here are set for a delay of 10 Δt.

8. An important feature of this particular arrangement is that regardless of what delays are selected, a signal always goes through the same number of switches. Consequently, the line's insertion loss doesn't vary with the switching.

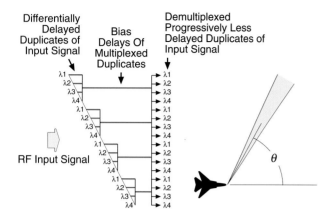

Differentially Delayed Duplicates of Input Signal

Bias Delays Of Multiplexed Duplicates

Demultiplexed Progressively Less Delayed Duplicates of Input Signal

RF Input Signal

26. One approach to wavelength multiplexing. Each of the three differential delays is produced by a separate BIFODEL, as is each of the three bias delays. By adding a fixed length to each bias delay line, the bulk of the hardware may be mounted remotely.

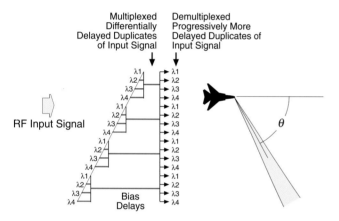

Multiplexed Differentially Delayed Duplicates of Input Signal

Demultiplexed Progressively More Delayed Duplicates of Input Signal

RF Input Signal

Bias Delays

27. Delays applied to steer the antenna's beam θ° to the right instead of the left. Delays are the same as shown in Fig. 26, but their order is reversed.

9. Up to 150 carriers having full 18 GHz bandwidths can be provided. The limiting factor: coupling of adjacent optical wavelengths.

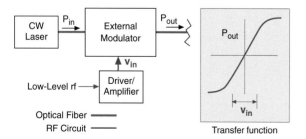

CW Laser — P_{in} — External Modulator — P_{out}

v_{in}

Low-Level rf — Driver/ Amplifier

Optical Fiber ▬▬▬
RF Circuit ▬▬▬

P_{out}

v_{in}

Transfer function

28. External modulator provides higher optical power than a photo diode; yields wider dynamic range, lower noise figure, and reduced RF input-to-output loss. But, is best employed remotely.

One approach to multiplexing is simplistically illustrated for transmission by a one-dimensional ESA in Fig. 26. This approach takes advantage of the fact that for any one look angle, *θ*, the difference in the delays that successive feeds must provide is the same from one end of the array of T/R-modules to the other. In other words, the required delay for feed (n + 1), differs from that for feed (n) by ΔT; the required delay for feed (n + 2) differs from that for feed (n) by 2 ΔT; the required delay for feed (n + 3), by 3 ΔT; and so on.

In this example, to provide the differential delays the feeds are grouped into four *subarrays* of four feeds each. The input signal is modulated on carriers having four optical wavelengths: λ1, λ2, λ3, and λ4 and applied in parallel to four BIFODELs, which delay it by 0, 1ΔT, 2ΔT, and 3ΔT, respectively.

The outputs of these BIFODELS are combined into a single wavelength-multiplexed signal. It is applied in parallel to four so-called "bias" BIFODELs, which provide the balance of the required delay for each subarray of feeds.

The signals output by each bias BIFODEL are wavelength demultiplexed, detected, and supplied to the corresponding subarray of T/R modules: the signal on carrier λ1, to the first module in the subarray; the signal on carrier λ2, to the second module in the subarray; and so on.

For negative values of *θ*—i. e., look angles to the right in Fig. 27—the sizes of both the differential delays and the bias delays are reversed: longest delay first, rather than last.

If the feeds are implemented entirely with optical components, the same hardware can be used for both transmission and reception. But to receive, a laser must be provided at each T/R module—adding, of course to its cost and complexity, and impacting performance.

Even so, the net reduction in hardware complexity and cost is substantial. With the multiplexing of many more frequencies,[9] it can be dramatic.

Another advantage of wavelength multiplexing is that, by adding a fixed increment (extension) to the length of each of the bias delay lines, the majority of the optical components may be mounted remotely from the antenna, thereby simplifying the installation. Also, with remoting, higher optical power can conveniently be provided by substituting an external modulator fed by light from a CW laser source, for each of the lower power directly modulated laser diodes (Fig. 28). With an external modulator, very much higher rf modulation frequencies may be used—up to 100 GHz, or so. And, with higher optical power, rf input-to-output loss can be reduced, and wider dynamic range and lower noise figures can be achieved.

Hybrid Implementation. Depending upon the application, the reduction in cost may be increased even further by providing the shorter delays with binary electronic delay lines. At L and S bands, for example, delay lines implemented with strip-line or microstrip circuitboards and gallium arsenide switches are every bit as suitable for short delays as fiber-optic delay lines. They are reversible, very small, and roughly two orders of magnitude cheaper.

The outputs of the electronic delay lines are converted to optical frequencies and applied to fiber-optic systems such as just described, which provide the longer delays for which electronic circuits are not suitable.

Potential Applications. Through advanced techniques such as those just described and others in the offing, the complexity and cost of TTD is gradually being reduced. As the high costs of suitable switches and other key optical components come down, photonic implementation of virtually all of the advanced features of the active ESA—including independently steered beams on different frequencies may become practical.

These capabilities, plus the wide instantaneous bandwidth achievable with photonic implementation, promise to make possible extremely broad situation awareness as well as to open the door to simultaneous shared use of the radar antenna for communications and electronic warfare.

The limiting factor then will be the cost of the T/R modules and their ability to support the wide rf bandwidths made available through TTD.

Another potential application which should not be overlooked is in long arrays—10 ft, or more—which may be required to transmit narrow pulses at look angles greater than about 30°. TTD then avoids problems of beam squint, or beam spread, which occur if the beam is steered with phase shifters, by ensuring that all of a pulse's energy arrives at the pulse's phase front at the same time (Fig.29).

Interferometric SAR (InSAR)

Just as an optical interferometer can measure variations in the thickness of a sheet of glass with precision approaching the wavelength of light, an interferometric SAR radar can measure the variations in the height of the terrain with precision approaching the wavelength of microwaves. Combined with conventional high-resolution SAR mapping, the interferometric height measurements enable the production of three-dimensional topographic maps.

Employed by satellite-borne radars, InSAR promises to provide the accurate, high-resolution global topographic maps required for geophysical applications.[10] Employed in

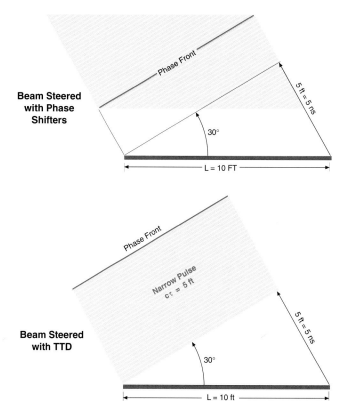

29. TTD ensures that all of a pulse's energy arrives at the pulse's phase front at the same time, which is important if the antenna is long, the look angle at all large, and the pulse narrow.

10. Spatial resolution on the order of a few tens of meters; height resolution on the order of a few meters and possibly as fine as 10 cm, for ice studies.

515

30. To obtain the values of elevation needed for three-dimensional mapping, the SAR radar measures the elevation angle, θ_e, of the line of sight to the center of each resolution cell in the swath or patch of ground being mapped.

airborne radars, InSAR promises to provide localized topographic maps of much finer resolution.

Basic Concept. An InSAR radar obtains the elevation data needed for three-dimensional mapping by determining the elevation angle of the line of sight to the center of each resolution cell in the swath or patch that is mapped (Fig. 30). From this angle (θ_e), the height (H) of the radar, and the slant range (r), to the cell, the cell's height and horizontal distance from the radar are computed.

The radar determines the elevation angle of a resolution cell in much the same way as a phase-comparison monopulse system determines a tracking error. As illustrated in Fig. 31 (below), radar returns from point **p** in the center of the cell are received by two antennas separated by a relatively short distance, **B**, on a cross-track baseline. The baseline is tilted a prescribed amount toward the area being mapped. Ranges r_1 and r_2 from the two antennas to **p** differ by an amount roughly equal to **B** times the sine of the angle, θ_L, between the line of sight to **p** and a line normal to the baseline.

The phases of the coherent radar returns received by the two antennas differ in proportion to the difference in the two ranges:

$$\phi = \frac{2\pi}{\lambda} \ (r_1 - r_2) \ \text{radians}$$

As with phase comparison monopulse, by measuring ϕ, the elevation angle (θ_L) between the line of sight to p and the normal to **B** may be computed.

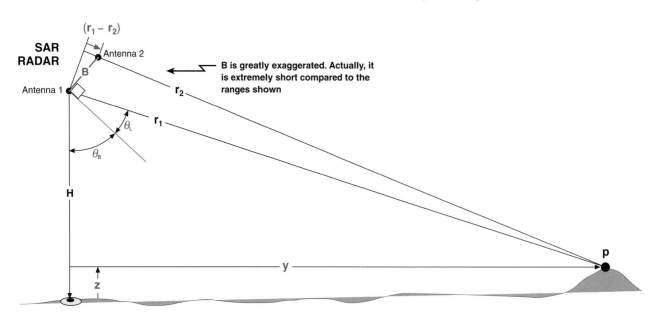

31. Parameters an InSAR radar measures to determine the elevation, z, and horizontal range, y, of a point, p, in the center of a resolution cell. Angle, θ_L, between the line of sight to p and the line normal to baseline, B, is determined by sensing the phase difference, ϕ, between the returns received by the two antennas as a result of the difference in slant range, $(r_1 - r_2)$, from p.

An equation for ϕ, in terms of r_1, λ, and the length and tilt of B, can be derived directly from the geometry illustrated in Fig. 31. Based upon that equation, an exact equation for θ_L can similarly be derived. To save you the trouble, both equations are presented in Fig. 32.

Having computed the value of θ_L, all that must be done to obtain the elevation angle, θ_e, is to add to θ_L the angle, θ_B, between the normal to the baseline and the vertical axis.

$$\theta_e = (\theta_L + \theta_B)$$

From this sum and the range r_1, the horizontal position (y) and elevation (z), of p may readily be computed.

$$y = r_1 \sin \theta_e$$

$$z = H - r_1 \cos \theta_e$$

Implementation. InSAR may be implemented by mapping the swath or patch in either of two different ways:

In one, a single pass is made with a radar having two antennas separated by the desired cross-track distance **B**. One antenna transmits and both bistatically receives.[11]

In the other implementation, two successive passes precisely separated by the desired cross-track baseline, **B**, are made with the same antenna on each pass operating monostatically.

Ambiguities and Their Resolution. As may be surmised from Fig. 32 in collecting returns from successive range increment across the full width of the swath being mapped, the range difference ($r_1 - r_2$), hence ϕ, increases continuously. Since the wavelength is comparatively short, the value of ϕ cycles repeatedly through 2π radians (360°) and so is ambiguous.

The ambiguities may be resolved by separately making a conventional SAR image (Fig. 33a) with the coherent returns received by each antenna. The two images are then coregistered and merged. Because of the phase difference, ϕ, between the images, the result is an interferogram (Fig. 33b).

By adding 2π to the value of ϕ each time a fringe in the interferogram is crossed (a process called *phase unwrapping*), the ambiguities are removed. The horizontal position, y, and elevation, z, of each cell are then accurately computed, and the map is topographically reconstructed (Fig. 33c).

The topographic accuracy depends critically, of course, on the accuracy with which the phase unwrapping is performed. Provided the signal-to-noise ratio is reasonably high and the fringes are not too close together, this is a straightforward process. It can become complicated, howev-

$$\phi = \frac{2\kappa\pi}{\lambda}\left[r_1 - (r_1{}^2 + B^2 + 2r_1 B \sin\theta)^{1/2}\right]$$

$$\theta_L = \sin^{-1}\left[\frac{\lambda^2\phi^2}{8(\kappa\pi)^2 r_1 B} - \frac{\lambda\phi}{2\kappa\pi B} - \frac{B}{2r_1}\right]$$

$\kappa = 1$ for bistatic (one pass) mapping

$\kappa = 2$ for monostatic (two pass) mapping

32. InSAR equations derived from the geometry shown in Fig. 31. For monostatic mapping, k = 2 since the phase shift, ϕ, corresponds to a difference in round-trip distance from the two antennas to point, **p**.

11. On a satellite, one antenna might be mounted on the end of a long pole.

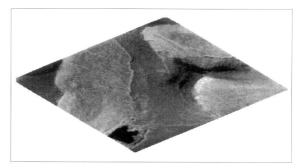

a. Basic SAR image made from returns received by one of the radar's two antennas.

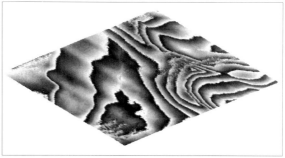

b. Interferogram produced by coregistering and merging the images produced with the returns received by the two antennas.

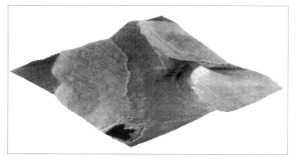

c. The reconstructed 3-D topographic map.

33. Images of a region in Wales obtained with DERA Malvern's C-band InSAR radar. (Crown copyright DERA Malvern)

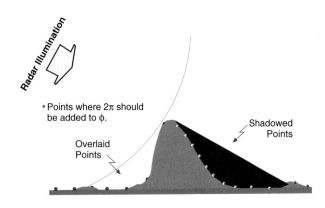

34. Possible sources of error. Steep slopes may result in returns from some points being overlaid by returns from a shorter slant range but longer horizontal range. Other points may be missed because they lie in shadows.

er if steep slopes or shadows are encountered (Fig. 34). For the slopes may result in some points at which fringes are crossed being overlaid by others, and the shadows may cause some points to be missed.

Summary

The advent of low-RCS aircraft and the growing threat of electronic countermeasures have spawned a number of advanced radar techniques.

For wideband multifrequency operation, SIMFAR conveniently generates multifrequency drive signals by phase modulating a microwave signal. To provide broad frequency coverage with 100% duty factor, STAR interleaves pulse trains of widely different radio frequencies.

To increase detection sensitivity, long coherent integration times have been made practical by techniques that compensate for target acceleration. Sequential detection techniques have further increased sensitivity by lowering detection thresholds and verifying target hits with detections made either with a high threshold or with a low threshold in several complete search frames.

To circumvent limitations on a fighter's power-aperture product and increase survivability, bistatic techniques have been perfected in which targets illuminated by one radar are detected by one or more passively operating radars

To reject external noise and noise jamming and compensate for motion-induced spreading of the doppler clutter spectrum in long-range surveillance radars, the received signals are passed through a joint angle-doppler filter that automatically adapts to changing clutter and jamming conditions.

To greatly broaden the instantaneous bandwidth of an active ESA, a fiber-optic feed is provided, and the phase shifts for beam steering are provided by selectively switching precisely cut segments of fiber into or out of the branches leading to the individual T/R modules.

To produce three-dimensional topographic maps, height is interferometrically measured by merging SAR maps made with returns received by two antennas separated by a relative short distance on a cross-track baseline.

Advanced Waveforms and Mode Control

41

T̲o enable the resolution of multiple targets at long ranges and to increase detection sensitivity against low-closing-rate targets, a number of new waveforms have been developed. In this chapter we'll take up three of these[1]:

- Range-gated high PRF

- Pulse burst

- Monopulse doppler

We'll also briefly consider a new search-while-track mode, which takes advantage of the ESA's extreme beam agility. We'll then be introduced to a mode-management software architecture for flexibly allocating the radar's resources and ensuring prompt response to high priority requirements in complex tactical situations.

Range-Gated High PRF

This mode overcomes the two chief limitations of high PRF range-while-search: reduced detection sensitivity against low closing rate targets and poor range resolution. Except at very low altitudes, performance of range-gated high PRF against both high-closing-rate and low-closing-rate targets is superior to that obtained with medium PRFs, and range measurement is more precise than that obtained at high PRFs with FM-ranging.

Range-gated high PRF differs from conventional high PRF waveforms in that the pulse width is narrowed somewhat and sufficient pulse compression is provided to enable resolution of closely spaced targets (Fig. 1). During the interpulse period, the radar returns are sampled at a high

1. All are essentially variations of the basic low-PRF and high-PRF waveforms described in detail in Chaps. 25 and 27.

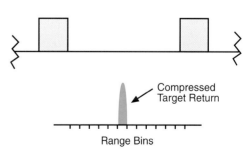

1. Range-gated high-PRF waveform differs from the conventional high-PRF waveform in that the pulsewidth is narrowed somewhat and sufficient pulse compression is provided to enable resolution of closely spaced targets.

rate and the samples are stored in range bins whose width corresponds to the compressed pulse width. From the contents of each range bin, a bank of narrowband doppler filters is formed, thereby also providing fine doppler resolution.

The PRF is made high enough to provide an unambiguous doppler clear region for the detection of high-closing-rate targets. The modest reduction in duty factor due to the shorter transmitted pulses is more than made up for by the commensurate reduction in eclipsing loss and the reduction in background noise provided by range gating. Elimination of doppler ambiguities and provision of fine range and doppler resolution minimize the amount of sidelobe clutter over which the echoes of low-closing-rate targets must be detected.

Because the maximum unambiguous range at high PRFs is extremely short, range ambiguities are, of course, severe. They may be resolved, however, by employing a combination of FM ranging, for coarse resolution, and PRF switching, for fine resolution.

The waveform is suitable for either cued or independent search operation. What makes it particularly attractive is the fine multiple-target resolution it provides. With 100-foot range bins, for example, the radar can individually display targets separated in range by as little as 300 feet (Fig. 2). Finer resolution can be obtained with narrower range bins.

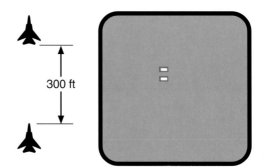

2.   With range-gated high PRF, a radar employing 100-foot range bins can individually display targets separated in range by as little as 300 feet.

When the waveform is applied to a STAR radar, two additional advantages may be gained over conventional high-PRF modes: increased average power without increased peak power, and spreading of the radiated power over a broad frequency band.

Pulse Burst

By transmitting high-PRF pulses in short bursts, this waveform goes a step further than range-gated high PRF in improving detection range against long-range all-aspects

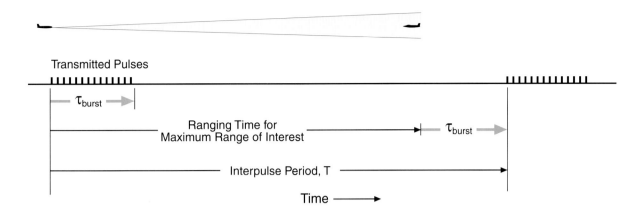

3. Pulse burst waveform. If the burst repetition period is made equal to the ranging time for the maximum range of interest plus the burst width, τ_{burst}, range ambiguities will be avoided. The burst width is added to keep the echoes of targets at maximum range from being eclipsed by the next burst of pulses.

targets. As illustrated in Fig. 3, if the repetition period of the bursts is made equal to the round-trip ranging time for the maximum range of interest plus the burst length, τ_{burst}, range ambiguities will be avoided. Moreover, returns from targets at ranges greater than half the burst length (0.5 $c\,\tau_{burst}$) will not be eclipsed by any of the transmitted pulses and will not be received simultaneously with any sidelobe return from shorter ranges.

Returns from targets at ranges less than half the burst length will, of course, be partially eclipsed and will be received with some sidelobe clutter from shorter ranges. But, because of the targets' short range, the loss of detection sensitivity is not particularly severe and becomes less and less so as the range decreases.

Monopulse Doppler

This waveform is essentially a low-PRF equivalent of pulse burst, in that a single long pulse is substituted for each pulse burst (Fig. 4). As a result, for the same peak power the average power is increased substantially.

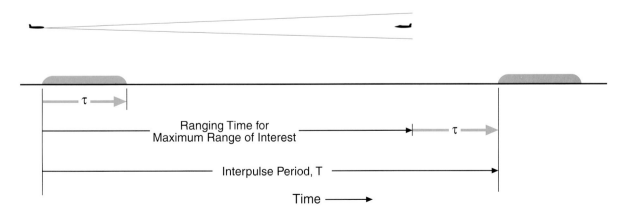

4. Monopulse doppler is essentially the same as pulse burst except that, instead of bursts of high-PRF pulses, single, long pulses are transmitted and the received signals are sampled at a high repetition rate.

For example, if the duty factor within the bursts were 40 percent, other factors being equal, monopulse doppler would provide two and a half times the average power of pulse burst.

Problems of doppler blind zones and returns from ground moving targets—which limit the effectiveness of conventional low-PRFs are avoided by sampling the radar return at a high enough rate to provide an adequately wide doppler-clear region between repetitions of the mainlobe clutter spectrum. Because of this and of range being unambiguous, coarse range as well as doppler resolution, may be obtained following the transmission of every pulse—hence the name, monopulse doppler. Finer range resolution may, of course, be obtained by employing pulse compression.

In one possible implementation, the samples of the radar return are fed in parallel to a set of range gates (Fig. 5, below). The opening of successive gates is staggered in time by an amount equal to the pulse width (compressed pulse width, if pulse compression is used). Each gate is left open for a time equal to the pulse width.

The samples passing through each gate are collected in range bins and coherently integrated to form a doppler filter bank for each range increment. Since the integration time of the doppler filters is limited to the duration of the transmitted pulse, the doppler resolution is fairly coarse.

If desired, finer doppler resolution can be obtained by forming a second bank of doppler filters with the outputs each coarse filter produces for several transmitted pulses.

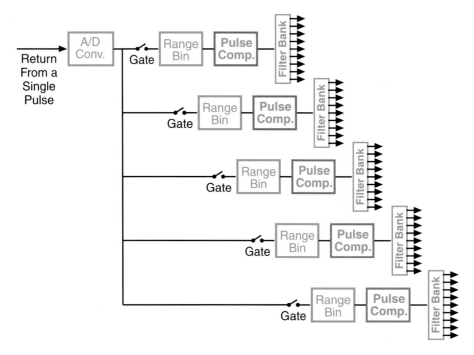

5. One possible implementation of monopulse doppler. Opening of successive gates—closing of the switches shown here—is staggered by the pulse width (compressed pulse width, if pulse compression is used). Gates are left open for a time equal to the pulse width.

Search-While-Track (SWT) Mode

A fighter's radar has three basic reasons to track more than one target simultaneously: (a) to individually monitor potentially threatening targets, (b) to provide periodic target illumination for semiactive guidance of missiles not requiring continuous illumination (e.g., Phoenix), and (c) to provide periodic target position updates for missiles employing command-inertial guidance (e.g., AMRAAM). At the same time, the radar may be required to search one or more narrow sectors for targets designated by offboard or other onboard sensors (e.g., IR search/track set) and to provide continuous situation awareness in a given sector.

While these requirements can be satisfied by the conventional track-while-scan mode,[2] it has a number of serious shortcomings. Not all of the targets must be tracked with the same revisit rates or the same dwell times. They may not all lie in the same sector or in the sector where designated targets must be searched for or where situation awareness is desired.

These limitations can all be surmounted by taking advantage of the extreme beam agility of the ESA (Fig. 7). Rather than refreshing target tracks each time the radar's beam sweeps over them in a continuous search scan, the beam of an ESA can jump almost instantaneously to any target as often and for as long a dwell as necessary to accurately track it. The beam can then jump back to whatever sector it was searching without appreciably increasing the scan frame time. While tracking targets in this way, the beam can simultaneously search specific narrow sectors for designated targets and provide situation awareness with selectable frame times in other sectors, or none at all.

To avoid confusion with conventional track-while-scan, this versatile new mode is called *search-while-track (SWT)*.

Mode Management

So far in this and earlier chapters, we've considered the various radar waveforms, modes, and techniques individually. But to carry out such functions as mode interleaving, adaptive dwell scheduling, multiple waveform utilization, and sensor fusion, the radar's front-end and processing resources must be successively allocated at the correct instants in time to each required internal operation.

A highly flexible and efficient answer to this requirement is a two-level mode-management software architecture outlined below.

In this architecture, the first level of management is performed by the avionic system's Sensor Manager. It receives requests for various radar operations from the flight crew's controls and other key subsystems and converts them into

6. One of several potential needs for SWT is to provide target position updates for AMRAAMs which may be in flight against widely separated targets outside the current search-scan sector.

2. Described in detail on pages 388–390.

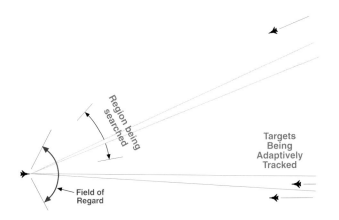

7. In search-while-track (SWT), a radar can search for targets while tracking targets anywhere within the field of regard, without materially increasing the scan frame time. Both track update intervals and dwell times are adaptively selected.

- **Perform search-while-track (SWT) in a volume centered at N, E, D of size ρ, φ, γ, with no LPI constraints and with a priority of 10.**

  Report any detections and track files found in this volume.

  Complete eight 2-second frames.

- **Perform an own-ship precision velocity update (PVU) with priority 20.**

- **Perform a long-range cued search about point xyz immediately, with LPI protocol #6.**

8. Sample prioritized sensor-level commands for basic radar functions in coarse time—order of seconds.

3. One such constraint is that formation of a SAR map cannot be interrupted.

prioritized commands in coarse time—on the order of seconds. Some representative commands are listed in Fig. 8.

The second level of management is performed by the Radar Manager. Upon receipt of the Sensor Manager's commands, it adjusts the coarse time line to account for the radar's current state of operation and system constraints.[3] It then allocates the radar's front-end and processing resources to the requisite tasks in fine time intervals—on the order of nanoseconds. Allocations typically include:

- Field of regard
- Length of individual dwells
- Waveform for each dwell
- Front-end hardware to transmit the waveform
- Processing resources to extract the required information from the collected data and report it to the Sensor Manager and the air crew or requesting avionics system

Thus, through simple prioritized radar-operation requests, the radar's resources are flexibly allocated so as to both avoid conflicts and assure prompt response to high priority requirements in complex tactical situations.

Summary

To increase detection sensitivity against tail-aspect targets, several advanced waveforms have been developed.

Range-gated high PRF resolves closely spaced targets and improves performance against low-closing-rate targets by employing pulse compression and a limited number of narrow range bins.

Pulse burst transmits high-PRF pulses in short bursts, to enable clutter-free detection of nose-aspect targets, and repeats the bursts at a low rate to simultaneously gain the advantage of low PRFs in avoiding sidelobe clutter.

Monopulse doppler accomplishes the same ends but provides higher average power by replacing the bursts with long pulses, and by sampling the returns at a high rate.

By taking full advantage of the ESA's extreme beam agility, a new search-while-track mode simultaneously tracks widely separated targets with interactively selected dwell times and revisit rates, searches narrow regions for designated targets, and selectively provides situation awareness.

By allocating the radar's front-end and processing resources to successive operations through prioritized requests, an advanced mode-selection software architecture enables the radar to flexibly and efficiently carry out such complex functions as mode interleaving, adaptive dwell scheduling, and multiple waveform utilization.

Low Probability of Intercept (LPI)

42

L*ow probability of intercept (LPI)* is the term used for there being a low probability that a radar's emissions will be usefully detected by an intercept receiver in another aircraft or on the ground.

For the air battle of the future, LPI is essential. In conventional aircraft the most important need for LPI is to avoid electronic countermeasures. In low observable aircraft, LPI additionally enhances the element of surprise and denies the enemy use of radar intercept queuing of its fighters. In aircraft of both types, LPI prevents successful attacks by antiradiation missiles.

In this chapter, we will review the generic types of intercept receivers and see what strategies may be used to defeat them. We'll then take up specific design features which may be incorporated in a radar to ensure a low probability of intercept. Finally, we'll very briefly assess the cost of LPI and consider possible future trends in LPI design.

Generic Intercept Systems

A combat aircraft may encounter any or all of four generic types of intercept receiving systems:

- Radar warning receivers (RWR)

- Intercept receiver sections of electronic countermeasures (ECM) systems

- Ground-based passive detection and tracking systems

- Antiradiation missiles (ARM)

The general capabilities of these systems are summarized in Table 1 and briefly outlined below.

TABLE 1. Generic Intercept Systems

System		Detects		Role
		Main-lobe	Side-lobe	
Airborne	RWR	X		• Warn air crew of potential attack • Cue evasive maneuvers & ECM.
	ECM Rcvr.	X	X	• Jammer turn-on, set-on, & pointing • Support sophisticated deception ECM.
Ground-based	DOA & EL		X	• Detect & locate intruding aircraft. • Cue attack or enable avoidance.
Missile	ARM		X	• Home on emissions. • Guide missile to emitter.

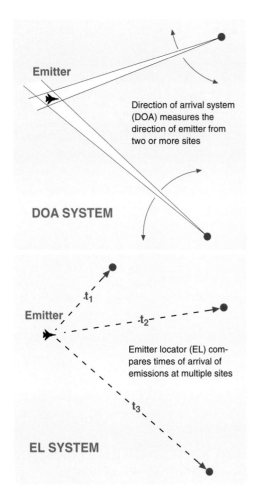

1. Generic ground-based passive intercept systems. Both primarily sense sidelobe emissions.

RWR. By sensing the mainlobe emissions from a radar in a potentially hostile aircraft, the RWR warns the air crew of imminent attack, enabling the pilot to maneuver evasively and to employ defensive countermeasures.

ECM System Receivers. Operating primarily in the radar's sidelobe regions, the intercept receiving portion of an airborne or ground-based ECM system provides the cueing necessary to concentrate jamming power at the radar's frequency and in the radar's direction, as well as to employ sophisticated deception countermeasures.

Ground-Based Systems. Intended to cue defending forces to the approach of intruding aircraft, these systems employ intercept receivers located at widely separated sites. With narrowbeam scanning antennas, the receivers simultaneously detect and track the sidelobe emissions from an aircraft's radar to determine its position. The systems (Fig. 1) are of two basic types. One, the *direction-of-arrival* system *(DOA)*, measures the direction of the source of the detected pulses and determines its location by triangulation. The other, the *emitter locator (EL)*, determines the emitter's location by measuring the time of arrival of its pulses.

Against low-observable aircraft, DOA systems may be used to provide lines of position to the aircraft. EL systems may then determine their actual positions, enabling fighters or ARMs to intercept them.

ARM. By detecting the sidelobe emissions of an aircraft's radar, the ARM homes in on the aircraft despite its evasive maneuvers, hence is a serious threat to any aircraft employing a radar.

All four of these passive "threats" may be defeated through a combination of (1) operational strategies of the air crew of the radar-bearing aircraft and (2) strategies of the radar designer.

Operational Strategies

The most effective LPI strategy of course is not to radiate at all. This strategy may be approached by limiting radar "on" time and operating with no higher power than absolutely necessary to achieve mission goals.

On stealthy interdiction missions, wherever possible the air crew should use collateral intelligence and reconnaissance information. Through careful mission planning they may be able to conduct an entire mission with only a few minutes—or even seconds—of radar operation.

In air-to-air combat situations, where continuous situation awareness is essential, the air crew should use onboard passive sensors—RWR or ESM system, IR search-track set, or forward-looking IR set. When a potentially hostile aircraft is detected, the radar may be used to measure range and possibly precise angle, which the passive sensors may not have provided. But it should be operated only in short bursts and then only to search the narrow sector in which the passive sensors indicate the target to be.

Design Strategies

Since the range at which a radar can detect a given target varies as the one-fourth power of the emitted signal power, whereas the range at which an intercept receiver can detect the radar varies only as the square root of the emitted power, the interceptor has a huge advantage over the radar. However, since signals from a multitude of other radars and electronic systems are inevitably present in a tactical environment, the radar designer has several opportunities to overcome this advantage.

Trade Integration for Reduced Peak Power. For a signal to be usefully detected by an intercept receiver, its source must be identified on the basis of such parameters as angle of arrival, radio frequency, PRF (obtained from times of arrival), and pulse width. To satisfy this requirement, the intercept receiver must detect individual pulses. Consequently, it can employ little or no signal integration; it is sensitive primarily to peak emitted power. The radar, on the other hand, is subject to no such requirement. By coherently integrating the echoes it receives over long periods, the peak power needed to detect a target can be greatly reduced, thereby reducing the detectability of the radar's signals (Fig. 2).

Trade Bandwidth for Reduced Peak Power. An intercept receiver must be able to separate overlapping signals which may be closely spaced in frequency. Consequently, the instantaneous bandwidth of each of its channels can be no wider than necessary to pass the narrowest pulses it can reasonably be expected to receive and measure their time and angle of arrival (Fig. 3). The radar, on the other hand, can be designed to spread its power over a much wider instantaneous-frequency band, thereby reducing the peak power the intercept receiver receives through any one of its channels by the ratio of the two bandwidths.

Trade Antenna Gain for Peak Power. Against an RWR, the radar has the advantage of being able to employ a large directional antenna, which the RWR cannot. During transmission, of course, the high gain of this antenna benefits the RWR as much as it benefits the radar. But during reception, the antenna's large intercept area enables the same

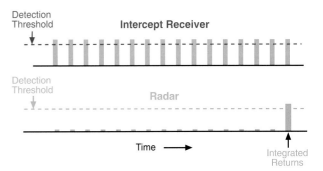

2. Because an intercept receiver must detect individual pulses, it is sensitive only to peak power. Because a radar can coherently integrate the returns it receives, it is sensitive to average power. Consequently, for LPI coherent integration time can be traded for reduced peak power.

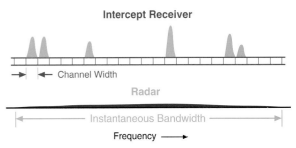

3. Because an intercept receiver must separate pulses closely spaced in frequency, its channels must be comparatively narrow. But a radar's bandwidth is limited only by its design. Consequently, for LPI, bandwidth can be traded for reduced peak power.

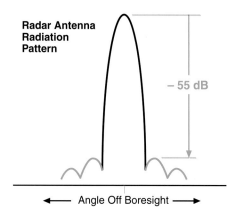

Radar Antenna Radiation Pattern

– 55 dB

← Angle Off Boresight →

4. Since most intercept receivers must rely upon detecting the radar's sidelobe emissions, for LPI the peak sidelobe gain should be down at least 55 dB.

detection sensitivity to be obtained with much lower peak power.

Against those intercept receivers which depend on sensing the radar antenna's sidelobe emissions—ECM system's receiver, ground-based DOA and EL systems, and ARMs— besides having a high gain and large intercept area, the radar antenna has the advantage of a very large difference between mainlobe and sidelobe gains. All of these characteristics can be traded for reduced peak power. While antenna size is generally limited by the dimensions of the aircraft, sidelobe reduction is not. For LPI, the peak sidelobe gain should be down at least 55 dB, relative to the peak mainlobe gain (Fig. 4).

Other Trades for Reduced Peak Power. Other features normally included in a radar to increase detection range that can correspondingly enable peak power to be reduced without reducing range include:

- High duty factor

- Low receiver noise figure

- Low receive losses

Low transmit losses, it might be noted, are of no advantage for LPI. For unless the radar is operating at maximum range, the peak emitted power can be set to the desired level for LPI regardless of these losses.

Special LPI-Enhancing Design Features

Special features which may be used to further enhance LPI include power management, use of wide instantaneous bandwidths, transmission of multiple antenna beams on different frequencies, randomizing waveform parameters, and mimicking the enemy's waveforms. Of these, power management is the most basic.

Power Management. The role of power management is to reduce the radar's peak radiated power to the absolute minimum needed to detect targets of interest at the *minimum* acceptable range, with *minimum* margin. As the radar's targets close to shorter range, the power management system must correspondingly reduce the emitted power (see panel, top of facing page).

The advantages of power management can best be appreciated by considering a simple example. Suppose that to detect a given target at a range of 80 miles, a certain radar must emit a peak power of 5,000 watts. To detect that same target at 5 miles, however, the radar would need to emit a peak power of only 0.076 watts!

POWER MANAGEMENT, PROBLEM 1

5,000 W

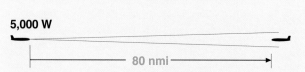

|← 80 nmi →|

Conditions: A certain radar can detect a given target at range R = 80 nmi by emitting a peak power P = 5,000 W

Question: How much power need the radar emit to detect the same target at 5 nmi?

Solution: The required peak power varies as the fourth power of the desired detection range. Therefore,

$$P_2 = P_1 \left(\frac{R_2}{R_1} \right)^4$$

$$P_2 = 5{,}000 \left(\frac{5}{80} \right)^4 = 0.076 \text{ W}$$

AVOIDING DETECTION, THROUGH POWER MANAGEMENT

Superficially, it seems impossible for a radar to avoid being detected by a target that the radar can detect. For the peak power which the radar must transmit to detect the target, P_{det}, is proportional to the fourth power of the target's range.

$$P_{det} = k_{det} \, R^4$$

Yet the peak power, P_{int}, which will enable an intercept receiver in the target to detect the radar is proportional only to the square of the target's range.

$$P_{int} = k_{int} \, R^2$$

However, by trading integration time, bandwidth, antenna gain, duty factor, and receiver sensitivity for peak emitted power, the factor k_{det} can be made very much smaller than k_{int}. As a result, a plot of P versus R for detection of the target by the radar is shifted down so that it intersects the plot of P versus R for detection of the radar by the intercept receiver at a reasonably long range.

The range, $R_{d_{max}}$, at which the two plots intersect—the range for which $P_{det} = P_{int}$—is the *LPI design range*.

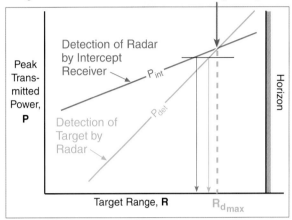

Avoiding Detection. By setting the radar's peak power just below a level corresponding to $R_{d_{max}}$ and progressively reducing it as the target closes to shorter ranges, the radar can avoid being detected by the intercept receiver.

Suppose now that, when the radar was emitting full power, a given intercept system could detect it at 300 miles. When the power was reduced to 0.076 watts, that same intercept system could detect the radar at only 1.2 miles.

Clearly, power management is essential for LPI. Also clear from the foregoing example: the power management system must be able to reduce the emitted power in small, precisely controlled steps over a very wide range—in this hypothetical example, nearly 50 dB.

One point to bear in mind: the interceptability of a given radar depends upon its mode of operation and on the capabilities of the intercept receiving system. Both may vary within any one mission, as well as from mission to mission.

In searching a narrow sector for a designated target at a given range, for example, the peak power may be set so that the radar detects the target without being detected by the target's intercept receiver. Yet in conducting broad area surveillance with the same power setting, the radar may be detected by the intercept receiver of a target of the same type before it closes sufficiently to be detected by the radar.[1]

POWER MANAGEMENT, PROBLEM 2

5,000 W

◀————— **300 nmi** —————

Conditions: When emitting a peak power of 5,000 W, the radar of Problem 1 can be detected by a given intercept receiver at 300 nmi.

Question: At what range can the radar be detected by the same intercept receiver, when emitting a peak power of only 0.076 W?

Solution: Since the signal travels only one way, the intercept range varies as the square root of the peak emitted power. Therefore,

$$R_2 = R_1 \left(\frac{P_2}{P_1} \right)^{0.5}$$

$$R_2 = 5,000 \left(\frac{0.076}{5,000} \right)^{0.5} = 1.2 \text{ nmi}$$

---

1. Detection range in this case is reduced because the radar's beam cannot dwell as long in the target's direction.

Or, a radar might operate at a low enough peak power that its signals would be below the detection threshold of an RWR in a target aircraft. Yet, with that same power setting, the radar might be detected by a ground-based intercept system having a large directional antenna and a highly sensitive receiver.

Wide Instantaneous Bandwidth. A radar's power can be spread uniformly over an extremely wide band of frequencies simply by transmitting extremely short pulses. But, with the desired low peak power, this would result in such low average power that the radar could not detect many targets.

A convenient solution to this dilemma is to transmit reasonably wide pulses and to phase modulate the transmitter with pulse-compression coding.

Pseudo-random codes spread a pulse's spectrum more uniformly than others. A large number of different pseudo-random codes can be easily generated. And they can be made virtually any length (see panel on page 532), enabling virtually any desired bandwidth to be obtained.

The 3-dB bandwidth of the central spectral line of a pulsed signal is:

$$BW_{3dB} \;=\; \frac{1}{\tau} \; \times \; (\text{Pulse Compression Ratio})$$

where τ is the uncompressed pulse width. With 1-ms wide pulses and 2000-to-1 pulse compression coding, for example, a bandwidth of 2 GHz may be obtained. By selecting a suitably high pulse compression ratio, therefore, the emitted signal can be spread over the radar's entire instantaneous bandwidth, which can be made quite broad.

Upon being received by the radar and decoded, target echoes are compressed into narrow pulses providing fine range resolution, and containing virtually all of the received power (Fig. 5). Yet, not knowing the pulse compression code used, an interceptor cannot similarly compress the radar's emitted pulses.

Multiple Beams on Different Frequencies. For any mode of operation in which the radar must search a solid angle of space, the ability to reduce peak power by increasing the coherent integration time is limited by the acceptable scan frame time. Within this limit, however, dwell times may be substantially increased by transmitting multiple beams on different radio frequencies.

Suppose, for example, that a volume, V, expressed in multiples of an angle equal to the radar's 3-dB beamwidth, is to be searched in the time, T. If the search were done

5. By modulating the radar's emitted pulses with pulse-compression coding, their power may be spread over the radar's entire instantaneous bandwidth. When the radar echoes are decoded, they are compressed into narrow pulses containing virtually all of the received power.

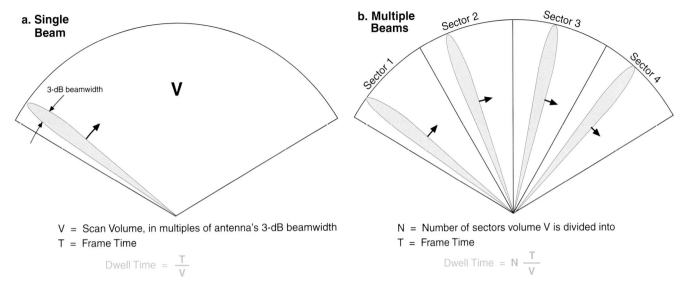

a. Single Beam

3-dB beamwidth

V

V = Scan Volume, in multiples of antenna's 3-dB beamwidth
T = Frame Time

$$\text{Dwell Time} = \frac{T}{V}$$

b. Multiple Beams

Sector 1   Sector 2   Sector 3   Sector 4

N = Number of sectors volume V is divided into
T = Frame Time

$$\text{Dwell Time} = N\frac{T}{V}$$

6. Increase in dwell time achievable by radiating multiple beams on different frequencies. For the same detection sensitivity, as the number, N, of beams is increased, peak power can be reduced by a factor of 1/N.

with a single beam, the maximum allowable dwell time would equal T/V (Fig. 6a).

On the other hand, if this same volume (V) were subdivided into N sectors and every sector were simultaneously searched by a different beam using a different radio frequency (Fig. 6b), the dwell time in each beam direction could be increased by a factor of N. Then, if the coherent integration time were increased to match the dwell time—now equal to NT/V—the peak power emitted in any one beam direction could be reduced by the factor 1/N.

In the extreme, provided adequate processor throughput is available, enough beams might be emitted to completely fill the scan volume (Fig. 7). No scanning would then be needed. Consequently, the coherent integration time could be made equal the total frame time, T.

Multiple beams may also be employed to advantage in other ways. They may, for example, be used to selectively search different portions of the total scan volume. Or, each beam may be used to scan the entire volume on a different frequency, thereby increasing detection sensitivity through frequency diversity rather than through increased integration time.

Random Waveform Parameters. For all practical purposes, in a dense signal environment a signal has not been *usefully* intercepted unless it has been successfully de-interleaved (sorted) and identified (Table 2). Consequently, besides reducing the probability that the radar's signals will be detected by an interceptor, the radar designer has opportunities for confounding the de-interleaving and identification processes, as well.

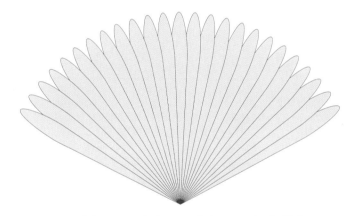

7. Enough beams might be provided to completely fill the scan volume. Then, no scanning would be needed, and the coherent integration time would equal the frame time.

TABLE 2   Basic Intercept Receiver Functions

Detection	Detect single pulses (peak power), with little or no integration.
De-interleaving (Sorting)	Separate pulses of individual emitters, in a dense signal environment.
Identification	Identify emitters by type; possibly even identify specific emitters.

PSEUDO-RANDOM PULSE COMPRESSION CODES

These are binary phase codes which appear to be entirely random in virtually every respect—except for being repeatable. Their advantages:

• A great many different codes can be generated easily and conveniently

• Codes can be made almost any length, hence provide extremely large compression ratios.

The codes are commonly generated in a shift register having two or more feedback paths.

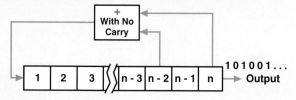

Filled initially with 1s or 1s and 0s, the register produces a code of 1s and 0s of length

$$N = 2^n - 1$$

where N is the number of digits output before the code repeats, and n is the number of digits the register holds.

An 11-digit register with the 9th and 11th digits fed back to the input, for example, produces a code 2,047 digits long. By changing the feedback connections, 176 different codes of that length can be produced .

The 0s and 1s in the code specify the relative phases—0° and 180°—for successive segments of the radar's transmitted pulse.

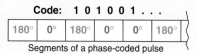

Segments of a phase-coded pulse

By shifting the register at intervals equal to the desired length of the segments, successive output digits can directly control the phase modulation of the radar signal.

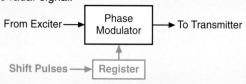

When the received pulse is decoded, the segments are superimposed, producing a pulse roughly N times the amplitude of the uncompressed pulse and only a little wider than the segments. The code generated by the 11-digit register of this example would thus yield a pulse compression ratio of roughly 2,000 to 1.

Among the waveform parameters typically used for both de-interleaving and classification are:

• Angle of arrival

• Radio Frequency

• PRF

Among those parameters typically used for classification alone are:

• Pulse width

• Scan rate

• Intrapulse modulation

• Interpulse modulation

• Beam width

• Signal polarization

Except for angle of arrival, all of the above-listed parameters can be varied randomly from one coherent integration period to the next.

Variations can be achieved without reducing detection sensitivity by taking advantage of the waveform agility available in modern airborne radars. Moreover, with two or more aircraft operating cooperatively—i.e., alternately providing target illumination for each other (Fig. 8)—even angle of arrival can be varied.

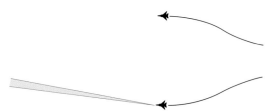

8. By cooperatively shifting radar transmission randomly between them, two or more aircraft can even vary the angle of arrival of their emissions.

Randomizing any of the parameters can confuse the classification process. That is particularly true for those intercept systems which classify signals by comparing their parameters with parameters stored in threat tables.

Mimicking Enemy Waveforms. Mimicking may also confuse signal classification. To be able to mimic an enemy's waveforms, though, the radar must not only have considerable waveform agility, but be able to operate over the full range of radio frequencies the enemy employs.

Cost of LPI

LPI techniques are not free; each of the LPI-enhancing features adds to the radar's cost. Most increase the costs of both software and hardware.

But by far the greatest cost of LPI is in digital processing throughput. As instantaneous bandwidth is increased, for instance, the required throughput goes up proportionately because of the increased number of range bins whose contents must be processed.

For, to the extent that bandwidth is increased through pulse compression coding, the wider the bandwidth, the narrower the compressed pulses will be, hence the more range bins required to cover the same range interval. Throughput similarly goes up with the number of simultaneous beams radiated.

To support a wide instantaneous bandwidth and a few simultaneous beams, the required throughput is staggering (Fig. 9). In fact, not until the 1990s were these features even deemed practical. With the dramatic advances being made in digital processor technology, however, the costs of these features are rapidly decreasing.

Be that as it may, in any discussion of costs one important fact must be borne in mind. With the exception of power management, virtually all of the LPI features maximize detection sensitivity greatly moderating the cost of LPI in performance.

Moreover, in those situations where the advantages of maximum detection range and situation awareness outweigh the advantages of LPI, the operator always has the option of overriding power management, operating the radar continuously, and searching the antenna's entire field of regard.

Possible Future Trends in LPI Design

Looking to the long-term future, one thing is certain: competition between radar designer and intercept receiver designer will never be static. For every improvement in LPI, improvements in intercept receiver design can be expected. LPI designers will continue to exploit coherent processing, which the intercept receiver cannot duplicate. And designers of intercept receivers will continue to exploit the R^2 advantage of one-way versus two-way propagation.

Probably, the most spectacular gains in both LPI and intercept receiver design will occur in signal processing, which is the subject of the next chapter.

Summary

There are four generic types of intercept receivers: radar warning receivers (RWR); intercept receivers of ECM systems; ground-based passive-detection systems (DOA and EL); and ARMs. RWRs typically detect only mainlobe radiation; the others, sidelobe radiation.

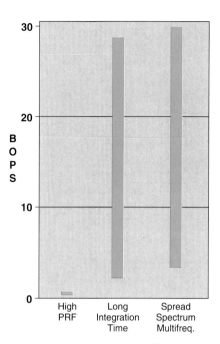

9. Ranges of throughputs in billions of computer operations per second (BOPS) required for long integration time and spread-spectrum/multifrequency operation. High PRF is shown for comparison.

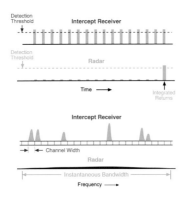

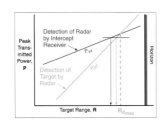

Operational strategies for LPI include limiting radar "on" time, using collateral intelligence and reconnaissance information wherever possible, relying heavily on onboard passive sensors, and searching only narrow sectors in which they indicate the target to be.

LPI design strategies capitalize primarily on the intercept receiver:

- Having to detect individual pulses, so that it can de-interleave them and identify their sources

- Having limited channel widths, so that the receiver can separate closely spaced signals.

Consequently, LPI can be enhanced by trading both long coherent integration time and wide instantaneous bandwidth for reduced peak power.

High antenna gain, reduced sidelobe levels, high duty factor, and increased receiver sensitivity can likewise be traded for reduced peak power.

LPI can be further enhanced by several special features. First among these is power management—keeping the peak emitted power just below the level at which it can be usefully detected by an intercept receiver in an approaching aircraft, yet just above the level at which the radar can detect the aircraft.

Added to this feature are (a) using extremely large amounts of pulse compression to spread the radar's signals over an exceptionally wide instantaneous bandwidth; (b) simultaneously transmitting multiple beams on different frequencies to reduce the constraint imposed on integration time by limits on scan-frame-time; (c) randomly changing waveform characteristics to confound the intercept receiver's signal de-interleaving and identification process; and (d) mimicking enemy waveforms.

The principal cost of LPI is greatly increased signal processing throughput.

Advanced Processor Architecture

43

Having read of the many advanced radar techniques in the offing, processor architecture may seem of little import. But the fact is that most of the advanced capabilities of airborne radars to date have only been made practical by substantial increases in digital processing throughput (Fig. 1).

In the 1970s, multimode operation was made possible in fighters by replacing the hardwired FFT processor with a *programmable signal processor (PSP)* having a throughput of around 130 MOPS.[1] In the 1980s, the addition of real-time SAR was made possible by quadrupling processing throughput. In the 1990s, the active ESA and other advanced capabilities of the F-22 were made possible by again quadrupling throughput.

Vastly higher throughputs will be needed to make practical some of the advanced radar capabilities currently envisioned. Spread spectrum, for example, is highly desirable for both ECCM and LPI. Yet, even a 500 MHz instantaneous bandwidth will require 500,000 MOPS.

In this chapter, we'll examine the key architectural features of the late 1990s-era processors: parallel processing, high throughput density, efficient modular design, fault tolerance, and integrated processing. We'll then take stock of a few technology advances which promise substantial throughput increases in the future.

Parallel Processing

To meet radar throughput requirements, two levels of parallel processing are typically employed: at the signal-processing *element* level, pipeline processing; at the processing *system* level, distributed processing.

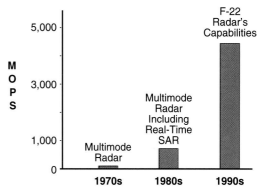

1. Growth of radar capabilities made possible by increases in processor throughput.

1. MOPS = Million operations per second.

PIPELINE PROCESSING

Each of the complex numbers to be processed, together with the multistep instruction for processing it, is sequentially loaded into a multistage register and shifted down, one stage at a time, by successive clock pulses. In each stage one step of the instruction is executed.

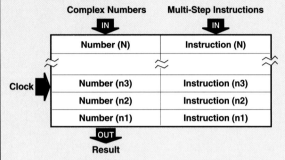

Once the pipeline is full, one butterfly or other algorithm is completed in every clock time.

Assuming a clock rate of 25 MHz and a 10 stage pipeline, such as might be provided for the FFT butterfly, the throughput would be:

10 operations x **25** million/sec. = **250 MOPS**

2. The technique is applicable to performing any series of additions, subtractions and multiplications of real, complex, or floating-point numbers.

3. The computations called for in the butterfly algorithm are detailed on pages 273 and 277.

Pipeline Processing. This technique was devised in the early days of digital signal processing to enable a *programmable* machine to perform the vast number of arithmetic operations required for doppler filtering fast enough to process radar returns in real time.[2]

The technique is implemented with a multistage register plus associated arithmetic elements (see panel, left). Keyed by successive clock pulses, each number to be processed—together with the multistep instruction for processing it—is sequentially loaded into the register's first stage.

The numbers are shifted down, a stage at a time, by successive clock pulses. In the first stage, the first step of the instruction is carried out. In the second stage, the second step; and so on.

The number of stages is the same as the number of individual processing operations necessary to execute the instruction for which the pipeline is designed. That number can vary from 2, for a very simple algorithm, to 8 or 10 for an FFT butterfly.[3] Once the pipeline is filled, one butterfly (or equivalent) may be computed in every clock time.

The increased throughput thus realized may be multiplied many times by distributing processing tasks among multiple processing elements (PEs) operating in parallel.

Distributed Processing. Both throughput and interconnect bandwidth increase directly with the number of PEs. Consequently, as throughput requirements have increased, the number of PEs used in airborne systems has been increased from three or four (see panel below) to a hundred or more, and no end is in sight.

SIMPLE DISTRIBUTED PROCESSING EXAMPLE

Distribution of processing tasks for parallel execution of a range-gated tracking mode by a processor employing one general purpose processing element—Array Controller—and three identical Signal Processing Elements.

Arrows indicate flow of data, e.g., Job 4 (assigned to Element 1) and Job 2 (assigned to Element 2) must be completed before Job 7 can be performed. Jobs 3 and 5 must be completed before Job 8 can be performed.

In one of many possible implementations (Fig. 2), communication between PEs is provided via two-dimensional mesh connections.

In another, the PEs are interconnected via nonblocking crossbar switches (Fig. 3). Large distributed systems of this type have been used to perform the billions of floating-point computations required in ultra-fine-resolution SAR applications.[4]

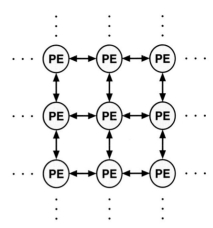

2.   One of many practical distribution schemes. Processing elements (PEs) are interconnected in a two-dimensional mesh. Pattern can be expanded in either dimension to accommodate more PEs.

4.   They have also been used in electro-optical applications.

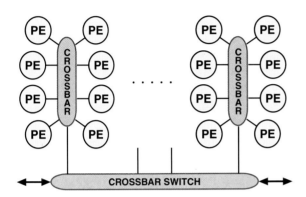

3.   Another distribution scheme. Same PEs as shown in Fig. 2 are clustered around crossbar switches. Number of PEs can be increased by adding more clusters.

Since the space available for avionic equipment in a high-performance military aircraft is limited, the maximum realizable throughput depends largely on how high the processor's throughput density is.

Achieving High-Throughput Density

Throughput density is a processor's maximum throughput divided by the volume of the processing hardware. High density is achieved primarily by implementing the processor with very large-scale integrated circuits (VLSIs).

Types of VLSIs Used. The VLSIs used are generally of three standard types:

- RISC[5] microprocessor chips

- Random-access memory (RAM) chips

- Programmable logic chips

5.   RISC stands for reduced instruction set computer.

plus custom designed signal processing and interface chips, called gate-arrays.

Gate Array Chips. Early in the era of VLSIs, the gate array was conceived as an economical means of (a) easing the limitation that defects impose on the maximum practical size of an integrated circuit and (b) producing affordable complex signal processing circuits for which there may be only a limited market. The basic building blocks of these circuits are logic gates.

THE GATE: Signal Processor Building Block

A gate is a digital circuit that performs a logic function. It may have one or more inputs, but only one output. Inputs and outputs are voltages of opposite polarities: "+" representing binary 1, and " – " representing binary 0.

Gates may be represented either by graphic symbols or by simple equations whose terms have values of 0 or 1 and whose connectives have special meanings., e.g., a "+" means "or"; a " • " means "and"; and a bar " ‾ " over a term means "not".

A gate's functions are defined by a truth table. It indicates what the gate's outputs will be for all possible combinations of inputs.

The most commonly used gates are the "and", "or", and "not" (inverter) gates.

An **AND** gate produces an output of 1 only if both inputs, A **and** B, are 1s. Otherwise, it produces an output of 0.

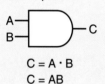

A	B	C
0	0	0
1	0	0
0	1	0
1	1	1

$$C = A \cdot B$$
$$C = AB$$

An **OR** gate, on the other hand, produces an output of 1 if A, or B, or both A and B are 1s.

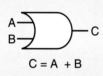

A	B	C
0	0	0
1	0	1
0	1	1
1	1	1

$$C = A + B$$

A **NOT**, (inverter) gate produces an output of 0 if its single input is 1 and an output of 1 if its single input is 0.

A	C
0	1
1	0

$$C = \overline{A}$$

There are many other gates. But all can be produced by combinations of just one: the NAND (not and) gate, commonly used in CMOS circuitry.

A **NAND** gate produces an output of 1 only if both inputs are **not** 1. Otherwise its output is 0.

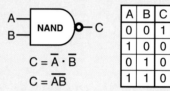

A	B	C
0	0	1
1	0	0
0	1	0
1	1	0

$$C = \overline{A} \cdot \overline{B}$$
$$C = \overline{AB}$$

Consequently, if both inputs are tied together, an input of 0 produces an output of 1, and an input of 1 produces an output of 0. The gate acts as a **NOT** gate, or inverter.

NOT
(Inverter)

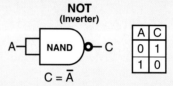

A	C
0	1
1	0

$$C = \overline{A}$$

If the output of a NAND gate is connected to both inputs of a NAND gate, therefore, the output is inverted. The two gates form an **OR** gate.

OR

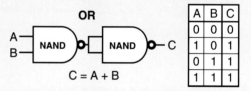

A	B	C
0	0	0
1	0	1
0	1	1
1	1	1

$$C = A + B$$

If the inputs to a NAND gate are similarly inverted, the three gates form an **AND** gate.

AND

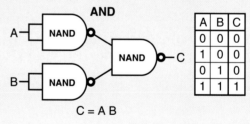

A	B	C
0	0	0
1	0	0
0	1	0
1	1	1

$$C = A B$$

As illustrated in the panel above, logic gates are of many different types. All possible types, however, may be produced with various combinations of just one: the "NAND" (not and) gate. Therefore, if an array of a great many NAND gates is produced on a single semiconductor chip, by appropriately interconnecting the usable gates when the interconnection layers are added, very large custom circuits can be economically produced.

Moreover, by implementing the gates with CMOS[6] technology (see panel facing page), relatively simple semiconductor processing can be employed. This leads to a low percentage of defects, hence high yields, making practical

6.   CMOS is a combination of MOSFETs (metal oxide silicon field effect transistors) having *complementary* characteristics.

CMOS: Key To Practicality Of Exceptionally Large Gate Arrays

CMOS is a combination of MOSFETs (Metal Oxide Silicon Field Effect Transistors) having complementary characteristics. Because MOS-FETs can be produced with relatively simple semiconductor processing, they make practical producing exceedingly large numbers of gates on a single semiconductor chip.

This panel explains what MOSFETs are, what their complementary characteristics are, and how they may be interconnected to form a NAND gate, with which all other gates may be produced.

A MOSFET is produced by heavily doping a lightly doped region on the surface of a silicon crystal substrate, to produce a channel of high conductivity. Centered over this channel is a tiny metal plate— called a "gate"— insulated from the crystal by an extremely thin layer of oxide.

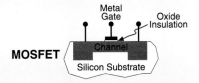

Terminals are provided at both ends of the channel; a terminal for a control voltage, on the gate.

Two complementary doping schemes are used. In one, N-type doping—which produces free negative charge carriers (electrons)—is used for the channel and P-type doping—which produces free positive charge carriers (holes)—is used for the substrate.

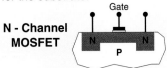

At the channel's lower edge, holes and electrons combine, depleting the number of free carriers there, and narrowing the channel .

If a negative voltage corresponding to a binary digit is applied to the gate, it attracts more holes from the substrate. They combine with more free electrons, narrowing the channel sufficiently to pinch it off, so no current can pass through.

If a positive voltage corresponding to a binary digit is applied to the gate, it repels the holes in the substrate, widening the channel and maximizing its conductivity.

The gate thus acts as a switch, which is closed by a positive control voltage and opened by a negative one.

Equivalent Circuit, N-Channel MOSFET

The other doping is P for the channel and N for the substrate.

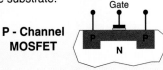

With it, the control voltage has the opposite effect. A positive voltage opens the switch; a negative, closes it.

Equivalent Circuit, P-Channel MOSFET

A NAND gate may be constructed by interconnecting 2 N-channel and 2 P-channel MOSFETs, as shown below. When inputs A and B are positive, both P-channel switches open, disconnecting the positive supply voltage.

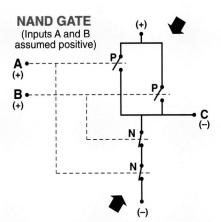

And the two N-channel switches close, connecting the negative supply voltage to the output, C.

When A **or** B is negative, at least one of the two P-channel switches closes, connecting the positive supply voltage to C.

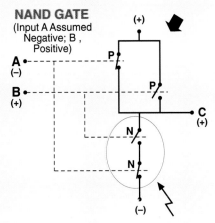

And at least one of the two N-channel switches opens, disconnecting the negative supply voltage.

the production of integrated-circuit chips containing hundreds of thousands of usable NAND gates.

The VLSIs used in a processor are mounted on plug-in circuit boards, called modules. Their makeup and electrical grouping are crucial to the processor's efficiency.

Efficient Modular Design

The goal here is to implement the processor with standard modules while— in the interest of low cost and ease of logistic support—keeping the number of different types of modules to a minimum. For a radar, most of the processor's modules are of four basic types:

- General-purpose processing
- Signal processing
- Global bulk memory
- Interface

To these may be added a relatively small number of special-purpose modules. Depending on the physical design standard chosen, one or more PEs of the same type may be included in a single module.

For operational simplicity and convenience in programming, the PEs are grouped electrically—though not necessarily physically—into clusters of various types.

One Approach to Clustering. A generic cluster of one general type is illustrated in Fig. 4. It consists of a cluster-control element and an appropriate mix of general-purpose PEs and signal-processing PEs, all sharing a multiport global bulk memory. The cluster controller, itself, is a general-purpose PE. A control bus provides low-latency control paths from the cluster controller to each PE and the bulk memory. Low-latency paths are also provided between each PE and the bulk memory.

Up to a limit imposed by electrical considerations, as many clusters of this sort may be included in a processor as are necessary to meet processing requirements (Fig. 5).

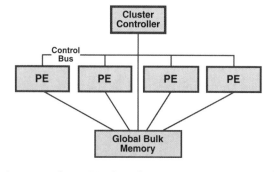

4. A generic cluster. Number of PEs may vary. Array controller is a general-purpose PE. Each of the others may be either general purpose or signal processing. Since all are not physically collocated, this is considered a virtual cluster.

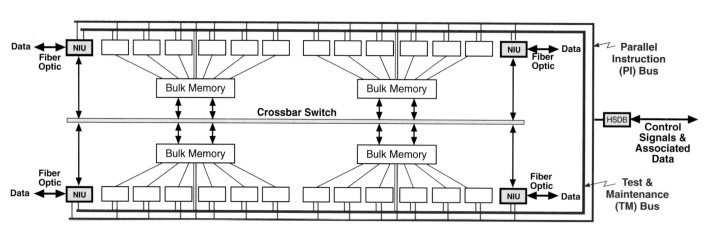

5.   Generic four-cluster processor. Crossbar is implemented with gate array chips and modularized.

Data, in the form of labeled messages, is exchanged between the bulk memories of all clusters through a non-blocking crossbar switch, providing point-to-point high-speed connections. The crossbar is implemented with gate-array chips and modularized.

Input and output data is transmitted between each cluster and the radar via a *high-speed fiber-optic data bus (HSDB)*. A module called a *network interface unit (NIU)* links each cluster to the bus.

All PEs are linked together by two dual-redundant buses. One is a parallel instruction (PI) bus, through which control signals and associated data are received from the other avionics subsystems. The other bus is a test and maintenance (TM) bus, through which system-status and self-test or reconfiguration instructions are conveyed.

Another Clustering Approach. Another approach to mixing general-purpose processing, signal processing, memory, and input/output elements is illustrated in Fig. 6. In it, a control bus is connected to all modules. A high-speed bus/crossbar switch is connected between the memory and the signal processing and input/output modules. This architecture is typically used in processors which are implemented with *commercial off-the-shelf (COTS)* hardware (see panel, below right).

Advantages. Whether custom or commercial hardware is used, modular architectures of both general types have three main advantages. They give multiple PEs ready access to the same stored data, simplify logistic support, and greatly facilitate achieving fault tolerance.

Fault Tolerance

Nothing could be more disconcerting to a flight crew in the crucial stage of an engagement than to have their radar abruptly shut down because of a failure in the radar's digital processor. Consequently, a processor is typically designed so that if failures occur the processor will continue to perform all its basic functions. This goal is achieved in three basic ways:

- Building into every module a comprehensive *built-in self-test (BIST)* capability that verifies the operation of every circuit and every connector in the module

- Employing a distributed operating system, which enables each module to continue its normal operation, including BIST, even if other modules fail

- Providing a processor-wide fault management system

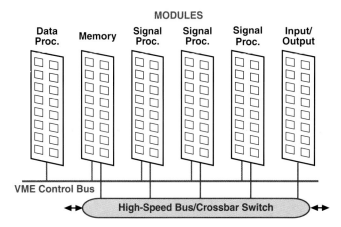

6. Cluster architecture typically used in processors implemented with COTS hardware.

COTS HARDWARE
Key to Lower Cost & Shorter Development Time

In the early days of digital processing, developers of airborne processors had little choice but to build their systems with medium-scale integrated circuits or custom-designed VLSIs. With the explosive growth of commercial digital applications, this is no longer so.

Today, a wide choice of high-quality commercial hardware is available with which to assemble compact, high-throughput distributed processors—VLSI microprocessors, crossbar switches, interfaces, buses, back-planes, etc.—all manufactured to the International Trade Association's exacting VME standards.

With COTS, orders of magnitude reductions may be achieved in cost and development time. But because of issues of operating temperature, ruggedization, and reliability, COTS has so far fallen short of meeting stringent military requirements—especially in processors for tactical aircraft.

In time, these issues may be resolved, and use of COTS hardware allowed in such applications.

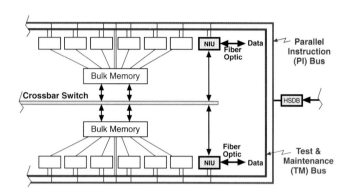

Fault management begins at startup, with the autonomous self-testing of each module. For this level of testing, a test control chip and a dedicated nonvolatile memory may be included in every module. Run at full speed, the tests generally can detect 99% of all possible faults.

In the architecture of Fig. 5 (repeated at left), according to a preprogrammed hierarchy of authority, one of the cluster controllers—having verified its operation—assumes mastership of the TM bus.

Thereafter, through an application program, that PE controls all processor-wide BIST, receives test results from all modules, filters them to eliminate false alarms, and permanently records the results.

During normal operation, status tests are performed continuously in all modules on a noninterference basis. If a failure occurs in any PE, the software automatically switches it out of its cluster and dynamically reallocates its tasks to another PE in the same or another cluster, selected so as to minimize degradation of performance.[7]

7. Greatly facilitating this task are the multiplicity of standard PEs, dual control paths, multiple memory ports, and separation of data and control paths.

Integrated Processing

The kind of digital processing required by a military aircraft's radar subsystem is virtually the same as that required by its electronic warfare (EW) subsystem and its electro-optical (EO) subsystems. Moreover, these subsystems have become increasingly interdependent.

There is little reason, therefore, to provide three separate processors for them—all employing the same kinds of control, data transfer, fault management, voltage-regulated power, cooling, and housing. By supporting the entire suit with a single integrated processor, the number of modules and external interconnections may be reduced, and a substantial saving in size and weight realized.

F-22 Example. An excellent example of integrated processing is the F-22 fighter's common integrated processor (CIP). Two CIPs serve the radar, electro-optical, and electronic warfare subsystems plus the balance of the avionics.

UAV Example. Another striking example is the integrated processor for the U.S. military's Global Hawk *unmanned aerial vehicle (UAV)*. This processor will serve a suite of radar, electro-optical, and IR sensors providing high-quality reconnaissance imagery in exploitable form via satellite directly to users in the field.

Employing 144 distributed processing elements, interconnected via a crossbar switch, the processor provides a throughput of 11.5 GFLOPS[8] for SAR imaging, plus eight BOPS for SAR and EO/IR image compression. In the inter-

7.   The Global Hawk UAV produces real-time high-resolution SAR, electro-optical, and IR imagery.

8.   1 GFLOPS $= 10^9$ floating point operations per second.

est of affordability, the processor is implemented entirely with ruggedized COTS hardware.

Providing Data Security. Along with its many advantages, integrated processing has created a new requirement: data security. The subsystems on a given platform invariably are subject to different military security restrictions. Unauthorized access to classified data in the processor's memory, therefore, must be prevented. This requires a secure operating system, as well as special hardware design features, primarily affecting memory access.

Advanced Developments

Projecting advances in digital-processing technology even a year or two ahead is risky. And projecting advances much beyond that borders on fanciful speculation. Yet, substantial increases in throughput may be realized through a combination of higher clock rates, higher density gate arrays for custom designs, and massive parallel processing, to name a few.

Higher Clock Rates. Throughput increases directly with clock rate. In the commercial world, clock rates have increased spectacularly. But in airborne military applications such as we're considering here, the increase has been far slower.

As a rule, these processors have employed synchronous data transfers,[9] which require all processing elements to operate at the same clock rate. That rate is limited by the time required for data to flow through the *slowest* of the thousands of different paths in the processor. This limitation may be avoided by operating asynchronously. Each processing element then has its own clock, and its rate can be made as high as the element's own circuitry will allow.[10]

Higher Density Gate Arrays. The number of gates that may reasonably be included in a single CMOS array is limited primarily by the feature size of the transistors. Over the years, feature size has been progressively decreased to small fractions of a micron. Yet, for that trend to extend much further, certain practical problems must be surmounted.

One of these is cooling. The denser the circuitry on the chip, the greater the amount of heat that must be removed to allow high processing speeds and ensure reliability. Several potential techniques are listed in Fig. 8.

Of these, the most efficient, but most expensive, is use of synthetic diamond either as a film deposited on a silicon substrate or as the substrate itself. Diamond is the best thermal conductor of any known electrically nonconducting material and has the added advantage of enhancing radiation hardening.

9. Synchronous operation circumvents several problems, such as instability due to electrical noise. But with technology advances, these problems have gradually disappeared.

10. This technique is commonly employed in commercial hardware, including products designed to comply with VME standards.

Conventional
- Air flow through cooling
- Liquid flow through cold plates
- Conduction cooled

More Efficient Possibilities
- Heat pipes
- Superconducting ceramic substrates
- Composite materials for heat sinks
- Localized thermoelectric cooling

Potentially Most Efficient
- Synthetic diamond films
- Synthetic diamond substrates

---

8. Cooling options. As circuit densities increase, more efficient cooling must be provided to ensure reliability. Going down the list, costs go up.

THE STORY ON SOFTWARE

The dramatic advances in digital processing hardware have been matched by similar advances in development of software for these machines.

In the late 1960s, when digital signal processing was introduced in radars for fighters, a general-purpose "data" processor was provided to control the radar's operation and so lighten the demands on the pilot. But processing speeds were so low and memory so limited that a hard-wired processor had to be provided for signal processing.

To run predictably in real time and fit within the available memory, software for the data processor had to be written in assembly language. The size of the program was limited by what could be executed in real time at the processor's limited speed and would fit in its memory. Consequently, even with a minimal tool set, the programming team could fully understand and readily develop the code.

By the mid 1970's, throughputs had increased enough to enable programmable signal processing, but only with a specialized pipeline processor.

Today, however, with dramatically increased throughput densities, reduced memory cost, and use of large-scale parallel processing and shared memories, it is possible to perform both signal and data processing in real-time with ruggedized commercial processors. Software for them can be developed in higher order languages, such as ADA and C. And the software can be compiled, checked out, and operated in a simulated environment with a full set of universally used, fully tested and supported, commercial software support tools.

Also available now are universally used operating systems—such as Unix-based systems—that can meet true real-time multiprocessing requirements.

Moreover, standardization of languages, tools, and operating systems has

- Facilitated the combining of multiple programs from different sources
- Allowed reuse of software from other applications
- Enabled insertion of commercial off-the-shelf (COTS) software packages

Currently emerging are automated software generating tools, which promise still further savings in software cost and development time.

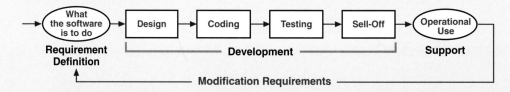

Typical software life cycle. *Today, the development process is automatically controlled by such tools as the Computer Aided Software Environment (CASE).*

Massive Parallel Processing. An order of magnitude increase in throughput may be realized with this technique, as demonstrated by a programmable module called SCAP designed to enable a processor to handle tasks requiring extraordinarily low latency and high throughput. SCAP is particularly adept at matrix operations, such as required for space time adaptive processing.

Comprised of a very large array of mesh-connected PEs, a single SCAP module has a throughput of 3.2 GFLOPS and a throughput density of 100 GFLOPS per cubic foot.

Conclusion

Add up the benefits of higher clock rates, deep submicron feature size, superior cooling and massive parallel processing, and it may be possible to meet the demanding processing requirements of the future.

PART X

Representative Radar Systems

GLOBAL HAWK, UNMANNED AERIAL VEHICLE (UAV)

This reconnaissance aircraft can fly 4,000 miles from its home base, loiter for 24 hours at an altitude of 65,000 feet, and return home without refueling. Equipped with a SAR radar capable of 1-foot resolution plus high-resolution electro-optical sensors, it can return sharp images virtually instantaneously either directly to ground forces or via satellite links to its home base. Missions can be completely preprogrammed or modified in flight.

Reconnaissance & Surveillance

E-2C Hawkeye

The APS-145 is the latest version of the Airborne Early Warning radar for the US Navy's carrier based E-2C Hawkeye. Early versions of the aircraft went into service in 1963. Since then, it and the radar have undergone numerous upgrades.

Designed for operation over both land or sea, the APS-145 can provide surveillance over 3,000,000 cubic miles of air space and can simultaneously monitor and track up to 2,000 targets. Looking beyond the horizon, it has a maximum detection range of 350 nmi.

Implementation. Operating at UHF frequencies (0.3 to 1.0 GHz) to minimize sea clutter, the radar employs a linear array of yaggi antennas, having monopulse sum and difference outputs. This array is housed in a 24-foot-diameter rotating radome, called a rotodome, which rotates at 5 rpm.

The transmitter is a high-power coherent master-oscillator power-amplifier (MOPA). It is switched through three different PRFs to eliminate doppler blind zones and employs linear frequency-modulation (chirp) pulse compression.

Adaptive Signal Processing. By means of DPCA and a double-delay AMTI clutter canceller, mainlobe clutter is eliminated, thereby avoiding the problem of low-closing (or opening) rate targets being obscured by spreading of the clutter spectrum due to the aircraft's advance when looking in broadside directions (normal to the

aircraft velocity). Clutter cancellation is followed by coherent signal integration with the FFT. To minimize false alarms, the detection threshold is adaptively adjusted—resolution-cell by resolution-cell—in accordance with the clutter level and the density of targets.

DPCA. The phase center of a side-looking planar array can be displaced forward or aft in the plane of the array by adding or subtracting a fraction of the monopulse difference signal, in quadrature, from/to the sum signal.

In the E-2C, following transmission of the first pulse of each successive pair of pulses, the phase center is displaced forward by the distance the aircraft will advance during the interpulse period. Following transmission of the second pulse, the phase center is displaced aft by the same amount. As a result, both pulses will travel the same round-trip distance to any point on the ground,

Carrier-based E-2C Hawkeye can monitor 3,000,000 cubic miles of airspace and simultaneously track 2,000 targets.

making possible complete cancellation of the ground return by the clutter canceller.

This process, of course, requires precise synchronization with the PRF, the velocity of the aircraft, and the rotation and look angle of the antenna.

Hawkeye's 24-foot diameter rotodome houses a monopulse connected linear array of UHF yaggi antennas, and rotates at 5 rpm.

AWACS installed in an E-3.

The AWACS antenna consists of a stacked array of 28 slotted waveguides, plus 28 reciprocal ferrite elevation-beam-steering phase shifters and 28 low-power nonreciprocal beam offset phase shifters.

E-3 AWACS Radar

The APY-2 is the radar for the U.S. Air Force E-3 Airborne Early Warning and Control System (AWACS). From an operational altitude of 30,000 ft, the radar can detect low altitude and sea-surface targets out to 215 nmi, coaltitude targets out to 430 nmi, and targets beyond the horizon at still greater ranges.

Implementation. Operating at S-band frequencies (nominally 3 GHz), the radar employs a 24-ft by 5-ft planar-array antenna, steered electronically in elevation, and housed in a rotodome which rotates at 6 rpm.

Besides phase shifters for elevation beam steering, phase shifters are also provided for offsetting the beam for reception during elevation scanning to compensate for the time delay between transmission of a pulse and reception of returns from long-range targets. The antenna has an extremely narrow azimuth beamwidth and is amplitude weighted for sidelobe reduction.

The transmitter chain consists of a solid-state predriver—whose output power is increased as a function of antenna elevation angle—a TWT intermediate power amplified, and a high-power pulse-modulated dual-klystron amplifier. For reliability, dual redundancy is employed throughout.

Following an extremely low-noise (HEMT) receiver preamplifier, two separate receive channels are provided: one for range-gated pulse-doppler operation; the other, for simple pulsed-radar operation.

Digital processing is performed by a signal processor, employing 534 pipeline gate arrays operating at 20 MHz; and a data processor, employing four RISC CPUs.

Modes of Operation. The radar has four primary modes of operation:

• *High-PRF pulse-doppler range-while-search*, for detecting targets in ground clutter;

• *High-PRF pulse-doppler range-while-search*, plus elevation scanning for additional elevation coverage and measurement of target elevation angles

• *Low-PRF pulsed radar search with pulse compression*, for detecting targets at long ranges beyond-the-horizon, where clutter is not a problem

• *Low-PRF pulsed radar search* for detecting surface ships, featuring extreme pulse compression and adaptive processing that adjusts for variations in sea clutter and blanks land returns on the basis of stored maps.

These modes can be interleaved to provide either all-altitude long-range aircraft detection or both aircraft and ship detection. A passive mode for detecting ECM sources is also provided.

Each 360° azimuth scan can be divided into up to 32 different sectors, in each of which a different operating mode and different conditions can be assigned or changed from scan to scan.

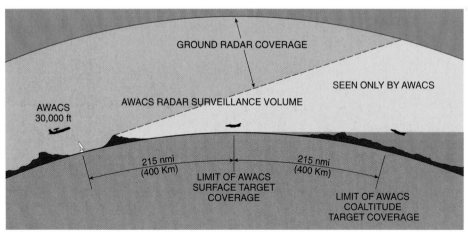

From an altitude of 30,000 ft., AWACS can detect sea and low-altitude targets out to 215 nmi and coaltitude targets out to 430 nmi.

Joint STARS

Joint STARS is a long-range, long-endurance, air-to-ground surveillance and battle management system carried aboard the U.S. Air Force E-8C aircraft. Operating at altitudes up to 42,000 feet, the system's high-power pulse-doppler radar is capable of looking deep behind hostile borders from a stand-off position and monitoring fixed and moving targets with a combination of high-resolution SAR mapping and moving-target indication (MTI) vehicle detection and tracking.

Implementation. The radar employs a 24-foot-long, roll-stabilized, slotted-waveguide, side-looking passive ESA, housed in a 26-foot-long radome carried under the forward section of the fuselage. The antenna is steered electronically in azimuth and mechanically in elevation.

Digital processing is performed by multiple signal processors. The radar data and signal processors are controlled by a VAX-based distributed processing system that includes individual digital processors at each of 17 operator work stations and one navigator/operator work station. All are readily accommodated in the E-8C's 140-ft-long cabin.

To separate targets having very low radial speeds from the accompanying mainlobe clutter, the displaced phase center technique described in Chap. 24 is used. To enable the targets' angular positions to be precisely determined, the antenna is subdivided lengthwise into three segments, also as described in Chap. 24.

Modes of Operation. The radar has three primary modes of operation:

- High-resolution SAR imaging, for detecting and identifying stationary targets

- Wide-area MTI surveillance, for situation awareness

- Sector MTI search, for battlefield reconnaissance.

As the name implies, the MTI modes are used to locate, identify, and track moving targets. When a vehicle that is being tracked stops, the radar can almost instantly produce a high-resolution SAR image of the vehicle and the surroundings.

All three modes may be flexibly selected or interleaved.

Targets detected with MTI are displayed as moving images. These can be superimposed on digitally stored maps or on the radar's SAR maps. And they can be stored and replayed at selectable speeds.

An operator can individually distinguish the vehicles in a convoy, even determine which vehicles are wheeled and which are tracked. If a vehicle stops, a SAR map showing it and the immediately surrounding area can be almost immediately produced.

Encrypted, the radar data may be relayed by a highly jam-resistant data link to an unlimited number of Army ground control stations.

Three-segment passive ESA, electronically steered in azimuth and manually steered in elevation, is housed in the 26-foot-long radome under the forward section of the fuselage.

Navigator's workstation (left) is one of 18 operator workstations. Each is equipped with a digital processor which is included in Joint STARS' VAX-based distributed processing system.

Fighter & Attack

F-22 Stealth Fighter

F-22 (APG-77)

The APG-77 is a multimode pulse-doppler radar meeting the air dominance and precision ground attack requirements of the F-22 stealth dual-role fighter. It may be armed with six AMRAAM missiles or two AMRAAMs plus two 1,000-pound GBU-33 glide bombs, two sidewinder IR missiles, and one 20-mm multi-barrel cannon—all of which are carried internally for low RCS. Four external stations are also available to carry additional weapons or fuel tanks

At present, very little can be said at an unclassified level about the radar other than that it employs an active ESA, that it incorporates extensive LPI features, and that its signal and data processing requirements are met by a common integrated processor (CIP).

The active ESA provides the frequency agility, low radar cross-section, and wide bandwidth required for the fighter's air dominance mission.

Two CIPs perform the signal and data processing for all of the F-22s sensors and mission avionics, with processor elements of just seven different types. One serves the radar, electro-optical, and electronic warfare subsystems; the other, the remaining avionics. Both have identical back planes and slots for 66 modules. Initially only 19 slots were filled in CIP 1 and 22 in CIP 2, leaving room for 200% growth in avionics capability.

The active ESA employed by the APG-77 to meet low RCS requirements provides extreme beam agility and supports enumerable growth features.

F-16 C/D (APG-68)

The APG-68 pulse-doppler radar meets the all-weather air superiority and air-to-ground strike requirements of the F-16 C/D fighter. Employing both head-up (HUD) and cockpit displays, it provides the easy hands-on, head-up operation essential for situation awareness in a one-man fighter.

Implementation. Consisting of four air-cooled line-replaceable units—antenna plus low-power RF unit, transmitter, and processor—it weighs 379 lbs., has a predicted mean time between failures of 250 hours, and a mean time to repair on the flight line of 30 minutes.

The antenna is a planar array, mounted in azimuth and elevation gimbals. Rotation about the roll axis is handled by suitably resolving the azimuth and elevation drive and position indicating signals.

A key feature of the transmitter is use of a dual mode TWT to meet the conflicting requirements of low peak power for high-PRFs and high peak power for medium PRFs.

The processor consists of a programmable signal processor and a radar data processor in a single unit.

Operation. A complete set of air-to-air, air-to-ground, and air-combat modes is provided.

The principal air-to-air search modes are a high-peak-power medium-PRF, and, an alert/confirm mode in which velocity search is used for alert. When a target is detected, it is confirmed on the next scan with an optimized medium-PRF waveform. If the target proves valid, it is presented in a range versus azimuth display. In both modes, the pilot can optionally restrict the search to a particular region of interest or request altitude data on a given target.

Also provided are track-while-scan for up to 10 targets, single-target track, and a situation awareness mode in which one or two pilot-selected targets are tracked continuously while the radar searches a pilot-selected volume. The radar also has a raid mode, which analyses possible multiple targets for differential velocities; a long-range up-look medium-PRF mode, optimized for low to moderate clutter environments; and a track retention capability for coasting through periods of single-target tracking when the signal drops below the clutter.

Air-to-ground modes include real-beam mapping, in which "hard" targets are sharpened with a monopulse technique; an expanded version of this mode optimized for maritime surveillance; and two doppler beam sharpening modes, providing 6:1 and 64:1 azimuth resolution improvement, respectively. Supplementing

The F-16 C/D Fighter

With air-to-air and air-to-ground radar displays presented on the head-up display (HUD) and all combat critical radar controls built into the throttle and side stick, the pilot never needs to take his eyes off a target or his hands off the aircraft controls.

these are fixed target tracking, ground-moving-target detection and tracking, and beacon modes.

Air-combat modes are automatically selected by pressing a "dog fight" switch on the throttle. Initially, the radar scans a 20° by 30° body-stabilized field of view and locks onto the first target detected within 10 nmi. The pilot also has the options of (a) selecting a 10° by 60° vertical scan, (b) steering to place the cursor of the HUD on the target and locking onto it by releasing a designate switch on the side stick, or (c) automatically acquiring a target anywhere within the antenna scan limits.

Growth. The APG-68 has sufficient throughput to support the addition of SAR, terrain following, terrain avoidance, PVU, PPU and other advanced modes.

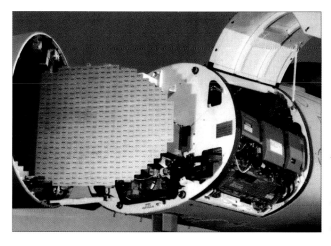

The radar's four air-cooled LRUs are organized for minimum interconnection and ease of maintenance. Each has its own power supply and BIT.

Twin-engine F/A-18 in a vertical climb.

Radar, in F/A-18, consists of five easily accessible LRUs, all having front panel connectors.

Transmitter employs a liquid-cooled, periodically-focussed permanent magnet TWT. Input power: 4.5 KW.

F/A-18 C/D (APG-73)

The APG-73 is an X-band pulse-doppler radar for the twin-engine F/A-18 C/D fighter/attack aircraft. Armed with AMRAAM and AIM-7 Sparrow missiles, the F/A 18 is tailored to carrier-based navy and marine applications.

Implementation. The radar consists of five flight-line replaceable units (LRUs)—antenna, transmitter, receiver/exciter, processor, and power supply.

The antenna is a monopulse planar array. Mounted in azimuth and elevation gimbals, it is directly positioned by electric torque motors, controlled by a servo electronics unit which plugs into the gimbal base. Aircraft roll rotation is accommodated by suitably resolving the azimuth and elevation drive and position indicating signals. To maximize ground coverage at steep look-down angles, the feed can be switched from producing a 3.3°-wide pencil beam to producing a wide fan beam. Included in the array are horns for a guard channel, for reducing the nulls in the radiation pattern during AIM-7 missile launches, and for providing flood illumination for AIM-7F visual launches.

The transmitter employs a liquid-cooled, gridded TWT, has a 4% bandwidth, and is capable of 13:1 Barker-code pulse compression.

The exciter provides a coherent transmitter drive signal of controllable amplitude for sensitivity time control and LPI power management. It provides local oscillator and reference signals for the receiver, and is capable of coherent PRI-to-PRI or noncoherent pulse-to-pulse frequency agility.

The receiver has two, triple down-conversion channels. During search, they carry the radar and guard signals; during monopulse operation, they carry the sum and difference signals and are time-shared for azimuth and elevation tracking. Zero to 45 dB of coarse AGC is provided in 15-dB increments; and 0 to 63 dB of fine AGC, in 1-dB increments. A/D converters sample the received signals at the following rates and resolutions.

A/D Sampling Rate	Resolution
10 MHz	11-Bit, Single
5 MHz	11-Bit, Dual
58 MHz	6-Bit, Dual or Single

The processor includes five mesh-connected processing elements (PEs): three identical pipe-line signal-processing elements (for range gating, doppler filtering and related functions) and two identical data processing elements (for loading programs for the selected modes of operation into the signal processing elements and performing overall control of the radar). Data word length is 32 bits; instruction word length, 64 bits; FFT filter banks, 2048-point.

The pipelines incorporate dedicated multiple intermediate memories and multiple high-speed parallel arithmetic units and are programmed so that no cycle time is devoted to nonproductive tasks, such as waiting for data or instructions from memory or for the incrementing of addresses.

All circuits are mounted on multilayer circuit boards, packaged in standard 5- by 9-inch, flow-through air-cooled plug-in modules.

The power-supply LRU rectifies the aircraft's prime power, converts it to the desired voltage levels, and conditions it. The unit is notable for having an overall efficiency of 82% and using programmable gate-array technology for control.

Air-To-Air Operation. A complete set of air-to-air search and track modes is provided.

For detecting high-closing-rate targets at maximum range, high-PRF velocity search is provided. For all aspect target detection, high-PRF with FM ranging is interleaved on alternate bars of the search scan with medium PRF employing 13:1 pulse compression and a guard channel for rejecting return from large-RCS point targets in the sidelobes.

In both modes, the high-PRF waveform alternates between two PRFs to minimize eclipsing, and a spotlight option provides a high update rate in a restricted pilot selected volume. Background tracks are initiated for all target "hits".

For target tracking, both track-while-scan (TWS) and single-target tracking modes are available. In TWS 10 targets can be tracked, up to eight of which, as prioritized by the pilot can be displayed. At the pilot's request, to facilitate multiple AMRAAM launches the radar automatically keeps the scan centered on the high-priority targets.

In single-target tracking, performance is optimize by automatically switching between high and medium PRF. To provide target illumination for AIM-7 launches, high PRF alone is used.

For situation awareness, TWS may be interleaved with single target tracking. To break out suspected multiple targets, finer than normal range and doppler resolution may be selected.

Four air-combat modes are provided. All employ medium PRFs, scan out in range, and automatically lock onto the first target detected.

Mode	Antenna
Gun	Scans HUD field of view
Vertical	Scans two vertical bars
Boresight	Fixed in boresight position
Wide Angle	Scans wide azimuth sector

At close ranges, a special medium-PRF track mode, employing CPI-to-CPI frequency agility to minimize glint, provides the high accuracy needed for the aircraft's gun-director.

Air-To-Ground Operation. The pilot has a wide choice of ground mapping modes.

- Real-beam, with sensitivity time control, 13:1 pulse compression in the longer range scales for increased clutter-to-noise ratio, and automatic switching from pencil to fan beam at steep depression angles.

- For navigation: wide-field-of-view, 8:1 DBS, with real beam mapping filling the ±5° forward blind sector.

- For finer resolution, a 45°-sector mode, with 19:1 DBS, 13:1 pulse compression, and four-look summing to reduce speckle.

- For still finer resolution, a similar mode maps a 12.6° wide patch with 67:1 beam sharpening.

- A SAR mode which maps a similar patch with the same medium resolution at all ranges.

- For detecting ships in high sea states, a noncoherent mode with pitch and roll compensation, 13:1 pulse compression, and pulse-to-pulse frequency agility for speckle suppression.

Ground-moving-target detection and tracking with coherent enhancement of slow moving targets may be interleaved with the mapping modes.

Air-to-ground navigation modes include fixed target tracking, coherent low-PRF air-to-ground ranging and its inverse, precision velocity update, and terrain avoidance for low-altitude penetration.

Reliability and Maintainability. The radar has a predicted mean-time-

Radar's vertically-polarized 26″ diameter planar array antenna showing corporate feed, monopulse networks, solid-state switches, and gimbal drives.

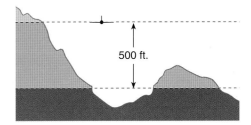

Scanning a ±35° sector out to 10 nmi ahead, a low-PRF non-coherent terrain avoidance mode senses terrain above the antenna's horizontal axis and terrain penetrating a plane 500 feet below it. A sector PPI display is used.

between failures of 208 hours. Through extensive built-in tests (BIT), it can detect 98% of all possible failures and isolate 99% of them to a single WRA.

Growth. Since 1998, two major improvements have been made to this radar:

1. The transmitter and antenna LRUs have been replaced with an active ESA.

2. A stretch generator module has been included in the receiver/exciter LRU enabling 1-foot-resolution SAR mapping.

Though originally developed for the F-4, the APG-76 is adaptable for installation in the nose, or wing and center-line pods on other aircraft.

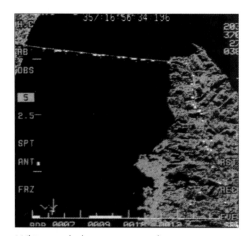

White symbols superimposed over ground map indicate precise locations of moving targets simultaneously detected with interferometric detection and clutter cancellation technique.

Weighing 625 pounds and including 7 LRUs, the APG-76 radar has a peak power of 12 Kw; receiver noise figure of 6.5 dB; antenna sum gain of 34.5 dB; and beamwidth of 2.2°.

F-4E (APG-76)

The APG-76 is a multimode Ku-band pulse-doppler radar originally developed by Westinghouse Norden Systems for Israel's F-4 Phantom 2000 fighters for air-to-air and air-to-ground precision targeting and weapon delivery. To date, 60 systems have been delivered.

Extended capability variants have been evaluated in simulated combat in wing tanks on the US Navy S-3 and US Air Force F-16.

Unique Capabilities. The radar is unique in being capable of simultaneous SAR mapping and ground moving target detection and tracking. Employing a three-segment mechanically steered planar array antenna and four low-noise receiver and signal processing channels, it features:

- Long-range multi-resolution SAR mapping

- All-speed ground moving target detection over the full, width of the forward sector

- Automatic tracking of ground moving and "did-move" targets

- Automatic detection and location of rotating antennas

The antenna has seven receive ports: sum, azimuth difference, elevation difference, guard, and three interferometer ports. In air-to-air modes, the sum, azimuth difference, elevation difference, and guard outputs are processed in parallel through the four receive channels. In air-to-ground modes, the sum signal is processed through one channel, and the three interferometer signals through the remaining three channels.

GMTI and GMTT. Employing the interferometric notching and tracking techniques described in Chap. 24, the radar can detect and precisely track ground moving targets having radial velocities of from 5 to 55 knots anywhere within the radar's ±60° azimuth field of view.

Ground clutter, meanwhile, is suppressed by subtracting the returns received by one interferometer antenna segment from the weighted returns received by another. This is done in the outputs of all of the doppler filters passing frequencies determined to be within the mainlobe of the two-way antenna pattern.

Adaptive CFAR detection thresholds are independently determined for clutter and clutter-free regions. Those targets satisfying an M-out-of-N detection criteria, are displayed as moving target symbols superimposed at the correct range and azimuth positions over the simultaneously produced SAR map.

Growth. As initially implemented, the radar employed five parallel operating vector pipeline processors and two scaler data processing elements. In a company funded program, these are being replaced with a COTS processor.

Also, as originally implemented the radar provided a wide selection of ground-map resolutions ranging from real beam, to doppler beam sharpening, on down to 10-foot-resolution SAR. The company has since developed and tested 3-foot and 1-foot resolution SAR modes plus a wide-area surveillance mode which combines high resolution SAR maps in a mosaic to facilitate continuous monitoring and tracking of moving targets.

Strategic Bombing

B-2 Bomber (APQ–181)

The APQ–181 is the multimode pulse-doppler radar for the B-2 long-range stealth bomber. It employs a low-RCS passive ESA antenna and incorporates advanced LPI features. Except for that and the fact that, like the APQ-164, it gives the aircraft the autonomous ability to navigate safely around hazards and use them to mask defensive systems, very little can yet be said about the radar at an unclassified level.

A tanker's view of the B-2 stealth bomber prior to refueling.

The B-1B Bomber

B-1B RADAR (APQ–164)

The APQ–164 is an X-band multimode pulse-doppler radar tailored to the requirements of the long-range strategic bomber. These include the ability to (a) penetrate deep into enemy territory at low altitudes night or day in fair or foul weather, undetected by enemy defenses; (b) detect, accurately identify, and destroy assigned targets; (c) immediately following a demanding 15-hour mission, be fully available for another mission.

Implementation. The radar features a 44 x 22 inch passive ESA, employing 1,526 phase control modules. Together with its beam-steering computer, the antenna is mounted in a roll gimbal having detents for locking it in forward, broadside, and vertical (down) positions. Besides being able to switch beam positions virtually instantaneously (order of 200 ms), the antenna provides extreme beam-steering accuracy, optimizes beam patterns for different modes, and offers a choice of either linear or circular polarization.

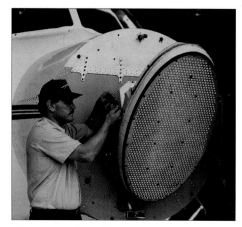

The radar's 44 x 22-inch passive ESA, together with the beam-steering computer, is mounted in a detented roll gimbal.

Most of the radar's units have a high degree of commonality with those of the F-16's APG-68. The transmitter employs the same dual-mode TWT (though it's liquid cooled in the APQ-164), and the receiver enables full two channel monopulse operation. To ensure a high degree of availability, except for the antenna, which is inherently fault tolerant, two independent chains of line-replaceable radar units are provided. In essence, two separate radars are carried in the B-1B: one, in operation; the other, in standby waiting to be switched in.

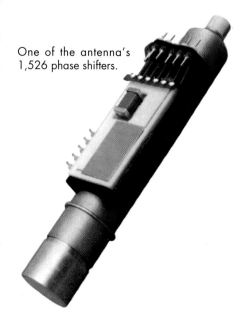

One of the antenna's 1,526 phase shifters.

Navigation Modes. For navigation, the primary mode is high-resolution SAR mapping. Typically, it's employed as follows: The B-1B's avionics give the radar the coordinates of a check point. The radar trains its antenna on the point, makes a patch map centered on it, and turns off. The map is stored and frozen on the display, giving the operator ample time to analyze it. Having located the check point, the operator designates it with a cursor, thereby updating the bomber's position and destination heading in the B-1B's INS.

Supplementing the SAR mode for navigation are real-beam mapping, weather detection, and velocity update modes. The weather mapping mode is essentially the same as real-beam ground mapping except that, if weather penetration is necessary, the antenna may be switched to circular polarization.

At altitudes up to 5,000 feet absolute, altitude is measured by a radar altimeter. From 5,000 to 50,000 feet, altitude updates may be obtained by moving the APQ-164 antenna to its vertical detent position.

For rendezvousing with tankers and other aircraft, an air-to-air beacon mode and a short-range air-to-air search mode are provided.

For penetration, automatic terrain following and terrain avoidance modes are provided. In terrain following, the radar supplies the B-1 avionics with a height versus range profile of a corridor centered on the projected flight path out to a range of 10 nmi, thereby enabling the automatic generation of appropriate climb and dive commands.

Through a unique azimuth and elevation extent algorithm, the radar differentiates between terrain and spurious returns from rain, towers, or electronic interference. By scanning in azimuth, terrain avoidance detects objects on either side of the flight path that are higher than a selected ground clearance plane, enabling the pilot to maintain the lowest possible overall altitude.

Weapon Delivery. For weapon delivery—both conventional and nuclear —high-resolution SAR mapping is used as just described. In addition, a ground-beacon tracking mode and a ground-moving-target-tracking mode enable precise targeting of both fixed and moving targets.

Because of the extreme flexibility of electronic beam steering, several radar modes can be sequentially time-shared, giving the pilot, copilot, and offensive systems officer (OSO) the equivalent of simultaneous independent use of different modes through their own displays.

Each of the bomber's three crew members has the equivalent of simultaneous independent use of different modes through their own displays.

The B-1B not only looks like a fighter, but its radar employs key hardware technology transferred from the radar for the F-16 C/D.

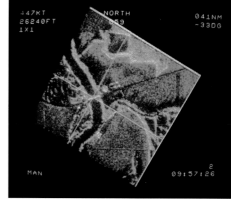

Given the coordinates of a check point, the radar makes a patch map centered on it and turns off.

Attack Helicopter

The AH-64D carries up to 16 RF or semi-active laser-guided Hellfire missiles and 76 70-mm folding fin aerial rockets or a combination of both, and up to 1,200 rounds of 30-mm ammunition

Lurking behind cover with only the radome of its millimeter wave radar showing, Longbow can quickly detect, classify, and prioritize more than 100 moving or stationary targets.

Flat, fully interchangeable color displays are provided in both cockpits.

AH-64D Apache Helicopter (Longbow Radar)

Longbow is a fast-reaction, low-exposure, high-resolution, millimeter-wave fire-control radar designed for the AH-D Apache attack helicopter. Mounted atop the main rotor mast to take advantage of terrain masking, the radar can pop up and, in seconds scan a 90° sector; then, drop down out of sight.

During that brief interval, it can detect, classify, and prioritize more than 100 moving and stationary ground targets, fixed wing aircraft, and both moving and hovering helicopters—discriminating between closely spaced targets of the same type with an extremely low false-alarm rate.

It then displays the 10 highest priority targets to the aircrew, and will automatically cue either an RF or a semi-active laser guided fire-and-forget Hellfire missile to the first target. Immediately after its launch, the system cues the next missile to the next priority target, and so on.

The radar also provides obstacle warning to alert the pilot to navigation hazards, including man-made structures, towers, etc.

Radar data is displayed on the pilot's night-vision helmet-mounted display and on two color-coded flat general purpose displays in each cockpit.

A derivative of the Longbow radar will be forthcoming for the RAH-66 Comanche helicopter. Using the same millimeter radar and the same Hellfire missiles as Apache, it will include a number of advanced features such as a smaller antenna.

Transport/Tanker Navigation

C-130 (APN-241)

Meeting all of the requirements of tanker/transport operations, the APN-241 is the baseline radar for the C-130J. It also can be employed in a number of other aircraft, when existing modules that interface with their avionics, controls, and displays are included.

Implementation. A light-weight X-band coherent pulse-doppler radar, the APN-241 consists of just two basic elements: an antenna and a receiver-transmitter-processor.

The antenna is a 26 by 32-inch, dual-channel, monopulse planar array. Stabilized about three axes, it provides ±135° of azimuth coverage and +10 to –25° of elevation coverage.

The receiver-transmitter-processor is all solid state. Operating at 9.3 to 9.41 GHz, it has a power output of 116 W peak; 9.5 W, average.

Modes of Operation. To meet all-weather delivery requirements, a variety of modes are provided:

- Weather—detects weather through weather, out to 320 nmi; turbulence out to 50 nmi

- Windshear detection—gives up to 90 seconds of warning of a microburst (probability of false alert $<10^{-4}$ per flight hour)

- Ground mapping—monopulse (2.5 to 10 times improvement over real beam); DBS, for higher resolution mapping off the nose

- Air-to-air detection—20 nmi against a C-130-sized target; indicates whether nose or tail aspect

Overlay Modes

- Beacon—interrogates both air and ground beacons

- Station-keeping

- Flight plan—navigation data from self-contained navigation system or other reference[1]

- Traffic collision avoidance system (TCAS)[1]

The radar is designed for two-person cockpit operation. Pilot and navigator can view and control different modes simultaneously.

Growth. A SAR mode has already been developed and is available as a software update. In development are terrain following, drop-zone wind measurement, and autonomous landing guidance modes.

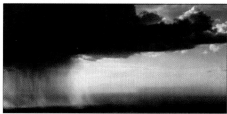

APN-241 can map weather out to 320 nmi, turbulence to 50 nmi, and give 90-second windshear warning.

1. Can be displayed autonomously or overlaid on any radar map.

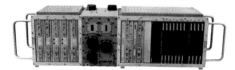

Receiver-Transmitter-Processor, includes interfaces for aircraft avionics, controls, and displays

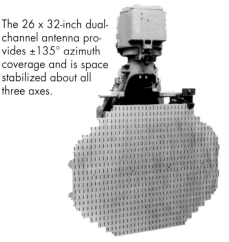

The 26 x 32-inch dual-channel antenna provides ±135° azimuth coverage and is space stabilized about all three axes.

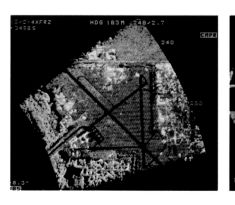

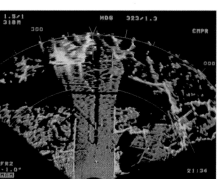

DBS ground map (far left) enables precise location of drop zones or target areas.

Monopulse ground map (left) enables blind radar approaches to small or unimproved landing sites.

Civil Applications

The RDR-4B is currently operating in nearly every type of commercial transport aircraft.

Antenna, radar electronics, controls, and display unit. *Pulse width:* radar modes, 6 and 18 μsecs (interlaced); weather, 2 μsecs. *PRF:* weather and maps, 380 Hz; turbulence mode, 1.6 kHz; windshear mode, 6kHz. *Peak transmitter power:* 125 W. Frequency agility for interference reduction.

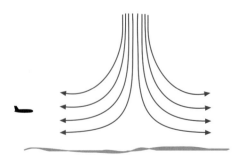

In windshear mode, antenna tilt is temporarily changed to the optimum for measuring the horizontal winds due to the out flow from a microburst.

RDR-4B Civil Weather Radar

The RDR-4B is a pulse-doppler forward-looking weather radar, operating in nearly every type of commercial transport aircraft. Besides the ability to penetrate weather systems and accurately map rainfall and turbulence, it meets all FAA requirements for stand-alone windshear detection.

At altitude, the radar provides a detailed ground-clutter-free color weather display of a ±40° forward sector out to a selectable maximum range of up to 320 nmi.

Whenever the absolute altitude is less than 2,300 feet, the windshear detection mode is automatically activated on alternate antenna scans.[1]

During clockwise scans, the radar continues to operate in the operator selected mode. But during counter-clockwise scans, it operates in windshear mode, with the antenna tilt tem-

porarily changed to the optimum angle for measuring horizontal winds caused by the outflow from a microburst core. The only apparent difference in the radar's operation, though, is that the display updates only on the clockwise scans.

If a windshear is detected at any range out to 5 nmi, an advisory icon appears on the display, and a chime or oral caution, "monitor radar display" sounds.

Below 1500 feet, if a windshear is detected within 3 nmi and ± 25° of the aircraft's heading, a windshear alert, sounds, and a warning appears on the radar display.

At 1.5 nmi, a windshear icon indicating the exact location of the microburst is superimposed over the normal display, and an oral warning, "windshear ahead, windshear ahead" sounds,[2] giving the pilot roughly 15 seconds to avoid the potential hazard.

---

1. If the radar isn't already on, it automatically turns itself on and operates continuously in windshear mode.

2. During landing, this would be changed to "go around, windshear ahead."

Standard Weather Display
■ Level 2 rain   ■ Level 4 & above
■ Level 3 rain   ■ Turbulence

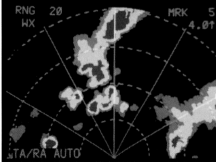

When windshear is encountered, red and yellow icon is superimposed over the weather display, indicating the microburst's exact position.

HISAR

This is an advanced high-resolution, real-time radar mapping system developed for a wide range of primarily civil uses. It can monitor environmental conditions such as flood damage, oil spills, and sea ice over vast regions in inclement weather. It can uncover illegal activities and detect illegal border crossings and threatening buildups without provoking a response.

Implementation. Suitable for use in small executive-class aircraft, HISAR consists of an X-band, multimode radar which looks out through a radome in the bottom of the aircraft, plus a computer workstation which controls the mission and displays the radar's outputs to the operator.

The radar has two planar array monopulse antennas mounted back-to-back. When aligned with the flight path, they enable surveillance to be switched quickly from one side of the aircraft to the other.

The workstation has two displays: one for mission planning and control; the other, for displaying SAR images and target data. A second workstation may be installed in a ground facility, enabling preflight mission planning and post-flight analysis to be performed there rather than in the aircraft.

Mission Control. For radar and mission control, the aircraft's GPS-aided INS provides HISAR with the aircraft's longitude, latitude, velocity, altitude, and attitude. To guide the aircraft on its mission, HISAR's workstation gives the autopilot flight commands. To collect the desired data at appropriate points in the preplanned mission, it cues the radar's selection of modes and controls their operation. At any point in the mission, of course, the radar can be redirected by the operator.

Radar Modes. Imaging is performed in three of HISAR's five radar modes:

- Wide-area search (DBS)—radar scans a 60° sector extending from 37 km to 110 km with resolution of 25 meters in range and 0.4 milliradian in azimuth

- SAR strip map—strip 37 km wide can be positioned in range anywhere between 20 and 110 km; resolution of 6 meters in both range and azimuth

- SAR spotlight—patch 3.5 km square, with 1.8 meters resolution in azimuth and elevation, can be placed anywhere in range between 37 and 110 km and in azimuth within ±45° off broadside

Supplementing these modes are (a) ground-moving target detection which can be interleaved with wide-area-search and strip-map modes, and (b) air-to-air search. The latter is a coherent low-PRF mode providing ±150° coverage in two elevation bars. It is capable of detecting helicopters and low-level, medium-speed aircraft out to a range of 70 km, enabling the operator to monitor the "air picture" over an immense area.

Multisensor Integration. HISAR is integrated with and exchanges cueing and data with a forward-looking infrared sensor (FLIR) and communication and electronic-intelligence sensors.

HISAR antenna looks out through radome in the bottom of this executive-class aircraft.

Radar has two planar array antennas mounted back-to-back on an azimuth gimbal.

The operator's workstation has two large displays. One (top) for mission planning and control; the other for SAR images and target data.

Long-wavelength FLIR images cued by HISAR: truck convoy (above, left), and tanker trucks (above, right). Note white image of wheels produced by heat of the tankers' tires.

RADAR SYSTEMS COVERED

RADAR	DEVELOPER
APS-145	Lockheed Martin
APY-1/2	Northrop Grumman ESSD* (Westinghouse)
JointSTARS	Northrop Grumman Norden Systems
APG-77	Joint Venture: Northrop Grumman Raytheon TI Systems
APG-68	Northrop Grumman ESSD* (Westinghouse)
APG-73	Raytheon (Hughes)
APG-76	Northrop Grumman Norden Systems
APQ-181	Raytheon (Hughes)
APQ-164	Northrop Grumman ESSD* (Westinghouse)
Longbow	Joint Venture: Lockheed Martin Northrop Grumman
APN-241	Northrop Grumman ESSD* (Westinghouse)
RDR-4B	Allied Signal
HISAR	Raytheon (Hughes)

*ESSD – Electronic Sensors & Systems Division

Appendix

Rules of Thumb

It is said that the instant way to become a radar expert is to learn a few rules of thumb. While that is absurd, it *is* true that most radar experts repeatedly use a number of very simple rules of thumb. And these can be just as useful to you.

Antenna Beamwidth

You can remember these rules most easily by keeping in mind (a) that the null on either side of the boresight line of a linear array occurs at an angle θ for which the difference in distance from the observer to the ends of the array equals one wavelength, λ, and (b) that the 3 dB beamwidth is roughly half the null-to-null.

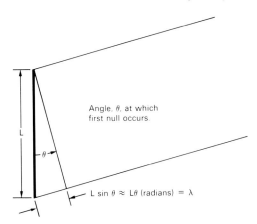

Angle, θ, at which first null occurs.

L sin θ ≈ Lθ (radians) = λ

Linear array

Real . . . . $\theta_{3\,dB} \sim \dfrac{\lambda}{L}$ radians

Synthetic . . . . $\theta_{3\,dB} \sim \dfrac{\lambda}{2L}$ radians

Note: λ and L must be in same units.

Circular aperture

Uniform illumination . . . . . $\theta_{3\,dB} \sim \dfrac{\lambda}{d}$ radians

Uniform illumination . . . . . $\theta_{3\,dB} \sim \dfrac{70}{d_{\,(inches)}}$ degrees
X-band

Tapered illumination . . . . . $\theta_{3\,dB} \sim \dfrac{85}{d_{\,(inches)}}$ degrees
X-band

To apply these rules to any wavelength, λ, multiply $\theta_{3\,dB}$ by λ/3.

Antenna Gain

You can remember these rules most easily by thinking of antenna gain as equal to the ratio of (a) the area of a sphere of unit radius (A = 4π) to (b) the area that is marked off on this sphere by a solid angle corresponding to the 3 dB width of the antenna beam.

Rectangular aperture . . . . $G \sim \dfrac{4\pi}{\theta_a \theta_e}$

3 dB az and el beamwidths, radians

Rectangular aperture . . . . $G \sim 1.4\,LW$
X-band

Length and width of aperture, cm

Circular aperture . . . . $G \approx d^2 \eta$
X-band   *Diameter, cm*   *Aperture efficiency*

To apply these rules to any wavelength, λ, multiply L, W, and d by 3/λ.

Bandwidth

Single Pulse . . . . $BW_{3\,dB} = \dfrac{1}{\tau}$

Pulse width

Spectral line . . . . $BW_{NN} \approx \dfrac{2}{NT}$

Number of pulses integrated
Interpulse period

Note. If τ or T are in seconds, BW is in hertz.

563

Doppler Shift (at X-band)

35 hertz per knot *(actually 34.30)*
30 hertz per statute mile per hour
19 hertz per kilometer per hour
20 kilohertz per 1000 feet per second

To apply these rules to any wavelength, λ, multiply by 3/λ.

Filter Passband

$$BW_{3\,dB} \sim \frac{1}{t_{int}}$$

$$BW_{3\,dB} \sim \frac{PRF}{(\text{Number of pulses integrated})}$$

Frequency, Wavelength, Period

$$f = \frac{30}{\lambda_{(cm)}} \text{ GHz}$$

$$\lambda = \frac{30}{f_{(GHz)}} \text{ cm}$$

$$T = \frac{1}{f_{(Hz)}} \text{ sec}$$

Pulse Length

1000 ft/μs of pulse width
500 radar ft/μs of pulse width

Round-Trip Ranging Times

12.4 μs per nautical mile
10.7 μs per statute mile
6.7 μs per kilometer

Range

Maximum unambiguous . . . . . . $R_u \approx \dfrac{80}{PRF_{(kHz)}}$ nmi

$$R_u \approx \frac{150}{PRF_{(kHz)}} \text{ km}$$

Resolution . . . . 500 ft/μs of
(compressed) pulse width

To radar horizon . . . . $R \sim \sqrt{2h_R}$
Statute miles   *Height of radar, ft*

To target beyond horizon . . . . $R_{max} \sim \sqrt{2\,h_R} + \sqrt{2\,h_T}$
Statute miles   *Height of radar, ft*
Height of target, ft

Units of Measure

1 radian = 57.3°
3 cm ≈ 0.1 ft
1 nmi ~ 6000 ft

Reference Data

UNITS OF SPEED

1 foot/sec = 0.59248 knot
= 0.681818 stat. mph
= 1.09728 kilometers/hr
1000 fps ~ 600 knots
1 kilometer/hr = 0.539957 knot
= 0.621371 stat. mph
= 0.277777 meters/sec
= 0.911344 ft/sec
1 mile/hr (stat.) = 0.868976 knot
= 1.60934 kilometers/hr
= 1.46667 ft/sec
1 knot* = 1.15078 stat. mph
= 1.85200 kilometers/hr
= 1.68781 feet/sec
*A knot is 1 nautical mile per hour.

APPROXIMATE SPEED OF SOUND (MACH 1)

Sea Level		36,000 to 82,000 ft*
1230 km/hr	Decreases Linearly To→	1062 km/hr
765 mph		660 mph
665 knots		573 knots

*Above 82,000 feet the speed increases linearly to 1215 km/hr (755 mph, 656 knots) at 154,000 ft.

SPEED OF LIGHT*

299,792.4562 kilometers/s
$\approx 300 \times 10^6$ meters/sec
$\cong$ 186,282 stat. miles/sec
$\cong$ 161,875 naut. miles/sec
$\cong$ 983.569 feet/μs
~ 1000 feet/μs
*In a vacuum.

RADIO FREQUENCY BANDS

Designation	Frequency	Wave Length
HF	3–30 MHz	100m–10m
VHF	30–300 MHz	10m–1m
UHF	300–3000 MHz	1m–10cm
SHF	3–30 GHz	10cm–1cm
EHF	30–300 GHz	1cm–1mm

Microwaves: $\sim$ 1m–1cm
Millimeter waves: $\sim$ 1cm–1mm

THE RADAR BANDS

Designation	Assigned Frequencies*
VHF	138–144 MHz 216–225 MHz
UHF	420–450 MHz 890–942 MHz
L	1215–1400 MHz
S	2300–2500 MHz 2700–3700 MHz
C	5250–5925 MHz
X	8500–10,680 MHz
K_u	13.4–14.0 GHz 15.7–17.7 GHz
K	24.05–24.25 GHz
K_a	33.4–36.0 GHz

*Bands specifically assigned for radar used by the International Telecommunications Union, ITU.

DECIMAL MULTIPLIER PREFIXES

Prefix	Symbol	Multiplier
giga	G	10^9
mega	M	10^6
kilo	k	10^3
milli	m	10^{-3}
micro	μ	10^{-6}
nano	n	10^{-9}

UNITS OF LENGTH

1 inch = 2.540 centimeters
0.1 foot $\cong$ 3 centimeters
1 foot = 30.4800 centimeters
1 yard = 0.914400 meter

1 meter = 1.09361 yards
= 3.28083 feet
= 39.3701 inches
= 10^6 microns
= 10^{10} Angstroms

1 kilometer = 0.539957 naut. mile
= 0.621371 stat. mile
= 1093.61 yards
= 3280.83 feet

1 statute mile = 0.868976 naut. mile
= 1.60934 kilometers
= 1760 yards
= 5280 feet

1 nautical mile = 1.15078 stat. miles
= 1.85200 kilometers
= 2025.37 yards
$\approx$ 2000 yards
= 6076.12 feet

BASIC TRIGONOMETRIC IDENTITIES

$\sin (A+B) = \sin A \cos B + \cos A \sin B$
$\sin (A-B) = \sin A \cos B - \cos A \sin B$
$\cos (A+B) = \cos A \cos B - \sin A \sin B$
$\cos (A-B) = \cos A \cos B + \sin A \sin B$

EQUIVALENCY SYMBOLS

Symbol	Meaning (as used in this book)
$\propto$	Proportional
$\sim$	Roughly equivalent
$\approx$	Approximately
$\cong$	Nearly equal
$=$	Equal

References

"A View of Current Status of Space-Time Processing Algorithm Research," H. Wang et al., IEEE International Radar Conference, pp. 789-791.

"Aircraft-Borne Interferometric SAR for 3-D High Resolution Radar Imaging," H. D. Griffiths, C. J. Baker, A. Currie, R. Voles, R. Bullock, and P. V. Brenna; IEE Colloquium Digest 1994.

"An Overview of Space-Time Adaptive Processing for Airborne Radars," Hong Wang, Dept of Electrical Engineering and Computer Science, Syracuse University, 1997.

"Low Probability of Intercept (LPI) Techniques," Phillips, E. V., Hughes Aircraft Company report P91000095 Chap. III-2, 1989. Paper from which Chapter 42 of this book was derived.

Radar Handbook, second edition, M. Skolnik, McGraw-Hill, 1990.

"Rapid Convergence Rate in Adaptive Arrays," I. S. Reed, J. D. Mallett, L. E. Brennan; IEEE Transactions on Aerospace and Electronic Systems, vol. AES-10, no. 6, Nov. 1974.

"Real-Time Adaptaive Airborne MTI, Part I: Space-Time Processing," R. Klemm; Proc. ICR 96, Oct. 1996.

"Switched Fiber-Optic Delay Architecutres," Akis P. Goutzoulis, D. Kenneth Davies; *Photonic Aspects of Modern Radar*, chap. 13, Artech House.

"Theory and Design of Interferometric Synthetic Aperture Radars," E. Rodriguez, J. M. Marin; IEEE Proceedings-F, vol. 139, no. 2, April 1992.

Index